Farwell's Rules of the Nautical Road

Farwell's Rules of the Nautical Road

Originally known as The Rules of the Nautical Road
by Captain Raymond F. Farwell

Sixth Edition prepared by
Frank E. Bassett, Commander, U.S. Navy
Richard A. Smith, Commander, Royal Navy

Naval Institute Press
Annapolis, Maryland

Library of Congress Cataloging in Publication Data

Farwell, Raymond Forrest, 1893-
Farwell's Rules of the nautical road.

Includes the complete text of the new U.S.
Inland Navigational Rules Act of 1980, along with
the 1972 International Regulations for Preventing
Collisions at sea.
Includes indexes.
1. Rule of the road at sea. 2. Inland navigation
—Law and legislation—United States. I. Bassett,
Frank E. II. Smith, Richard Arthur, 1940-
III. International Regulations for Preventing
Collisions at Sea (1972). 1982. IV. United
States. Inland Navigational Rules Act of 1980.
1982. V. Title. VI. Title: Farwell's Rules of
the nautical road.

K4184.F37 1982 343'.0966 82-14410
ISBN 0-87021-180-3 342.3966

Sixth Edition

The opinions and assertions of the authors of this
book are personal ones and do not necessarily reflect the views
of either the Navy or the naval service as a whole.

Contents

Illustrations

Preface to the Sixth Edition

This book was first published in 1941 as *The Rules of the Nautical Road,* by the late Captain Raymond F. Farwell, U. S. Naval Reserve. Since its first publication, numerous statutory and regulatory changes have been made in both local and international nautical rules. Such changes automatically necessitate revisions of any authoritative texts pertaining to this subject, and hence the book has been revised whenever occasion demanded.

Over the years, common usage has made Captain Farwell's name synonymous with the book itself; accordingly, since the Fourth Edition it has been titled *Farwell's Rules of the Nautical Road.*

Among the specific statutory and regulatory actions resulting in changes are the following:

Executive Order 9083, effective 1 March 1942, which transferred the functions of the former Bureau of Marine Inspection and Navigation to the U. S. Coast Guard. The authority reposing in the Board of Supervising Inspectors, Bureau of Marine Inspection and Navigation, to promulgate Pilot Rules to supplement and have co-jurisdiction with the local statutory rules was thereby temporarily given to the Commandant, U. S. Coast Guard.

Reorganization Plan No. 3 of 1946, effective 16 July 1946, which made permanent the transfer of the functions of the former Bureau of Marine Inspection and Navigation to the U. S. Coast Guard.

The renumbering and reorganization of the Code of Federal Regulations, effective 30 November 1948, which resulted in the renumbering of the respective Pilot Rules.

Public Law 544, 80th Congress, i.e., Act of May 21, 1948, effective 1 January 1949, relating to and revising the statutory rules for the prevention of collisions on certain inland waters of the United States and on the western rivers.

The 1948 International Conference on Safety of Life at Sea at London, which, in addition to revising the 1929 International Convention for the Safety of Life at Sea, revised the International Rules drawn up in 1889 at a Conference of twenty-six maritime nations, adopted by this country in 1890, and put into effect in 1897.

Public Law 172, 82nd Congress, i.e., Act of October 11, 1951, adopting the 1948 revision of the International Rules and authorizing the President to fix their effective date by Presidential Proclamation.

The announcement by the United Kingdom on 19 December 1952, in accordance with the provisions of the Final Act of the 1948 International Conference on Safety of Life at Sea, of the effective date of the revised International Rules as 1 January 1954.

Public Law 232, 83rd Congress, i.e., Act of August 8, 1953, extending the applicability of the Inland Rules and the Pilot Rules published pursuant thereto to the Mobile River and its tributaries above Choctaw Point.

Public Law 552, 84th Congress, i.e., Act of June 4, 1956, revising the lights required to be carried by motorboats when in inland waters, the western rivers, or the Great Lakes.

Public Law 350, 85th Congress, i.e., Act of March 28, 1958, extending Great Lakes Rules to foreign vessels and increasing penalties for violation of Great Lakes Rules.

Public Law 635, 85th Congress, i.e., Act of August 14, 1958, amending rules regarding towing lights and stern lights in inland waters and western rivers.

Public Law 658, 85th Congress, i.e., Act of August 14, 1958, clarifying authority to establish pilot rules in inland waters and western rivers and increasing penalties for violation of Inland Rules.

The 1960 International Conference on Safety of Life at Sea, at London, which proposed numerous changes to International Rules revised by the 1948 Conference, as well as making various recommendations as to how to use radar as an aid in avoiding collisions.

Public Law 84, 88th Congress, i.e., Act of August 5, 1963, revising anchorage regulations for vessels at anchor in inland waters.

Public Law 131, 88th Congress, i.e., Act of September 24, 1963, adopting the 1960 revision of the International Rules, and authorizing the President to proclaim same effective at a subsequent date.

Public Law 163, 88th Congress, i.e., Act of October 30, 1963, providing for special regulations for vessels passing under low bridges built on navigable waters of the United States.

Proclamation 3632 by the President on 29 December 1964, that the 1948 International Rules would be rescinded and the 1960 International Rules made effective on 1 September 1965.

Public Law 670, 89th Congress, i.e., Act of October 15, 1966, transferring functions, powers, and duties of the Secretary of the Army re water vessel anchorage in *special anchorage areas* to the Secretary of Transportation.

Public Law 764, 89th Congress, effective 3 February 1967, requiring small sail vessels and power boats under 65 feet to keep clear of larger vessels when in narrow channels.

The transfer of the Coast Guard on 1 April 1967 from the Department of the Treasury *to the Department of Transportation.*

Miscellaneous revisions and amendments to the Pilot Rules since the first edition, dated February 1941.

The 1972 International Convention on Revision of the International Regulations for Preventing Collisions at Sea, at London, that proposed numerous changes, including a major editorial reorganization, to the International Rules revised by the 1960 Conference.

The IMCO Recommendation on Navigational Watchkeeping, which gives guidance on standards pertinent to the rules of the road, particularly on the subject of lookout.

Public Law 340, 92nd Congress, i.e., Act of July 10, 1972, known as the Ports and Waterways Safety Act of 1972, which authorizes the establishment of vessel traffic services in the navigable waters of the United States.

The Vessel Bridge-to-Bridge Radiotelephone Act, effective 1 January 1973, requiring certain vessels to maintain watch on a radiotelephone while under way on the navigable waters of the United States.

The International Maritime Consultative Organization (IMCO) Resolution of 20 November 1973 adopting Routing Systems on the high seas, in harmony with the International Regulations for Preventing Collisions at Sea, 1972.

The announcement by IMCO in June 1976 in accordance with the provisions of Article IV of the 1972 Convention on the International Regulations for Preventing Collisions at Sea, of the effective date of the revised International Rules as 15 July 1977.

The acceptance by the President of the United States on 12 December 1975 of the 1972 Convention on the International Regulations for Preventing Collisions at Sea, and the *deposit of this acceptance with IMCO,* effective 23 November 1976.

Public Law 95–75, 95th Congress, i.e., Act of July 27, 1977, known as the International Navigational Rules Act of 1977, implementing the International Regulations for Preventing Collisions at Sea, 1972.

Public Law 95–474, 95th Congress, i.e., Act of October 17, 1978, known as the Port and Tanker Safety Act of 1972.

Executive Order 12234 of September 3, 1980, implementing the International Convention for the Safety of Life at Sea, 1974.

Public Law 96–591, 96th Congress, i.e., Act of December 24, 1980, known as the Inland Navigational Rules Act of 1980, unifying the rules for preventing collisions on the inland waters of the United States, the effective date being set as December 24, 1981.

Adoption by IMCO, in November 1981, of the proposed amendments to the International Regulations for Preventing Collisions at Sea, 1972, with the recommendation that the amendments should enter into force 18 months after adoption (May 1983).

The major impetus for the publication of this edition was due to the significant changes in the rules of the road on the inland waters of the United States as contained in the Inland Navigational Rules Act of 1980, which went into effect on December 24, 1981.

The authors are indebted to the mariners who wrote expressing their views on the fifth edition of *Farwell's*. Their assistance in helping to eliminate errors for future editions is greatly appreciated. Additional comments are welcomed for the sixth edition.

This book is dedicated to the proposition that obedience to the rules is the surest way to avoid collision.

Frank E. Bassett
Commander, U.S. Navy

Richard A. Smith
Commander, Royal Navy

Preface to the First Edition

Marine Collision Law has too long been a specialty of judges on the admiralty bench and of a very limited number of admiralty lawyers at the bar. It should, of course, be instead a specialty of the mariner on the bridge. The present book is planned to satisfy the needs of classes in seamanship such as those at the Naval Academy and at Naval R.O.T.C. colleges, and at the same time to serve as a useful handbook on the subject for the officers at sea in the actual practice of navigation. To this two-fold end an adequate amount of case material has been included, and in addition an index sufficiently comprehensive so that ready references may be made to any desired rule.

To the man on the bridge it may be superfluous to say that a collision situation is neither the time nor the place for him to look up the law that determines his proper action. But it must be pointed out that the alternative is such a thorough understanding of his duty in every situation that his action will amost instinctively be the proper one from the standpoint of both law and seamanship. It is said that Admiral Knight required his officers to read through the rules of the road each time before getting under way. Certain it is that the complexity of rules effective under American law makes frequent study of them both desirable and necessary.

As teacher and practical mariner of some experience, the writer believes that a clear, definite, positive knowledge of the principles here presented should be an essential part of the professional equipment of every man on the bridge engaging in the practice of navigation. Court decisions in an overwhelming number prove that nearly all marine collisions follow violations of the rules of the road. The inference is that the rules, if implicitly obeyed, are practically collision-proof. It is the writer's opinion that most of this seeming disregard of the law is due to misunderstanding of the rules as interpreted by the courts. The rules will not be better obeyed until they are better understood. Such a better understanding by the mariner is the primary purpose of this book.

University of Washington
December 15, 1940

Raymond F. Farwell
Captain, U. S. Naval Reserve

Farwell's Rules of the Nautical Road

Part I

The Rules of the Nautical Road

1

Summary of Changes in Inland Rules*

Since 15 July 1977, all mariners on the high seas have sailed under the revised International Rules of the Road, technically known as the International Regulations for Preventing Collisions at Sea, 1972 (72 COLREGS). These rules apply to all international waters and to local waters of those countries having adopted them that do not have different local rules of their own. For United States waters, however, under the authority of International Rule 1(b), separate rules have long existed. This hodge-podge of different rules for inland waters, the Western Rivers, and the Great Lakes, plus the associated pilot rules and interpretive rulings, created a great deal of confusion for those mariners who sailed in different local waters and particularly for those who sailed in international waters as well. The new inland rules have practically eliminated the proliferation of separate rules for different inland waters, and they are now in close conformity with the International Regulations.

Since the signing of the 72 COLREGS in 1972, and particularly since the enactment of implementing legislation in 1977, the Coast Guard has been actively pursuing unification of the inland rules and regulations into a single and coherent system. To assist in developing the best possible set of rules, that agency established the Rules of the Road Advisory Committee, composed of knowledgeable men and women representing a cross section of maritime interests that use or are familiar with our various inland waterway systems. After deliberations and discussions over a

*The reader will note variations in spelling and usage throughout this book. For instance, certain words are spelled in different ways: colored and coloured, maneuver and manoeuvre, draft and draught. These differences are due in part to the British style in which the international regulations were written, and the American style in which the U.S. rules are phrased. It is not permissible to alter spelling and usage to bring the style of the various rules into common agreement.

period of eight years, the Advisory Committee developed a proposal to unify the inland navigational rules. The outcome of that proposal was the Inland Navigational Rules Act of 1980, which was signed into law on 24 December, 1981 (Public Law 96–591, 94 Stat. 3415, Title 33, U.S. Code).

The new Inland Rules supersede the old Inland Rules, the Western Rivers Rules, the Great Lakes Rules, their respective pilot rules and interpretive rulings, and parts of the Motorboat Act of 1940. The new rules went into effect on all United States waters except on the Great Lakes on 24 December 1981. The effective date for the Great Lakes was delayed to allow the Canadian government to review the new rules and permit both the U. S. and Canada to adopt them simultaneously. As of this writing, this was expected to occur on 1 March 1983.

It is necessary for mariners who sail upon the waters of the United States to familiarize themselves with all the changes brought about by the revised rules. It is the purpose of these pages to draw attention to the principal changes in order that those familiar with the old rules may quickly and readily determine the nature and the extent of the changes. For a detailed consideration of any point in question, it is suggested that the reader refer to the index.

In addition to substantive alterations to the rules, major editorial changes have been made. The Steering and Sailing Rules now are gathered together and appear before the rules prescribing lights and shapes. Radar, heretofore unmentioned in the old rules, is now prominently incorporated into the main body of the rules. Technical material, of more interest to the shipbuilder than to the mariner, is placed in the annexes, which are regulatory as opposed to statutory, for the thirty-eight rules themselves. The many old pilot rules have been winnowed down to a noncontradictory few and appear as Annex V. What should be of major interest to the mariner who is familiar with, or has occasion to sail under, the International Regulations, is the near alignment of the two sets of rules.

The secondary purpose of this chapter is to compare the 72 COLREGS and the new Inland Rules and point out the differences that remain. In certain instances where the 72 COLREGS were considered inappropriate for specific waters of the United States, the committee deemed it necessary, in the interests of navigational safety, to adopt some of the existing domestic rules and practices. There are also a number of other deviations from the existing 72 COLREGS, incorporating technical and interpretive modifications adopted by the Inter-Governmental Maritime Consultative Organization (IMCO), in proposed amendments to the International Regulations. (See Appendix D.) These amendments are expected to enter into force in midsummer 1983. A brief presentation of the major differences between the two sets of rules follows, and the reader is invited to refer to the detailed discussions later in this chapter and in the applicable sections of Part I and Part II for more comparisons.

Whistle signals The Inland Rules retain from the old inland rules the concept of "signals of intent" rather than "signals of action" used on the high seas. It was felt by the advisory committee that the existing proposal and reply whistle signals provide a greater margin of safety on our inland waters. On the high seas, the one- or two-blast signal is sounded to indicate a change of course. In inland waters, particularly on many of our winding rivers, this would lead to a cacophony of whistle blowing. Thus, the one- or two-blast signal is used by a proposing vessel to indicate her intent in crossing, overtaking, or meeting another vessel regardless of a change in heading; the other vessel signifies her agreement by sounding a like signal.

Narrow channels Under the new inland rules, a vessel downbound with the current, in certain specified waters, has the right-of-way over an upbound vessel and proposes the manner of passage. This provision is not found in the International Regulations. The Inland Navigational Rules also dispense with the complicated intent-to-pass and assent-to-pass overtaking signals required by the international rules for vessels in a narrow channel, retaining instead the old one- and two-blast passing signals.

Crossing situation Inland Rule 15(b), which requires a crossing vessel to keep out of the way of an ascending or descending power-driven vessel operating in certain designated rivers, is not found in the International Regulations.

Vessels constrained by their draft Such vessels are not mentioned in the Inland Rules, either in the rules outlining responsibilities between vessels or the rules describing special lights and day shapes.

Towing lights and lights for tows The Inland Rules require: a special flashing yellow light to be placed forward on a tow or tows being pushed ahead; white lights to mark a partially submerged tow; and a second yellow towing light for vessels towing ahead or pushing alongside. They exempt vessels in certain specified waters from carrying any white towing lights when pushing ahead or towing alongside, and permit placement of the white towing lights on either the forward or after mast. The International Rules are presently silent on these provisions, but upon entry into force of the proposed amendments in 1983, the white lights for a partially submerged tow will be required, and the placing of the white towing lights on either mast will be authorized.

Optional all-round white light A power-driven vessel operating on the Great Lakes may carry an all-round white light in lieu of the second masthead light and sternlight.

Use of radiotelephone Under Inland Rule 34(h), a vessel that reaches agreement with another vessel in a meeting, crossing, or overtaking situation by using the radiotelephone is *not* obliged to sound the prescribed whistle signals for these situations.

Rule-By-Rule Analysis

This analysis draws extensively from a report published by the House of Representatives Committee on Merchant Marine and Fisheries, which in turn was based on an analysis prepared by the Rules of the Road Advisory Committee dated April 1980.

As has been noted, the Inland Navigational Rules are, in many instances, identical (with minor editorial changes) to the 1972 International Regulations. They are divided into five parts containing 38 rules (as are the International Regulations) as follows:

Part A—General

Contains rules on applicability, responsibility, and definitions. This part includes Rules 1 through 3.

Part B—Steering and sailing rules

Contains three subparts for the conduct of vessels in any condition of visibility, in sight of one another, and in restricted visibility. Subparts I, II, and III include Rules 4 through 10, 11 through 18, and 19, respectively.

Part C—Lights and shapes

Contains rules on applicability, definitions, visibility of lights; and lights for power-driven vessels underway, vessels towing and pushing, sailing vessels underway, vessels under oars, fishing vessels, vessels not under command, anchored vessels, vessels aground, and seaplanes. This part includes Rules 20 through 31.

Part D—Sound and light signals

Contains rules and definitions concerning equipment for sound signals, maneuvering and warning signals, sound signals in restricted visibility, signals to attract attention, and distress signals. This part includes Rules 32 through 37.

Part E—Exemptions

Provides certain exemptions to permit a smooth transition period. This part consists of Rule 38.

In addition, there are five annexes, the equivalent of Annex V not being found in the International Regulations, as follows:

Annex I—Positioning and Technical Details of Lights and Shapes
Annex II—Additional Signals for Fishing Vessels Fishing In Close Proximity
Annex III—Technical Details of Sound Signal Appliances
Annex IV—Distress Signals
Annex V—Pilot Rules

PART A—GENERAL

Application

RULE 1 1. This rule is similar to the rules establishing the old Inland Rules and is consistent with the approach of the 72 COLREGS. The Inland Rules apply to all vessels of all nations when on the navigable waters of the United States, shoreward of the COLREGS demarcation lines (see Appendix A) including the United States side of the Great Lakes international boundary. These rules also apply to U.S. vessels when on the Canadian side of the boundary line to the extent that they do not conflict with Canadian law. The Great Lakes were exempt from the enactment date of 24 December 1981 as was previously mentioned and as of this writing were expected to come under the new rules in March 1983.

2. Rule 1. b(ii) states that vessels complying with the construction and equipment requirements of the COLREGS are also in compliance with the Inland Rules.

3. A new provision of the rules parallels the COLREGS and recognizes that vessel traffic service (VTS) regulations may be in effect in certain areas.

4. A provision, similar to old Inland Rule 13, permits special rules for additional lights or whistle signals for warships, convoys, and vessels fishing as a fleet. Authority is also granted to permit vessels that cannot comply fully with the light, shape, and sound-signaling requirements of the rules to comply as closely as possible. Certificates of alternate compliance for naval vessels may be issued by the Secretary of the Navy.

Responsibility

RULE 2 The same as the 72 COLREGS, this rule takes in both the old "Rule of Good Seamanship" and the "General Prudential Rule" as found in Articles 29 and 27 of the old Inland Rules. The former rule has been expanded. It now states that no vessel shall be exempt from failure to comply with all of the rules rather than from the simple neglect to carry lights or signals. The latter is almost identical to the old rule, adding the "limitation of the vessels involved" as a consideration, and leaves no doubt that it also applies to all the rules. Excluded is the requirement to keep a proper lookout, which is now covered in Rule 5.

General Definitions

RULE 3 1. This rule contains general definitions that are necessary to carry out the purposes of the rules, some not appearing in the old rules. Most of the definitions are the same as those found in the 72 COLREGS, except that the term "vessel constrained by her draft" is not included, and definitions 3(l) through 3(q) are peculiar only to U.S. inland waterways.

2. The word "vessel" now includes nondisplacement craft and seaplanes. Thus, surface effect ships, hovercraft, hydrofoils, and seaplanes are considered as ordinary vessels within the rules.

3. The definition of "sailing vessel" is now expressed in more positive terms, stressing the nonuse of propelling machinery.

4. The term "engaged in fishing" has been expanded to include any fishing apparatus that restricts maneuverability. Trolling lines remain excluded.

5. The term "vessel not under command" is defined for the first time. The reason why a vessel may be considered not under command is clarified.

6. A new definition is a "vessel restricted in her ability to maneuver," which encompasses cable and pipe layers, dredgers, ships engaged in replenishing underway, flight operations, or minesweeping vessels and vessels engaged in difficult towing operations, which are unable to keep out of the way of other vessels. The rule makes it clear that this list is not all inclusive.

7. Defining vessels in sight of one another as being "when one can be observed visually from the other" may at first glance appear to be superfluous until one remembers that radar observations are not intended to be substituted for visual observations.

PART B—STEERING AND SAILING RULES

Application

RULE 4 A new rule that states that the rules in this subpart, which relates to lookouts, safe speed, risk of collision, action to avoid collision, narrow channels, and VTS, apply in both clear and restricted visibility. This rule in fact recognizes that ships do navigate in conditions of restricted visibility.

Lookout

RULE 5 This rule expands upon the old Article 29 requirement to "keep a proper lookout" and now further defines a proper lookout to include the use of hearing, as well as sight, and the use of all available means, including the appropriate use of radar.

Safe speed

RULE 6 This is a completely new rule that replaces Article 16 of the old Inland Rules, which admonished a vessel to "go at a moderate speed." The term "safe speed" replaces "moderate speed" and, unlike the latter, must be maintained at all times and in all visibilities. The rule gives a list of factors to be taken into account in determining a safe speed, the first of which is the state of visibility. There is no longer any reference to stopping engines as in the old rules or reducing speed to bare steerage-way. However, "safe speed" may include these or other actions in good or restricted visibility.

Risk of collision

RULE 7 This rule is the same as for the 72 COLREGS and amplifies the old rules and recognizes the wide use of radar. It requires that all available appropriate means must be used to determine risk of collision and that proper use of operational radar, including plotting, is mandatory. Manual plotting by radar is not required, however, when the information would be meaningless, as in a winding channel or river. In discussing change of bearing, warning is given that risk of collision can still exist with vessels at close range, even though an appreciable bearing change is evident. This rule applies in all visibilities.

Action to Avoid Collision

RULE 8 1. Again, this rule is the same as for the 72 COLREGS and is essentially a new rule. It advises the mariner to take positive and timely action to avoid collision with another vessel and directs that such action should result in passing that vessel at a safe distance. The mariner is also advised that an alteration of course alone, with due regard for special circumstances, may be the best action to take to avoid collision.

2. Alteration of course and/or speed have to be readily apparent to another vessel, whether she is observing visually or by radar.

In Narrow Channels

RULE 9 1. Rule 9(a)(i) is the same as Rule 9(a) of the 72 COLREGS and is similar to Article 25 of the old inland rules, except that it now requires *all* vessels to keep to the starboard side of the channel.

2. Inland Rule 9(a)(ii), not found in the 72 COLREGS, applies in narrow channels or fairways of the Great Lakes, Western Rivers, or on waters specified by the Secretary and gives the right-of-way and the initiative for proposing the manner of passing to power-driven vessels proceeding downbound with the current. This rule was not in the old Inland Rules but is similar to the superseded Western Rivers Rule 19 and Great Lakes Rule 24.

3. Rules 9(b) and (c) are the same as the 72 COLREGS and are similar to Articles 20 and 26 of the old rules. Small vessels and sailing vessels are enjoined in strong language not to impede the passage of a vessel restricted to navigating in a narrow channel or fairway. Fishing vessels engaged in fishing are similarly instructed not to impede the passage of such a vessel.

4. Rule 9(d) is essentially a new rule and forbids a vessel from crossing a narrow channel or fairway so as to impede the passage of a vessel limited

to navigating safely only in the channel. This rule is similar to the 72 COLREGS rule except that it replaces the word "may" with the word "shall" when discussing the use of the doubt/danger signal and makes its use mandatory by the vessel restricted to the channel if in doubt as to the intentions of the crossing vessel. This rule does not shift the right-of-way.

5. Rule 9(e)(i) maintains that overtaking whistle signals from the old Pilot Rules anytime a vessel intends to overtake in a narrow channel: one short blast to overtake to port, two short blasts to overtake to starboard. The overtaken vessel signals her assent by sounding the same signal. The 72 COLREGS, however, require signals in an overtaking situation only if the vessel being overtaken must maneuver to permit a safe passing. The overtaking vessel sounds the signals prescribed in COLREG Rule 34 (c) (different signals from those described above); the overtaken vessel must then maneuver to permit a safe passage and must sound the prescribed signal of Rule 34(c). Again, for vessels being overtaken in doubt as to the intentions of an overtaking vessel, the doubt signal is mandatory in inland waters, while only recommended on the high seas.

6. The bend signal, the same for both sets of rules, is a carry-over from the Pilot Rules but is now required by all vessels approaching a bend or fairway where another vessel may be obscured by an intervening obstruction.

7. Rule 9(g) prohibits anchoring in a narrow channel, as does the COLREG Rule, "if the circumstances of the case admit."

Vessel Traffic Services

RULE 10 This completely new rule requires compliance with federal vessel traffic service regulations that may be established in a given port or geographic area. It is equivalent to COLREG Rule 10, which deals with traffic separation schemes, but is less detailed—leaving the specifics to the actual regulations for each area.

Subpart II—Conduct of Vessels in Any Condition of Visibility

Application

RULE 11 This rule is the same as that of the 72 COLREGS and is consistent with the old rules. It specifies that the rules in Subpart II apply only when each vessel is in sight of the other. These rules do not apply to vessels operating in restricted visibility and relying on radar.

Sailing Vessels

RULE 12 This rule is the same as that of the 72 COLREGS. Article 17 of the old Inland Rules has been updated to reflect the prevalence of modern-day fore-and-aft rigged vessels instead of the old square-riggers. The reference to "running free" has been dropped. The rule now conforms more closely to yacht-racing rules. A vessel with the wind on the port side that cannot determine on which side a vessel to windward has the wind must consider herself the give-way vessel.

Overtaking

RULE 13 This rule is very similar to Article 24 of the old Inland Rules and is applicable to vessels in sight of one another in any condition of visibility. At the time of this writing, it differs from the 72 COLREGS only in the applicability. The proposed amendments to the COLREGS, expected to go into force in midsummer 1983, will remove this difference.

Head-on Situation

RULE 14 This new rule is similar to the old Inland Rules and various Pilot Rules, but greatly clarifies what used to be called the "meeting situation." The rule, the same as in the 72 COLREGS, specifies an alteration of course to starboard for a port-to-port passage, and directs a vessel in doubt as to whether or not a head-on situation exists, to assume so and act accordingly.

Crossing Situation

RULE 15 1. Rule 15(a) combines Article 19 and 22 of the old Inland Rules and is the same in the 72 COLREGS. The vessel that has the other to starboard is the give-way vessel and should avoid crossing ahead.

2. Rule 15(b) was not in the old Inland Rules and is not found in the COLREGS. Similar to the superseded Western Rivers Rule 19, it gives a power-driven vessel ascending or descending a river the right-of-way over a crossing vessel on the Western Rivers, the Great Lakes, and other waters designated by the Secretary. This rule is consistent with Inland Rule 9(d).

Action by Give-way Vessel

RULE 16 This rule replaces Article 23 of the old Inland Rules and is the same as the 72 COLREGS. The term "burdened" has been replaced by "give-way" and the rule enjoins "early and substantial action" for the give-way vessel.

Action by Stand-on Vessel

RULE 17 1. Rule 17 is the same as the 72 COLREGS. Rule 17 (a)(i) is the same as Article 21 of the old Inland Rules, replacing "privileged" with "stand-on," but Rule 17(a)(ii) is a new provision in that the stand-on vessel is given freedom to take action prior to extremis "as soon as it becomes apparent that the vessel required to keep out of the way is not taking appropriate action." This is a major change. The reasoning is that a stand-on vessel, cannot know the quality of alertness prevailing in the give-way vessel, nor her capabilities, and as a result, may have in the past unnecessarily stood on until a dangerous situation arose. Now she can take action to avoid collision by her maneuvers alone—providing, if the circumstances permit, she does not alter course to port for a vessel on her own port side. Notwithstanding this, the give-way vessel is not relieved of her responsibilities.

2. As before, a stand-on vessel finding herself in extremis is required to take "such action as will best aid to avoid collision."

Responsibilities Between Vessels

RULE 18 1. This is a new rule that gathers together all those vessels that are restricted in their ability to maneuver for various reasons. Clear details are given of those vessels required to keep clear of certain other types of vessels.

2. This rule is the same as the 72 COLREGS except that reference to "a vessel constrained by her draft" is not included.

3. The rule concerning seaplanes is new and is the same as the 72 COLREGS.

Subpart III—Conduct of Vessels in Restricted Visibility

Conduct of Vessels in Restricted Visibility

RULE 19 1. This new rule contains a number of modifications from the old rules and is the same as that of the 72 COLREGS. The rule clearly applies to all vessels out of sight of one another, not only to those in an area of restricted visibility but also to any near such an area.

2. The mariner is strictly reminded of the importance of maintaining a safe speed in all visibilities, with specific emphasis on the need to take into account the circumstances of restricted visibility. In addition, a new requirement is that a power-driven vessel must have her engines ready for immediate maneuvering.

3. Clear advice is offered on taking avoiding action when a developing close-quarters situation is detected by radar alone—i.e., before the other vessel is seen or her fog signal heard. To be able to take the necessary action under this rule, it is axiomatic that systematic radar observation and plotting, whether manual or automatic, is required. If such plotting reveals that a close-quarters situation is developing, then avoiding action is mandatory.

4. The action to be taken by a vessel on hearing a fog signal apparently forward of her beam has been changed. If risk of collision or a close-quarters situation has been deemed not to exist, presumably by the use of radar, there is no requirement to "stop engines" as under the old rules. More flexibility is therefore given to a vessel capable of making efficient

use of radar. However, if this exception proviso does not apply, then a vessel must reduce her speed to the minimum at which she can effectively steer. If this action is insufficient, then the vessel must take all way off and navigate with extreme caution until risk of collision is over. This applies to all vessels and not just power-driven vessels as in the past. In effect, the new rule requires all vessels to slow down and to maintain a much lower speed than did the previous requirement for vessels so fitted to merely "stop engines" for an unspecified period of time.

PART C—LIGHTS AND SHAPES

Application

RULE 20 1. Although basically similar to Article 1 of the old rules, the new rule requires, rather than permits, vessels to show lights in restricted visibility between sunrise and sunset. It is the same as the 72 COLREGS.

2. Technical details of lights and shapes are omitted in this part. The requirements for their size, color, and shape are now consolidated in Annex I of the Rules.

Definitions

RULE 21 1. This new rule is the same as the 72 COLREGS with two exceptions: one, it states that a vessel less than 12 meters in length may place her masthead light and sidelight as close to the centerline as possible; and two, it adds a special yellow flashing light, to be placed at the head of a tow being pushed ahead.

2. The repetition in various old rules of describing the arcs, ranges, and position of various lights is avoided in the new rules. This particular rule defines the meaning, color, and arcs of lights. Technical details of positioning such lights are relegated to Annex I. Only the information essential to the mariner is contained here.

3. A new color has been added to the spectrum in the towing lights for vessels being towed. It is a yellow light shown in addition to, and having the same characteristics as, a sternlight.

4. For the first time a flashing light for use on vessels is introduced and defined. This is the 120-flashes-per-minute light displayed by surface effect ships operating in the nondisplacement mode.

Visibility of Lights

RULE 22 The visibility standards of this rule are essentially the same as in the old Inland Rules. This rule lists the range of visibility of lights required for vessels, according to their length in meters. In general, for large vessels ranges have been increased by one mile. For smaller vessels the minimum visibility ranges generally remain the same, except for the sidelights in one class and the masthead light in another. The rule is the same in the 72 COLREGS except for the addition of the special flashing light and the white lights to mark a partially submerged tow. The latter lights will also appear in the COLREGS when the proposed amendments go into effect.

Power-driven Vessels Underway

RULE 23 1. This rule, similar to the 72 COLREGS, neatly gathers together the requirements for lights to be shown by a power-driven vessel. It is far less cumbersome to read and understand than the old rules. The requirements are basically the same as before.

2. A new, all-round flashing yellow light is prescribed for air-cushion vessels operating in the nondisplacement mode.

3. Small power-driven vessels that do not have the capability for the usual lights or experience difficulty in showing them, are accommodated in part (c) of this rule, although the Inland Rule establishes a length of 12 meters—vice 7 meters in the 72 COLREGS—as the dividing line for such vessels.

4. Not found in the 72 COLREGS, Rule 23(d), a continuation of superseded Great Lakes Rule 3, permits the use of an all-round white light in lieu of the second masthead light and sternlight for power-driven vessels operating on the Great Lakes. This exception was adopted because such a light could be seen more easily above the low-lying fogs prevalent in the area.

Towing and Pushing

RULE 24 1. This rule, dealing with the lights for vessels towing or pushing, replaces Article 3 of the old Inland Rules, but contains some significant differences. It is essentially the same as the 72 COLREGS, with modifications considered necessary to retain some consistency with the old rules. One major departure from the international rule is the option given to vessels on inland waters to carry the two or three white towing lights on either the forward or after mast. This difference is expected to be eliminated when the proposed amendments to the 72 COLREGS, which will give vessels towing on the high seas the same option, enter into force in midsummer 1983. Other differences from the international rule include the addition of: a yellow towing light in place of the sternlight for a power-driven vessel pushing ahead or towing alongside; a special flashing light at the head of a tow being pushed ahead; white lights to mark a partly submerged object (expected to be added to the COLREGS); and the removal of the requirement for vessels pushing ahead or towing alongside on the Western Rivers and other specified waters to display any white masthead lights.

2. For a vessel towing astern, the superseded inland rule has been modified to require the display of two white masthead lights, or three when the length of the tow exceeds 200 meters. The optional "small white light abaft the funnel . . . for the tow to steer by" no longer exists, and a white sternlight is now mandatory. A yellow towing light is carried directly above the sternlight and covers the same arc.

3. A vessel pushing ahead that is rigidly connected to the vessel being pushed is regarded as a composite unit with the same maneuverability as a power-driven vessel and is therefore required to be lighted as a power-driven vessel.

4. When pushing ahead or towing alongside, a vessel displays the two masthead lights, sidelights, and two yellow towing lights in a vertical line.

5. The requirement for a towing or pushing vessel 50 meters or more in length to display a second masthead light is explicit; it remains optional for a towing vessel under 50 meters in length.

6. One significant change from the prior U.S. rules is that vessels being towed show sidelights and a sternlight, except that any number of vessels towed alongside or pushed ahead are lighted as one vessel. A vessel being pushed ahead also shows the special flashing light at the forward end.

7. Rule 24(g) directs that an inconspicuous partly submerged tow be marked by white lights. Efforts should be made to indicate the presence of such a tow, or especially a tow that cannot be lighted at all, by some means. The towing vessel can illuminate such an unlighted or poorly lighted tow with her searchlight to warn an approaching vessel. In an emergency, when a vessel has taken another in tow and cannot display the normal towing lights, she should also make use of the searchlight to illuminate the tow.

Sailing Vessels Underway and Vessels under Oars

RULE 25 1. This rule is basically the same as the 72 COLREGS. The basic lighting requirements for sailing vessels in the old Inland Rules have not been changed by Rule 25, with the following exceptions:

(a) in vessels under 20 meters in length the sternlight may also be incorporated into a combined lantern, which may be displayed near the top of the mast.

(b) the optional red over green masthead lights for a sailing vessel can be exhibited at or near the top of the mast, but are not permitted in a sailing vessel showing the combined lantern in (a) above.

2. Small vessels of less than 7 meters are encouraged to exhibit normal navigation lights, but if they cannot, they are required to show a white light in sufficient time to prevent collision.

3. A cone, apex down, is required for a sailing vessel proceeding under power.

Fishing Vessels

RULE 26 1. This rule is the same as the 72 COLREGS, adopting a uniform system of lights for fishing vessels. The red over white all-round lights for a vessel engaged in fishing, other than trawling, is retained from Article 9(c) of the old Inland Rules, but now must be shown in a vertical line. A vessel trawling shows green over white all-round lights in a vertical line. Both show sidelights and a sternlight when making way.

2. The day shape for both fishing vessels and trawlers consists of two cones, apexes together. A vessel less than 20 meters in length may show a basket instead.

3. Signals are provided to mark the direction of gear extending more than 150 meters horizontally from a fishing vessel.

4. Vessels engaged in fishing frequently operate in close proximity to each other. There are new provisions in Annex II for additional signals, to enable vessels, not necessarily only fishing vessels, to identify vessels engaged in special fishing operations—e.g., pair trawling or purse seining.

Vessels Not Under Command or Restricted in Their Ability to Maneuver

RULE 27 1. This rule is the same as the 72 COLREGS, with minor modifications. This is a new rule, as there were no rules for a vessel not under command or restricted in her ability to maneuver in the old Inland Rules. The former shows two red all-round lights or two black balls in a vertical line, while the latter—except for a minesweeper—shows three lights, red over white over red, or three black balls. When making way, they both show sidelights and a sternlight, and a vessel restricted in her ability to maneuver shows the required anchor lights or shapes when at anchor, in addition to the special lights or shapes.

2. A vessel engaged in a towing operation that renders her unable to deviate from her course shows, in addition to her normal towing lights, the lights or shapes of a vessel restricted in her ability to maneuver.

3. Vessels engaged in dredging or underwater operations show the same lights and shapes as a vessel restricted in her ability to maneuver. In addition they must, when an obstruction exists, indicate the clear and foul side by use of special lights and shapes. When at anchor, however, vessels engaged in dredging or underwater operations only display their special identification, and when appropriate, obstruction lights and no anchor lights.

4. For small vessels engaged in diving operations where it is impracticable to show the proper lights and shapes, an alternate signal, adopted from the International Code of Signals, is provided—a rigid replica of the code flag "A."

5. Minesweepers are required to carry three green lights or black balls regardless of the side on which operations are being conducted, one near the foremast and one on each foreyard.

6. A vessel less than 12 meters in length is not required to show any of the lights or shapes of Rule 27 except when divers are operating, and then she should display the code flag "A."

RULE 28 Not used. It only appears to preserve the continuity of the numbering system between the two sets of rules. International Rule 28 addresses vessels constrained by their draft.

Pilot Vessels

RULE 29 Essentially the same as the 72 COLREGS, this rule retains the white over red lights from Article 8 of the old Inland Rules. Otherwise, the old rules have been streamlined; there is no longer mention of sailing pilot vessels, and the requirement for a flare-up light has been dropped.

Anchored Vessels and Vessels Aground

RULE 30 1. This rule is the same as the 72 COLREGS with two additions: vessels less than 12 meters in length are exempt from exhibiting special lights or shapes when aground; and vessels less than 20 meters in length are not required to show anchor lights or shapes when anchored in designated anchorage areas.

2. The light and shape requirements for anchored vessels are basically the same as Article 11 of the old Inland Rules. However, lengths are now expressed in meters, and a vessel less than 50 meters in length may now show her all-round white light "where it can best be seen," and not just in the fore part of the vessel as in the old rule.

3. Vessels at anchor, over 100 meters in length, are required to illuminate their decks. Although this is a new provision in the rules, the practice of exhibiting deck lights is common. It aids considerably in identifying vessels at anchor.

4. Recognition is given to small vessels under 7 meters in length that have difficulty in carrying the lights and shapes in this rule. Notwithstanding this, if such a vessel does anchor or ground" "in or near a narrow channel, fairway or anchorage, or where other vessels normally navigate," she must show the required lights and shapes.

5. Rule 30(d), requiring the display of two red lights or three black balls when aground, is a new rule having no precedent in the prior rules.

Seaplanes

RULE 31 This rule sets forth the requirement that seaplanes comply as closely as possible to the rules in their exhibition of lights and shapes. This is the same as the 72 COLREGS rule and is similar to the prior rules.

PART D—SOUND AND LIGHT SIGNALS

Definitions

RULE 32 The definition of "whistle" has been revised to take into account the new technical requirements laid down in Annex III of the Rules. No mention is made of the siren or foghorn as in the old rules. The duration of the sound signals is unchanged, but the long blast, or bend signal, of old Article 18 no longer exists in the rules. Rule 32 is the same as the 72 COLREGS.

Equipment for Sound Signals

RULE 33 1. All vessels over 12 meters, not merely power-driven ones, are now required to carry a whistle and a bell. If the vessel is over 100 meters, a gong is also required. The term "foghorn" has been removed. Automatic sounding equipment is permitted, but must always be capable of manual operation. Technical specifications are relegated to Annex III.

2. Vessels under 12 meters must be able, through other means, to make efficient sound signals if they cannot carry the equipment above.

3. This rule is similar to the old rules and is the same as the 72 COLREGS.

Maneuvering and Warning Signals

RULE 34 1. As has already been noted, the provisions for maneuvering and warning signals differ considerably from the 72 COLREGS. The signals of "intent" and "assent" have been retained from the old Inland Rules. These proposal and reply signals were considered by the Rules of the Road Advisory Committee to be safer and more convenient than the action signals used in COLREGS waters.

2. As before, when two vessels are in sight of one another in a meeting or crossing situation, one short blast by the proposing vessel signals her intent to leave the other on her own port side while two short blasts signal her intent to leave the other to starboard. The answering vessel, if agreeing to the proposal, replies with the same signal. Cross signals, answering one signal with two or two with one, are prohibited. A clarification from the old rules requires that these signals be exchanged when the vessels will meet or cross within a half mile of each other.

3. Rule 34(c), relating to the overtaking situation, retains the signals and procedures of the old rules. The overtaking vessel sounds one short blast to overtake to port and two short blasts to overtake to starboard. The overtaken vessel replies with the same signal, if she assents.

4. The danger, or in-doubt signal, is now five or more short blasts, instead of four or more in the earlier rules. The doubt signal is now *mandatory* for any vessel that doubts the action or intentions of the other vessel in an approaching situation when the vessels are in sight of one another. Three short blasts means a vessel is operating astern propulsion.

6. A new provision is that any one of the whistle signals in Rule 35 may be supplemented by light signals, but they must be synchronized with the whistle signals.

7. The bend signal—one prolonged blast—is sounded by a vessel approaching a blind bend. The prolonged blast is also sounded by a vessel leaving her berth or dock. The latter requirement is not found in the 72 COLREGS.

8. Also not found in the 72 COLREGS are the provisions of Rule 34(h), which permit a vessel that reaches an agreement by radiotelephone with another vessel in a meeting, crossing, or passing situation to dispense with the prescribed whistle signals.

Sound Signals in Restricted Visibility

RULE 35 1. There are significant differences between the signals required by this rule and those required by the old Inland Rules to be sounded by vessels in restricted visibility. Additionally, the maximum interval between sounding fog signals has increased from the one minute interval to an interval of not more than two minutes. Rule 35 is the same as the 72 COLREGS except that there is no mention of vessels constrained by their draft, and the requirement of small vessels in designated special anchorages to sound fog signals is deleted.

2. A power-driven vessel sounds one prolonged blast at intervals of not more than two minutes when underway making way and two prolonged blasts at the same interval when underway not making way.

3. Vessels hampered in some way—i.e., vessels not under command, restricted in their ability to maneuver, engaged in towing or pushing, engaged in fishing, or sailing vessels—all sound the same signals of one prolonged blast followed by two short blasts. Sailing vessels no longer sound a specific signal indicating their point of sail, which could be confused with maneuvering signals for vessels in sight of one another.

4. The signal for a manned vessel being towed, or the last of a series of vessels towed, is one prolonged followed by three short blasts made immediately after the signal of the towing vessel, if practicable.

5. All anchored vessels sound the required fog signals at intervals of not more than one minute. Apart from expressing lengths in meters and requiring the gong, where needed, to be sounded immediately after the bell, there is no change from Article 15 of the old Inland Rules. When aground, a vessel gives the bell signal, and the gong if required, plus three distinct strokes on the bell immediately before and after the rapid ringing

of the bell. Optional signals, with no required interval, are permitted for vessels at anchor or aground to warn approaching vessels of their position (one short, one prolonged, one short blast). A pilot vessel, in addition to the appropriate required fog signals, may sound an identity signal of four short blasts.

6. Vessels of less than 12 meters, whether underway or not, are released from the obligation to sound the prescribed signals, but must, instead, make an efficient sound signal at not more than two-minute intervals.

Signals to Attract Attention

RULE 36 The same as the 72 COLREGS and similar to the old rules, a searchlight to indicate the direction of danger is authorized for all vessels.

Distress Signals

RULE 37 This is a brief reference to use the signals in Annex IV of the rules if in distress and requiring assistance. The distress signals were not placed in the main body of the rules because they are not directly connected with preventing collisions at sea.

PART E—EXEMPTIONS

Exemptions

RULE 38 1. This new rule is similar to the 72 COLREGS. It allows exemptions for vessels built, or whose keels had been laid, prior to 24 December 1980. The change of units from feet to meters for the placing of lights and the new technical requirements for lights, shapes, and sound-signaling apparatus could result in costly alterations for existing ships.

ANNEX I—POSITIONING AND TECHNICAL DETAILS OF LIGHTS AND SHAPES

1. This annex is based on Annex I to the 72 COLREGS, but has been changed to reflect the special conditions found on United States inland waters. It also incorporates minor amendments to Annex I to the 72 COLREGS, which are expected to become effective in midsummer 1983. Annex I supplements the Inland Navigational Rules covering lights and shapes by specifying the following: the vertical and horizontal positioning and spacing of lights; details of the location of direction-indicating lights for fishing vessels, dredgers, and vessels engaged in underwater operations; screens for sidelights; color and dimensions of shapes; color of lights; intensity of lights; horizontal and vertical sectors; and details of the optional maneuvering light.

2. Items of interest to the mariner are:

(a) allowance must be made for all normal conditions of trim in the vertical separation of masthead lights. On a long voyage it is possible for the after (range) light not to be visible above, or separately from, the forward (masthead) light.

(b) masthead lights must be placed above and clear of all other lights except as noted in Section 84.03f(2).

(c) horizontal spacing of lights for power-driven vessels must be such as to give a more accurate indication of the approximate length of the vessel.

(d) for vessels 20 meters or more in length, the sidelights must not be placed in front of the forward masthead light.

(e) screens for sidelights must be painted mat black.

(f) all shapes are now black.

(g) notwithstanding Item 2 (b) above, the maneuvering light should be placed on the same fore-and-aft vertical plane as the masthead lights, and where practicable, at a minimum height of 2 meters above the forward masthead light, provided that it is not placed less than 2 meters above or below the after masthead light.

ANNEX II—ADDITIONAL SIGNALS FOR FISHING VESSELS FISHING IN CLOSE PROXIMITY

1. This new annex provides additional signals for vessels fishing in the proximity of each other. The signals are optional, but as fishing vessels do frequently operate at close quarters, they provide the opportunity to apprise other vessels of the particular stage of fishing operations in progress. As these signals are identical to those found in Annex II of the 72 COLREGS, they will help to provide standardized signals for fishing vessels on a worldwide basis and hopefully remove much of the conflict, confusion, and entanglement of nets that occur when fishing vessels of different types or nationalities are fishing the same grounds.

2. While of primary interest to the fishing industry and those agencies that enforce the regulations, it behooves every mariner to know these signals. In the event of an inadvertent or necessary approach to a fleet of fishing vessels, he will be better able to extract himself with minimum danger to all.

ANNEX III—TECHNICAL DETAILS OF SOUND SIGNAL APPLIANCES

Details of primary interest to equipment manufacturers are found in this annex. They include frequencies, audible ranges, signal intensity, directional properties, and positioning. The object is to ensure minimum standards among all vessels. Even though his vessel may be fitted with automatic sound-signal apparatuses, the mariner is required to be able to operate them in a manual mode. Annex III is based on Annex III of the 72 COLREGS, but has been changed in response to special situations that occur on inland waters.

ANNEX IV—DISTRESS SIGNALS

1. The majority of signals are not new and are almost identical to those designated for use on international waters.
2. The relevance of the International Code of Signals and the Merchant Ship Search and Rescue Manual is brought to the attention of the mariner.

ANNEX V—PILOT RULES

These supplemental rules primarily concern navigation lights for special applications. This annex is not found in the 72 COLREGS. The rules are not new but have been rewritten to make them easier to read and understand. In keeping with the policy of adhering as closely as possible to the 72 COLREGS and in attempting to cut down on excessive additional regulations, the number of rules in this annex have been kept to a minimum. New rules may be added, however, as they become necessary.

NOTES

There long has been a need to modernize, unify, and simplify the navigational rules for vessels on the high seas and on the inland waters of the United States. With the enactment of the International Navigational Act of 1977, the United States became a party to the International Regulations for Preventing Collisions at Sea, 1972 (COLREGS). That left the inland waters of the United States with various sets of inland navigational rules nowhere near in conformity with the International Regulations as admonished by Rule 1(b). With the enactment of the Inland Navigational Rules Act of 1980, and its subsequent entry into force on 24 December 1981, the U.S. rules closely parallel the international rules, and mariners are relieved of the responsibility to learn several sets of rules when plying between the two jurisdictions. Some differences remain where it was felt the special circumstances or common practices of operating on U.S. waters dictated the retention of some existing rules.

The most significant difference between the two sets of rules concerns whistle signals. The existing inland signals of intent and reply are retained, while the 72 COLREGS retain the international action signals. This dichotomy can still lead to problems when two vessels are approaching each other on either side of the demarcation line marking the boundary of the two jurisdictions.

The enacting legislation for the 1980 Inland Navigational Rules provided for the establishment of a Rules of the Road Advisory Council (RORAC), a continuation of the Advisory Committee, composed of the same or similarly experienced mariners who originally developed the new Inland Rules. The establishment of this council will ensure continued public participation as these new rules are refined over the years with appropriate amendments and additions to Annex V.

2

Applicability, Scope, and Definitions

International Rules

Application

RULE 1 (a) These Rules shall apply to all vessels upon the high seas and in all waters connected therewith navigable by seagoing vessels.

(b) Nothing in these Rules shall interfere with the operation of special rules made by an appropriate authority for roadsteads, harbours, rivers, lakes or inland waterways connected with the high seas and navigable by seagoing vessels. Such special rules shall conform as closely as possible to these rules.

(c) Nothing in these Rules shall interfere with the operation of any special rules made by the Government of any State with respect to additional station or signal lights or whistle signals for ships of war and vessels proceeding under convoy, or with respect to additional station

Inland Rules

Application

RULE 1 (a) These Rules apply to all vessels upon the inland waters of the United States, and to vessels of the United States on the Canadian waters of the Great Lakes to the extent that there is no conflict with Canadian law.

(b)(i) These Rules constitute special rules made by an appropriate authority within the meaning of Rule 1(b) of the International Regulations.

(ii) All vessels complying with the construction and equipment requirements of the International Regulations are considered to be in compliance with these Rules.

(c) Nothing in these Rules shall interfere with the operation of any special rules made by the Secretary of the Navy with respect to additional station or signal lights and shapes or whistle signals for ships of war and vessels proceeding under convoy, or by the Secretary

International Rules

or signal lights for fishing vessels engaged in fishing as fleet. These additional station or signal lights or whistle signals shall, so far as possible, be such that they cannot be mistaken for any light or signal authorized elsewhere under these Rules.

(d) Traffic separation schemes may be adopted by the Organization for the purpose of these Rules.

(e) Whenever the Government concerned shall have determined that a vessel of special construction or purpose cannot comply fully with the provisions of any of these rules with respect to the number, position, range or arc of visibility of lights or shapes, as well as to the disposition and characteristics of sound-signalling appliances, without interfering with the special function of the vessel, such vessel shall comply with such other provisions in regard to the number, position, range or arc of visibility of lights or shapes, as well as to the disposition and characteristics of sound-signaling appliances, as her Government shall have determined to be the closest possible compliance with these rules in respect to that vessel.

Inland Rules

with respect to additional station or signal lights and shapes for fishing vessels engaged in fishing as a fleet. These additional station or signal lights and shapes or whistle signals shall, so far as possible, be such that they cannot be mistaken for any light, shape, or signal authorized elsewhere under these Rules. Notice of such special rules shall be published in the Federal Register and, after the effective date specified in such notice, they shall have effect as if they were a part of these Rules.

(d) Vessel traffic service regulations may be in effect in certain areas.

(e) Whenever the Secretary determines that a vessel or class of vessels of special construction or purpose cannot comply fully with the provisions of any of these Rules with respect to the number, position, range, or arc of visibility of lights or shapes, as well as to the disposition and characteristics of sound-signaling appliances, without interfering with the special function of the vessel, the vessel shall comply with such other provisions in regard to the number, position, range, or arc of visibility of lights or shapes, as well as to the disposition and characteristics of sound signaling appliances, as the Secretary shall have determined to be the closest possible compliance with these Rules. The Secretary may issue a certificate of alternative compliance for a vessel or class of

International Rules

Responsibility

RULE 2 (a) Nothing in these rules shall exonerate any vessel, or the owner, master or crew thereof, from the consequences of any neglect to comply with these Rules or of the neglect of any precaution which may be required by the ordinary practice of seamen, or by the special circumstances of the case.

(b) In construing and complying with these Rules due regard shall be had to all dangers of navigation and collision and to any special circumstances, including the limitations of the vessels involved, which may make a departure from these Rules necessary to avoid immediate danger.

General Definitions

RULE 3 For the purpose of

Inland Rules

vessels specifying the closest possible compliance with these Rules. The Secretary of the Navy shall make these determinations and issue certificates of alternative compliance for vessels of the Navy.

(f) The Secretary may accept a certificate of alternative compliance issued by a contracting party to the International Regulations if he determines that the alternative compliance standards of the contracting party are substantially the same as those of the United States.

Responsibility

RULE 2 (a) Nothing in these Rules shall exonerate any vessel, or the owner, master, or crew thereof, from the consequences of any neglect to comply with these Rules or of the neglect of any precaution which may be required by the ordinary practice of seamen, or by the special circumstances of the case.

(b) In construing and complying with these Rules due regard shall be had to all dangers of navigation and collision and to any special circumstances, including the limitations of the vessels involved, which may make a departure from these Rules necessary to avoid immediate danger.

General Definitions

RULE 3 For the purpose of

International Rules

these Rules, except where the context otherwise requires:

(a) The word "vessel" includes every description of water craft, including non-displacement craft and seaplanes, used or capable of being used as a means of transportation on water.

(b) The term "power-driven vessel" means any vessel propelled by machinery.

(c) The term "sailing vessel" means any vessel under sail provided that propelling machinery, if fitted, is not being used.

(d) The term "vessel engaged in fishing" means any vessel fishing with nets, lines, trawls or other fishing apparatus which restrict manoeuvrability, but does not include a vessel fishing with trolling lines or other fishing apparatus which do not restrict manoeuvrability.

(e) The word "seaplane" includes any aircraft designed to manoeuvre on the water.

(f) The term "vessel not under command" means a vessel which through some exceptional circumstance is unable to manoeuvre as required by these rules and is therefore unable to keep out of the way of another vessel.

(g) The term "vessel restricted in her ability to manoeuvre" means a vessel which from the nature of her work is restricted in her ability to manoeuvre as required by these rules and is therefore unable to

Inland Rules

these Rules and this Act, except where the context otherwise requires:

(a) The word "vessel" includes every description of water craft, including nondisplacement craft and seaplanes, used or capable of being used as a means of transportation on water;

(b) The term "power-driven vessel" means any vessel propelled by machinery;

(c) The term "sailing vessel" means any vessel under sail provided that propelling machinery, if fitted, is not being used;

(d) The term "vessel engaged in fishing" means any vessel fishing with nets, lines, trawls, or other fishing apparatus which restricts maneuverability, but does not include a vessel fishing with trolling lines or other fishing apparatus which do not restrict maneuverability;

(e) The word "seaplane" includes any aircraft designed to maneuver on the water;

(f) The term "vessel not under command" means a vessel which through some exceptional circumstance is unable to maneuver as required by these Rules and is therefore unable to keep out of the way of another vessel;

(g) The term "vessel restricted in her ability to maneuver" means a vessel which from the nature of her work is restricted in her ability to maneuver as required by these Rules and is therefore unable to

International Rules

keep out of the way of another vessel.

The following vessels shall be regarded as vessels restricted in their ability to manoeuvre:

(i) a vessel engaged in laying, servicing or picking up a navigation mark, submarine cable or pipeline;

(ii) a vessel engaged in dredging, surveying or underwater operations;

(iii) a vessel engaged in replenishment or transferring persons, provisions or cargo while underway;

(iv) a vessel engaged in the launching or recovery of aircraft;

(v) a vessel engaged in minesweeping operations;

(vi) a vessel engaged in a towing operation such as severely restricts the towing vessel and her tow in their ability to deviate from their course.

(h) The term "vessel constrained by her draught" means a power-driven vessel which because of her draught in relation to the available depth of water is severely restricted in her ability to deviate from the course she is following.

(i) The word "underway" means that a vessel is not at anchor, or made fast to the shore, or aground.

(j) The words "length" and "breadth" of a vessel mean her length overall and greatest breadth.

(k) Vessels shall be deemed to be in sight of one another only when

Inland Rules

keep out of the way of another vessel; vessels restricted in their ability to maneuver include, but are not limited to:

(i) a vessel engaged in laying, servicing, or picking up a navigation mark, submarine cable, or pipeline;

(ii) a vessel engaged in dredging, surveying, or underwater operations;

(iii) a vessel engaged in replenishment or transferring persons, provisions, or cargo while underway;

(iv) a vessel engaged in the launching or recovery of aircraft;

(v) a vessel engaged in minesweeping operations; and

(vi) a vessel engaged in a towing operation such as severely restricts the towing vessel and her tow in their ability to deviate from their course.

(h) The word "underway" means that a vessel is not at anchor, or made fast to the shore, or aground;

(i) The words "length" and "breadth" of a vessel mean her length overall and greatest breadth;

(j) Vessels shall be deemed to be in sight of one another only when

International Rules

one can be observed visually from the other.

(l) The term "restricted visibility" means any condition in which visibility is restricted by fog, mist, falling snow, heavy rainstorms, sandstorms or any other similar causes.

Inland Rules

one can be observed visually from the other;

(k) The term "restricted visibility" means any condition in which visibility is restricted by fog, mist, falling snow, heavy rainstorms, sandstorms, or any similar causes;

(m) "Great Lakes" means the Great Lakes and their connecting and tributary waters including the Calumet River as far as the Thomas J. O'Brien Lock and Controlling Works (between mile 326 and 327), the Chicago River as far as the east side of the Ashland Avenue Bridge (between mile 321 and 322), and the Saint Lawrence River as far east as the lower exit of Saint Lambert Lock;

(n) "Secretary" means the Secretary of the department in which the Coast Guard is operating;

(o) "Inland Waters" means the navigable waters of the United States shoreward of the navigational demarcation lines dividing the high seas from harbors, rivers, and other inland waters of the United States and the waters of the Great Lakes on the United States side of the International Boundary;

(p) "Inland Rules" or "Rules" means the Inland Navigational Rules and the annexes thereto, which govern the conduct of vessels and specify the lights, shapes, and sound signals that apply on inland waters; and

(q) "International Regulations"

International Rules

Inland Rules

means the International Regulations for Preventing Collisions at Sea, 1972, including annexes currently in force for the United States.

Safe Speed

RULE 6 Every vessel shall at all times proceed at a safe speed so that she can take proper and effective action to avoid collision and be stopped within a distance appropriate to the prevailing circumstances and conditions.

In determining a safe speed the following factors shall be among those taken into account:

(a) By all vessels:

(i) the state of visibility;

(ii) the traffic density including concentrations of fishing vessels or any other vessels;

(iii) the manoeuvrability of the vessel with special reference to stopping distance and turning ability in the prevailing conditions;

(iv) at night the presence of background light such as from shore lights or from back scatter of her own lights;

(v) the state of wind, sea and current, and the proximity of navigational hazards;

(vi) the draught in relation to the available depth of water.

(b) Additionally, by vessels with operational radar:

(i) the characteristics, efficiency and limitations of the radar equipment;

Safe Speed

RULE 6 Every vessel shall at all times proceed at a safe speed so that she can take proper and effective action to avoid collision and be stopped within a distance appropriate to the prevailing circumstances and conditions.

In determining a safe speed the following factors shall be among those taken into account:

(a) By all vessels:

(i) the state of visibility;

(ii) the traffic density including concentrations of fishing vessels or any other vessels;

(iii) the maneuverability of the vessel with special reference to stopping distance and turning ability in the prevailing conditions;

(iv) at night the presence of background light such as from shore lights or from back scatter of her own lights;

(v) the state of wind, sea, and current, and the proximity of navigational hazards;

(vi) the draft in relation to the available depth of water.

(b) Additionally, by vessels with operational radar:

(i) the characteristics, efficiency and limitations of the radar equipment;

International Rules

(ii) any constraints imposed by the radar range scale in use;

(iii) the effect on radar detection of the sea state, weather and other sources of interference;

(iv) the possibility that small vessels, ice and other floating objects may not be detected by radar at an adequate range;

(v) the number, location and movement of vessels detected by radar;

(vi) the more exact assessment of the visibility that may be possible when radar is used to determine the range of vessels or other objects in the vicinity.

Risk of Collision

RULE 7 (a) Every vessel shall use all available means appropriate to the prevailing circumstances and conditions to determine if risk of collision exists. If there is any doubt such risk shall be deemed to exist.

(b) Proper use shall be made of radar equipment if fitted and operational, including long-range scanning to obtain early warning of risk of collision and radar plotting or equivalent systematic observation of detected objects.

(c) Assumptions shall not be made on the basis of scanty information, especially scanty radar information.

(d) In determining if risk of collision exists the following considerations shall be among those taken into account:

Inland Rules

(ii) any constraints imposed by the radar range scale in use;

(iii) the effect on radar detection of the sea state, weather, and other sources of interference;

(iv) the possibility that small vessels, ice and other floating objects may not be detected by radar at an adequate range;

(v) the number, location, and movement of vessels detected by radar; and

(vi) the more exact assessment of the visibility that may be possible when radar is used to determine the range of vessels or other objects in the vicinity.

Risk of Collision

RULE 7 (a) Every vessel shall use all available means appropriate to the prevailing circumstances and conditions to determine if risk of collision exists. If there is any doubt such risk shall be deemed to exist.

(b) Proper use shall be made of radar equipment if fitted and operational, including long-range scanning to obtain early warning of risk of collision and radar plotting or equivalent systematic observation of detected objects.

(c) Assumptions shall not be made on the basis of scanty information, especially scanty radar information.

(d) In determining if risk of collision exists the following considerations shall be among those taken into account:

International Rules

(i) such risk shall be deemed to exist if the compass bearing of an approaching vessel does not appreciably change;

(ii) such risk may sometimes exist even when an appreciable bearing change is evident, particularly when approaching a very large vessel or a tow or when approaching a vessel at close range.

Inland Rules

(i) such risk shall be deemed to exist if the compass bearing of an approaching vessel does not appreciably change; and

(ii) such risk may sometimes exist even when an appreciable bearing change is evident, particularly when approaching a very large vessel or a tow or when approaching a vessel at close range.

International Rules

Application

RULE 20 (a) Rules in this Part shall be complied with in all weathers.

(b) The rules concerning lights shall be complied with from sunset to sunrise, and during such times no other lights shall be exhibited, except such lights as cannot be mistaken for the lights specified in these Rules or do not impair their visibility or distinctive character, or interfere with the keeping of a proper look-out.

(c) The lights prescribed by these Rules shall, if carried, also be exhibited from sunrise to sunset in restricted visibility and may be exhibited in all other circumstances when it is deemed necessary.

(d) The Rules concerning shapes shall be complied with by day.

(e) The lights and shapes specified in these Rules shall comply with the provisions of Annex I to these Regulations

Inland Rules

Application

RULE 20 (a) Rules in this Part shall be complied with in all weathers.

(b) The Rules concerning lights shall be complied with from sunset to sunrise, and during such times no other lights shall be exhibited, except such lights as cannot be mistaken for the lights specified in these Rules or do not impair their visibility or distinctive character, or interfere with the keeping of a proper lookout.

(c) The lights prescribed by these Rules shall, if carried, also be exhibited from sunrise to sunset in restricted visibility and may be exhibited in all other circumstances when it is deemed necessary.

(d) The Rules concerning shapes shall be complied with by day.

(e) The lights and shapes specified in these Rules shall comply with the provisions of Annex I of these Rules.

International Rules

Definitions

RULE 21 (a) "Masthead light" means a white light placed over the fore and aft centreline of the vessel showing an unbroken light over an arc of the horizon of 225 degrees and so fixed as to show the light from right ahead to 22.5 degrees abaft the beam on either side of the vessel.

(b) "Sidelights" means a green light on the starboard side and a red light on the port side each showing an unbroken light over an arc of the horizon of 112.5 degrees and so fixed as to show the light from right ahead to 22.5 degrees abaft the beam on its respective side. In a vessel of less than 20 metres in length the sidelights may be combined in one lantern carried on the fore and aft centreline of the vessel.

(c) "Sternlight" means a white light placed as nearly as practicable at the stern showing an unbroken light over an arc of the horizon of 135 degrees and so fixed as to show the light 67.5 degrees from right aft on each side of the vessel.

Inland Rules

Definitions

RULE 21 (a) "Masthead light" means a white light placed over the fore and aft centerline of the vessel showing an unbroken light over an arc of the horizon of 225 degrees and so fixed as to show the light from right ahead to 22.5 degrees abaft the beam on either side of the vessel, except that on a vessel of less than 12 meters in length the masthead light shall be placed as nearly as practicable to the fore and aft centerline of the vessel.

(b) "Sidelights" means a green light on the starboard side and a red light on the port side each showing an unbroken light over an arc of the horizon of 112.5 degrees and so fixed as to show the light from right ahead to 22.5 degrees abaft the beam on its respective side. On a vessel of less than 20 meters in length the sidelights may be combined in one lantern carried on the fore and aft centerline of the vessel, except that on a vessel of less than 12 meters in length the sidelights when combined in one lantern shall be placed as nearly as practicable to the fore and aft centerline of the vessel.

(c) "Sternlight" means a white light placed as nearly as practicable at the stern showing an unbroken light over an arc of the horizon of 135 degrees and so fixed as to show the light 67.5 degrees from right aft on each side of the vessel.

International Rules

(d) "Towing light" means a yellow light having the same characteristics as the "sternlight" defined in paragraph (c) of this Rule.

(e) "All-round light" means a light showing an unbroken light over an arc of the horizon of 360 degrees.

(f) "Flashing light" means a light flashing at regular intervals at a frequency of 120 flashes or more per minute.

Definitions

RULE 32 (a) The word "whistle" means any sound signalling appliance capable of producing the prescribed blasts and which complies with the specifications in Annex III to these regulations.

(b) The term "short blast" means a blast of about one second's duration.

(c) The term "prolonged blast" means a blast of from four to six seconds' duration.

(d) "Towing light" means a yellow light having the same characteristics as the "sternlight" defined in paragraph (c) of this Rule.

(e) "All-round light" means a light showing an unbroken light over an arc of the horizon of 360 degrees.

(f) "Flashing light" means a light flashing at regular intervals at a frequency of 120 flashes or more per minute.

(g) "Special flashing light" means a yellow light flashing at regular intervals at a frequency of 50 to 70 flashes per minute, placed as far forward and as nearly as practicable on the fore and aft centerline of the tow and showing an unbroken light over an arc of the horizon of not less than 180 degrees nor more than 225 degrees and so fixed as to show the light from right ahead to abeam and no more than 22.5 degrees abaft the beam on either side of the vessel.

Definitions

RULE 32 (a) The word "whistle" means any sound signaling appliance capable of producing the prescribed blasts and which complies with specifications in Annex III to these Rules.

(b) The term "short blast" means a blast of about one second's duration.

(c) The term "prolonged blast" means a blast of from four to six seconds' duration.

NOTES

International Rules The International Rules are based on an international agreement of maritime nations of the world. They are given force by separate statutes enacted by the member countries. The present International Rules are the outgrowth of the International Conference on Safety of Life at Sea in London, England, in 1972. Officially, the rules are known as "The International Regulations for Preventing Collisions at Sea, 1972"; the Senate gave its advice and consent of 28 October 1975, the President approved acceptance on 12 December 1975, and the acceptance was deposited with IMCO (the Intergovernmental Maritime Consultative Organization) effective 23 November 1976. The Secretary General of IMCO declared the 1972 Rules to be in effect from 1200 local time, 15 July 1977. They were proclaimed by the President in January 1977 and put into effect without enabling legislation by Congress. On 27 July 1977, Congress enacted Public Law 95–75, the International Navigational Rules Act of 1977, implementing the 72 COLREGS.

Inland Rules The Inland Navigational Rules Act of 1980 was signed into law on 24 December 1980 and went into effect on 24 December 1981. The new rules supersede the old Inland Rules, the Western Rivers Rules, the Great Lakes Rules, and all their associated pilot rules, and parts of the Motorboat Act of 1940. The Inland Rules are effective on all the inland waters of the United States except the Great Lakes, pending discussions with the Canadian government. At the time of this writing, the effective date for the Great Lakes is expected to be 1 March 1983. The Inland Rules are statutory and codified under Title 33, United States Code, Section 2001 through 2038. The Annexes are regulatory and can be found in Title 33, Code of Federal Regulations, Parts 84 through 88.

Application Every vessel and every seaplane on the water, regardless of flag, ownership, or service, navigates under the International Regulations on the high seas and their approaches outside prescribed inland waters. Similarly, every vessel is under Inland Rules on the inland waters of the United States.

Specific demarcation lines separate inland waters from the waters where the COLREGS are in effect. These may be found in Appendix A.

The Inland Rules constitute the special rules permitted by Rule 1(b) of the International Regulations and now, finally, conform closely to them. Inland Rule 1(b)(ii) states that vessels complying with the International Regulations with regard to construction and equipment requirements are in compliance with the Inland Rules also.

Rule 1(c) continues the international and inland practice of permitting special rules with respect to additional station or signal lights and shapes or whistle signals for ships of war or fishing vessels. And both sets of rules provide authority to permit vessels that cannot fully comply with the light, shape, or sound-signaling requirements to comply as closely as possible. The Secretary of Transportation and the Secretary of the Navy may issue certificates of alternate compliance for vessels of the Coast Guard and vessels of the Navy, respectively. The Secretary of Transportation may accept a certificate of alternative compliance issued by a contracting party to the 72 COLREGS.

TSS/VTS International and Inland Rule 1(d) speaks to the IMCO adopted traffic separation schemes (TSS) and the inland vessel traffic services (VTS), respectively, in effect in certain areas. Currently, vessel traffic services are in effect in Puget Sound, Washington; San Francisco, California; Houston Ship Channel, Texas; and New Orleans and Berwick Bay, Louisiana.

Penalty The International Rules do not contain a monetary penalty, as such a penalty would be more or less unenforceable. Compliance with the rules is based on liability in a collision and, in some countries, action against the licenses of mariners involved. In Inland Rules, where enforcement is more feasible, various penalties are provided for in Rule 38. Certain countries have also made it an offense for vessels of their flag to violate traffic separation schemes—e.g., in the English Channel where both France and the United Kingdom police the straits.

Definitions Rule 3 contains the general definitions that are necessary to carry out the purposes of the rules. The definitions in the Inland Rules are now the same as, or similar to, the definitions in the International Regulations, the significant exception being the absence of the term "vessel constrained by her draft" in the Inland Rules. The definitions found in Inland Rules 3(l) through 3(q) define terms that are peculiar to United States waters.

Responsibility Rule 2, identical for both jurisdictions, contains the concepts of the "Rule of Good Seamanship" and the "Rule of Special Circumstances," often called the "General Prudential Rule," which are firmly embedded in the seasoned mariner's mind. The precepts of Rules 2(a) and (b) extend to the entire body of each set of rules.

Safe speed Special notice should be given to the way in which both sets of rules define safe speed. Now in agreement with the International Regulations, the Inland Rules direct vessels to proceed at a safe speed at *all times*. The old term "moderate speed," just during reduced visibility, has left us. Rule 6 lists a number of factors to be considered in determining a safe speed and also provides a number of considerations for a vessel equipped with an operational radar.

Risk of collision Under both sets of rules, continual observations of the bearing of an approaching vessel are required to determine risk of collision, but specifically under all conditions of visibility and using all appropriate means. Whereas the use of compass bearing is stressed, so, too, is the proper use of radar equipment. The rules later set down separate courses of action to avoid a close-quarters situation, dependent upon whether a vessel is in sight of another or not.

IMCO IMCO, the Intergovernmental Maritime Consultative Organization, is an outgrowth of an international conference in Geneva in 1948, whose purpose was to create an organization to disseminate, coordinate, and modernize shipping matters heretofore dealt with by intermittent international conferences. IMCO was established in 1958. Its membership now includes nearly all maritime countries of the world. In May 1982 IMCO changed its name to IMO—International Maritime Organization.

3

Running and Anchor Lights

International Rules

PART C—LIGHTS AND SHAPES

Application

RULE 20 (a) Rules in this Part shall be complied with in all weathers.

(b) The Rules concerning lights shall be complied with from sunset to sunrise, and during such times no other lights shall be exhibited, except such lights as cannot be mistaken for the lights specified in these Rules or do not impair their visibility or distinctive character, or interfere with the keeping of a proper look-out.

(c) The lights prescribed by these Rules shall, if carried, also be exhibited from sunrise to sunset in restricted visibility and may be exhibited in all other circumstances when it is deemed necessary.

(e) The lights . . . specified in these Rules shall comply with the

Inland Rules

PART C—LIGHTS AND SHAPES

Application

RULE 20 (a) Rules in this Part shall be complied with in all weathers.

(b) The Rules concerning lights shall be complied with from sunset to sunrise, and during such times no other lights shall be exhibited, except such lights as cannot be mistaken for the lights specified in these Rules or do not impair their visibility or distinctive character, or interfere with the keeping of a proper lookout.

(c) The lights prescribed by these Rules shall, if carried, also be exhibited from sunrise to sunset in restricted visibility and may be exhibited in all other circumstances when it is deemed necessary.

(e) The lights . . . specified in these Rules shall comply with the

International Rules

provisions of Annex I to these Regulations.

Definitions

RULE 21 (a) "Masthead light" means a white light placed over the fore and aft centreline of the vessel showing an unbroken light over an arc of the horizon of 225 degrees and so fixed as to show the light from right ahead to 22.5 degrees abaft the beam on either side of the vessel.

(b) "Sidelights" means a green light on the starboard side and a red light on the port side each showing an unbroken light over an arc of the horizon of 112.5 degrees and so fixed as to show the light from right ahead to 22.5 degrees abaft the beam on its respective side. In a vessel of less than 20 metres in length the sidelights may be combined in one lantern carried on the fore and aft centreline of the vessel.

(c) "Sternlight" means a white light placed as nearly as practicable at the stern showing an unbroken light over an arc of the horizon of 135 degrees and so fixed as to show

Inland Rules

provisions of Annex I of these Rules.

Definitions

RULE 21 (a) "Masthead light" means a white light placed over the fore and aft centerline of the vessel showing an unbroken light over an arc of the horizon of 225 degrees and so fixed as to show the light from right ahead to 22.5 degrees abaft the beam on either side of the vessel, except that on a vessel of less than 12 meters in length the masthead light shall be placed as nearly as practicable to the fore and aft centerline of the vessel.

(b) "Sidelights" means a green light on the starboard side and a red light on the port side each showing an unbroken light over an arc of the horizon of 112.5 degrees and so fixed as to show the light from right ahead to 22.5 degrees abaft the beam on its respective side. On a vessel of less than 20 meters in length the sidelights may be combined in one lantern carried on the fore and aft centerline of the vessel, except that on a vessel of less than 12 meters in length the sidelights when combined in one lantern shall be placed as nearly as practicable to the fore and aft centerline of the vessel.

(c) "Sternlight" means a white light placed as nearly as practicable at the stern showing an unbroken light over an arc of the horizon of 135 degrees and so fixed as to show

International Rules

the light 67.5 degrees from right aft on each side of the vessel.

(e) "All-round light" means a light showing an unbroken light over an arc of the horizon of 360 degrees.

(f) "Flashing light" means a light flashing at regular intervals at a frequency of 120 flashes or more per minute.

Visibility of Lights

RULE 22 The lights prescribed in these Rules shall have an intensity as specified in Section 8 of Annex I to these regulations so as to be visible at the following minimum ranges:

(a) In vessels of 50 metres or more in length:

a masthead light, 6 miles;

a sidelight, 3 miles;

a sternlight, 3 miles;

a towing light, 3 miles

a white, red, green or yellow all-round light, 3 miles.

(b) In vessels of 12 metres or more in length but less than 50 metres in length:

a masthead light, 5 miles; except that where the length of the vessel is less than 20 metres, 3 miles;

a sidelight, 2 miles;

a sternlight, 2 miles;

a towing light, 2 miles

a white, red, green or yellow all-round light, 2 miles.

(c) In vessels of less than 12 metres in length:

Inland Rules

the light 67.5 degrees from right aft on each side of the vessel.

(e) "All-round light" means a light showing an unbroken light over an arc of the horizon of 360 degrees.

(f) "Flashing light" means a light flashing at regular intervals at a frequency of 120 flashes or more per minute.

Visibility of Lights

RULE 22 The lights prescribed in these Rules shall have an intensity as specified in Annex I to these Rules, so as to be visible at the following minimum ranges:

(a) In a vessel of 50 meters or more in length:

a masthead light, 6 miles;

a sidelight, 3 miles;

a sternlight, 3 miles;

a towing light, 3 miles;

a white, red, green or yellow all-round light, 3 miles; and

a special flashing light, 2 miles.

(b) In a vessel of 12 meters or more in length but less than 50 meters in length:

a masthead light, 5 miles; except that where the length of the vessel is less than 20 meters, 3 miles;

a sidelight, 2 miles;

a sternlight, 2 miles;

a towing light, 2 miles;

a white, red, green or yellow all-round light, 2 miles; and

a special flashing light, 2 miles.

(c) In a vessel of less than 12 meters in length:

International Rules

a masthead light, 2 miles;
a sidelight, 1 mile;
a sternlight, 2 miles;
a towing light, 2 miles;
a white or yellow all-round light, 2 miles.

Power-driven Vessels Underway[1]

RULE 23 (a) A power-driven vessel underway shall exhibit:

(i) a masthead light forward;

(ii) a second masthead light abaft of and higher than the forward one; except that a vessel of less than 50 metres in length shall not be obliged to exhibit such light but may do so;

(iii) sidelights;[2]

(iv) a sternlight.

(b) An air-cushion vessel when operating in the nondisplacement mode shall, in addition to the lights prescribed in paragraph (a) of this Rule, exhibit an all-round flashing yellow light.

[1]Heights of light, including vertical and horizontal separations, are found in Annex I to the International Regulations and Inland Rules.

[2]Construction details of screens are found in Annex I to the International Regulations and Inland Rules.

Inland Rules

a masthead light, 2 miles;
a sidelight, 1 mile;
a sternlight, 2 miles;
a towing light, 2 miles;
a white, red, green or yellow all-round light, 2 miles; and
a special flashing light, 2 miles.

(d) In an inconspicuous, partly submerged vessel or object being towed:

a white all-round light, 3 miles.

Power-driven Vessels Underway[1]

RULE 23 (a) A power-driven vessel underway shall exhibit:

(i) a masthead light forward; except that a vessel of less than 20 meters in length need not exhibit this light forward of amidships but shall exhibit it as far forward as is practicable.

(ii) a second masthead light abaft of and higher than the forward one, except that a vessel of less than 50 meters in length shall not be obliged to exhibit such light but may do so;

(iii) sidelights;[2] and

(iv) a sternlight.

(b) An air-cushion vessel when operating in the nondisplacement mode shall, in addition to the lights prescribed in paragraph (a) of this Rule, exhibit an all-round flashing yellow light where it can best be seen.

Fig. 1. Power-driven vessel underway, with after masthead light, high seas or inland waters. International and Inland Rule 23(a).

Fig. 2. An air-cushion vessel operating in the non-displacement mode, showing all-round flashing yellow light. International and Inland Rule 23(b).

Fig. 3. Power-driven vessel underway, with fixed stern light, high seas or inland waters. International and Inland Rule 23(a).

Fig. 4. Power-driven vessel underway, less than 50 meters in length, not showing after masthead light. International and Inland Rule 23(a).

International Rules

(c) A power-driven vessel of less than 7 metres in length and whose maximum speed does not exceed 7 knots may, in lieu of the lights prescribed in paragraph (a) of this Rule, exhibit an all-round white light. Such vessel shall, if practicable, also exhibit sidelights.

Sailing Vessel Underway and Vessels Under Oars

RULE 25 (a) A sailing vessel underway shall exhibit:

(i) sidelights;

(ii) a sternlight.

(b) In a sailing vessel of less than 12 metres in length the lights prescribed in paragraph (a) of this Rule may be combined in one lantern carried at or near the top of the mast where it can best be seen.

(c) A sailing vessel underway may, in addition to the lights prescribed in paragraph (a) of this Rule, exhibit at or near the top of the mast, where they can best be seen, two all-round lights in a vertical line, the upper being red and the lower green, but these lights shall not be exhibited in conjunction with the combined lantern permitted by paragraph (b) of this Rule.

(d)(i) A sailing vessel of less

Inland Rules

(c) A power-driven vessel of less than 12 meters in length may, in lieu of the lights prescribed in paragraph (a) of this Rule, exhibit an all-round white light and sidelights.

(d) A power-driven vessel when operating on the Great Lakes may carry an all-round white light in lieu of the second masthead light and sternlight prescribed in paragraph (a) of this Rule. The light shall be carried in the position of the second masthead light and be visible at the same minimum range.

Sailing Vessels Underway and Vessels Under Oars

RULE 25 (a) A sailing vessel underway shall exhibit:

(i) sidelights; and

(ii) a sternlight.

(b) In a sailing vessel of less than 20 meters in length the lights prescribed in paragraph (a) of this Rule may be combined in one lantern carried at or near the top of the mast where it can best be seen.

(c) A sailing vessel underway may, in addition to the lights prescribed in paragraph (a) of this Rule, exhibit at or near the top of the mast, where they can best be seen, two all-round lights in a vertical line, the upper being red and the lower green, but these lights shall not be exhibited in conjunction with the combined lantern permitted by paragraph (b) of this Rule.

(d)(i) A sailing vessel of less than

Fig. 5. Power-driven vessel, less than 12 meters in length (7 meters in international waters) showing all-round white light and sidelights. International and Inland Rule 23(c).

Fig. 6. Sailing vessel underway, showing optional foremast lights, high seas and inland waters. International and Inland Rule 25(c).

Fig. 7. Sailing vessel, less than 20 meters in length (12 meters in international waters) showing combined lantern at or near top of mast. International and Inland Rule 25(b).

Fig. 8. Sailing vessel underway, stern view, inland waters or high seas. International and Inland Rule 25(a).

International Rules

than 7 metres in length shall, if practicable, exhibit the lights prescribed in paragraph (a) or (b) of this Rule, but if she does not, she shall have ready at hand an electric torch or lighted lantern showing a white light which shall be exhibited in sufficient time to prevent collision.

(ii) A vessel under oars may exhibit the lights prescribed in this Rule for sailing vessels, but if she does not, she shall have ready at hand an electric torch or lighted lantern showing a white light which shall be exhibited in sufficient time to prevent collision.

Anchored Vessels and Vessels Aground

RULE 30 (a) A vessel at anchor shall exhibit where it can best be seen:

(i) in the fore part, an all-round white light . . . ;

(ii) at or near the stern and at a lower level than the light prescribed in subparagraph (i), an all-round white light.

(b) A vessel of less than 50 metres in length may exhibit an all-round white light where it can best be seen instead of the lights prescribed in paragraph (a) or this Rule.

(c) A vessel at anchor may, and a vessel of 100 metres and more in length shall, also use the available working or equivalent lights to illuminate her decks.

(d) A vessel aground shall ex-

Inland Rules

7 meters in length shall, if practicable, exhibit the lights prescribed in paragraph (a) or (b) of this Rule, but if she does not, she shall have ready at hand an electric torch or lighted lantern showing a white light which shall be exhibited in sufficient time to prevent collision.

(ii) A vessel under oars may exhibit the lights prescribed in this Rule for sailing vessels, but if she does not, she shall have ready at hand an electric torch or lighted lantern showing a white light which shall be exhibited in sufficient time to prevent collision.

Anchored Vessels and Vessels Aground

RULE 30 (a) A vessel at anchor shall exhibit where it can best be seen:

(i) in the fore part, an all-round white light . . . ;

(ii) at or near the stern and at a lower level than the light prescribed in subparagraph (i), an all-round white light.

(b) A vessel of less than 50 meters in length may exhibit an all-round white light where it can best be seen instead of the lights prescribed in paragraph (a) of this Rule.

(c) A vessel at anchor may, and a vessel of 100 metres and more in length shall, also use the available working or equivalent lights to illuminate her decks.

(d) A vessel aground shall ex-

Fig. 9. Power-driven vessel of less than 20 meters, showing combined lantern in lieu of sidelights. International and Inland Rules 21(b) and 23(a).

Fig. 10. A vessel under oars (or less than 7 meters in length, under sail) with lighted lantern (or flashlight) showing a white light. International and Inland Rule 25(d).

Fig. 11. Vessel less than 50 meters in length at anchor. International and Inland Rule 30(b).

Fig. 12. Vessel 50 meters or more in length at anchor, inland waters or high seas. International and Inland Rule 30(a).

International Rules

hibit the lights prescribed in paragraph (a) or (b) of this Rule and in addition, where they can best be seen:

(i) two all-round red lights in a vertical line;

(e) A vessel of less than 7 metres in length, when at anchor or aground, not in or near a narrow channel, fairway or anchorage, or where other vessels normally navigate, shall not be required to exhibit the lights . . . prescribed in paragraphs (a), (b), or (d) of this Rule.

Inland Rules

hibit the lights prescribed in paragraph (a) or (b) or this Rule and in addition, if practicable, where they can best be seen:

(i) two all-round red lights in a vertical line;

(e) A vessel of less than 7 meters in length, when at anchor, not in or near a narrow channel, fairway, anchorage, or where other vessels normally navigate, shall not be required to exhibit the lights . . . prescribed in paragraph (a) and (b) of this Rule.

(f) A vessel of less than 12 meters in length when aground shall not be required to exhibit the lights . . . prescribed in subparagraphs (d)(i) . . . of this Rule.

(g) A vessel of less than 20 meters in length, when at anchor in a special anchorage area designated by the Secretary, shall not be required to exhibit the anchor lights . . . required by this Rule.

International Rules

Seaplanes

RULE 31 Where it is impracticable for a seaplane to exhibit lights . . . of the characteristics or in the positions prescribed in the Rules of this Part she shall exhibit lights . . . as closely similar in characteristics and position as is possible.

Inland Rules

Seaplanes

RULE 31 Where it is impracticable for a seaplane to exhibit lights . . . of the characteristics or in the positions prescribed in the Rules of this Part she shall exhibit lights . . . as closely similar in characteristics and position as is possible.

International Rules

Exemptions—Navy and Coast Guard Vessels

RULE 1 (e) Whenever the Government concerned shall have determined that a vessel of special

Inland Rules

Exemptions—Navy and Coast Guard Vessels

RULE 1 (e) Whenever the Secretary determines that a vessel or class of vessels of special con-

Fig. 13. Vessel less than 50 meters in length aground, inland waters and high seas. International and Inland Rule 30(d).

Fig. 14. Vessel aground, high seas and inland waters. International and Inland Rule 30(d).

International Rules

construction or purpose cannot comply fully with the provisions of any of these Rules with respect to the number, position, range or arc of visibility of lights . . . without interfering with the special function of the vessel or seaplane, such vessel shall comply with such other provisions in regard to the number, position, range or arc of visibility of lights . . . as her Government shall have determined to be the closest possible compliance with these Rules in respect to that vessel.

*Navy and Coast Guard Vessel Exceptions**
(33 U.S.C. 1052)

Any requirement of such regulations in respect of the number, position, range of visibility, or arc of visibility of the lights required to be displayed by vessels shall not apply to any vessel of the Navy or of the Coast Guard whenever the Secretary of the Navy or the Secretary of Transportation, in the case of Coast Guard vessels operating under the Department of Trans-

*(See Title 33, Code of Federal Regulations, Part 135 and Title 32, Code of Federal Regulations, Parts 706 and 707.)

Inland Rules

struction or purpose cannot comply fully with the provisions of any of these Rules with respect to the number, position, range, or arc of visibility of lights . . . without interfering with the special function of the vessel, the vessel shall comply with such other provisions in regard to the number, position, range, or arc of visibility of lights . . . as the Secretary shall have determined to be the closest possible compliance with these Rules. The Secretary may issue a certificate of alternative compliance for a vessel or class of vessels specifying the closest possible compliance with these Rules. The Secretary of the Navy shall make these determinations and issue certificates of alternative compliance for vessels of the Navy.

International Rules

portation, or such official as either may designate, shall find or certify that, by reason of special construction, it is not possible for such vessel or class of vessels to comply with such regulations. The lights of any such exempted vessel or class of vessels, however, shall conform as closely to the requirements of the applicable regulations as the Secretary or such official shall find or certify to be feasible. Notice of such findings or certification and of the character and position of the lights prescribed to be displayed on such exempted vessel or class of vessels shall be published in the Federal Register and in the Notice to Mariners and, after the effective date specified in such notice, shall have effect as part of such regulations.

PART E—EXEMPTIONS

Exemptions

RULE 38 Any vessel (or class of vessels) provided that she complies with the requirements of the International Regulations for Preventing Collisions at Sea, 1960, the keel of which is laid or which is at a corresponding stage of construction before the entry into force of these Regulations may be exempted from compliance therewith as follows:

(a) The installation of lights with ranges prescribed in Rule 22, until four years after the date of entry into force of these Regulations.

Inland Rules

PART E—EXEMPTIONS

Exemptions

RULE 38 Any vessel or class of vessels, the keel of which is laid or which is at a corresponding stage of contruction before the date of enactment of this Act, provided that she complies with the requirements of—

(a) The Act of June 7, 1897 (30 Stat. 96), as amended (33 U.S.C. 154–232) for vessels navigating the waters subject to that statute;

International Rules

(b) The installation of lights with colour specifications as prescribed in Section 7 of Annex I to these Regulations, until four years after the date of entry into force of these Regulations.

(c) The repositioning of lights as a result of conversion from Imperial to metric units and rounding off measurement figures, permanent exemption.

(d)(i) The repositioning of masthead lights on vessels of less than 150 metres in length, resulting from the prescriptions of Section 3(a) of Annex I, permanent exemption.

(ii) The repositioning of masthead lights on vessels of 150 metres or more in length, resulting from the prescriptions of Section 3(a) of Annex I to these Regulations, until nine years after the date of entry into force of these Regulations.

(e) The repositioning of masthead lights resulting from the prescriptions of Section 2(b) of Annex I, until nine years after the date of entry into force of these Regulations.

(f) The repositioning of sidelights resulting from the prescriptions of Sections 2(g) and 3(b) of

Inland Rules

(b) Section 4233 of the Revised Statutes (33 U.S.C. 301–356) for vessels navigating the waters subject to that statute;

(c) The Act of February 8, 1895 (28 Stat. 645), as amended (33 U.S.C. 241–295) for vessels navigating the waters subject to that statute; or

(d) Sections 3, 4, and 5 of the Act of April 25, 1940 (54 Stat. 163), as amended (46 U.S.C. 526 b, c, and d) for motorboats navigating the waters subject to that statute; shall be exempted from compliance with the technical Annexes to these Rules as follows:

(i) the installation of lights with ranges prescribed in Rule 22, until 4 years after the effective date of these Rules, except that vessels of less than 20 meters in length are permanently exempt;

(ii) the installation of lights with color specifications as prescribed in Annex I to these Rules, until 4 years after the effective date of these Rules, except that vessels of less than 20 meters in length are permanently exempt;

(iii) the repositioning of lights as a result of conversion to metric units and rounding off measurement figures, are permanently exempt; and

(iv) the horizontal repositioning of masthead lights prescribed by Annex I to these Rules:

(1) on vessels of less than 150

International Rules

Annex I, until nine years after the date of entry into force of these Regulations.

Inland Rules

meters in length, permanent exemption.

(2) on vessels of 150 meters or more in length, until 9 years after the effective date of these Rules.

(v) the restructuring or repositioning of all lights to meet the prescriptions of Annex I to these Rules, until 9 years after the effective date of these Rules;

(vi) power-driven vessels of 12 meters or more but less than 20 meters in length are permanently exempt from the provisions of Rule 23(a)(i) and 23(a)(iv) provided that, in place of these lights, the vessel exhibits a white light aft visible all round the horizon;

NOTES

The definitions of various lights and visibilities required on the high seas and in inland waters are contained in Rules 21 and 22, International Regulations and Inland Rules. The technical provisions of Annex I of both sets of rules are concerned with the placement and specifications of lights. These details are provided to ensure uniform construction of navigation lights by all manufacturers. In the past, operating under definitions such as "visible on a dark night with a clear atmosphere," it was possible for manufacturers to use different parameters for calculating ranges and chromaticities, leading to varying color intensities on board vessels.

Running lights differences With the passing of the Inland Navigational Rules Act of 1980, the basic differences between the running and anchor lights to be displayed by vessels in inland and international waters have largely disappeared. The application, definitions, and visibility and annexes are nearly identical. The inland rules mention an all-round white light for a partly submerged object and a special flashing light, which the international regulations do not. There are some differences with respect to the position of, or requirements for, lights for vessels less than 20, 12, or 7 meters in length that bear noting by operators of the vessels concerned, and the inland rules add Rule 23(d), which allows power-driven vessels operating on the Great Lakes to exhibit an all-round white light in lieu of the second masthead light and sternlight prescribed in Rule 24(a).

Power-driven vessels underway Rule 23, both sets of rules, requires a power-driven vessel underway to exhibit a masthead light forward, a second masthead light aft of and higher than the forward one, sidelights and a sternlight. Vessels less than 50 meters in length need not show the second masthead light. Vessels less than 12 meters (7 meters for international waters) may, in lieu of these lights, show an all-round white light and sidelights.

Sailing vessels In accordance with Rule 25, both sets of rules, in addition to sidelights and a sternlight, a sailing vessel *may* carry at or near the top of the mast two all-round lights in a vertical line, the upper red and the lower green. For vessels less than 20 meters (12 meters on the high seas), the sidelights and sternlights may be combined in one lantern at or near the top of the mast. If such a combined lantern is carried, the optional red and green cannot be. Sailing vessels under 7 meters and vessels under oars should have handy a flashlight or lantern to indicate their presence to other vessels if they are not carrying the prescribed lights of Rules 25(a) and (b), which they are not required to do.

Air-cushion vessels Where such vessels are operating in the nondisplacement mode, it must be borne in mind that their heading, as indicated by their lights, may be considerably different from their track over the seas. This is caused by drift, due to strong winds and the unique handling features of the craft. The flashing all-round yellow light required by Rule 23(b), International Regulations and Inland Rules, should alert mariners to the need for caution when meeting these high-speed craft, which have difficulty in indicating their true aspect.

Anchor lights International and Inland Rules 30(a) and (b), require a vessel at anchor, 50 meters in length or greater, to exhibit two all-round white lights—one forward and a lower one, aft. A vessel less than 50 meters in length need show only one anchor light where it can best be seen. The two red lights shown by a vessel aground, both in international and inland waters, must be shown along with the proper anchor lights. Rule 30(e), International Regulations and Inland Rules, exempts, in certain circumstances, vessels of less than 7 meters from the necessity to show the anchor and aground lights of Rule 30(a) (b) or (d). In inland waters, any vessel less than 12 meters in length need not show the two red lights when aground, and any vessel less than 20 meters in length at anchor in a special anchorage area established pursuant to Rule 30(g) need not show any light whatsoever. These anchorages may be found in numerous places along the coasts of the United States. Descriptions of the anchorages are published separately in Title 33, Code of Federal Regulations, and in local harbor guides.

Exemptions Rule 38 discusses exemptions for vessels or classes of vessels for which the keel had been laid before entry into force of the International Regulations or enactment of the Inland Rules, for each jurisdiction. The latter set of rules was enacted on 24 December 1980 and provides for vessels to be exempt from complying with certain sections of the technical annexes for specified periods of time. Power-driven vessels of less than 20 meters, but more than 12 meters in length, are permanently exempt from showing the masthead light and sternlight of Rule 23, provided they exhibit an all-round white light.

Exemptions—Naval and Coast Guard vessels Naval and Coast Guard vessels of special construction are authorized some variation in the carriage of lights required by the rules, but such vessels are required to be in the *closest possible compliance* with the lighting requirements prescribed. The extent of the exemptions authorized such vessels by the rules is clear under Rule 1(e). The Secretary of Transportation for Coast Guard vessels and the Secretary of the Navy for naval vessels may issue a certificate of alternate compliance for a vessel or class of vessels *specifying* the closest possible compliance with the Inland Rules.

Submarines are among the exempted vessels. Pursuant to Rule 1(c) International Regulations and Inland Rules, submarines are authorized to display a flashing amber-colored light in addition to the modified navigation lights as a special identifying signal.

Positioning and technical details of lights The reader is invited to refer to Annex I of both sets of rules (Appendixes B and E) for positioning and technical details of lights.

4

Towing Lights

International Rules

Definitions

RULE 21 (d) "Towing light" means a yellow light having the same characteristics as the "sternlight" defined in paragraph (c) of this rule.

Visibility of Lights

RULE 22 The lights prescribed in these Rules shall have an intensity as specified in Annex I to

Inland Rules

Definitions

RULE 21 (d) "Towing light" means a yellow light having the same characteristics as the "sternlight" defined in paragraph (c) of this rule.

(g) "Special flashing light" means a yellow light flashing at regular intervals at a frequency of 50 to 70 flashes per minute, placed as far forward and as nearly as practicable on the fore and aft centerline of the tow and showing an unbroken light over an arc of the horizon of not less than 180 degrees nor more than 225 degrees and so fixed as to show the light from right ahead to abeam and no more than 22.5 degrees abaft the beam on either side of the vessel.

Visibility of Lights

RULE 22 The lights prescribed in these Rules shall have an intensity as specified in Annex I to

International Rules

these Rules, so as to be visible at the following minimum ranges:

(a) In vessels of 50 metres or more in length:

a towing light, 3 miles;

(b) In vessels of 12 metres or more in length but less than 50 metres in length:

a towing light, 2 miles;

(c) In vessels of less than 12 metres in length:

a towing light, 2 miles;

Power-Driven Vessels When Towing or Pushing

RULE 24 (a) A power-driven vessel when towing shall exhibit:

(i) Instead of the light prescribed in Rule 23 (a)(i), two masthead lights forward in a vertical line. When the length of the tow, measuring from the stern of the towing vessel to the after end of the tow exceeds 200 metres, three such lights in a vertical line;

(ii) sidelights;

(iii) a sternlight;

(iv) a towing light in a vertical line above the sternlight;

(b) When a pushing vessel and a vessel being pushed ahead are rigidly connected in a composite unit they shall be regarded as a power-driven vessel and exhibit the lights prescribed in Rule 23.

(c) A power-driven vessel when pushing ahead or towing alongside, except in the case of a composite unit, shall exhibit:

Inland Rules

these Rules, so as to be visible at the following minimum ranges:

(a) In a vessel of 50 meters or more in length:

a towing light, 3 miles;

(b) In a vessel of 12 meters or more in length but less than 50 meters in length:

a towing light, 2 miles;

(c) In a vessel of less than 12 meters in length:

a towing light, 2 miles;

Power-Driven Vessels When Towing or Pushing

RULE 24 (a) A power-driven vessel when towing astern shall exhibit:

(i) instead of the light prescribed either in Rule 23 (a)(i) or 23(a)(ii), two masthead lights in a vertical line. When the length of the tow, measuring from the stern of the towing vessel to the after end of the tow exceeds 200 meters, three such lights in a vertical line;

(ii) sidelights;

(iii) a sternlight;

(iv) a towing light in a vertical line above the sternlight;

(b) When a pushing vessel and a vessel being pushed ahead are rigidly connected in a composite unit they shall be regarded as a power-driven vessel and exhibit the lights prescribed in Rule 23.

(c) A power-driven vessel when pushing ahead or towing alongside, except as required by paragraphs (b) and (i) of this Rule, shall exhibit:

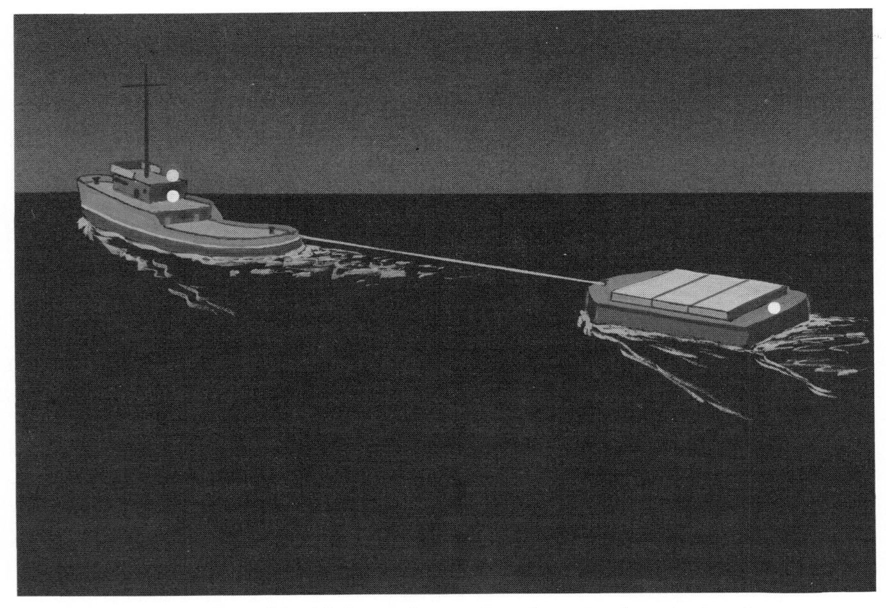

Fig. 15. Tug (any length) with barge in tow (any length of tow), view from astern. International and Inland Rules 24(a) and (e).

Fig. 16. Tug with large power-driven vessel in tow, length of tow over 200 meters, high seas or inland waters. After range light may be omitted if towing vessel is under 50 meters. International and Inland Rules 24(a) and (e).

International Rules

(i) instead of the light prescribed in Rule 23(a)(i), two masthead lights forward in a vertical line;
(ii) sidelights;
(iii) a sternlight.

(d) A power-driven vessel to which paragraphs (a) and (c) of this rule apply shall also comply with Rule 23(a)(ii).

Vessels Being Towed and Vessels Being Pushed Ahead

(e) A vessel or object being towed shall exhibit:

(i) sidelights;
(ii) a sternlight;

(f) Provided that any number of vessels being towed alongside or pushed in a group shall be lighted as one vessel,

(i) a vessel being pushed ahead, not being part of a composite unit, shall exhibit at the forward end, sidelights;

(ii) a vessel being towed alongside shall exhibit a sternlight and at the forward end sidelights.

Inland Rules

(i) instead of the light prescribed either in Rule 23 (a)(i) or 23(a)(ii), two masthead lights in a vertical line;
(ii) sidelights; and
(iii) two towing lights in a vertical line.

(d) A power-driven vessel to which paragraphs (a) or (c) of this Rule apply shall also comply with Rule 23(a)(i) and 23(a)(ii).

Vessels Being Towed and Vessels Being Pushed Ahead

(e) A vessel or object other than those referred to in paragraph (g) of this Rule being towed shall exhibit:

(i) sidelights;
(ii) a sternlight;

(f) Provided that any number of vessels being towed alongside or pushed in a group shall be lighted as one vessel:

(i) a vessel being pushed ahead, not being part of a composite unit, shall exhibit at the forward end sidelights, and a special flashing light; and

(ii) a vessel being towed alongside shall exhibit a sternlight and at the forward end sidelights.

(g) An inconspicuous, partly submerged vessel or object being towed shall exhibit:

(i) if it is less than 25 meters in breadth, one all-round white light at or near each end;

(ii) if it is 25 meters or more in breadth, four all-round white

Fig. 17. Tug with barge, canal boat, or scow alongside. International and Inland Rules 24(c) and (e).

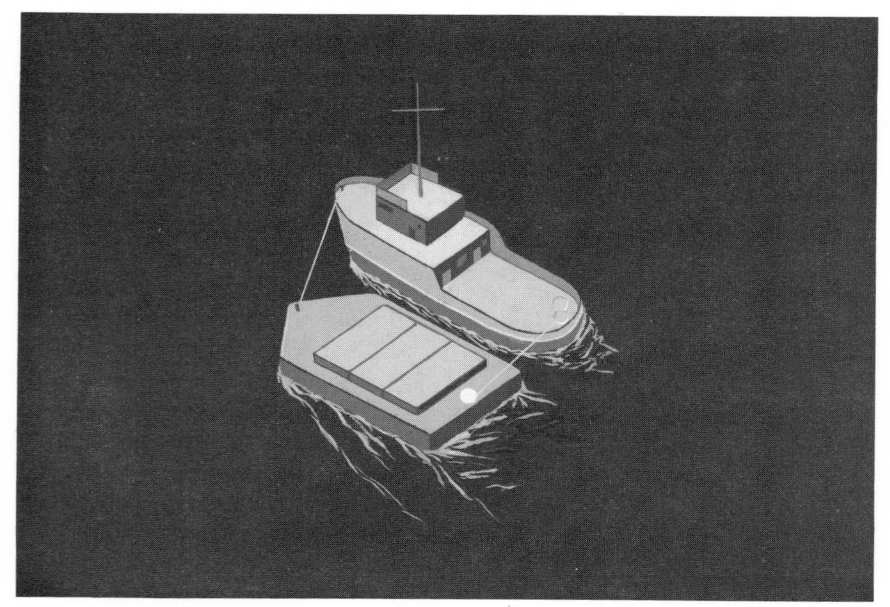

Fig. 18. Tug towing alongside, inland waters. Inland Rules 24(c) and (e).

International Rules

Towed Object Difficult to Light

(g) Where from any sufficient cause it is impracticable for a vessel or object being towed to exhibit the lights prescribed in paragraph (e) of this rule, all possible measures shall be taken to light the vessel or object towed or at least to indicate the presence of the unlighted vessel or object.

Inland Rules

lights to mark its length and breadth;

(iii) if it exceeds 100 meters in length, additional all-round white lights between the lights prescribed in subparagraphs (i) and (ii) so that the distance between the lights shall not exceed 100 meters; *Provided*, That any vessels or objects being towed alongside each other shall be lighted as one vessel or object;

(v) the towing vessel may direct a searchlight in the direction of the tow to indicate its presence to an approaching vessel.

Towed Object Difficult to Light

(h) Where from any sufficient cause it is impracticable for a vessel or object being towed to exhibit the lights prescribed in paragraph (e) or (g) of this Rule, all possible measures shall be taken to light the vessel or object towed or at least to indicate the presence of the unlighted vessel or object.

(i) Notwithstanding paragraph (c), on the Western Rivers and on waters specified by the Secretary, a power-driven vessel when pushing ahead or towing alongside, except as paragraph (b) applies, shall exhibit:

(i) sidelights; and

(ii) two towing lights in a vertical line.

(j) Where from any sufficient cause it is impracticable for a vessel not normally engaged in towing operations to display the light pre-

International Rules | *Inland Rules*

scribed by paragraph (a), (c) or (i) of this Rule, such vessel shall not be required to exhibit those lights when engaged in towing another vessel in distress or otherwise in need of assistance. All possible measures shall be taken to indicate the nature of the relationship between the towing vessel and the vessel being assisted. The searchlight authorized by Rule 36 may be used to illuminate the tow.

NOTES

Towing lights When the *Inland Navigation Rules Act of 1980* went into effect on 24 December 1981, the wide disparity between towing lights on inland waters and on the high seas was greatly reduced. Further alignment will take place in midsummer 1983 when the proposed amendments to the 1972 COLREGS are scheduled to go into effect. Be that as it may, both sets of rules require a power-driven vessel towing astern, alongside, or pushing ahead, to exhibit two 225° white lights in a vertical line, and if 50 meters or more in length a second masthead light aft of and higher than the ones forward. When the length of the tow, measured from the stern of the towing vessel to the stern of the last vessel towed, is more than 200 meters, the towing vessel shows three such lights in a vertical line. These lights must be of the same construction and character as the masthead light, and one of them must be in the same position as the masthead light. The lower towing light may be below the masthead light. Inland Rule 24, however, permits the towing lights to be carried on the forward or after mast. As noted above, when the proposed amendments go into effect, International Rule 24 will allow the same latitude.

In addition to the 225° white lights discussed above, the towing vessel will show her regular sidelight, a sternlight, and a yellow 135° towing light placed above the white sternlight. In inland waters, when pushing ahead or towing alongside, a power-driven towing vessel will show two yellow towing lights placed to show the same arc, the second yellow light replacing the sternlight.

Composite unit Advances in ship design have resulted in the capability of a pushing vessel being rigidly connected to a vessel she is pushing ahead. This composite unit is usually designed with sufficient power to

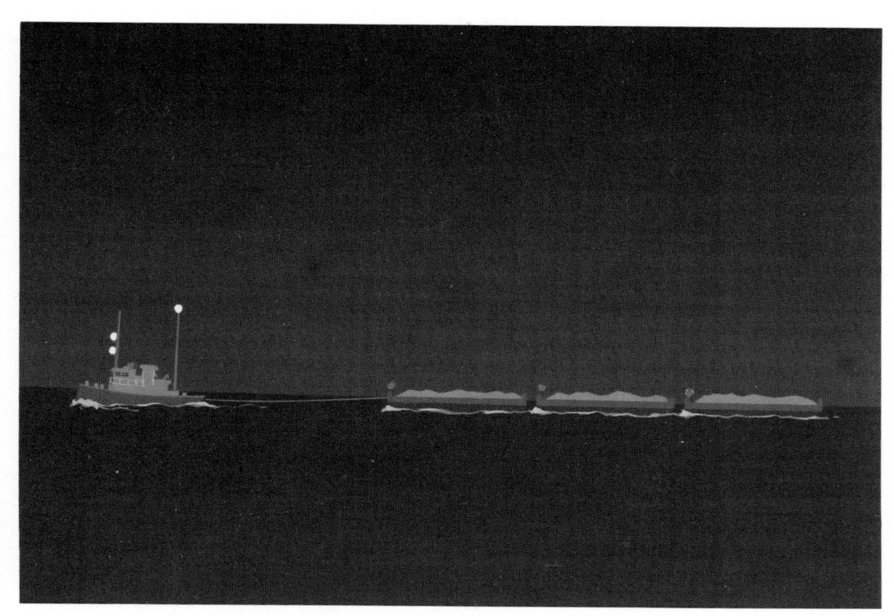

Fig. 19. Tug with barges, length of tow not over 200 meters. After masthead light may be omitted if tug is under 50 meters. International and Inland Rules 24(a) and (e).

Fig. 20. Tug with barges, length of tow over 200 meters. After masthead light may be omitted if tug is under 50 meters. International and Inland Rules 24(a) and (e).

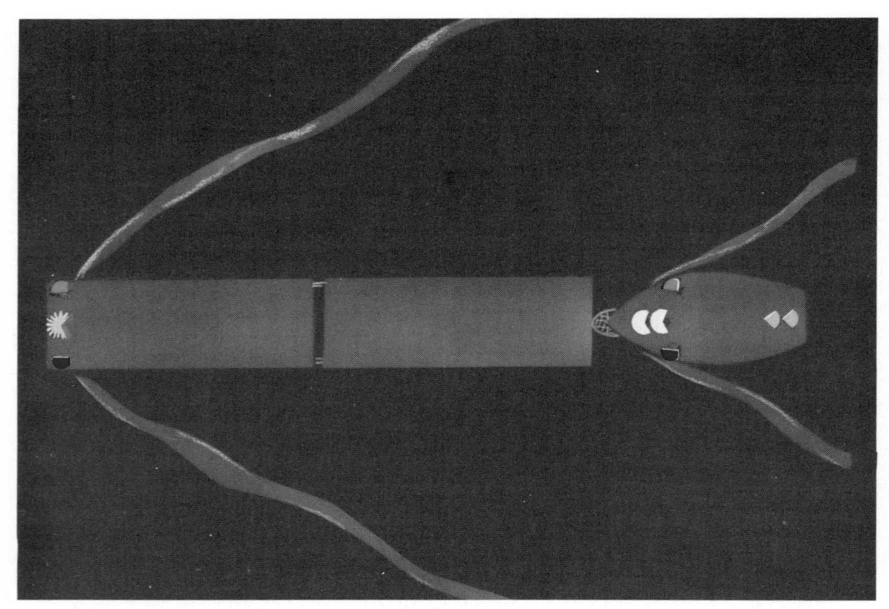

Fig. 21. Towboat under 50 meters pushing barges or other vessels. Inland Rules 24(c) and (f).

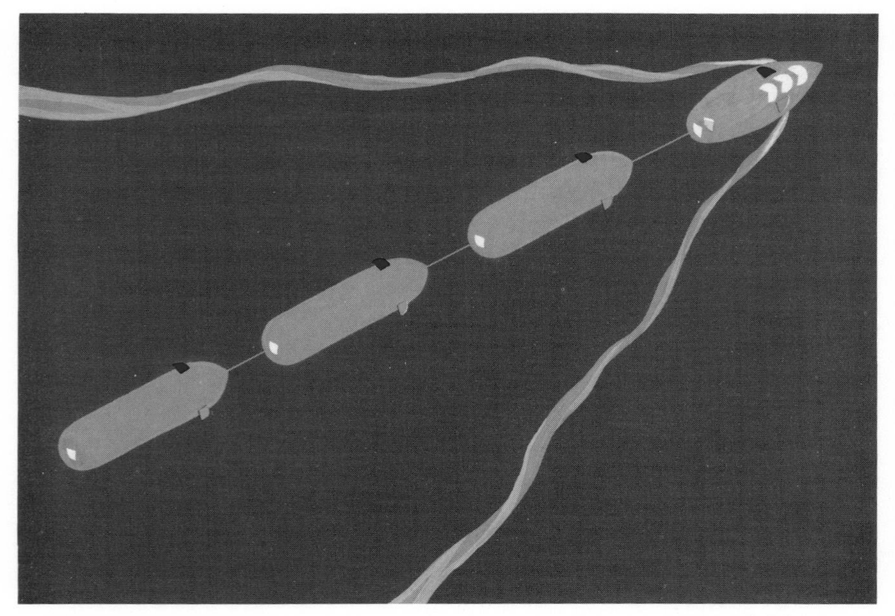

Fig. 22. Tug with barges in tandem, length of tow over 200 meters, high seas or inland waters. International and Inland Rules 24(a).

Fig. 23. Tug towing partly submerged object, over 200 meters in length and 25 meters in breadth, inland waters only. Inland Rules 24(a) and (g).

Fig. 24. Tug with barges alongside. International and Inland Rules 24(c) and (f).

operate in the open sea. It is regarded as a power-driven vessel and shows, on the high seas and in inland waters, the lights as such (International and Inland Rule 24(b)).

Vessels towed and being pushed ahead International and Inland Rules both require a vessel or object being towed alongside or astern to exhibit sidelights and a sternlight. The inland rules except an inconspicuous or partly submerged object from this requirement, specifying a number of all-round white lights to be shown instead, depending on the length and breadth of the object. Rule 24(f) for both jurisdictions provides that any number of vessels being towed alongside or pushed in a group are to be lighted as one vessel. Vessels being pushed ahead, not being part of a composite unit, exhibit sidelights at the forward end. In inland waters, a special flashing yellow light is also carried by such vessels. On western rivers and waters specified by the Secretary, vessels towing alongside or pushing ahead are exempt from the requirement to display masthead lights.

Towed object difficult to light All possible measures should be taken to indicate the presence of any underwater tow or any tow which is awash or nearly so, such as a dracone, upon which it is impossible to display lights. A vessel with such a tow would show the two or three vertical masthead lights and the red-white-red occupation lights of Rule 27(b), as appropriate, unless it were a minesweeper, in which case it would show the special lights of Rule 27(f). When showing two or three vertical masthead lights, the required yellow towing light for a tow astern will assist other vessels in identifying the towing vessel and alerting them to the possibility of an unlighted tow. Nevertheless, it would be prudent for a towing vessel to direct the searchlight authorized by Rule 36 in the direction of the tow to illuminate it or to indicate its presence to an approaching vessel.

5

Special Lights

International Rules

Visibility of Lights

RULE 22 The lights prescribed in these rules shall have an intensity as specified in Section 8 of Annex I to these regulations so as to be visible at the following minimum ranges:

(a) In vessels of 50 metres or more in length:
a white, red, green or yellow all-round light, 3 miles;

(b) In vessels of 12 metres or more in length but less than 50 metres in length:
a white, red, green or yellow all-round light, 2 miles;

(c) In vessels of less than 12 metres in length:
a white, red, green or yellow all-round light, 2 miles.

Lights for Fishing Vessels

RULE 26 (a) A vessel en-

Inland Rules

Visibility of Lights

RULE 22 The lights prescribed in these Rules shall have an intensity as specified in Annex I to these Rules, so as to be visible at the following minimum ranges:

(a) In a vessel of 50 meters or more in length:
a white, red, green or yellow all-round light, 3 miles; and
a special flashing light, 2 miles.

(b) In a vessel of 12 meters or more in length but less than 50 meters in length:
a white, red, green or yellow all-round light, 2 miles; and
a special flashing light, 2 miles.

(c) In a vessel of less than 12 meters in length:
a white, red, green or yellow all-round light, 2 miles; and
a special flashing light, 2 miles.

Lights for Fishing Vessels

RULE 26 (a) A vessel en-

International Rules

gaged in fishing, whether underway or at anchor, shall exhibit only the lights . . . prescribed in this rule.

(b) A vessel when engaged in trawling, by which is meant the dragging through the water of a dredge net or other apparatus used as a fishing appliance, shall exhibit:

(i) two all-round lights in a vertical line, the upper being green and the lower white, . . .

(ii) a masthead light abaft of and higher than the all-round green light; a vessel of less than 50 metres in length shall not be obliged to exhibit such a light but may do so;

(iii) when making way through the water, in addition to the lights prescribed in this paragraph, sidelights and a sternlight.

(c) A vessel engaged in fishing, other than trawling shall exhibit:

(i) two all-round lights in a vertical line, the upper being red and the lower white, . . .

(ii) When there is outlying gear extending more than 150 metres horizontally from the vessel, an all-round white light . . . in the direction of the gear;

(iii) when making way through the water, in addition to the lights prescribed in this paragraph, sidelights and a sternlight.

(d) A vessel engaged in fishing in close proximity to other vessels engaged in fishing may exhibit the additional signals described in Annex II to these regulations.

(e) A vessel when not engaged in

Inland Rules

gaged in fishing, whether underway or at anchor, shall exhibit only the lights . . . prescribed in this Rule.

(b) A vessel when engaged in trawling, by which is meant the dragging through the water of a dredge net or other apparatus used as a fishing appliance, shall exhibit:

(i) two all-round lights in a vertical line, the upper being green and the lower white, . . .

(ii) a masthead light abaft of and higher than the all-round green light; a vessel of less than 50 metres in length shall not be obliged to exhibit such a light but may do so; and

(iii) when making way through the water, in addition to the lights prescribed in this paragraph, sidelights and a sternlight.

(c) A vessel engaged in fishing, other than trawling shall exhibit:

(i) two all-round lights in a vertical line, the upper being red and the lower white, . . .

(ii) when there is outlying gear extending more than 150 meters horizontally from the vessel, an all-round white light . . . in the direction of the gear; and

(iii) when making way through the water, in addition to the lights prescribed in this paragraph, sidelights and a sternlight.

(d) A vessel engaged in fishing in close proximity to other vessels engaged in fishing may exhibit the additional signals described in Annex II to these regulations.

(e) A vessel when not engaged in

Fig. 25. Vessel trawling, making way, greater in length than 50 meters. International and Inland Rule 26(b).

Fig. 26. Vessel engaged in trawling, not making way, with net fast upon an obstruction. International and Inland Rule 26(b) and Annex II.

International Rules

fishing shall not exhibit the lights . . . prescribed in this Rule, but only those prescribed for a vessel of her length.

ANNEX II—ADDITIONAL SIGNALS FOR FISHING VESSELS FISHING IN CLOSE PROXIMITY

1. *General*

The lights mentioned herein shall, if exhibited in pursuance of Rule 26(d), be placed where they can best be seen. They shall be at least 0.9 metre apart but at a lower level than lights prescribed in Rule 26(b)(i) and (c)(i). The lights shall be visible all round the horizon at a distance of at least 1 mile but at a lesser distance than the lights prescribed by these rules for fishing vessels.

2. *Signals for trawlers*

(a) Vessels when engaged in trawling, whether using demersal or pelagic gear, may exhibit:

(i) when shooting their nets: two white lights in a vertical line;

(ii) when hauling their nets: one white light over one red light in a vertical line;

(iii) when the net has come fast upon an obstruction: two red lights in a vertical line.

(b) Each vessel engaged in pair trawling may exhibit:

(i) by night, a searchlight directed forward and in the direction of the other vessel of the pair;

Inland Rules

fishing shall not exhibit the lights . . . prescribed in this Rule, but only those prescribed for a vessel of her length.

PART 85—ANNEX II—ADDITIONAL SIGNALS FOR FISHING VESSELS FISHING IN CLOSE PROXIMITY

§ 85.1 General. The lights mentioned herein shall, if exhibited in pursuance of Rule 26(d), be placed where they can best be seen. They shall be at least 0.9 meter apart but at a lower level than lights prescribed in Rule 26(b)(i) and (c)(i) contained in the Navigational Rules Act of 1980. The lights shall be visible all around the horizon at a distance of at least 1 mile but at a lesser distance from the lights prescribed by these Rules for fishing vessels.

§ 85.3 Signals for trawlers. (a) Vessels when engaged in trawling, whether using demersal or pelagic gear, may exhibit:

(1) When shooting their nets: two white lights in a vertical line;

(2) When hauling their nets: one white light over one red light in a vertical line;

(3) When the net has come fast upon an obstruction: two red lights in a vertical line.

(b) Each vessel engaged in pair trawling may exhibit:

(1) By night, a search light directed forward and in the direction of the other vessel of the pair;

Fig. 27. Vessel fishing with nets or lines extending more than 150 meters horizontally, trolling lines excepted, underway making way. International and Inland Rule 26(c).

Fig. 28. Vessel trawling, making way, less than 50 meters in length. International and Inland Rule 26(b)(i).

Fig. 29. Vessels engaged in pair trawling, showing searchlights forward and in direction of other vessel. International and Inland Rule 26(b) and Annex II.

Fig. 30. Vessel fishing with nets or lines extending over 150 meters, underway but not making way. International and Inland Rule 26(c)(i) and (ii)

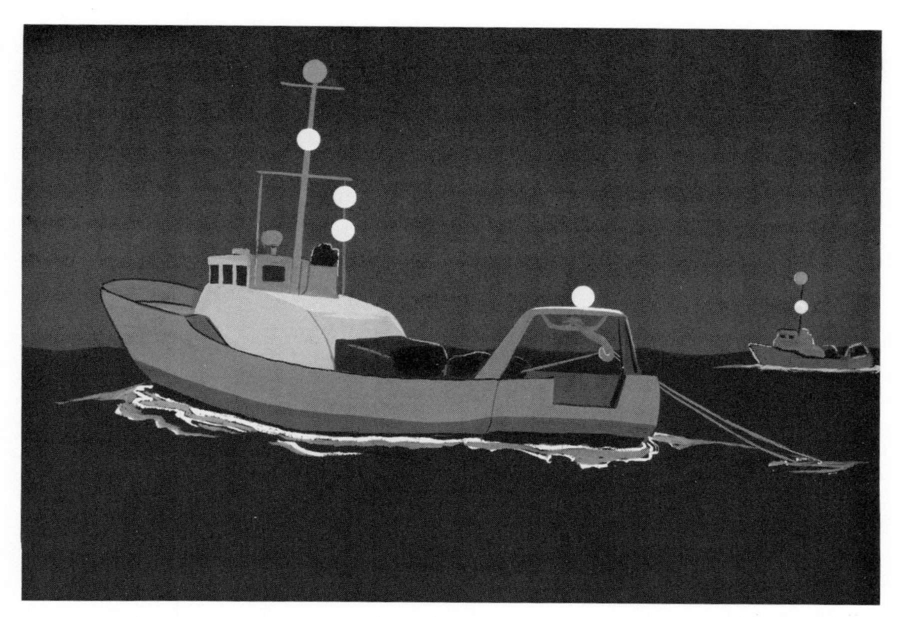

Fig. 31. Vessel engaged in trawling, underway, making way, and shooting her nets. International and Inland Rule 26(b) and Annex II.

Fig. 32. Vessel engaged in purse seining, hampered by her fishing gear, showing optional, flashing, alternate yellow lights. International and Inland Rule 26(c) and Annex II.

International Rules

(ii) when shooting or hauling their nets or when their nets have come fast upon an obstruction, the lights prescribed in 2(a) above.

3. *Signals for purse seiners*

Vessels engaged in fishing with purse seine gear may exhibit two yellow lights in a vertical line. These lights shall flash alternately every second and with equal light and occultation duration. These lights may be exhibited only when the vessel is hampered by its fishing gear.

Inland Rules

(2) When shooting or hauling their nets or when their nets have come fast upon an obstruction, the lights prescribed in paragraph (a) above.

§ 85.5 Signals for purse seiners. Vessels engaged in fishing with purse seine gear may exhibit two yellow lights in a vertical line. These lights shall flash alternately every second and with equal light and occultation duration. These lights may be exhibited only when the vessel is hampered by its fishing gear.

International Rules

Vessels Not Under Command or Restricted in Their Ability to Manoeuvre

RULE 27 (a) A vessel not under command shall exhibit:

(i) two all-round red lights in a vertical line where they can best be seen;

(iii) when making way through the water, in addition to the lights prescribed in this paragraph, sidelights and a sternlight.

(b) A vessel restricted in her ability to manoeuvre, except a vessel engaged in minesweeping operations, shall exhibit:

(i) three all-round lights in a vertical line where they can best be seen. The highest and lowest of these lights shall be red and the middle light shall be white;

Inland Rules

Vessels Not Under Command or Restricted in Their Ability to Maneuver

RULE 27 (a) A vessel not under command shall exhibit:

(i) two all-round red lights in a vertical line where they can best be seen;

(iii) when making way through the water, in addition to the lights prescribed in this paragraph, sidelights and a sternlight.

(b) A vessel restricted in her ability to maneuver, except a vessel engaged in minesweeping operations, shall exhibit:

(i) three all-round lights in a vertical line where they can best be seen. The highest and lowest of these lights shall be red and the middle light shall be white;

Fig. 33. Vessel engaged in trawling (less than 50 meters), underway but not making way, and hauling her nets. Second vessel same, but underway, not hauling her nets. International and Inland Rule 26(b) and Annex II.

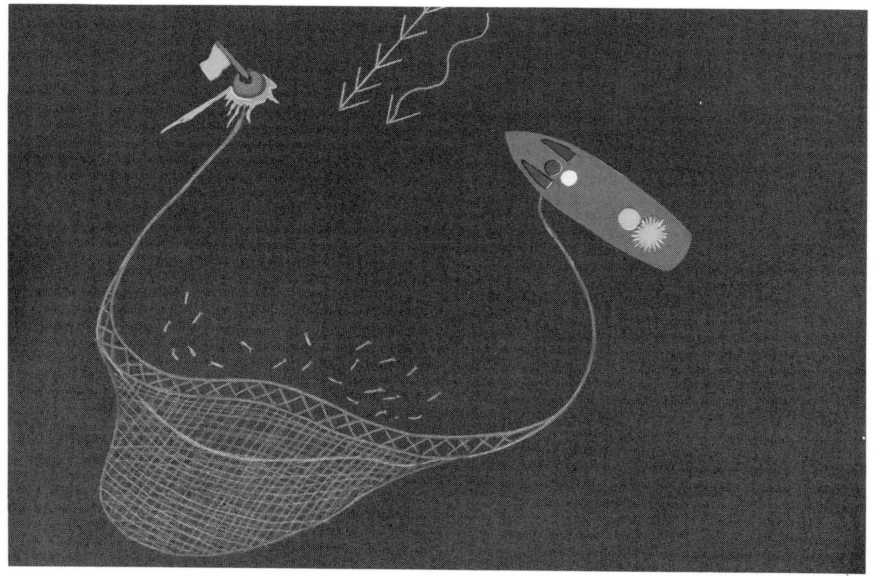

Fig. 34. Vessel purse seine fishing, underway with way on, hampered by her fishing gear, showing optional alternate yellow flashing light. International and Inland Rule 26(c) and Annex II.

International Rules

(iii) when making way through the water, masthead lights, sidelight and a sternlight, in addition to the lights prescribed in subparagraph (i);

(iv) when at anchor, in addition to the lights . . . prescribed in subparagraphs (i) and (ii), the light, lights . . . prescribed in Rule 30.

(c) A vessel engaged in a towing operation such as severely restricts the towing vessel and her tow in their ability to deviate from their course shall, in addition to the lights . . . prescribed in subparagraph (b)(i) and (ii) of this Rule, exhibit the lights . . . prescribed in Rule 24(a).

(d) A vessel engaged in dredging or underwater operations, when restricted in her ability to manoeuvre, shall exhibit the lights . . . prescribed in paragraph (b) of this rule and shall in addition, when an obstruction exists, exhibit:

(i) two all-round red lights . . . in a vertical line to indicate the side on which the obstruction exists;

(ii) two all-round green lights . . . in a vertical line to indicate the side on which another vessel may pass;

(iii) when making way through the water, in addition to the lights prescribed in this paragraph, masthead lights, sidelights, and a sternlight;

(iv) a vessel to which this paragraph applies when at anchor shall exhibit the lights . . . prescribed in

Inland Rules

(iii) when making way through the water, masthead lights, sidelights and a sternlight, in addition to the lights prescribed in subparagraph (b)(i); and

(iv) when at anchor, in addition to the lights . . . prescribed in subparagraphs (b)(i) and (ii), the light, lights . . . prescribed in Rule 30.

(c) A vessel engaged in a towing operation which severely restricts the towing vessel and her tow in their ability to deviate from their course shall, in addition to the lights . . . prescribed in subparagraphs (b)(i) and (ii) of this Rule, exhibit the lights . . . prescribed in Rule 24.

(d) A vessel engaged in dredging or underwater operations, when restricted in her ability to maneuver, shall exhibit the lights . . . prescribed in subparagraphs (b)(i), (ii), and (iii) of this Rule and shall in addition, when an obstruction exists, exhibit:

(i) two all-round red lights . . . in a vertical line to indicate the side on which the obstruction exists;

(ii) two all-round green lights . . . in a vertical line to indicate the side on which another vessel may pass; and

(iii) when at anchor, the lights . . . prescribed by this paragraph, instead of the lights . . . prescribed

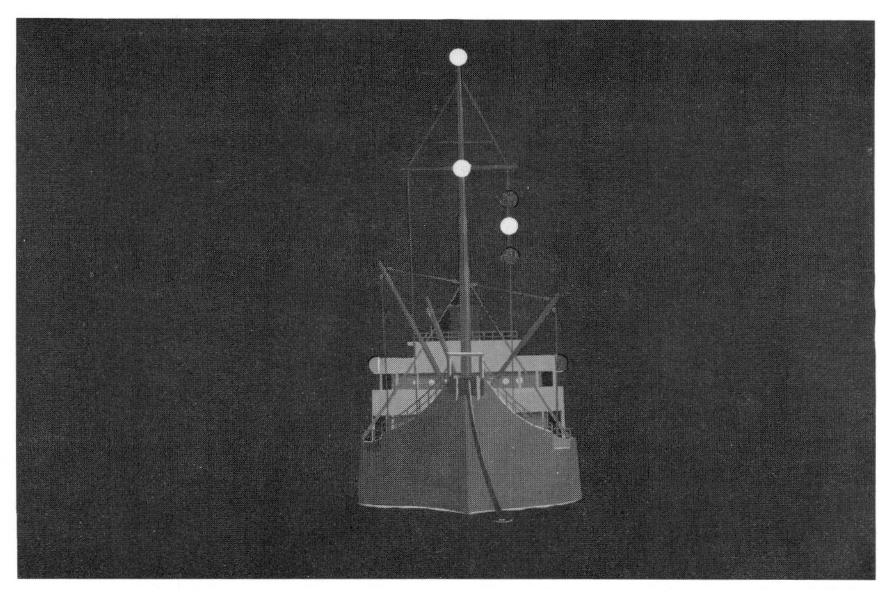

Fig. 35. Cable ship at work, underway and making way. After masthead light optional for vessels less than 50 meters. Same for vessel servicing navigation marks, surveying, replenishment, or launching or recovering aircraft. International and Inland Rule 27(b).

Fig. 36. Vessel not under command, underway but not making way. International and Inland Rule 27(a).

International Rules

subparagraphs (i) and (ii) instead of the lights . . . prescribed in Rule 30.

(f) A vessel engaged in minesweeping operations shall, in addition to the lights prescribed for a power-driven vessel in Rule 23, exhibit three all-round green lights . . . One of these lights . . . shall be exhibited at or near the foremast head and one at each end of the fore yard. These lights . . . indicate that it is dangerous for another vessel to approach closer than 1,000 metres astern or 500 metres on either side of the minesweeper.

(g) Vessels of less than 7 metres in length shall not be required to exhibit the lights prescribed in this rule.

(h) The signals prescribed in this rule are not signals of vessels in distress and requiring assistance. Such signals are contained in Annex IV to these Regulations.

Vessels Constrained by Their Draught

RULE 28 A vessel con-

Inland Rules

in Rule 30 for anchored vessels.

(e) Whenever the size of a vessel engaged in diving operations makes it impracticable to exhibit all lights . . . prescribed in paragraph (d) of this Rule, the following shall instead be exhibited:

(i) Three all-round lights in a vertical line where they can best be seen. The highest and lowest of these lights shall be red and the middle light shall be white.

(f) A vessel engaged in minesweeping operations shall, in addition to the lights prescribed for a power-driven vessel in Rule 23, exhibit three all-round green lights . . . One of these lights . . . shall be exhibited near the foremast head and one at each end of the fore yard. These lights . . . indicate that it is dangerous for another vessel to approach closer than 1,000 meters astern or 500 meters on either side of the minesweeper.

(g) A vessel of less than 12 meters in length, except when engaged in diving operations, is not required to exhibit the lights . . . prescribed in this Rule.

(h) The signals prescribed in this Rule are not signals of vessels in distress and requiring assistance. Such signals are contained in Annex IV to these Rules.

RULE 28 [Reserved]

Fig. 37. Vessel at anchor, restricted in her ability to maneuver. International and Inland Rule 27(b)(i) and (iv).

Fig. 38. Vessel engaged in dredging, restricted in her ability to maneuver, underway with way on, with no obstruction. International and Inland Rule 27(b)(i) and (iii).

Fig. 39. Vessel engaged in dredging or underwater operations, restricted in her ability to maneuver, at anchor with an obstruction existing. International and Inland Rule 27(d).

Fig. 40. Cable ship at work, underway but not making way. Same for vessel servicing navigation mark, surveying, replenishing, or launching or recovering aircraft. International and Inland Rule 27(b).

Fig. 41. Vessel engaged in towing operation, length of tow over 200 meters such as renders her unable to deviate from her course. International and Inland Rules 24(a) and 27(c).

Fig. 42. Minesweeper with gear out. International and Inland Rule 27(f).

International Rules

strained by her draught may, in addition to the lights prescribed for power-driven vessels in Rule 23, exhibit where they can best be seen three all-round red lights in a vertical line, . . .

Pilot Vessels

RULE 29 (a) A vessel engaged on pilotage duty shall exhibit:

(i) at or near the masthead, two all-round lights in a vertical line, the upper being white and the lower red;

(ii) when underway, in addition, sidelights and a sternlight;

(iii) when at anchor, in addition to the lights prescribed in subparagraph (i), the anchor light, lights . . .

(b) A pilot vessel when not engaged on pilotage duty shall exhibit the lights . . . prescribed for a similar vessel of her length.

Signals to Attract Attention

RULE 36 If necessary to attract the attention of another vessel, any vessel may make light . . . signals that cannot be mistaken for any signal authorized elsewhere in these Rules, or may direct the beam of her searchlight in the direction of the danger, in such a way as not to embarrass any vessel.

Station and Signal Lights

RULE 1 (c) Nothing in these Rules shall interfere with the op-

Inland Rules

Pilot Vessels

RULE 29 (a) A vessel engaged on pilotage duty shall exhibit:

(i) at or near the masthead, two all-round lights in a vertical line, the upper being white and the lower red;

(ii) when underway, in addition, sidelights and a sternlight;

(iii) when at anchor, in addition to the lights prescribed in subparagraph (i), the anchor light, lights, . . . prescribed in Rule 30 for anchored vessels.

(b) A pilot vessel when not engaged on pilotage duty shall exhibit the lights . . . prescribed for a vessel of her length.

Signals to Attract Attention

RULE 36 If necessary to attract the attention of another vessel, any vessel may make light . . . signals that cannot be mistaken for any signal authorized elsewhere in these Rules, or may direct the beam of her searchlight in the direction of the danger, in such a way as not to embarrass any vessel.

Station and Signal Lights

RULE 1 (c) Nothing in these Rules shall interfere with the op-

Fig. 43. Vessel constrained by her draught, high seas only. International Rule 28.

Fig. 44. Power-driven pilot vessel underway on station, high seas or inland waters. International and Inland Rule 29(a).

International Rules

eration of any special rules made by the Government of any State with respect to additional station or signal lights . . . for ships of war and vessels proceeding under convoy, or with respect to additional station or signal lights for fishing vessels engaged in fishing as a fleet. These additional station or signal lights . . . shall, so far as possible, be such that they cannot be mistaken for any light . . . authorized elsewhere under these Rules.

(e) Whenever the Government concerned shall have determined that a vessel of special construction or purpose cannot comply fully with the provisions of any of these Rules with respect to the number, position, range or arc of visibility of lights . . . without interfering with the special function of the vessel, such vessel shall comply with such other provisions in regard to the number, position, range or arc of visibility of lights . . . as her Government shall have determined to be the closest possible compliance with these Rules in respect to that vessel.

Inland Rules

eration of any special rules made by the Secretary of the Navy with respect to additional station or signal lights . . . for ships of war and vessels proceeding under convoy, or by the Secretary with respect to additional station or signal lights . . . for fishing vessels engaged in fishing as a fleet. These additional station or signal lights . . . shall, so far as possible, be such that they cannot be mistaken for any light . . . authorized elsewhere under these Rules. Notice of such special rules shall be published in the Federal Register and, after the effective date specified in such notice, they shall have effect as if they were a part of these Rules.

(e) Whenever the Secretary determines that a vessel or class of vessels of special construction or purpose cannot comply fully with the provisions of any of these Rules with respect to the number, position, range, or arc of visibility of lights . . . , without interfering with the special function of the vessel, the vessel shall comply with such other provisions in regard to the number, position, range, or arc of visibility of lights . . . , as the Secretary shall have determined to be the closest possible compliance with these Rules. The Secretary may issue a certificate of alternative compliance for a vessel or class of vessels specifying the closest possible compliance with these Rules. The Secretary of the Navy shall make these determinations and

International Rules

Inland Rules

issue certificates of alternative compliance for vessels of the Navy.

Navy and Coast Guard Vessel Exceptions (33 U.S.C. 1052)[1]

Any requirement of such regulations in respect of the number, position, range of visibility, or arc of visibility of the lights required to be displayed by vessels shall not apply to any vessel of the Navy or of the Coast Guard whenever the Secretary of the Navy or the Secretary of Transportation, in the case of Coast Guard vessels operating under the Department of Transportation, or such official as either may designate, shall find or certify that, by reason of special construction, it is not possible for such vessel or class of vessels to comply with such regulations. The lights of any such exempted vessel or class of vessels, however, shall conform as closely to the requirements of the applicable regulations as the Secretary or such official shall find or certify to be feasible. Notice of such findings or certification and of the character and position of the lights prescribed to be displayed on such exempted vessel or class of vessels shall be published in the Federal Register and in the Notice to Mariners and, after the effective date specified in such notice, shall

[1](See Title 33, Code of Federal Regulations, Part 135 and Title 32, Code of Federal Regulations, Parts 706 and 707.)

International Rules

have effect as part of such regulations.

Inland Rules

PART 88—ANNEX V, PILOT RULES

§ 88.01 Purpose and applicability. This part applies to all vessels operating on United States inland waters and to United States vessels operating on the Canadian waters of the Great Lakes to the extent there is no conflict with Canadian law.

§ 88.09 Temporary exemptions from light and shape requirements when operating under bridges. A vessel's navigation lights and shapes may be lowered if necessary to pass under a bridge.

§ 88.11 Law enforcement vessels. (a) Law enforcement vessels may display a flashing blue light when engaged in direct law enforcement activities. This light shall be located so that it does not interfere with the visibility of the vessel's navigation lights.

(b) The blue light described in this section may be displayed by law enforcement vessels of the United States and their political subdivisions.

§ 88.13 Lights on barges at bank or dock. (a) The following barges shall display at night and if practicable in periods of restricted visibility the lights described in paragraph (b) of this section—

(1) Every barge projecting into a buoyed or restricted channel.

(2) Every barge so moored that

International Rules

Inland Rules

it reduces the available navigable width of any channel to less than 80 meters.

(3) Barges moored in groups more than two barges wide or to a maximum width of over 25 meters.

(4) Every barge not moored parallel to the bank or dock.

(b) Barges described in paragraph (a) of this section shall carry two unobstructed white lights of an intensity to be visible for at least one mile on a clear dark night, and arranged as follows:

(1) On a single moored barge, lights shall be placed on the two corners farthest from the bank or dock.

(2) On barges moored in group formation, a light shall be placed on each of the upstream and downstream ends of the group, on the corners farthest from the bank or dock.

(3) Any barge in a group, projecting from the main body of the group toward the channel, shall be lighted as a single barge.

(c) Barges moored in any slip or slough which is used primarily for mooring purposes are exempt from the lighting requirements of this section.

(d) Barges moored in well-illuminated areas are exempt from the lighting requirements of this section.

§88.15 *Lights on dredge pipelines.* Dredge pipelines that are floating or supported on trestles shall display the following lights at night

International Rules

Inland Rules

and in periods of restricted visibility.

(a) One row of yellow lights. The lights must be—

(1) Flashing 50 to 70 times per minute,

(2) Visible all around the horizon,

(3) Visible for at least 2 miles on a clear dark night,

(4) Not less than 1 and not more than 3.5 meters above the water,

(5) Approximately equally spaced, and

(6) Not more than 10 meters apart where the pipeline crosses a navigable channel. Where the pipeline does not cross a navigable channel the lights must be sufficient in number to clearly show the pipeline's length and course.

(b) Two red lights at each end of the pipeline, including the ends in a channel where the pipeline is separated to allow vessels to pass (whether open or closed). The lights must be—

(1) Visible all around the horizon, and

(2) Visible for at least 2 miles on a clear dark night, and

(3) One meter apart in a vertical line with the lower light at the same height above the water as the flashing yellow light.

NOTES

Fishing vessels In order to establish what special lights they must carry, fishing vessels are divided into two broad categories: trawlers, and vessels fishing by any method other than trawling. Both the International Regulations and the Inland Rules are now consistent on the lights necessary for both classes of fishing vessels.

Fig. 45. Barges moored as a group. Inland Rules, Annex V, Section 88.13.

Fig. 46. Dredge pipelines extending across a navigable channel, pipeline separated to allow vessels to pass. Inland Rules, Annex V, Section 88.15.

Trawlers drag behind them, at varying depths, a dredge net or similar apparatus. Normally, it descends into the sea at a short distance from the trawler. Trawlers show a green all-round light over a white all-around light to indicate the comparative safety with which they may be approached. In addition, except when handling gear or fast to an obstruction, they are normally making way. However, the development of trawlers working together with a single apparatus, i.e., pair trawling, requires more caution from the mariner, specifically in not passing between two such vessels. The side-by-side maneuvers of a pair, plus the searchlights authorized by Annex II, should help mariners, and not just fishermen, to recognize the situation. The other lights shown by trawlers, when fishing in company with other vessels also fishing, are primarily of interest to their fellow fishermen. Nevertheless, it behooves the mariner to understand these extra signals in the not too unlikely event of his meeting a fleet of fishermen on his approach to land. By knowing what the trawlers are doing, there is less risk of causing damage to nets.

Other methods of fishing are diverse but sometimes involve the use of extensive lengths of lines or nets, which hamper a vessel's maneuverability far more than a trawler's dredge. The gear used in such methods often lies close to the surface, can be several miles in length, is usually unlighted, and is vulnerable to the screws of a power-driven vessel. The all-round red light over an all-around white light serves to warn the prudent mariner of the need to give such vessels a wide berth, particularly when they are fishing in close proximity to each other.

In both inland waters and on the high seas, fishing vessels engaged in fishing while at anchor do not show anchor lights. It is not, therefore, possible to distinguish by lights alone if a vessel engaged in fishing is at anchor. International Regulations and Inland Rules, however, both require vessels engaged in fishing to show sidelights and sternlights if making way through the water.

Should any vessel fishing, other than trawling, have gear out that extends over 150 meters horizontally, she is required by Rule 26(c)(ii) to show an additional white light in the direction of the gear. A trawler does not show such a light, due to the nature of her gear. However, a trawler 50 meters in length or greater is required to show a second masthead light higher and abaft the the all-round green light. The second masthead light is optional for trawlers under 50 meters in length.

Not under command or restricted in ability to maneuver Here again, both sets of rules are nearly identical for vessels in these categories, although no special lights for vessels not under command existed in the old Inland Rules. Vessels not under command on the high seas and now on inland waters carry the two all-around red lights specified in Rule 27(a)(i). Vessels not under command do not display masthead lights but do show sidelights and sternlights when making way. Vessels restricted in their ability to maneuver *do* show masthead lights as well as sidelights and sternlights, in addition to the three all-around red-white-red lights, when making way. In the event that vessels restricted in their ability to maneuver drop anchor, they display anchor lights as well as the red-white-red occupational lights, with the exception of a vessel engaged in dredging or underwater operations, which does not show anchor lights. A vessel engaged in dredging or underwater operations is also required to display two all-round red lights and two all-round green lights to indicate the unsafe and safe passing side, respectively.

Vessels constrained by their draft There is no equivalent to this international rule in the inland rules. Such a vessel in inland waters would need to invoke Rule 2, the Responsibility Rule, and use the danger signal or radiotelephone to signify her constraint.

Minesweepers Rule 27(f) International Regulations and Inland Rules requires a vessel engaged in minesweeping to show three all-round green lights in addition to the masthead, side, and stern lights.

Pilot vessels The use of the white-over-red identification lights, the sidelights and sternlights when underway, and the required anchor lights when at anchor, is the same for pilot vessels on duty in international and inland waters.

Naval and Coast Guard vessels Rule 1(c), International Regulations and Inland Rules, permits naval and Coast Guard vessels to carry speed lights and other special lights in addition to their prescribed lights. Rule 1(e) provides exemptions regarding lights to be carried by naval and Coast Guard vessels of *special construction*. Note that *closest possible compliance* with the rules is required by vessels of special construction. In general, naval and Coast Guard vessels are in no way relieved from the obligation to carry the lights prescribed by the respective rules.

Inland Rules special lights Annex V, Pilot Rules to the Inland Rules, provides for special lights for barges moored to a bank or dock in certain situations. Such lights are white and should be placed on the corners of the barges farthest from the bank.

The annex also provides for a flashing blue light for law-enforcement vessels and flashing yellow lights marking a dredge pipeline to clearly show its length and course. When the pipeline is separated in a navigable channel to allow vessels to pass, the ends are marked by two vertical red lights.

Positioning and technical details of lights The reader is invited to refer to Annex I of both sets of rules for positioning and spacing of lights.

6

Day Shapes

International Rules

PART C—LIGHTS AND SHAPES

Application

RULE 20 (a) Rules in this Part shall be complied with in all weathers.

(d) The Rules concerning shapes shall be complied with by day.

(e) The . . . shapes specified in these Rules shall comply with the provisions of Annex I to these Regulations.

Towing and Pushing

RULE 24 (a) A power-driven vessel when towing shall exhibit:

(v) when the length of the tow exceeds 200 metres, a diamond shape where it can best be seen.

(e) A vessel or object being towed shall exhibit:

Inland Rules

PART C—LIGHTS AND SHAPES

Application

RULE 20 (a) Rules in this Part shall be complied with in all weathers.

(d) The Rules concerning shapes shall be complied with by day.

(e) The . . . shapes specified in these Rules shall comply with the provisions of Annex I of these Rules.

Towing and Pushing

RULE 24 (a) A power-driven vessel when towing astern shall exhibit:

(v) when the length of the tow exceeds 200 meters, a diamond shape where it can best be seen.

(e) A vessel or object other than those referred to in paragraph (g) of this Rule being towed shall exhibit:

International Rules

(iii) when the length of the tow exceeds 200 metres, a diamond shape where it can best be seen.

Sailing Vessels Underway

RULE 25 (e) A vessel proceeding under sail when also being propelled by machinery shall exhibit forward where it can best be seen a conical shape, apex downwards.

Fishing Vessels

RULE 26 (a) A vessel engaged in fishing, whether underway or at anchor, shall exhibit only the . . . shapes prescribed in this Rule.

(b) A vessel when engaged in trawling, by which is meant the dragging through the water of a dredge net or other apparatus used as a fishing appliance, shall exhibit:

(i) . . . a shape consisting of two cones with their apexes together in a vertical line one above the other; a vessel of less than 20 metres in length may instead of this shape exhibit a basket;

(c) A vessel engaged in fishing, other than trawling, shall exhibit:

(i) . . . a shape consisting of two cones with apexes together in a ver-

Inland Rules

(iii) when the length of the tow exceeds 200 meters, a diamond shape where it can best be seen.

(g) An inconspicuous, partly submerged vessel or object being towed shall exhibit:

(iv) a diamond shape at or near the aftermost extremity of the last vessel or object being towed.

Sailing Vessels Underway

RULE 25 (e) A vessel proceeding under sail when also being propelled by machinery shall exhibit forward where it can best be seen a conical shape, apex downward. A vessel of less than 12 meters in length is not required to exhibit this shape, but may do so.

Fishing Vessels

RULE 26 (a) A vessel engaged in fishing, whether underway or at anchor, shall exhibit only the . . . shapes prescribed in this Rule.

(b) A vessel when engaged in trawling, by which is meant the dragging through the water of a dredge net or other apparatus used as a fishing appliance, shall exhibit:

(i) . . . a shape consisting of two cones with their apexes together in a vertical line one above the other; a vessel of less than 20 meters in length may instead of this shape exhibit a basket;

(c) A vessel engaged in fishing, other than trawling, shall exhibit:

(i) . . . a shape consisting of two cones with apexes together in a ver-

International Rules

tical line one above the other; a vessel of less than 20 metres in length may instead of this shape exhibit a basket;

(ii) When there is outlying gear extending more than 150 metres horizontally from the vessel, . . . a cone apex upwards in the direction of the gear;

(e) A vessel when not engaged in fishing shall not exhibit the . . . shapes prescribed in this rule, but only those prescribed for a vessel of her length.

Inland Rules

tical line one above the other; a vessel of less than 20 meters in length may instead of this shape exhibit a basket:

(ii) when there is outlying gear extending more than 150 meters horizontally from the vessel, . . . a cone apex upward in the direction of the gear;

(e) A vessel when not engaged in fishing shall not exhibit the . . . shapes prescribed in this Rule, but only those prescribed for a vessel of her length.

International Rules

Vessels Not Under Command or Restricted in Their Ability to Manoeuvre

RULE 27 (a) A vessel not under command shall exhibit:

(ii) two balls or similar shapes in a vertical line where they can best be seen;

(b) A vessel restricted in her ability to manoeuvre, except a vessel engaged in minesweeping operations, shall exhibit:

(ii) Three shapes in a vertical line where they can best be seen. The highest and lowest of these shapes shall be balls and the middle one a diamond;

(iv) when at anchor, in addition to the . . . shapes . . . prescribed in subparagraph . . . (ii), the . . . shape prescribed in Rule 30.

(c) A vessel engaged in a towing operation such as renders her unable to deviate from her course

Inland Rules

Vessels Not Under Command or Restricted in Their Ability to Maneuver

RULE 27 (a) A vessel not under command shall exhibit:

(ii) two balls or similar shapes in a vertical line where they can best be seen;

(b) A vessel restricted in her ability to maneuver, except a vessel engaged in minesweeping operations, shall exhibit:

(ii) three shapes in a vertical line where they can best be seen. The highest and lowest of these shapes shall be balls and the middle one a diamond;

(iv) when at anchor, in addition to the . . . shapes prescribed in subparagraph (ii), . . . shapes prescribed in Rule 30.

(c) A vessel engaged in a towing operation which severely restricts the towing vessel and her tow in

Fig. 47. Vessel towing, vessel towed, length of tow over 200 meters. International and Inland Rules 24(a) and (e).

Fig. 48. Vessel at anchor. International and Inland Rule 30(a)(i).

International Rules

shall, in addition to the . . . shapes prescribed in subparagraph (b) . . . (ii) of this rule, exhibit the . . . shape prescribed in Rule 24(a).

(d) A vessel engaged in dredging or underwater operations, when restricted in her ability to manoeuvre, shall exhibit the . . . shapes prescribed in paragraph (b) of this rule and shall in addition, when an obstruction exists, exhibit:

(i) . . . two balls in a vertical line to indicate the side on which the obstruction exists;

(ii) . . . two diamonds in a vertical line to indicate the side on which another vessel may pass;

(iv) a vessel to which this paragraph applies when at anchor shall exhibit the . . . shapes prescribed in subparagraphs (i) and (ii) instead of the . . . shapes prescribed in Rule 30.

(e) Whenever the size of a vessel engaged in diving operations makes it impracticable to exhibit the shapes prescribed in paragraph (d) of this rule, a rigid replica of the International Code flag "A" not less than 1 metre in height shall be exhibited. Measures shall be taken to ensure all-round visibility.

(f) A vessel engaged in minesweeping operations shall exhibit . . . three balls. One of these . . . shall be exhibited at or near the foremast head and one at each end of

Inland Rules

their ability to deviate from their course shall, in addition to the . . . shapes prescribed in subparagraph . . . (b)(ii) of this Rule, exhibit the . . . shape prescribed in Rule 24.

(d) A vessel engaged in dredging or underwater operations, when restricted in her ability to maneuver shall exhibit the . . . shapes prescribed in subparagraph . . . (b)(ii) . . . of this Rule and shall in addition, when an obstruction exists, exhibit

(i) . . . two balls in a vertical line to indicate the side on which the obstruction exists;

(ii) . . . two diamonds in a vertical line to indicate the side on which another vessel may pass;

(iii) when at anchor, the . . . shape prescribed by this paragraph, instead of the . . . shapes prescribed in Rule 30 for anchored vessels.

(e) Whenever the size of a vessel engaged in diving operations makes it impracticable to exhibit all . . . shapes prescribed in paragraph (d) of this Rule, the following shall instead be exhibited:

(ii) A rigid replica of the international Code flag "A" not less than 1 meter in height. Measures shall be taken to insure its all-round visibility.

(f) A vessel engaged in minesweeping operations shall . . . exhibit . . . three balls. One of these . . . shapes shall be exhibited near the foremast head and one at each end

Fig. 49. Vessel under sail and power by day. International and Inland Rule 25(e).

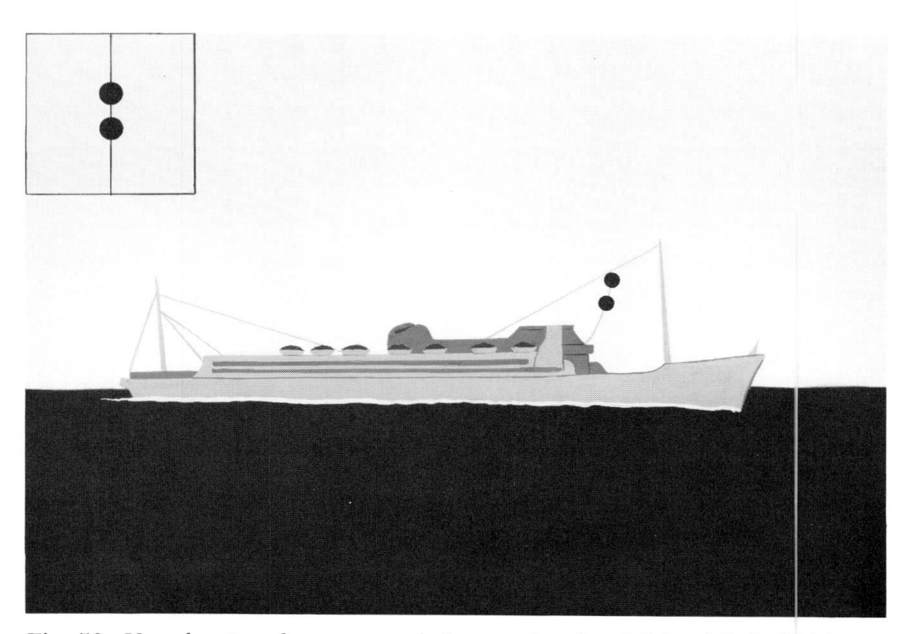

Fig. 50. Vessel not under command. International and Inland Rule 27(a).

International Rules

the fore yard. These . . . shapes indicate that it is dangerous for another vessel to approach closer than 1,000 metres astern or 500 metres on either side of the minesweeper.

(h) The signals prescribed in this rule are not signals of vessels in distress and requiring assistance. Such signals are contained in Annex IV to these Regulations.

Inland Rules

of the fore yard. These . . . shapes indicate that it is dangerous for another vessel to approach closer than 1,000 meters astern or 500 meters on either side of the minesweeper.

(g) A vessel of less than 12 meters in length, except when engaged in diving operations, is not required to exhibit the . . . shapes prescribed in this Rule.

(h) The signals prescribed in this Rule are not signals of vessels in distress and requiring assistance. Such signals are contained in Annex IV to these Rules.

Vessels Constrained by Their Draught

RULE 28 A vessel constrained by her draught may, . . . exhibit where they can best be seen . . . a cylinder.

RULE 28

[Reserved]

Anchored Vessels and Vessels Aground

RULE 30 (a) A vessel at anchor shall exhibit where it can best be seen:

(i) in the fore part, . . . one ball;

(d) A vessel aground shall exhibit . . . where they can best be seen:

(ii) three balls in a vertical line.

(e) A vessel of less than 7 metres in length, when at anchor or aground, not in or near a narrow channel, fairway or anchorage, or

Anchored Vessels and Vessels Aground

RULE 30 (a) A vessel at anchor shall exhibit where it can best be seen:

(i) in the fore part, . . . one ball;

(d) A vessel aground shall exhibit . . . where they can best be seen:

(ii) three balls in a vertical line.

(e) A vessel of less than 7 meters in length, when at anchor, not in or near a narrow channel, fairway, anchorage, or where other vessels

Fig. 51. Vessel restricted in her ability to maneuver. When at anchor, one ball would be shown in addition. International and Inland Rules 27(b)(ii) and (iv).

Fig. 52. Vessel engaged in towing, unable to deviate from her course and length of tow over 200 meters. International and Inland Rule 27(c).

Fig. 53. Vessel engaged in dredging or underwater operations, restricted in her ability to maneuver when underway, or at anchor with an obstruction existing to one side. International and Inland Rule 27(d).

Fig. 54. Minesweeper at work. International and Inland Rule 27(f).

Fig. 55. Vessel fishing, 20 meters or more in length, gear out more than 150 meters. International and Inland Rule 26(c).

Fig. 56. Fishing vessel with nets, lines, or trawls out, less than 20 meters. International and Inland Rule 26(c)(i).

International Rules

where other vessels normally navigate, shall not be required to exhibit the . . . shapes prescribed in paragraphs (a), . . . or (d) of this Rule.

Inland Rules

normally navigate, shall not be required to exhibit the . . . shape prescribed in paragraph (a) . . . of this Rule.

(f) A vessel of less than 12 meters in length when aground shall not be required to exhibit the . . . shapes prescribed in subparagraphs (d) . . . (ii) of this Rule.

(g) A vessel of less than 20 meters in length, when at anchor in a special anchorage area designated by the Secretary, shall not be required to exhibit the anchor . . . shapes required by this Rule.

International Rules

Seaplanes

RULE 31 Where it is impracticable for a seaplane to exhibit . . . shapes of the characteristics or in the positions prescribed in the Rules of this Part she shall exhibit . . . shapes as closely similar in characteristics and position as is possible.

Inland Rules

Seaplanes

RULE 31 Where it is impracticable for a seaplane to exhibit . . . shapes of the characteristics or in the positions prescribed in the Rules of this Part she shall exhibit . . . shapes as closely similar in characteristics and position as is possible.

International Rules

ANNEX I

6. *Shapes*

(a) Shapes shall be black and of the following sizes:

(i) a ball shall have a diameter of not less than 0.6 metre;

(ii) a cone shall have a base diameter of not less than 0.6 metre and a height equal to its diameter;

(iii) a cylinder shall have a diameter of at least 0.6 metre and a height of twice its diameter;

Inland Rules

ANNEX I

§ 84.11 Shapes. (a) Shapes shall be black and of the following sizes:

(1) A ball shall have a diameter of not less than 0.6 meter;

(2) A cone shall have a base diameter of not less than 0.6 meter and a height equal to its diameter;

(3) A diamond shape shall consist of two cones (as defined in paragraph (a)(2) of this section) having a common base.

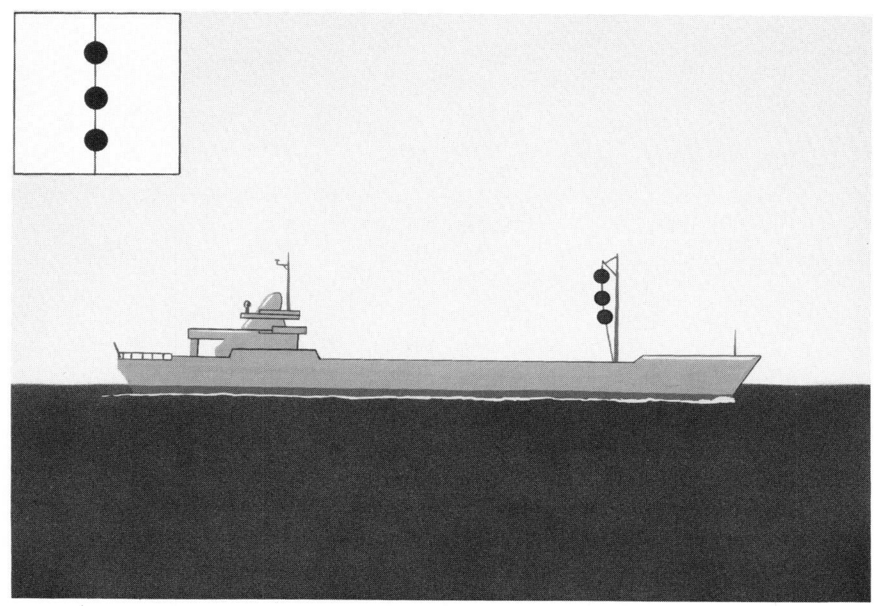

Fig. 57. Vessel aground. International and Inland Rule 30(d)(ii).

Fig. 58. Vessel constrained by her draught, high seas only. International Rule 28.

International Rules	*Inland Rules*
(iv) a diamond shape shall consist of two cones as defined in (ii) above having a common base.	
(b) The vertical distance between shapes shall be at least 1.5 metre.	(b) The vertical distance between shapes shall be at least 1.5 meter.
(c) In a vessel of less than 20 metres in length shapes of lesser dimensions but commensurate with the size of the vessel may be used and the distance apart may be correspondingly reduced.	(c) In a vessel of less than 20 meters in length shapes of lesser dimensions but commensurate with the size of the vessel may be used and the distance apart may be correspondingly reduced.

NOTES

Application As with lights, the technical details concerning the color, size, and placement of shapes on the high seas are consolidated in Annex I of the International Regulations and the Inland Rules. All day shapes are black.

Tows over 200 meters Both sets of rules require the same day shape under Rule 24(a). A vessel with a tow extending more than 200 meters astern (measured from the stern of the towing vessel to the stern of the last vessel of the tow) is required to display a diamond shape. The vessel or object towed also shows the same signal. This day signal is applicable only to tows astern. In inland waters, a submerged or partly submerged object being towed displays the diamond shape at the aftermost end of the object.

Sailboats under sail and power The conical shape required by Rule 25(e) is mandatory on the high seas and in inland waters, but a vessel less than 12 meters in length in inland waters is not required to exhibit the shape.

Fishing vessels Rule 26 requires vessels engaged in fishing on the high seas and in inland waters to display by day, whether at anchor or underway, a shape consisting of two cones point to point. A vessel less than 20 meters in length may substitute a basket for this special black shape, but any vessel of any size with gear extending more than 150 meters horizontally must also show one cone, point up, in the direction of the gear. No difference is made in the shapes required of vessels engaged in trawling and vessels engaged in fishing by other means, as was done in the lights.

Vessels not under command or restricted in ability to maneuver The day signal for a vessel not under command, two black balls, is applicable to vessels on the high seas and in inland waters. Similarly, Rule 27(b) provides for a special shape for vessels restricted in their ability to maneuver, except minesweeping vessels, consisting of one black ball above a diamond and one black ball below. If such a vessel were to anchor, she would show this shape plus the normal anchor shape. A vessel towing, which for any reason is unable to deviate from its course, displays the same signal for vessels restricted in their ability to maneuver in addition to a diamond shape when the length of tow is greater than 200 meters. Vessels, when dredging or engaged in underwater operations, also display the shape for a vessel restricted in its ability to maneuver, as well as special signals to indicate on which side an obstruction may exist and on which side it is safe to pass. However, when at anchor they do *not* exhibit an anchor shape. Rule 27(e) requires a day shape consisting of a rigid replica of the International Code Flag "A" for small vessels engaged in diving operations that are unable to show the shape of Rule 27(d).

Exemptions Under Inland Rule 27(g), a vessel less than 12 meters in length, except when engaged in diving operations, is not required to exhibit day shapes. The International Regulations are presently silent on this point, but the proposed amendments to the International Regulations, expected to take effect in midsummer 1983, will add the same exemption.

Minesweepers Minesweepers must show a day shape of one black ball at the foremast head and one at each end of the foreyard regardless of the side on which the danger exists. However, such shapes would undoubtedly be impractical on craft such as air-cushion vehicles or rotary-wing aircraft, which are increasingly being used in mine countermeasures operations.

Vessels constrained by their draft Rule 28, International Rules, provides for a situation increasingly seen on the shallow seas and straits of the world—a large deep-draft ship that is restricted in her ability to maneuver because of shallow waters. The depth of water is not the only factor, however, which determines whether or not a vessel is constrained by her draft. The availability of sufficiently deep water on all sides for the vessel to navigate safely must also be considered. A vessel navigating with small under-keel clearance but with adequate sea room to take avoiding action should not be regarded as a vessel constrained by her draft. The vessel may show by day a black cylinder shape. The Inland Rules do not mention this recent outgrowth of the modern development of very large crude carriers (VLCCs).

Vessels at anchor or aground Rule 30(a), Inland Rules, requires vessels at anchor or aground, except vessels engaged in fishing, to exhibit one black ball or three black balls, respectively. The same is true on the high seas. Vessels dredging and engaged in underwater operations are not permitted to display the single black ball or the aground shapes. Vessels less than seven meters in length are not required to display the black ball when anchored away from a fairway. Vessels less than 12 meters in length aground in inland waters need not display any day shapes. In special designated anchorage areas in inland waters, vessels under 20 meters in length are not required to show the anchor ball.

7

Sound, Light, and Distress Signals for All Vessels

International Rules

RULE 1 (c) Nothing in these Rules shall interfere with the operation of any special rules made by the Government of any State with respect to additional station or signal lights or whistle signals for ships of war and vessels proceeding under convoy, or with respect to additional station or signal lights for fishing vessels engaged in fishing as a fleet. These additional station or signal lights or whistle signals shall, so far as possible, be such that they cannot be mistaken for any light or signal authorized elsewhere under these Rules.

Inland Rules

RULE 1 (c) Nothing in these Rules shall interfere with the operation of any special rules made by the Secretary of the Navy with respect to additional station or signal lights and shapes or whistle signals for ships of war and vessels proceeding under convoy, or by the Secretary with respect to additional station or signal lights and shapes for fishing vessels engaged in fishing as a fleet. These additional station or signal lights and shapes or whistle signals shall, so far as possible, be such that they cannot be mistaken for any light, shape, or signal authorized elsewhere under these Rules. Notice of such special rules shall be published in the Federal Register and, after the effective date specified in such notice, they shall have effect as if they were a part of these Rules.

International Rules

PART D—SOUND AND LIGHT SIGNALS

Definitions

RULE 32 (a) The word "whistle" means any sound signalling appliance capable of producing the prescribed blasts and which complies with the specifications in Annex III to these Regulations.

(b) The term "short blast" means a blast of about one second's duration.

(c) The term "prolonged blast" means a blast of from four to six seconds' duration.

Equipment for Sound Signals

RULE 33 (a) A vessel of 12 metres or more in length shall be provided with a whistle and a bell and a vessel of 100 metres or more in length shall, in addition, be provided with a gong, the tone and sound of which cannot be confused with that of the bell. The whistle, bell and gong shall comply with the specifications in Annex III to these Regulations. The bell or gong or both may be replaced by other equipment having the same respective sound characteristics, provided that manual sounding of the required signals shall always be possible.

(b) A vessel of less than 12 metres in length shall not be obliged to carry the sound-signalling appliances prescribed in paragraph (a) of this Rule but if she does not, she shall be provided with

Inland Rules

PART D—SOUND AND LIGHT SIGNALS

Definitions

RULE 32 (a) The word "whistle" means any sound signalling appliance capable of producing the prescribed blasts and which complies with specifications in Annex III to these Rules.

(b) The term "short blast" means a blast of about one second's duration.

(c) The term "prolonged blast" means a blast of from four to six seconds' duration.

Equipment for Sound Signals

RULE 33 (a) A vessel of 12 meters or more in length shall be provided with a whistle and a bell and a vessel of 100 meters or more in length shall, in addition, be provided with a gong, the tone and sound of which cannot be confused with that of the bell. The whistle, bell and gong shall comply with the specifications in Annex III to these Rules. The bell or gong or both may be replaced by other equipment having the same respective sound characteristics, provided that manual sounding of the prescribed signals shall always be possible.

(b) A vessel of less than 12 meters in length shall not be obliged to carry the sound signaling appliances prescribed in paragraph (a) of this Rule but if she does not, she shall be provided with

International Rules

some other means of making an efficient sound signal.

Manoeuvring and Warning Signals

RULE 34 (a) When vessels are in sight of one another, a power-driven vessel underway, when manoeuvring as authorized or required by these Rules, shall indicate that manoeuvre by the following signals on her whistle:

—one short blast to mean "I am altering my course to starboard";

—two short blasts to mean "I am altering my course to port";

—three short blasts to mean "I am operating astern propulsion."

(b) Any vessel may supplement the whistle signals prescribed in paragraph (a) of this Rule by light signals, repeated as appropriate, whilst the manoeuvre is being carried out:

Inland Rules

some other means of making an efficient sound signal.

Maneuvering and Warning Signals

RULE 34 (a) When power-driven vessels are in sight of one another and meeting or crossing at a distance within half a mile of each other, each vessel underway, when maneuvering as authorized or required by these Rules:

(i) shall indicate that maneuver by the following signals on her whistle: one short blast to mean "I intend to leave you on my port side"; two short blasts to mean "I intend to leave you on my starboard side"; and three short blasts to mean "I am operating astern propulsion".

(ii) upon hearing the one or two blast signal of the other shall, if in agreement, sound the same whistle signal and take the steps necessary to effect a safe passing. If, however, from any cause, the vessel doubts the safety of the proposed maneuver, she shall sound the danger signal specified in paragraph (d) of this Rule and each vessel shall take appropriate precautionary action until a safe passing agreement is made.

(b) A vessel may supplement the whistle signals prescribed in paragraph (a) of this Rule by light signals:

International Rules

(i) these light signals shall have the following significance:

—one flash to mean "I am altering my course to starboard";

—two flashes to mean "I am altering my course to port";

—three flashes to mean "I am operating astern propulsion";

(ii) the duration of each flash shall be about one second, the interval between flashes shall be about one second, and the interval between successive signals shall be not less than ten seconds;

(iii) the light used for this signal shall, if fitted, be an all-round white light visible at a minimum range of 5 miles, and shall comply with the provisions of Annex I.

(c) When in sight of one another in a narrow channel or fairway:

(i) a vessel intending to overtake another shall in compliance with Rule 9(e)(i) indicate her intention by the following signals on her whistle:

—two prolonged blasts followed by one short blast to mean "I intend to overtake you on your starboard side";

—two prolonged blasts followed by two short blasts to mean "I intend to overtake you on your port side."

(ii) the vessel about to be overtaken when acting in accordance with Rule 9(e)(i) shall indicate her agreement by the following signal on her whistle:

Inland Rules

(i) These signals shall have the following significance: one flash to mean "I intend to leave you on my port side"; two flashes to mean "I intend to leave you on my starboard side"; three flashes to mean "I am operating astern propulsion.";

(ii) The duration of each flash shall be about one second; and

(iii) The light used for this signal shall, if fitted, be one all-round white or yellow light, visible at a minimum range of 2 miles, synchronized with the whistle, and shall comply with the provisions of Annex I to these Rules.

(c) When in sight of one another:

(i) a power-driven vessel intending to overtake another power-driven vessel shall indicate her intention by the following signals on her whistle: one short blast to mean "I intend to overtake you on your starboard side"; two short blasts to mean "I intend to overtake you on your port side"; and

(ii) the power-driven vessel about to be overtaken shall, if in agreement, sound a similar sound signal. If in doubt she shall sound the danger signal prescribed in

International Rules

—one prolonged, one short, one prolonged and one short blast, in that order.

(d) When vessels in sight of one another are approaching each other and from any cause either vessel fails to understand the intentions of the other, or is in doubt whether sufficient action is being taken to avoid collision, the vessel in doubt shall immediately indicate such doubt by giving at least five short and rapid blasts on the whistle. Such signal may be supplemented by a light signal of at least five short and rapid flashes.

(e) A vessel nearing a bend or an area of a channel or fairway where other vessels may be obscured by an intervening obstruction shall sound one prolonged blast. Such signal shall be answered with a prolonged blast by any approaching vessel that may be within hearing around the bend or behind the intervening obstruction.

(f) If whistles are fitted on a vessel at a distance apart of more than 100 metres, one whistle only shall be used for giving manoeuvring and warning signals.

Inland Rules

Paragraph (d).

(d) When vessels in sight of one another are approaching each other and from any cause either vessel fails to understand the intentions or actions of the other, or is in doubt whether sufficient action is being taken by the other to avoid collision, the vessel in doubt shall immediately indicate such doubt by giving at least five short and rapid blasts on the whistle. This signal may be supplemented by a light signal of at least five short and rapid flashes.

(e) A vessel nearing a bend or an area of a channel or fairway where other vessels may be obscured by an intervening obstruction shall sound one prolonged blast. This signal shall be answered with a prolonged blast by any approaching vessel that may be within hearing around the bend or behind the intervening obstruction.

(f) If whistles are fitted on a vessel at a distance apart of more than 100 meters, one whistle only shall be used for giving maneuvring and warning signals.

(g) When a power-driven vessel is leaving a dock or berth, she shall sound one prolonged blast.

(h) A vessel that reaches agreement with another vessel in a meeting, crossing, or overtaking situation by using the radiotelephone as prescribed by the Bridge-to-Bridge Radiotelephone Act (85 Stat. 165; 33 U.S.C. 1207), is not obliged to

International Rules

Sound Signals in Restricted Visibility

RULE 35 In or near an area of restricted visibility, whether by day or night, the signals prescribed in this Rule shall be used as follows:

(a) A power-driven vessel making way through the water shall sound at intervals of not more than 2 minutes one prolonged blast.

(b) A power-driven vessel underway but stopped and making no way through the water shall sound at intervals of not more than 2 minutes two prolonged blasts in succession with an interval of about 2 seconds between them.

(c) A vessel not under command, a vessel restricted in her ability to manoeuvre, a vessel constrained by her draught, a sailing vessel, a vessel engaged in fishing and a vessel engaged in towing or pushing another vessel shall, instead of the signals prescribed in paragraphs (a) or (b) of this rule, sound at intervals of not more than 2 minutes three blasts in succession, namely one prolonged followed by two short blasts.

(d) A vessel towed or if more than one vessel is towed the last vessel of the tow, if manned, shall at intervals of not more than 2 min-

Inland Rules

sound the whistle signals prescribed by this Rule, but may do so. If agreement is not reached, then whistle signals shall be exchanged in a timely manner and shall prevail.

Sound Signals in Restricted Visibility

RULE 35 In or near an area of restricted visibility, whether by day or night, the signals prescribed in this Rule shall be used as follows:

(a) A power-driven vessel making way through the water shall sound at intervals of not more than two minutes one prolonged blast.

(b) A power-driven vessel underway but stopped and making no way through the water shall sound at intervals of not more than two minutes two prolonged blasts in succession with an interval of about two seconds between them.

(c) A vessel not under command; a vessel restricted in her ability to maneuver, whether underway or at anchor; a sailing vessel; a vessel engaged in fishing, whether underway or at anchor; and a vessel engaged in towing or pushing another vessel shall, instead of the signals prescribed in paragraphs (a) or (b) of this Rule, sound at intervals of not more than two minutes, three blasts in succesion; namely, one prolonged followed by two short blasts.

(d) A vessel towed or if more than one vessel is towed the last vessel of the tow, if manned, shall at intervals of not more than two min-

International Rules

utes sound four blasts in succession, namely one prolonged followed by three short blasts. When practicable, this signal shall be made immediately after the signal made by the towing vessel.

(e) When a pushing vessel and a vessel being pushed ahead are rigidly connected in a composite unit they shall be regarded as a power-driven vessel and shall give the signals prescribed in paragraphs (a) or (b) of this Rule.

(f) A vessel at anchor shall at intervals of not more than one minute ring the bell rapidly for about 5 seconds. In a vessel of 100 metres or more in length the bell shall be sounded in the forepart of the vessel and immediately after the ringing of the bell the gong shall be sounded rapidly for about 5 seconds in the after part of the vessel. A vessel at anchor may in addition sound three blasts in succession, namely one short, one prolonged and one short blast, to give warning of her position and of the possibility of collision to an approaching vessel.

(g) A vessel aground shall give the bell signal and if required the gong signal prescribed in paragraph (f) of this Rule and shall, in addition, give three separate and distinct strokes on the bell immediately before and after the rapid ringing of the bell. A vessel aground may in addition sound an appropriate whistle signal.

(h) A vessel of less than 12 metres in length shall not be

Inland Rules

utes sound four blasts in succession; namely, one prolonged followed by three short blasts. When practicable, this signal shall be made immediately after the signal made by the towing vessel.

(e) When a pushing vessel and a vessel being pushed ahead are rigidly connected in a composite unit they shall be regarded as a power-driven vessel and shall give the signals prescribed in paragraphs (a) or (b) of this Rule.

(f) A vessel at anchor shall at intervals of not more than one minute ring the bell rapidly for about five seconds. In a vessel of 100 meters or more in length the bell shall be sounded in the forepart of the vessel and immediately after the ringing of the bell the gong shall be sounded rapidly for about five seconds in the after part of the vessel. A vessel at anchor may in addition sound three blasts in succesion; namely, one short, one prolonged and one short blast, to give warning of her position and of the possibility of collision to an approaching vessel.

(g) A vessel aground shall give the bell signal and if required the gong signal prescribed in paragraph (f) of this Rule and shall, in addition, give three separate and distinct strokes on the bell immediately before and after the rapid ringing of the bell. A vessel aground may in addition sound an appropriate whistle signal.

(h) A vessel of less than 12 meters in length shall not be

International Rules

obliged to give the above-mentioned signals but, if she does not, shall make some other efficient sound signal at intervals of not more than 2 minutes.

(i) A pilot vessel when engaged on pilotage duty may in addition to the signals prescribed in paragraphs (a), (b) or (f) of this Rule sound an identity signal consisting of four short blasts.

Inland Rules

obliged to give the above-mentioned signals but, if she does not, shall make some other efficient sound signal at intervals of not more than two minutes.

(i) A pilot vessel when engaged on pilotage duty may in addition to the signals prescribed in paragraphs (a), (b) or (f) of this Rule sound an identity signal consisting of four short blasts.

(j) The following vessels shall not be required to sound signals as prescribed in paragraph (f) of this Rule when anchored in a special anchorage area designated by the Secretary:

(i) a vessel of less than 20 meters in length; and

(ii) a barge, canal boat, scow, or other nondescript craft.

International Rules

Signals to Attract Attention

RULE 36 If necessary to attract the attention of another vessel, any vessel may make light or sound signals that cannot be mistaken for any signal authorized elsewhere in these Rules, or may direct the beam of her searchlight in the direction of the danger, in such a way as not to embarrass any vessel.

Distress Signals

RULE 37 When a vessel is in distress and requires assistance she shall use or exhibit the signals prescribed in Annex IV to these Regulations.

Inland Rules

Signals to Attract Attention

RULE 36 If necessary to attract the attention of another vessel, any vessel may make light or sound signals that cannot be mistaken for any signal authorized elsewhere in these Rules, or may direct the beam of her searchlight in the direction of the danger, in such a way as not to embarrass any vessel.

Distress Signals

RULE 37 When a vessel is in distress and requires assistance she shall use or exhibit the signals described in Annex IV to these Rules.

International Rules

ANNEX IV—DISTRESS SIGNALS

1. The following signals, used or exhibited either together or separately, indicate distress and need of assistance:

(a) a gun or other explosive signal fired at intervals of about a minute;

(b) a continuous sounding with any fog-signalling apparatus;

(c) rockets or shells, throwing red stars fired one at a time at short intervals;

(d) a signal made by radiotelegraphy or by any other signalling method consisting of the group . . . --- . . . (SOS) in the Morse Code;

(e) a signal sent by radiotelephony consisting of the spoken word "Mayday";

(f) the International Code Signal of distress indicated by N.C.;

(g) a signal consisting of a square flag having above or below it a ball or anything resembling a ball;

(h) flames on the vessel (as from a burning tar barrel, oil barrel, etc.)

Inland Rules

PART 87—ANNEX IV, DISTRESS SIGNALS[1]

§ 87.1 Need of assistance. The following signals, used or exhibited either together or separately, indicate distress and need of assistance:

(a) A gun or other explosive signal fired at intervals of about a minute.

(b) A continuous sounding with any fog-signaling apparatus;

(c) Rockets or shells, throwing red stars fired one at a time at short intervals;

(d) A signal made by radiotelegraphy or by any other signaling method consisting of the group . . . ——— . . . (SOS) in the Morse Code,

(e) A signal sent by radiotelephony consisting of the spoken word "Mayday";

(f) The International Code Signal of distress indicated by N.C.

(g) A signal consisting of a square flag having above or below it a ball or anything resembling a ball:

(h) Flames on the vessel (as from a burning tar barrel, oil barrel, etc.);

[1]Authority: Sec. 3, Pub. L. 96–591, 33 U.S.C. 2071, 49 CFR 1.46(n)(14)

International Rules

(i) a rocket parachute flare or a hand flare showing a red light;

(j) a smoke signal giving off orange-coloured smoke;

(k) slowly and repeatedly raising and lowering arms outstretched to each side;

(l) the radiotelegraph alarm signal;

(m) the radiotelephone alarm signal;

(n) signals transmitted by emergency position-indicating radio beacons.

Inland Rules

(i) A rocket parachute flare or a hand flare showing a red light;

(j) A smoke signal giving off orange-colored smoke;

(k) Slowly and repeatedly raising and lowering arms outstretched to each side;

(l) The radiotelegraph alarm signal;

(m) The radiotelephone alarm signal;

(n) Signals transmitted by emergency position-indicating radio beacons;

(o) A high intensity white light flashing at regular intervals from 50 to 70 times per minute.

2. The use or exhibition of any of the foregoing signals except for the purpose of indicating distress and need of assistance and the use of other signals which may be confused with any of the above signals is prohibited.

§ 87.3 Exclusive use. The use or exhibition of any of the foregoing signals except for the purpose of indicating distress and need of assistance and the use of other signals which may be confused with any of the above signals is prohibited.

3. Attention is drawn to the relevant sections of the International Code of Signals, the Merchant Ship Search and Rescue Manual and the following signals:

(a) a piece of orange-coloured canvas with either a black square and circle or other appropriate symbol (for identification from the air);

(b) a dye marker.

§ 87.5 Supplemental signals. Attention is drawn to the relevant sections of the International Code of Signals, the Merchant Ship Search and Rescue Manual and the following signals:

(a) A piece of orange-colored canvas with either a black square and circle or other appropriate symbol (for identification from the air);

(b) A dye marker.

Fig. 59. Firing a gun. (Day or night distress signal.) International and Inland Rules, Annex IV.

Fig. 60. Red rockets at night indicate distress, high seas or inland waters. International and Inland Rules, Annex IV.

Fig. 61. Distress signal (NC), high seas or inland waters. International and Inland Rules, Annex IV.

Fig. 62. Distant distress signal, high seas or inland waters. International and Inland Rules, Annex IV.

Fig. 63. Flames at night indicate distress, high seas or inland waters. International and Inland Rules, Annex IV.

Fig. 64. Distress signal, inverted ensign, high seas or inland waters. There is no applicable rule; this practice is based on custom.

Fig. 65. Distress signal, orange smoke, high seas or inland waters. International and Inland Rules, Annex IV.

Fig. 66. Distress signal, person raising and lowering arms outstretched to each side, high seas or inland waters. International and Inland Rules, Annex IV.

Fig. 67. Submarine distress signal, high seas or inland waters.

Fig. 68. Airplane with red Very's signal. "Send assistance to save personnel," high seas or inland waters.

International Rules

PART E—EXEMPTIONS

Exemptions

RULE 38 Any vessel (or class of vessels) provided that she complies with the requirements of the International Regulations for Preventing Collisions at Sea, 1960, the keel of which is laid or which is at a corresponding stage of construction before the entry into force of these Regulations may be exempted from compliance therewith as follows:

(g) The requirements for sound signal appliances prescribed in Annex III, until nine years after the date of entry into force of these Regulations.

Inland Rules

PART E—EXEMPTIONS

Exemptions

RULE 38 Any vessel or class of vessels, the keel of which is laid or which is at a corresponding stage of construction before the date of enactment of this Act, provided that she complies with the requirements of—

(a) The Act of June 7, 1897 (30 Stat. 96), as amended (33 U.S.C. 154–232) for vessels navigating the waters subject to that statute;

(b) Section 4233 of the Revised Statutes (33 U.S.C. 301–356) for vessels navigating the waters subject to that statute;

(c) The Act of February 8, 1895 (28 Stat. 645), as amended (33 U.S.C. 241–295) for vessels navigating the waters subject to that statute; or

(d) Sections 3, 4, and 5 of the Act of April 25, 1940 (54 Stat. 163), as amended (46 U.S.C. 526 b, c, and d) for motorboats navigating the waters subject to that statute; shall be exempted from compliance with the technical Annexes to these Rules as follows:

(vii) the requirements for sound signal appliances prescribed in Annex III to these Rules, until 9 years after the effective date of these Rules.

NOTES

Additional signals Both sets of rules authorize additional light signals for men-of-war or other vessels proceeding in company and for vessels fishing as a fleet, and for all categories, both sets of rules permit the use of additional whistle signals.

Definitions In both sets of rules, the short blast is defined as being of about one second duration, the prolonged blast being of from four to six seconds' duration. The long blast is no longer mentioned in either rules.

Sound signal apparatus The term "whistle" is comprehensively specified in Annex III to both the International Regulations and the Inland Rules. There is no longer any mention of a foghorn under either set of rules. Vessels 12 meters or more in length on the high seas or in inland waters are required to carry a whistle and a bell whether power-driven or under sail, while vessels over 100 meters must also carry a gong. Vessels of less than 12 meters in both jurisdictions are not required to carry a whistle or bell but must be provided with an efficient sound signaling device.

Signals when vessels are in sight of one another International Rule 34(a) makes the one- and two-blast signals for power-driven vessels purely rudder signals to be given whenever course is changed in accordance with the rules, but not to be given *unless* course is changed. The proper signal must be given with a change in course whether the other vessel is a power-driven or a sailing vessel. No means of acknowledging such a signal is provided, but if the other vessel makes a change in course she would, if power-driven, be required to make the proper signal.

In inland waters, when power-driven vessels are in sight of one another and meeting or crossing, the one- or two-blast signal must be given and *answered* whenever the vessels approach within half a mile, regardless of a change in course by either or both vessels. This does not mean that vessels should always wait until within one-half mile before signaling, but it does mean that in inland waters they must never approach that close without exchanging one- or two-blast signals. Excepting the three-short-blast signal, which has the same meaning in both sets of rules, these signals are not action signals but are signals of intent and agreement. When sounded by a proposing vessel they must be answered. If the answering vessel is not in agreement, then she must sound the in-doubt signal. Cross signals, answering one blast with two, or two blasts with one, are not permitted.

Overtaking Under Rule 34, International Regulations, the signals for overtaking differ depending upon whether the situation occurs on the high seas or in a narrow channel. On the high seas, no signals are required by either vessel if no change of course is necessitated by the overtaking situation, the one- and two-blast signal only being required when a course change is required to effect a safe passing. However, in a narrow channel, when overtaking can take place only if the vessel to be overtaken has to take action to permit safe passing, the vessel intending to overtake must sound the appropriate signals: two prolonged blasts followed by one short blast to overtake to starboard, or two prolonged blasts followed by two short blasts to overtake to port. The vessel to be overtaken indicates her agreement by sounding one prolonged, one short, one prolonged, and one short blast. Either vessel, if in doubt, must sound the in-doubt signal. In inland waters, under Rule 34, a power-driven vessel in sight signals her intent by sounding one short blast to overtake to starboard, and two short blasts to overtake to port. The vessel ahead signals her assent by sounding a like signal. Again, if either vessel has any doubt about what the other is doing, the doubt signal is required. It should be noted that both Inland and International Rules 34(a), (b), and (c), exclude the one-, two-, and three-blast signals in fog before the vessels sight each other and limits their application to cases where each vessel can see the other or her lights.

Whistle light Under Rule 34(b), vessels operating in international or inland waters may further supplement the "in-sight" whistle signals of Rule 34(a) by means of an all-round white light, such that the number of flashes corresponds to the number of short blasts. Under the International Regulations, the white light does not have to be synchronized, and the light signal can be repeated without necessarily repeating the whistle signal.

In-doubt signal The five-short-blast signal of Rule 34(d) for both jurisdictions is mandatory and applies to all vessels that have doubt about another vessel's actions to avoid collision, if the vessels are in sight of one another. If the vessels are not in sight of one another due to restricted visibility, then only the signals prescribed in Rule 35 are required. However, if a vessel thought that a collision could be avoided by sounding the in-doubt signal even though the other vessel was not in sight, the "responsibility" requirement of Rule 2 would prevail. A vessel failing to use the in-doubt signal when circumstances require it is almost certain to be held at least partly liable for a collision.

Vessel Bridge-to-Bridge Radiotelephone Act This important act went into effect on 1 January 1973 and has the force of law upon inland waters. (See Appendix F.) It applies to the following categories of vessels in any condition of visibility:

(1) power-driven vessels of 300 gross tons and upward;
(2) vessels of 100 gross tons and upward carrying one or more passengers for hire;
(3) commercial towing vessels of 26 feet or over in length;
(4) manned dredges and floating plants working in or near a channel or fairway.

All the preceding vessels must guard frequency 156.65 MHz (Channel 13) for bridge-to-bridge communications. The frequency is for the exclusive use of the person in charge of the vessel for the purpose of transmitting and confirming the intentions of his vessel and any other information necessary for the safe navigation of vessels. It has become common practice to exchange maneuvering signals verbally between vessels guarding the VHF channel used for Bridge-to-Bridge Radio Telephone communications. In fact, under Rule 34(h) of the Inland Rules, vessels that come to an agreement by radiotelephone in a meeting, crossing, or overtaking situation are not required to exchange the prescribed whistle signals. However, if agreement is not reached, then whistle signals must be exchanged in a timely manner.

Bend signal The bend signal is now the same for both international and inland waters—one prolonged blast. This signal is answered in like manner by an approaching vessel that may be obscured. The inland rules also specify a prolonged blast by a power-driven vessel leaving her berth.

Sound signals in restricted visibility These signals are required under Rule 35, International Regulations and Inland Rules, in or near an area of restricted visibility, whether by day or night. Both sets of rules exempt vessels under 12 meters from the obligation to make precisely the laid-down signals, but if they do not or cannot, they must make some other equally effective sound at the required interval.

Intervals The intervals between required sound signals in restricted visibility should be regarded as maximum intervals only. In conditions of heavy traffic, or more probably when another vessel is known to be near, it may well be prudent that the signals be given more frequently. The rules differentiate between vessels underway, which make required signals at intervals of not more than two minutes, and certain vessels not underway that make required signals at intervals of not more than one minute.

Power-driven vessel underway with no way on in restricted visibility Rule 35(b) provides a signal to distinguish between a power vessel underway, making way, and one underway, not making way. When using such signals, care should be taken to see that the vessel has lost all headway (or sternway) before the two prolonged blasts are started.

Vessels at anchor in fog The bell signal for a vessel at anchor on the high seas or in inland waters must be given at least once a minute not only by a single vessel but, when several vessels are anchored together, by every vessel in the nest. This rule is rigidly enforced, however large the number of vessels. The rapid ringing of a ship's bell is a standardized signal, and the substitution of miscellaneous noises such as the beating of a dishpan or the sound of a pneumatic drill is not permitted by the courts, even though such a noise might be heard by the approaching vessel.

Rule 35(f), International Regulations and Inland Rules, also requires a vessel 100 meters or more in length anchored in fog to sound a gong, the sound of which cannot be confused with the bell. Such vessels must sound the bell in the forepart of the vessel in the manner required and immediately thereafter sound the gong in the after part.

It will be noted that all vessels at anchor in fog may sound a special sound signal of three blasts (one short, one prolonged, one short) to give an approaching vessel more definite warning of its position and to indicate the possibility of a collision. This signal, which is the letter R in the Morse Code, can be regarded as a special application of Rule 36, and is in addition to the required signals.

In inland waters all vessels less than 20 meters in length and nondescript vessels of any size that are anchored in special anchorage areas are exempt from sounding the anchor signal. Special anchorage areas are established by the Secretary of Transportation. They may be found in practically any part of any coast of the United States. The descriptions are not included in this text due to their bulk and limited relationship to the rules for preventing collisions.

Vessels engaged in special activities in restricted visibility The signal of one prolonged blast followed by two short blasts is used in inland waters and on the high seas for a vessel towing in fog, a sailing vessel, a vessel fishing, a vessel underway that is not under command, and a vessel restricted in her ability to maneuver. The last category includes vessels engaged in dredging, surveying, underwater operations, minesweeping, underway replenishment, launching or recovery of aircraft, and the laying, servicing, or picking up of a navigation mark, submarine cable, or pipeline. In the event of some of these activities being carried out in fog while at anchor—particularly for vessels engaged in underwater operations—it could be assumed that the three-blast signal supersedes the regular anchor signal, though the rules do not specifically regulate on this point.

International Rule 35(c) also provides for a vessel constrained by her draft to sound the "restricted in her ability to maneuver" signal in restricted visibility. A vessel pushing another ahead that is rigidly connected to it gives the signal for a power-driven vessel and not that of a vessel towing or pushing.

A vessel towed at sea and in inland waters, if manned, must sound, in restricted visibility, the signal in Rule 35(d)—namely, one prolonged blast followed by three short blasts. If there is more than one vessel towed, only the last vessel of the tow should give this signal, and where practicable, it should immediately follow the signal made by the towing vessel.

Vessels aground Vessels aground on the high seas and in inland waters, in addition to the normal bell and gong signals for a vessel at anchor, must also sound three strokes of the bell immediately before and after the rapid ringing of the bell. A vessel aground can, in addition to the required signal, make an appropriate whistle signal that could be a distress signal or a suitable letter from the International Code of Signals.

Pilot vessels in fog Pilot vessels on pilotage duty in inland waters and in international waters may, in addition to the fog signals prescribed in the rules, sound an identity signal of four short blasts.

Sound signal appliances The reader is invited to refer to Annex III of both sets of rules for the technical details of sound signal appliances.

8

Conduct of Vessels in Any Condition of Visibility

International Rules

PART B—STEERING AND SAILING RULES

Section 1—Conduct of Vessels in Any Condition of Visibility

Application

RULE 4 Rules in this Section apply in any condition of visibility.

Look-out

RULE 5 Every vessel shall at all times maintain a proper look-out by sight and hearing as well as by all available means appropriate in the prevailing circumstances and conditions so as to make a full appraisal of the situation and of the risk of collision.

Safe Speed

RULE 6 Every vessel shall at all times proceed at a safe speed so that she can take proper and effective action to avoid collision and be stopped within a distance

Inland Rules

PART B—STEERING AND SAILING RULES

Subpart I—Conduct of Vessels in Any Condition of Visibility

Application

RULE 4 Rules in this subpart apply in any condition of visibility.

Look-out

RULE 5 Every vessel shall at all times maintain a proper look-out by sight and hearing as well as by all available means appropriate in the prevailing circumstances and conditions so as to make a full appraisal of the situation and of the risk of collision.

Safe Speed

RULE 6 Every vessel shall at all times proceed at a safe speed so that she can take proper and effective action to avoid collision and be stopped within a distance

International Rules

appropriate to the prevailing circumstances and conditions.

In determining a safe speed the following factors shall be among those taken into account:

(a) By all vessels:

(i) the state of visibility;

(ii) the traffic density including concentrations of fishing vessels or any other vessels;

(iii) the maneuverability of the vessel with special reference to stopping distance and turning ability in the prevailing conditions;

(iv) at night the presence of background light such as from shore lights or from back scatter of her own lights;

(v) the state of wind, sea, and current, and the proximity of navigational hazards;

(vi) the draught in relation to the available depth of water.

(b) Additionally, by vessels with operational radar:

(i) the characteristics, efficiency and limitations of the radar equipment;

(ii) any constraints imposed by the radar range scale in use;

(iii) the effect on radar detection of the sea state, weather and other sources of interference:

(iv) the possibility that small vessels, ice and other floating objects may not be detected by radar at an adequate range;

(v) the number, location and movement of vessels detected by radar;

(vi) the more exact assessment

Inland Rules

appropriate to the prevailing circumstances and conditions.

In determining a safe speed the following factors shall be among those taken into account:

(a) By all vessels:

(i) the state of visibility;

(ii) the traffic density including concentration of fishing vessels or any other vessels;

(iii) the maneuverability of the vessel with special reference to stopping distance and turning ability in the prevailing conditions;

(iv) at night the presence of background light such as from shore lights or from back scatter of her own lights;

(v) the state of wind, sea, and current, and the proximity of navigational hazards;

(vi) the draft in relation to the available depth of water.

(b) Additionally, by vessels with operational radar:

(i) The characteristics, efficiency and limitations of the radar equipment;

(ii) any constraints imposed by the radar range scale in use;

(iii) the effect on radar detection of the sea state, weather, and other sources of interference;

(iv) the possibility that small vessels, ice and other floating objects may not be detected by radar at an adequate range;

(v) the number, location, and movement of vessels detected by radar;

(vi) the more exact assessment

International Rules

of the visibility that may be possible when radar is used to determine the range of vessels or other objects in the vicinity.

Risk of Collision

RULE 7 (a) Every vessel shall use all available means appropriate to the prevailing circumstances and conditions to determine if risk of collision exists. If there is any doubt such risk shall be deemed to exist.

(b) Proper use shall be made of radar equipment if fitted and operational, including long-range scanning to obtain early warning of risk of collision and radar plotting or equivalent systematic observation of detected objects.

(c) Assumptions shall not be made on the basis of scanty information, especially scanty radar information.

(d) In determining if risk of collision exists the following considerations shall be among those taken into account:

(i) such risk shall be deemed to exist if the compass bearing of an approaching vessel does not appreciably change;

(ii) such risk may sometimes exist even when an appreciable bearing change is evident, particularly when approaching a very large vessel or a tow or when approaching a vessel at close range.

Action to Avoid Collision

RULE 8 (a) Any action taken

Inland Rules

of the visibility that may be possible when radar is used to determine the range of vessels or other objects in the vicinity.

Risk of Collision

RULE 7 (a) Every vessel shall use all available means appropriate to the prevailing circumstances and conditions to determine if risk of collision exists. If there is any doubt such risk shall be deemed to exist.

(b) Proper use shall be made of radar equipment if fitted and operational, including long-range scanning to obtain early warning of risk of collision and radar plotting or equivalent systematic observation of detected objects.

(c) Assumptions shall not be made on the basis of scanty information, especially scanty radar information.

(d) In determining if risk of collision exists the following considerations shall be among those taken into account:

(i) such risk shall be deemed to exist if the compass bearing of an approaching vessel does not appreciably change; and

(ii) such risk may sometimes exist even when an appreciable bearing change is evident, particularly when approaching a very large vessel or a tow or when approaching a vessel at close range.

Action to Avoid Collision

RULE 8 (a) Any action taken

International Rules

to avoid collision shall, if the circumstances of the case admit, be positive, made in ample time and with due regard to the observance of good seamanship.

(b) Any alteration of course and/or speed to avoid collision shall, if the circumstances of the case admit, be large enough to be readily apparent to another vessel observing visually or by radar; a succession of small alterations of course and/or speed should be avoided.

(c) If there is sufficient sea room, alteration of course alone may be the most effective action to avoid a close-quarters situation provided that it is made in good time, is substantial and does not result in another close-quarters situation.

(d) Action taken to avoid collision with another vessel shall be such as to result in passing at a safe distance. The effectiveness of the action shall be carefully checked until the other vessel is finally past and clear.

(e) If necessary to avoid collision or allow more time to assess the situation, a vessel shall slacken her speed or take all way off by stopping or reversing her means of propulsion.

Narrow Channels

RULE 9 (a) A vessel proceeding along the course of a narrow channel or fairway shall keep as near to the outer limit of the

Inland Rules

to avoid collision shall, if the circumstances of the case admit, be positive, made in ample time and with due regard to the observance of good seamanship.

(b) Any alteration of course or speed to avoid collision shall, if the circumstances of the case admit, be large enough to be readily apparent to another vessel observing visually or by radar; a succession of small alterations of course or speed should be avoided.

(c) If there is sufficient sea room, alteration of course alone may be the most effective action to avoid a close-quarters situation provided that it is made in good time, is substantial and does not result in another close-quarters situation.

(d) Action taken to avoid collision with another vessel shall be such as to result in passing at a safe distance. The effectiveness of the action shall be carefully checked until the other vessel is finally past and clear.

(e) If necessary to avoid collision or allow more time to assess the situation, a vessel shall slacken her speed or take all way off by stopping or reversing her means of propulsion.

Narrow Channels

RULE 9 (a)(i) A vessel proceeding along the course of a narrow channel or fairway shall keep as near to the outer limit of the

International Rules

channel or fairway which lies on her starboard side as is safe and practicable.

(b) A vessel of less than 20 metres in length or a sailing vessel shall not impede the passage of a vessel which can safely navigate only within a narrow channel or fairway.

(c) A vessel engaged in fishing shall not impede the passage of any other vessel navigating within a narrow channel or fairway.

(d) A vessel shall not cross a narrow channel or fairway if such crossing impedes the passage of a vessel which can safely navigate only within such channel or fairway. The latter vessel may use the sound signal prescribed in Rule 34(d) if in doubt as to the intention of the crossing vessel.

(e)(i) In a narrow channel or fairway when overtaking can take place only if the vessel to be over-

Inland Rules

channel or fairway which lies on her starboard side as is safe and practicable.

(ii) Notwithstanding paragraph (a)(i) and Rule 14(a), a power-driven vessel operating in narrow channels or fairways on the Great Lakes, Western Rivers, or waters specified by the Secretary, and proceeding downbound with a following current shall have the right-of-way over an upbound vessel, shall propose the manner and place of passage, and shall initiate the maneuvering signals prescribed by Rule 34(a)(i), as appropriate. The vessel proceeding upbound against the current shall hold as necessary to effect safe passage.

(b) A vessel of less than 20 meters in length or a sailing vessel shall not impede the passage of a vessel that can safely navigate only within a narrow channel or fairway.

(c) A vessel engaged in fishing shall not impede the passage of any other vessel navigating within a narrow channel or fairway.

(d) A vessel shall not cross a narrow channel or fairway if such crossing impedes the passage of a vessel which can safely navigate only within that channel or fairway. The latter vessel shall use the danger signal prescribed in Rule 34(d) if in doubt as to the intention of the crossing vessel.

(e)(i) In a narrow channel or fairway when overtaking, the vessel intending to overtake shall indicate

International Rules

taken has to take action to permit safe passing, the vessel intending to overtake shall indicate her intention by sounding the appropriate signal prescribed in Rule 34(c)(i). The vessel to be overtaken shall, if in agreement, sound the appropriate signal prescribed in Rule 34(c)(ii) and take steps to permit safe passing. If in doubt she may sound the signals prescribed in Rule 34(d).

(ii) This Rule does not relieve the overtaking vessel of her obligation under Rule 13.

(f) A vessel nearing a bend or an area of a narrow channel or fairway where other vessels may be obscured by an intervening obstruction shall navigate with particular alertness and caution and shall sound the appropriate signal prescribed in Rule 34(e).

(g) Any vessel shall, if the circumstances of the case admit, avoid anchoring in a narrow channel.

Traffic Separation Schemes

RULE 10 (a) This rule applies to traffic separation schemes adopted by the organization.

(b) A vessel using a traffic separation scheme shall:

(i) proceed in the appropriate traffic lane in the general direction of traffic flow for that lane;

(ii) so far as practicable keep clear of a traffic separation line or separation zone;

(iii) normally join or leave a

Inland Rules

her intention by sounding the appropriate signal prescribed in Rule 34(c) and take steps to permit safe passing. The overtaken vessel, if in agreement, shall sound the same signal. If in doubt she shall sound the danger signal prescribed in Rule 34(d).

(ii) This Rule does not relieve the overtaking vessel of her obligation under Rule 13.

(f) A vessel nearing a bend or an area or fairway where other vessels may be obscured by an intervening obstruction shall navigate with particular alertness and caution and shall sound the appropriate signal prescribed in Rule 34(e).

(g) Every vessel shall, if the circumstances of the case admit, avoid anchoring in a narrow channel.

Vessel Traffic Services

RULE 10 Each vessel required by regulation to participate in a vessel traffic service shall comply with the applicable regulations.

International Rules *Inland Rules*

traffic lane at the termination of the lane, but when joining or leaving from the side shall do so at as small an angle to the general direction of traffic flow as practicable.

(c) A vessel shall so far as practicable avoid crossing traffic lanes, but if obliged to do so shall cross as nearly as practicable at right angles to the general direction of traffic flow.

(d) Inshore traffic zones shall not normally be used by through traffic which can safely use the appropriate traffic lane within the adjacent traffic separation scheme.

(e) A vessel, other than a crossing vessel, shall not normally enter a separation zone or cross a separation line except:

(i) in cases of emergency to avoid immediate danger;

(ii) to engage in fishing within a separation zone.

(f) A vessel navigating in areas near the terminations of traffic separation schemes shall do so with particular caution.

(g) A vessel shall so far as practicable avoid anchoring in a traffic separation scheme or in areas near its terminations.

(h) A vessel not using a traffic separation scheme shall avoid it by as wide a margin as is practicable.

(i) A vessel engaged in fishing shall not impede the passage of any vessel following a traffic lane.

(j) A vessel of less than 20 metres in length or a sailing vessel shall not impede the safe passage of

International Rules	*Inland Rules*
a power-driven vessel following a traffic lane.	

NOTES

Application Rule 4 for both inland and international waters specifies the applicability of the rules contained in the subpart (section), and recognizes the fact that vessels do navigate in clear and restricted visibility.

Lookout In Rule 5, identical in both sets of rules, the important subject of a proper lookout is discussed. The rule specifies that a proper lookout by sight and hearing shall be maintained by all vessels and at all times. In addition, a lookout shall be kept by all available means appropriate to the prevailing circumstances. The term "all available means" would certainly include, but not be limited to, intelligent and systematic use of operational radar to detect the location and movement of vessels in the vicinity. The U.S. federal courts have defined the person who is a proper lookout to be an experienced seaman, alert and vigilant, without other duties, and properly stationed in the vessel for the prevailing circumstances. In many cases the last definition may mean that the lookout should be as low down and as far forward on the vessel as circumstances permit.

Safe Speed Rule 6, International Regulations, and Rule 6, Inland Rules, are identical and require all vessels to proceed at a safe speed in all conditions of visibility in order to allow them the maximum possible time to take effective action to avoid a collision. It further specifies factors to be considered by all vessels in determining a safe speed, including those vessels with operational radar. It is relevant to note that the term "moderate speed" has disappeared entirely from the language as has its application, which was limited to periods of restricted visibility. Safe speed is required in clear weather as well and must be determined by the prevailing circumstances. It is obvious that the state of visibility is a primary consideration for arriving at a safe speed. When there is little or no visibility, safe speed becomes bare steerageway, usually not over 3 or 4 knots. In very thick weather, in crowded harbors or other regions of dense traffic, vessels should, if practicable, find an anchorage. Vessels have been held at fault in collision cases for getting underway or for failing to come to anchor under conditions of poor visibility. Although a safe speed is a matter for individual judgment, it must be a speed slow enough to allow time to prevent a collision.

Risk of collision Rule 7 in both sets of rules is the same and requires the use of all available means to determine if risk of collision exists, and specifically includes the use of radar. The basic method of determining whether risk of collision exists is still by means of compass bearings. The bearings of approaching vessels should be frequently observed and the bearing drift, if any, ascertained as a means of determining risk of collision. If the compass bearing of the other vessel is steady or nearly steady and the range is decreasing, there is risk of collsion. Even when there is a marked bearing change, in the case of a large vessel or a tow at close range, such risk may sometimes exist.

Action to avoid collision Here again, in Rule 8 the two sets of rules contain identical provisions. Specific and sound advice is presented to the mariner regarding his actions to avoid collision, and such actions are applicable to the stand-on vessel as well as to the give-way vessel. Any avoiding action shall be positive, made in ample time, and follow the dictates of good seamanship. Any change of course or speed should be large enough to be readily detected by another vessel either visually or by radar. Assuming there is sufficient sea room, a substantial course change away from a close-quarters situation is often the best avoiding action. In any case, the rule requires that the mariner check the effectiveness of his action continuously until a safe passage has occurred.

Narrow Channels Rule 9(a) of the International Regulations and Rule 9(a)(i) of the Inland Rules apply to all vessels, not just power-driven vessels, and require them to keep as near to the starboard limit of the channel as is safe and practicable. In identical wording, both sets of rules go on to admonish small vessels and sailing vessels not to interfere with the safe passage of a vessel restricted to following a narrow channel; nor is a fishing vessel to impede the passage of any vessel navigating a channel. Rule 9 also recognizes the problems experienced by large ships in narrow channels by specifying that vessels shall not cross a channel such that the passage of a large ship is impeded. If she is unsure of the intentions of a crossing vessel, under Inland Rule 9 (d) the vessel restricted to navigating within the channel is mandated to use the doubt signal, but is only invited to under International Rule 9(d). Rule 9(a)(ii) Inland Rules, not found in the International Regulations, abrogates the starboard side of the channel requirement for vessels operating in certain specified waters and downbound with the current. It recognizes the limited maneuverability of a downbound vessel and is considered essential for the safety of navigation in our narrow inland channels and fairways. Such vessels, in addition to having the right-of-way over an upbound vessel, specify the manner of passage and initiate the appropriate whistle signals. Both sets of rules require the one-prolonged-blast "bend" signal and prohibit anchoring in a narrow channel.

Overtaking in a narrow channel Here, in Rule 9, the two sets of rules depart from each other somewhat. Although both require signals of proposal and agreement for the overtaking situation in a narrow channel, International Rule 9(e)(i) requires them only if the vessel to be overtaken has to take action to permit safe passing. Referring back to the sound signals of Rule 34, it should be noted that the prescribed signals of proposal and assent are vastly different. In either case, the vessel to be overtaken, if in doubt, must sound the five-short-blast doubt signal.

Traffic separation schemes Higher concentrations of traffic in various bodies of water have led to the development of traffic separation schemes to confine vessels to traffic lanes flowing in the same general direction. International Rule 10 provides the requirements for vessels using those traffic separation schemes (TSS) adopted by IMCO. Provisions are made in this rule for local traffic, as well as for vessels that must enter the separation zone in an emergency or to engage in fishing. Other measures affecting the workability and safety of traffic separation schemes are included. Inland Rule 10 addresses vessel traffic services (VTS), the U.S. equivalent to a traffic separation scheme. These services are established for certain waterways of the United States, and Inland Rule 10 requires compliance with the applicable regulations.

9

Conduct of Vessels in Sight of One Another

International Rules

SECTION II—CONDUCT OF VESSELS IN SIGHT OF ONE ANOTHER

Application

RULE 11 Rules in this section apply to vessels in sight of one another.

Sailing Vessels

RULE 12 (a) When two sailing vessels are approaching one another, so as to involve risk of collision, one of them shall keep out of the way of the other as follows:

(i) when each has the wind on a different side, the vessel which has the wind on the port side shall keep out of the way of the other;

(ii) when both have the wind on the same side, the vessel which is to windward shall keep out of the way of the vessel which is to leeward;

(iii) if a vessel with the wind on the port side sees a vessel to windward and cannot determine with

Inland Rules

SUBPART II—CONDUCT OF VESSELS IN SIGHT OF ONE ANOTHER

Application

RULE 11 Rules in this subpart apply to vessels in sight of one another.

Sailing Vessels

RULE 12 (a) When two sailing vessels are approaching one another, so as to involve risk of collision, one of them shall keep out of the way of the other as follows:

(i) when each has the wind on a different side, the vessel which has the wind on the port side shall keep out of the way of the other;

(ii) when both have the wind on the same side, the vessel which is to windward shall keep out of the way of the vessel which is to leeward;

(iii) if a vessel with the wind on the port side sees a vessel to windward and cannot determine with

International Rules

certainty whether the other vessel has the wind on the port or on the starboard side, she shall keep out of the way of the other.

(b) For the purposes of this rule the windward side shall be deemed to be the side opposite to that on which the mainsail is carried or, in the case of a square-rigged vessel, the side opposite to that on which the largest fore-and-aft sail is carried.

Overtaking

RULE 13 (a) Notwithstanding anything contained in the rules of this Section any vessel overtaking any other shall keep out of the way of the vessel being overtaken.

(b) A vessel shall be deemed to be overtaking when coming up with another vessel from a direction more than 22.5 degrees abaft her beam, that is, in such a position with reference to the vessel she is overtaking, that at night she would be able to see only the sternlight of that vessel but neither of her sidelights.

(c) When a vessel is in any doubt as to whether she is overtaking another, she shall assume that this is the case and act accordingly.

(d) Any subsequent alteration of the bearing between the two vessels shall not make the overtaking vessel a crossing vessel within the meaning of these rules or relieve her of the duty of keeping clear of the overtaken vessel until she is finally past and clear.

Inland Rules

certainty whether the other vessel has the wind on the port or on the starboard side, she shall keep out of the way of the other.

(b) For the purpose of this Rule the windward side shall be deemed to be the side opposite to that on which the mainsail is carried or, in the case of a square-rigged vessel, the side opposite to that on which the largest fore-and-aft sail is carried.

Overtaking

RULE 13 (a) Notwithstanding anything contained in Rules 4 through 18, any vessel overtaking any other shall keep out of the way of the vessel being overtaken.

(b) A vessel shall be deemed to be overtaking when coming up with another vessel from a direction more than 22.5 degrees abaft her beam; that is, in such a position with reference to the vessel she is overtaking, that at night she would be able to see only the sternlight of that vessel but neither of her sidelights.

(c) When a vessel is in any doubt as to whether she is overtaking another, she shall assume that this is the case and act accordingly.

(d) Any subsequent alteration of the bearing between the two vessels shall not make the overtaking vessel a crossing vessel within the meaning of these Rules or relieve her of the duty of keeping clear of the overtaken vessel until she is finally past and clear.

International Rules

Head-on Situation

RULE 14 (a) When two power-driven vessels are meeting on reciprocal or nearly reciprocal courses so as to involve risk of collision each shall alter her course to starboard so that each shall pass on the port side of the other.

(b) Such a situation shall be deemed to exist when a vessel sees the other ahead or nearly ahead and by night she could see the masthead lights of the other in a line or nearly in a line and/or both sidelights and by day she observes the corresponding aspect of the other vessel.

(c) When a vessel is in any doubt as to whether such a situation exists she shall assume that it does exist and act accordingly.

Crossing Situation

RULE 15 When two power-driven vessels are crossing so as to involve risk of collision, the vessel which has the other on her own starboard side shall keep out of the way and shall, if the circumstances of the case admit, avoid crossing ahead of the other vessel.

Action by Give-way Vessel

RULE 16 Every vessel which is directed to keep out of the way of

Inland Rules

Head-on Situation

RULE 14 (a) When two power-driven vessels are meeting on reciprocal or nearly reciprocal courses so as to involve risk of collision each shall alter her course to starboard so that each shall pass on the port side of the other.

(b) Such a situation shall be deemed to exist when a vessel sees the other ahead or nearly ahead and by night she could see the masthead lights of the other in a line or nearly in a line or both sidelights and by day she observes the corresponding aspect of the other vessel.

(c) When a vessel is in any doubt as to whether such a situation exists she shall assume that it does exist and act accordingly.

Crossing Situation

RULE 15 (a) When two power-driven vessels are crossing so as to involve risk of collision, the vessel which has the other on her own starboard side shall keep out of the way and shall, if the circumstances of the case admit, avoid crossing ahead of the other vessel.

(b) Notwithstanding paragraph (a), on the Great Lakes, Western Rivers, or water specified by the Secretary, a vessel crossing a river shall keep out of the way of a power-driven vessel ascending or descending the river.

Action by Give-way Vessel

RULE 16 Every vessel which is directed to keep out of the way of

International Rules

another vessel shall, so far as possible, take early and substantial action to keep well clear.

Action by Stand-on Vessel

RULE 17 (a)(i) Where one of two vessels is to keep out of the way, the other shall keep her course and speed.

(ii) The latter vessel may however take action to avoid collision by her manoeuvre alone, as soon as it becomes apparent to her that the vessel required to keep out of the way is not taking appropriate action in compliance with these Rules.

(b) When, from any cause, the vessel required to keep her course and speed finds herself so close that collision cannot be avoided by the action of the give-way vessel alone, she shall take such action as will best aid to avoid collision.

(c) A power-driven vessel which takes action in a crossing situation in accordance with sub-paragraph (a)(ii) of this rule to avoid collision with another power-driven vessel shall, if the circumstances of the case admit, not alter course to port for a vessel on her own port side.

(d) This Rule does not relieve the give-way vessel of her obligation to keep out of the way.

Responsibility Between Vessels

RULE 18 Except where Rules 9, 10 and 13 otherwise require:

(a) A power-driven vessel underway shall keep out of the way of:

Inland Rules

another vessel shall, so far as possible, take early and substantial action to keep well clear.

Action by Stand-on Vessel

RULE 17 (a)(i) Where one of two vessels is to keep out of the way, the other shall keep her course and speed.

(ii) The latter vessel may, however, take action to avoid collision by her maneuver alone, as soon as it becomes apparent to her that the vessel required to keep out of the way is not taking appropriate action in compliance with these Rules.

(b) When, from any cause, the vessel required to keep her course and speed finds herself so close that collision cannot be avoided by the action of the give-way vessel alone, she shall take such action as will best aid to avoid collision.

(c) A power-driven vessel which takes action in a crossing situation in accordance with subparagraph (a)(ii) of this Rule to avoid collision with another power-driven vessel shall, if the circumstances of the case admit, not alter course to port for a vessel on her own port side.

(d) This Rule does not relieve the give-way vessel of her obligation to keep out of the way.

Responsibilities Between Vessels

RULE 18 Except where Rules 9, 10, and 13 otherwise require:

(a) A power-driven vessel underway shall keep out of the way of:

International Rules

(i) a vessel not under command;
(ii) a vessel restricted in her ability to manoeuvre;
(iii) A vessel engaged in fishing;
(iv) a sailing vessel.

(b) A sailing vessel underway shall keep out of the way of:
(i) a vessel not under command;
(ii) a vessel restricted in her ability to manoeuver;
(iii) a vessel engaged in fishing.

(c) A vessel engaged in fishing when underway shall, so far as possible, keep out of the way of:
(i) a vessel not under command;
(ii) a vessel restricted in her ability to manoeuvre.

(d)(i) Any vessel other than a vessel not under command or a vessel restricted in her ability to manoeuvre shall, if the circumstances admit, avoid impeding the safe passage of a vessel constrained by her draught, exhibiting the signals in Rule 28.

(ii) A vessel constrained by her draught shall navigate with particular caution having full regard to her special condition.

(e) A seaplane on the water shall, in general, keep well clear of all vessels and avoid impeding their navigation. In circumstances, however, where risk of collision exists, she shall comply with the Rules of this Part.

Inland Rules

(i) a vessel not under command;
(ii) a vessel restricted in her ability to maneuver;
(iii) a vessel engaged in fishing;
(iv) a sailing vessel.

(b) A sailing vessel underway shall keep out of the way of:
(i) a vessel not under command;
(ii) a vessel restricted in her ability to maneuver;
(iii) a vessel engaged in fishing.

(c) A vessel engaged in fishing when underway shall, so far as possible, keep out of the way of:
(i) a vessel not under command;
(ii) a vessel restricted in her ability to maneuver.

(d) A seaplane on the water shall, in general, keep well clear of all vessels and avoid impeding their navigation. In circumstances, however, where risk of collision exists, she shall comply with the Rules of this Part.

NOTES

Application Rule 11 specifies the applicability of Rules 12 through 18 in Section II of the International Rules and Subpart II of the Inland

Rules that apply to vessels in sight of one another. As defined in Rule 3(j), vessels are deemed to be in sight of one another only when they can be observed visually one from the other. Except for differences in Rules 13, 15, and 18, discussed below, the Inland Rules in Subpart II are now identical to the corresponding International Rules.

Sailing vessels Rule 12 describes the actions required of two sailing vessels approaching one another at such an angle that there is risk of collision. It specifies which vessel is to keep out of the way of the other: the right-of-way is determined first by point of sail and then, if two vessels have the wind on the same side, by which vessel is to windward of the other. In the situation where a sailing vessel with the wind on the port side cannot determine on which side another sailing vessel has the wind, the rule directs the former to keep out of the way of the latter.

Overtaking Rule 13, International Regulations, and Rule 13, Inland Rules, are virtually the same. One significant exception is that the former rule applies only to vessels in sight of one another while the latter extends the applicability of the overtaking rules to any condition of visibility. Obviously, this extension falls short of including vessels not in sight of one another due to restricted visibility. An overtaking vessel is defined as one coming up from more than 22.5 degrees abaft the beam of another. The overtaking vessel must keep out of the way of the vessel to be overtaken and has the option of passing on either side, provided it is safe. The rule mentions the case where a vessel is in doubt whether she is overtaking or crossing, a common dilemma at sea; in such case the overtaking situation is assumed to exist. The obligation of the overtaking vessel to keep clear lasts not only until she has passed, but until she is "finally past and clear."

Head-on situation Both Rule 14, International Regulations, and Rule 14, Inland Rules, describe a "head-on" or meeting situation in the same terms and require that vessels pass port to port when they are in sight and there is risk of collision. This holds true whether the vessels are already port to port, or are exactly head-on, or are a little starboard to starboard but not far enough to pass well clear on that side. In the head-on situation, it is clear that a course change to port is contrary to the intent of both sets of rules although it is not mentioned in either. Rule 14(c) requires a vessel, when in doubt, to consider a situation a head-on one and alter course to starboard to avoid collision. This should eliminate the possibility of conflicting maneuvers when two vessels are meeting almost end-on. Nevertheless, the normal requirements in this rule for the head-on situation are superseded by Inland Rule 9(a)(ii) in the case of a power-driven vessel proceeding downbound with a following current in certain specified waters. When meeting an upbound vessel in this case, the downbound vessel has the right-of-way and shall propose the manner of passage.

Crossing situation Rule 15, International Regulations, and Rule

15(a), Inland Rules, are identical. Both require the power-driven vessel that is to port of another crossing power-driven vessel to keep out of the way and avoid, where possible, passing ahead of the other. A crossing situation has usually been defined in terms of what it is not—i.e., neither overtaking nor meeting—but it can generally be considered to be two power-driven vessels in visual sight of one another approaching at right angles or obliquely so as to involve risk of collision. The give-way vessel is the one with the other on her own starboard side within an arc from dead ahead to 22.5 degrees abaft the beam. Inland Rule 15(b) directs a vessel crossing a river in certain designated areas to stay out of the way of a power-driven vessel ascending or descending the river.

Action by the give-way vessel Under Rule 16 in both jurisdictions, the give-way vessel is directed to take early and substantial action to keep well clear. The vessel to port in a crossing situation is allowed the following actions: turning to starboard, turning to port, or reducing speed by stopping or backing engines. The give-way vessel is enjoined to avoid crossing ahead.

Action by the stand-on vessel Again, the rules governing the actions of the stand-on vessel are the same for a vessel on the high seas and in inland waters. She is directed to hold course and speed. If and when she finds herself in extremis, that is, so close that collision would be inevitable if she took no action, she is required to take "such action as will best aid to avoid collision." However, a major new provision in the 1972 Regulations and now in the Inland Navigational Rules Act of 1980, allows the stand-on vessel the freedom to take action prior to extremis as soon as it becomes apparent to her that the give-way vessel is not taking appropriate action. If the situation is a crossing one, the stand-on vessel is warned against a turn to port for a vessel on her own port side. The signal of doubt (five or more short blasts) should always be used by the stand-on vessel if she is in doubt that the other vessel is paying attention to her duties.

Responsibilities between vessels Rule 18, International Regulations and Rule 18, Inland Waters, govern all vessels that are not under command or are restricted in their ability to maneuver and admonishes other vessels to keep well clear of them. It also clearly defines a hierarchy among vessels and lists their mutual responsibilities. The new Inland Rules do not provide for vessels constrained by their drafts, but the COLREGS Rules 18(d)(i) and (ii) include provisions for a "vessel constrained by her draught," which take into account vessels that are restricted to narrow channels or fairways due to their deep draft and cannot readily maneuver outside them.

10

Conduct of Vessels in Restricted Visibility

International Rules

SECTION III—CONDUCT OF VESSELS IN RESTRICTED VISIBILITY

Conduct of Vessels in Restricted Visibility

RULE 19 (a) This Rule applies to vessels not in sight of one another when navigating in or near an area of restricted visibility.

(b) Every vessel shall proceed at a safe speed adapted to the prevailing circumstances and conditions of restricted visibility. A power-driven vessel shall have her engines ready for immediate manoeuvre.

(c) Every vessel shall have due regard to the prevailing circumstances and conditions of restricted visibility when complying with the Rules of Section I of this Part.

(d) A vessel which detects by radar alone the presence of another vessel shall determine if a close-quarters situation is developing and/or risk of collision exists. If

Inland Rules

SUBPART III—CONDUCT OF VESSELS IN RESTRICTED VISIBILITY

Conduct of Vessels in Restricted Visibility

RULE 19 (a) This Rule applies to vessels not in sight of one another when navigating in or near an area of restricted visibility.

(b) Every vessel shall proceed at a safe speed adapted to the prevailing circumstances and conditions of restricted visibility. A power-driven vessel shall have her engines ready for immediate maneuver.

(c) Every vessel shall have due regard to the prevailing circumstances and conditions of restricted visibility when complying with Rules 4 through 10.

(d) A vessel which detects by radar alone the presence of another vessel shall determine if a close-quarters situation is developing or risk of collision exists. If so,

International Rules

so, she shall take avoiding action in ample time, provided that when such action consists of an alteration of course, so far as possible the following shall be avoided:

(i) an alteration of course to port for a vessel forward of the beam, other than for a vessel being overtaken;

(ii) an alteration of course towards a vessel abeam or abaft the beam.

(e) Except where it has been determined that a risk of collision does not exist, every vessel which hears apparently forward of her beam the fog signal of another vessel, or which cannot avoid a close-quarters situation with another vessel forward of her beam, shall reduce her speed to the minimum at which she can be kept on her course. She shall if necessary take all her way off and in any event navigate with extreme caution until danger of collision is over.

Inland Rules

she shall take avoiding action in ample time, provided that when such action consists of an alteration of course, so far as possible the following shall be avoided:

(i) an alteration of course to port for a vessel forward of the beam, other than for a vessel being overtaken; and

(ii) an alteration of course toward a vessel abeam or abaft the beam.

(e) Except where it has been determined that a risk of collision does not exist, every vessel which hears apparently forward of her beam the fog signal of another vessel, or which cannot avoid a close-quarters situation with another vessel foward of her beam, shall reduce her speed to the minimum at which she can be kept on course. She shall if necessary take all her way off and, in any event, navigate with extreme caution until danger of collision is over.

NOTES

Conduct of vessels in restricted visibility Rule 19 applies to vessels not in sight of one another due to some degree of restricted visibility. The rule also applies to vessels navigating near an area of restricted visibility. It reminds the mariner to proceed at a "safe" speed in restricted visibility with careful regard to the circumstances involved. The term "safe speed" is defined more exactly in Rule 6, which is identical for both international and inland waters, and the mariner should refer back to it for specific requirements. He should also refer to the discussion of safe speed in the notes to chapter 8. In addition to proceeding at a safe speed, the engines of a vessel should be ready for immediate maneuver, and the mariner is reminded to be particularly aware of existing conditions of restricted visibility when considering: the posting of a proper lookout; whether or not a risk of collision exists, as well as action to be taken to avoid collision; conduct in narrow channels; and participation in vessel traffic services. The new Inland Rules, now nearly identical to the COLREGS, reinforce the positive approach to the use of radar under conditions of restricted visibility. In this regard, Rule 7(b) states that an operating radar should be used for long-range scanning for early warning as well as for plotting a detected object. The rules spell out the action to be taken by a vessel in restricted visibility, when she detects by radar alone another vessel with whom a risk of collision exists. Vessels are to take avoiding action when their radar indicates that a close-quarters situation is developing. This action must be taken in ample time, once the situation is assessed. In taking avoiding action that involves a change of course, specific guidance is given in the rules. The guidelines are such that if two vessels detect each other by radar, their avoiding action will be complementary. To be able to act under this rule it is axiomatic that systematic radar plotting, either manual or automated, is required.

In the new Inland Rules, a vessel that hears a fog signal apparently forward of the beam now has the same flexibility as was previously enjoyed only by vessels on the high seas. There is no requirement to stop engines if the radar plot shows no risk of collision. If the close-quarters situation cannot be avoided, a vessel must reduce her speed to the minimum at which she can hold steerageway, 3-4 knots for most vessels. The vessel is directed, if necessary, to take all her way off, which could include stopping or backing her engines, and then to navigate only with extreme caution.

The use of radar should not encourage a liberal interpretation of Rule 19. The courts are not like to agree with such an interpretation. In the cases settled to date, the courts have not been sympathetic to arguments that radar justifies the omission of taking any precaution either stated or implied in the rules. The sound signals to be used under conditions of restricted visibility are to be found in chapter 7.

Examples of avoiding action taken under International and Inland Rules 19(d)(i) and (ii) Some simple examples of avoiding action, involving change of course only, that could be taken in restricted visibility follow. Vessels are considered to be effectively using their radar. All PPI scopes are compass-stabilized with own ship at the center (i.e., a relative picture) and show a heading marker. No attempt has been made to illustrate more complicated situations, or the effect of speed changes, which more properly belong in radar and collision-avoidance manuals.

(1) Both vessels determine from systematic observation that a close-quarters situation is developing, and that neither is overtaking the other.

(2) Both vessels elect to maneuver by making a course change.

(3) The circumstances in this case allow them to comply with Rule

Example #1

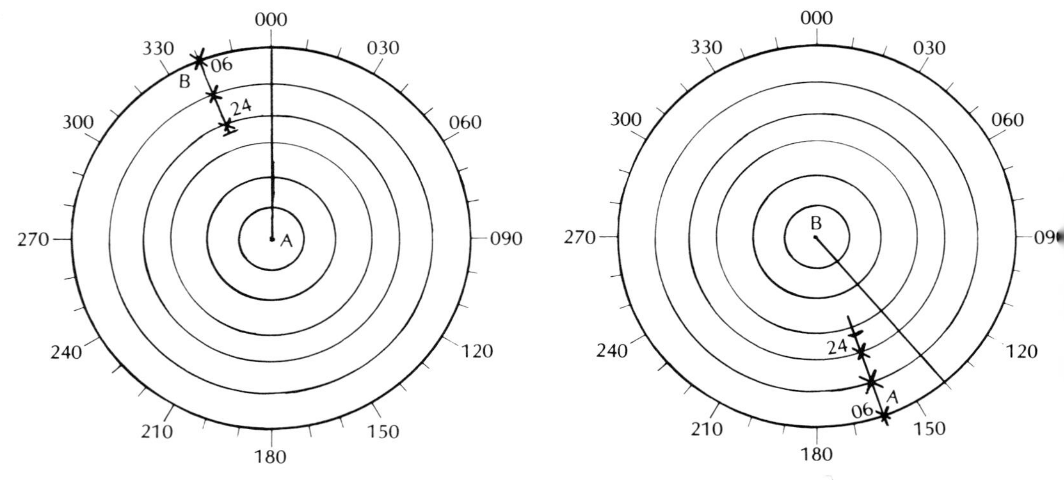

A, on course 000°T, speed 10kn, holds B ahead of her beam.

B, on course 140°T, speed 10 kn, holds A ahead of her beam.

19(d)(i) for both inland and international waters. Therefore they both turn boldly to starboard, though not necessarily at the same moment.

(4) Both continue with systematic observation to ensure their action is effective.

A true-motion plot would indicate the situation and the avoiding action taken as shown below:

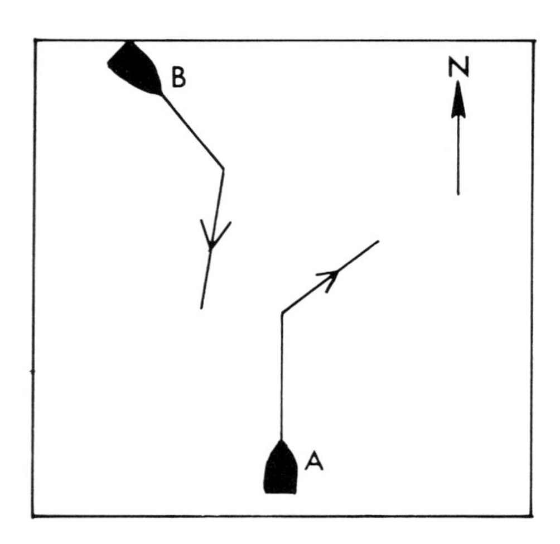

Example #2

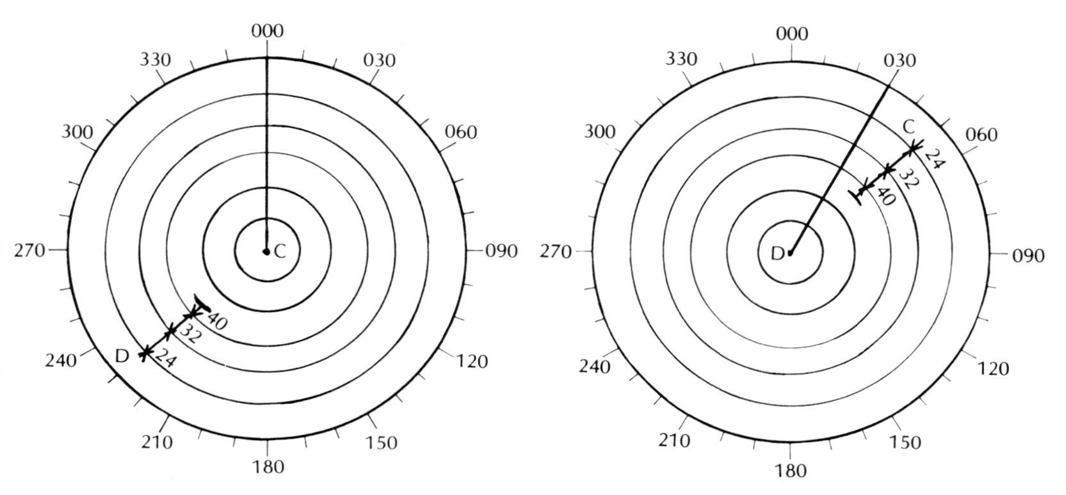

C, on course 000°T, speed 5 kn, holds D abaft her beam.

D, on course 030°T, speed 11 kn, holds C ahead of her beam.

(1) Both vessels determine from systematic observation that a close-quarters situation is developing.

(2) C elects to alter course boldly away from D, i.e., to starboard.

(3) D determines that she is overtaking C. To avoid C she may alter to port or to starboard, there being no preferred direction of alteration specified in Rule 19 (d)(i) in this case.

A true-motion plot would indicate the situation and the avoiding action taken as shown below.

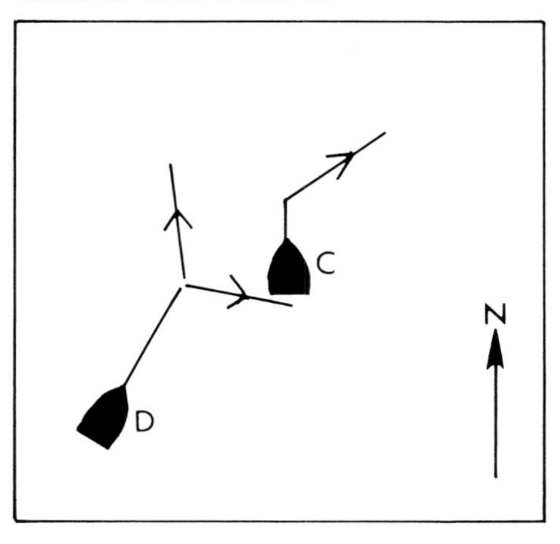

In example #2 it was stated that D was overtaking C. The term *overtaking*, although used, is not defined in the restricted visibility portion of the International Rules.

From Rules 19(d)(i) and (ii), taken together, it *might* be inferred that the term means when coming up generally from abaft the beam of another vessel. The precise meaning laid down in Rule 13(b) of each set of rules, of more than 22.5 degrees abaft the beam, is strictly applicable only when vessels are in sight of each other, though it might be considered a reasonable extension to assume it also applies in restricted visibility.

Part II

The Law of the Nautical Road

11

Principles of Marine Collision Law

Jurisdiction in Collision Cases

Jurisdiction over cases of collision between vessels on public navigable waters is placed by the Constitution of the United States in the hands of federal courts, sitting as courts of admiralty. Public navigable waters may be defined as waters used, or capable of being used, in interstate commerce. The definition excludes from federal jurisdiction collisions occurring on a lake wholly within a state, but includes cases on a navigable river that flows in or between two states or empties into the sea or an inlet of the sea. A collision case ordinarily begins in the federal district court, from which it may be appealed to the circuit court of appeals and on proper grounds to the United States Supreme Court.[1]

State courts have exclusive jurisdiction over a collision on a lake completely surrounded by territory of the state and concurrent jurisdiction over one occurring on any portion of public navigable waters within the state. Thus, a collision that takes place on Puget Sound may by mutual consent of the parties be adjudicated in a damage suit between the vessels' owners in the Superior Court of the State of Washington. However, such cases are actually tried, in an overwhelming majority, in the federal courts, because usually at least one of the litigants regards certain admiralty principles peculiar to these courts as favorable to his side of the case. Cases of collision on public navigable waters, unless action has already

[1]In June 1982, the U.S. Supreme Court, in *Foremost Insurance Company* v. *Richardson*, ruled that a jurisdiction test should not be tied to the commercial use of the vessels involved in a collision. In other words, "there is no requirement that the maritime activity be an exclusively commercial one" in order to properly state a claim within the admiralty jurisdiction of the federal courts.

been started in a state court, may always be taken to a federal court by either party.

Investigation of a collision that occurred in the waters of a foreign country is governed by the law of that country, as also, said Judge Krupansky of the District Court of Ohio, are "the rights and liabilities arising as a consequence of the collision."[2]

After a collision has occurred on the high seas, an action may be brought in the country of either of the plaintiffs or the defendants, or in any other country where the law permits such action to be brought. Generally speaking, most courts will allow an action to be brought if the defendant vessel is in a port of their country at the time, in which case the proceedings and findings are in accordance with the law of that land.

A valid judgment by a foreign court, delivered abroad before action is brought in another country and given in the presence of both parties, is final and conclusive and a bar to further proceedings between the two same parties in the other country, except possibly for the enforcement, extension, or registration of that judgment.

Legal Personality of a Vessel

In American admiralty cases, an important principle applied by the federal courts is the legal personality of the vessel, which is assumed to make the vessel herself the wrongdoer when collision follows a violation of the rules of the road. The action may be, and commonly is, between John Doe and the steamer So-and-So, or it may be between the two vessels. The vessel sued or "libeled" is held until the claims against her are satisfied, unless the owners obtain her release by paying into the court an amount equal to her appraised valuation, or post a bond double the amount of existing liens. If a judgment is obtained against her, the vessel may be sold at public auction by the marshal in order that the proceeds of the sale may place funds in the registry of the court for satisfaction of the judgment. A final recognition of the vessel's personality is seen in the fact that a marshal's sale, properly conducted, divests her of all maritime liens against her, and starts her out with her new owners free of any old claims against her. She thus literally receives a new lease on life.

Division of Damages

Originally a principle of admiralty courts was the doctrine of equal responsibility for unequal fault. When two vessels were in collision and both of them had violated a rule, the liability of each vessel was 50 percent of the total loss, regardless of her degree of guilt. One vessel may have been guilty of what seemed to be only a minor infraction and the other

[2]The *Steelton*, [1979] 1 Lloyd's Rep. 431.

may have violated several rules and been flagrantly and deliberately negligent. The courts were concerned with the *fact* of fault but not, in most cases, with its amount:

> Damages from collisions between vessels both at fault must be equally divided, irrespective of degree of fault.[3]

This rule was an ancient form of rough justice, a means of apportioning damages where it was difficult to measure which party was more at fault.

The Brussels Collision Liability Convention of 1910 subsequently provided for the apportionment of damages on the basis of degree of fault whenever it was possible to do so. The United States was virtually alone among the world's major maritime nations in not adhering to the Convention—a fact that encouraged transoceanic court forum shopping. While the lower federal courts followed the old equally divided damages rule, they did so only grudgingly, and in 1975 the Supreme Court divested themselves of this vestigial relic and substituted a rule of comparative negligence:

> We hold that when two or more parties have contributed by their fault to cause property damage in a maritime collision or stranding, liability for such damage is to be allocated among the parties proportionately to the comparative degree of their fault, and that liability for such damages is to be allocated equally only when the parties are equally at fault or when it is not possible fairly to measure the comparative degree of their fault.[4]

If one vessel is solely at fault, she is liable for the total damage to the other, subject only to the provisions of the Limited Liability Acts, which limit the liability of a vessel of her value after the collision, plus earnings for the voyage, collected or collectible. If neither vessel is at fault, we then have that extremely rare species of collision characterized by the courts as inevitable accident, and each vessel, of course, bears her own loss, be it great or small. It is significant that only about 1 percent of all the cases in the books come under this category. It is evident, therefore, that in a collision between vessel *A* and *B*, we have four possibilities as to liability as determined by our admiralty courts: *A* solely liable; *B* solely liable: *A* and *B* both liable, in which case the damages are proportional to degree of fault; and neither liable, in practice a very rare occurrence.

Limited Liability of Vessel

The principle of limited liability is a very old one, and was originally based on the high degree of risk that was inherent in any maritime venture. The theory was, and to some extent still is, that capital would be

[3]The *Marian* (1933) 66 F(2d) 354.
[4]U.S.V. Reliable Transfer Co., Inc., 1975, 44 L. Ed. 251.

discouraged from investing in a business where an absentee agent, the master, might by his negligence involve his owners in enormous losses through disaster to a ship and her cargo, unless the amount of that loss could be limited somewhat in accordance with the old common-law principle. As described by Justice Holmes in the Supreme Court in a case just after the First World War:

> The notion, as applicable to a collision case, seems to us to be that if you surrender the offending vessel you are free, just as it was said by a judge in the time of Edward III: "If my dog kills your sheep and I, freshly after the fact, tender you the dog, you are without recourse against me."[5]

Under the original statute of 1851 and its later amendments, the law in the United States now provides that where a vessel is at fault without the privity or knowledge of her owners the limit of her liability is the value of the vessel at the expiration of the voyage, plus any earnings that have accrued or are collectible for the transportation of passengers or cargo. Under a usual form of contract providing that, once the voyage has begun, freight is payable, ship lost or not lost, it frequently happens in the case of a commercial vessel that although the vessel herself is a total loss, these earnings, technically known as "pending freight," may amount to a considerable item. Pending freight is added to the value of the vessel at the end or breaking-up of a voyage, and therefore her value after the accident causing the liability determines the limit, and not her value before the accident as in English law. This value, including pending freight, may be levied upon for faulty damage to the other vessel or her cargo or for injury to personnel on either vessel. The Harter Act and the later Carriage of Goods by Sea Act, however, excuse the vessel in a faulty collision from liability for damage to her own cargo. It will be seen that where a vessel that is solely at fault in a collision is totally lost and there is no pending freight, the injured vessel can recover nothing for damage either to herself or to her cargo.

An amendment to the Limited Liability Acts, patterned after the English law and passed 29 August 1935, changes the liability for death or personal injury to a maximum of $60 per gross register ton where the remaining value of the ship is less than that amount. The liability for damage to vessels or cargoes was not affected by this amendment. However, the Convention on Limitation of Liability for Maritime Claims, 1976, held under the auspices of IMCO (Inter-Governmental Maritime Consultative Organization) provides for the calculation of general limits in accordance with a rather complicated version of the English law. In essence the size of the "limitation fund" does not increase in the same

[5]Liverpool, Brazil, and River Plate Steam Navigation Co. vs. Brooklyn Eastern District Terminal (1919) 64 L. Ed. 130.

amount as the tonnage of the ship does, but is progressively reduced instead. The result would be that owners of larger ships would only provide a smaller fund than the tonnages of their vessels warrant under the present straightforward English application of a multiplier of the registered tonnage. There are other innovations agreed upon by the convention which, if adopted and made part of national laws, could align practice in this branch of the law in various countries.

One fundamental principle that is likely to remain unchanged in all jurisdictions is that the privilege of limiting liability is available only to the shipowner who is in no way to blame for the damage. The term "privity or knowledge of the owners" refers to cases in which damage and consequent liability are incurred through circumstances not beyond the control of the owners or managers. If such guilty knowledge can be implied to the owners or managers, limited liability cannot be invoked, and the injured vessel or cargo owners may bring all the offending owners' resources into that suit. One excellent illustration is the case of the USS *Chicago* and the British freighter, *Silver Palm*, which were in a collision in a fog off the California coast in October 1933. The vessels sighted each other a minute before the collision, and both commanding officers ordered their engines full speed astern. The *Chicago* was practically brought to a standstill, but the *Silver Palm*, unknown to her master, had an early type of diesel engine that could not be reversed at speeds higher than 6 knots until the fuel was shut off and the vessel had lost a substantial part of her headway, the result being that she struck the *Chicago* while still making excessive speed. When the government showed that this defect was known to her owners, but they had not warned the master, who was making his first voyage on the vessel, of the conditions, both the district court and the circuit court of appeals denied the *Silver Palm's* petition for limitation of liability, thus opening the way for the United States to levy the entire fleet of her owners, if necessary, to make up the difference between the value of the damaged *Silver Palm* and the amount of the judgment.[6]

Recent cases have reinforced this rule in other countries. On 27 June 1970, a collision occurred between a racing dinghy and an unmanned barge that was being towed by the tug *Kathy K* in English Bay, Vancouver. The crew of the sailboat were killed. Although both vessels were at fault, the owner and operator of the tug was unable to limit his liability because it was shown that he had failed to do the following:

(i) discuss the length of tow ropes to be used in Vancouver Harbor;
(ii) supply the *Kathy K* with copies of the Collision Regulations or the harbor regulations;

[6]Decided October 28, 1937. The circuit court found both vessels at fault, thus halving the damage that could be collected under the district court decision. Certiorari denied (1938) 82 L. Ed. 1539.

(iii) see that she had the minimum crew of three persons as required by her inspection certificate;
(iv) see that the master was sufficiently experienced;
(v) see that the vessel was equipped with a whistle control on the flying bridge; and
(vi) see that she had an alternative means of sounding the whistle if the electrical system failed.

The Canadian Supreme Court found that the owners "were actually privy to the fault and negligence" and that the operator's acquiescence "in the tug and tow being left in such inexperienced hands created a situation exposing other traffic in the bay to potential dangers which in fact ensued."[7]

On another occasion the Supreme Court of Canada found that "no actual fault" could be attributed to the owners because the experienced master of a tug was "so unpredictably careless as to overestimate the height of the tide."[8]

Failure to provide copies of local regulations was also the cause for the English Court of Appeal to deny limitation of costs to the owners of a motor vessel involved in a collision in the River Thames:

> The managing owner ought to have foreseen that, without specific instructions, the Master, however competent, might fail to have the Port of London River By-laws available.[9]

One of the more interesting problems of limitation of liability arises when by negligent navigation a tug causes her tow to collide with another vessel. The tug's limitation fund may be trifling compared to that of her tow. The tug owner is able, in some countries, to limit liability to a sum based on the tonnage of the tug alone, which may not be sufficient to satisfy the claim in full.

In some cases, an American judge may put together the tonnage of the tug and tow to raise the amount of the limitation fund. In one case judged in the United States a tug, a barge, and a derrick lighter working together had anchored separately at a good distance from each other. The barge broke ground during a storm and collided with a gas well structure. The tonnage of all three vessels was used when calculating the limitation fund. However, a vessel hiring a tug is not regarded as being involved in such a "common enterprise," and if the tug's action alone causes damage, American practice would be in line with English practice where the liability would be limited to the tonnage of the tug. As was said by Lord Denning,

[7]The *Kathy K* [1976] 1 Lloyd's Rep. 154.
[8]The *Chugaway II* [1973] 2 Lloyd's Rep. 159.
[9]Rederij Erven, H. Groen & Groen v. the *England* owners and others [1973] 1 Lloyd's Rep. 373.

"there is not much room for justice in this rule. . ."! The rule has its critics on both sides of the Atlantic, and although English judges cannot get away from it—but do try to mitigate its harshness—in some circumstances, such as that outlined above, American judges can be more flexible.[10]

Liability of the Mariner

Apart from the owners, the master, officers and crew of a vessel are clearly liable for their actions. Until recently, prosecutions for infringement of the International Regulations for the Prevention of Collision at Sea were few. Judgment on members of ships' crews was usually confined to those arising from collision cases that ended up in court, though inquiries by national boards of inquiry could and did result in the loss or suspension of qualifications. However, for those infringements of the collision regulations that did not result in a collision, it was difficult to establish intentional or careless conduct on the part of shipmasters. The advent of traffic separation schemes governed by the international rules has removed some of this difficulty. A vessel navigating contrary to the rules simply needs to be plotted and identified. The surveillance of traffic separation schemes has led to an increase in the number of masters brought before the courts for contravening the rules. Offenders who are not under the jurisdiction of the littoral states are reported to their flag nation, who follow-up substantiated allegations in their own court or disciplinary system. The penalties can be steep, involving heavy fines and, on occasion, the arrest of the offending vessel.

Naval Vessel Not Subject to Lien

The doctrine of personality of the vessel is modified with respect to navy and other publicly owned vessels, in that it is contrary to public policy to permit such vessels to be libeled and taken into custody of the marshal and thus held out of service. Nevertheless, the old theory that "the King can do no wrong" is never used as a defense in cases of collision between a ship of the United States Navy and a merchant vessel, even in time of war. Instead, under a special statutory provision, the government permits itself to be sued as an owner *in personam*, and the damage sustained by both vessels is adjudicated exactly as it would be done were the action *in rem* between the vessels themselves. Indeed when the collision occurred because the naval vessel was patrolling without light and fog signals, due to actual war conditions, so scrupulously has the United States accepted its responsibility that it has paid the losses in full.[11]

[10]Synopsis from "Mariner at the Bar," *Seaways Bulletin* Oct 1977.
[11]Watts vs. U.S. (1903) 123 F 105.

Rules of the Road Are Mandatory

At the beginning, it will be well to have in mind certain general principles that govern the action of the courts in determining collision liability. The first of these is that the rules of the road applicable to a particular case are in no sense optional, but are for the most part absolutely mandatory. To avoid liability for a collision, the requirements *must* be obeyed. The courts will excuse a departure from the rules on two grounds only, and one of these—to avoid immediate danger—is provided for by the rules themselves.[12] Departure from the rules for any other reason, or for no reason at all, must be justified on the ground that while there was a technical violation, the circumstances were such that it could not possibly have contributed to the collision. As said by the Supreme Court, not only once but in substance many times:

> But when a ship at the time of collision is in actual violation of a statutory rule intended to prevent collision, it is no more than a reasonable presumption that the fault, if not the sole cause, was at least a contributory cause of the disaster. In such a case the burden rests upon the ship of showing, not merely that her fault might not have been one of the causes, or that it probably was not, but that it could not have been.[13]

It will be recognized that a disregard of any rule on the basis of convenience, courtesy, good nature, or disbelief in its efficacy places the navigator under a burden of proof that it is almost impossible for him to carry.

Obedience Must Be Timely

In the second place, not only must the rules be obeyed, but the action prescribed by them must be taken in ample time to carry out their purpose. It must be remembered that the rules are intended not only to prevent collision but to prevent serious and imminent risk of collision. This precept applies with particular force to the use of sound signals, which should always be given in time to be corrected if misunderstood, or as one judge expressed it "in time to maneuver out of a misunderstanding." Obedience to the letter of the rule is not obedience to the spirit of the rule unless it is rendered *before* the vessels are in dangerous proximity, in a sufficiently timely manner so that each vessel is aware of the other's intentions in time to conduct herself in accordance with them and aid in carrying them out.

Rules Apply Alike to All Vessels

A third principle that must be remembered is that the rules apply with equal force to all vessels on public navigable waters, without regard to

[12]Special Circumstances, Rule 2, International Regulations and Inland Rules.
[13]The *Pennsylvania* (1875) 19 Wall 125.

flag, ownership, service, size, or speed. A so-called "stand-on" vessel is as much under the obligation to hold course and speed as is the "give-way" vessel to keep out of her way. No rights or exemptions, except those conferred by the rule, or amendments thereto, apply under American law to naval vessels, passenger liners, ferries, or towboats with tows, and the same steering and sailing rules govern the USS *America* and a 30-foot trawler. To give the rules their maximum effectiveness, this is, of course, precisely as it should be.

Rules Modified by Court Interpretation

A fourth principle of the rules too often overlooked by the mariner in his seagoing practice of collision law is that to avoid liability he must know not only what the rules applicable to a given situation provide but what the federal courts have interpreted them to mean. Judicial interpretation has, in the history of the rules, performed three important functions. First, it has determined the legal meaning of certain phrases not defined in the rules themselves, such as efficient whistle or siren, flare-up light, proper lookout, special circumstances, immediate danger, ordinary practice of seamen, and risk of collision; it is in accordance with the meanings thus established that these terms are construed in collision cases. Second, it has filled certain gaps in the rules, sometimes modifying the statute to do this. For example, Article 28, the old Inland Rules, provided that three short blasts, when vessels were in sight of each other, meant, "my engines are going at full speed astern," while the courts required the same signal to be given when the engines were going at less than full speed astern or when one engine was going ahead and the other astern, with the vessel actually making sternboard.[14] Again, the courts have determined the proper signals when vessels approach each other in a collision situation stern first, a point on which the rules are silent. Third, judicial interpretation has been used not only to eliminate the old Pilot Rules found contradictory to the old Inland Rules, but to reconcile occasional inconsistencies or conflicts in the latter. This is illustrated in the treatment by the courts of the apparently inconsistent sections of old Article 18, Inland Rules, in which Rule III required the danger signal by an approaching vessel when she failed to understand the course or intention of the other *from any cause*, and Rule IX of the same article provided that in fog, when vessels cannot see each other, *fog signals only must be given*. The courts have decided, in effect, that the danger signal in inland waters must be included as a fog, as well as a clear weather, signal.[15]

[14]The *Sicilian Prince* (CCA NY 1903) 144 F 951; the *Deutschland* (CCA NY 1904) 137 F 1018; the *San Juan* (Calif 1927) A.M.C. 384.
[15]The *Celtic Monarch* (Wash 1910) 17 F. 1006.

Whatever the mariner thinks of the legal setup, which has the effect of giving the courts more authority over the rules of the road than the Commandant, U.S. Coast Guard, who enforces them through the local inspectors, the mariner must obey the law as he finds it—and that means in practice, as the admiralty judges interpret it. Notwithstanding the fact that in this country we do not have special admiralty courts, but any federal judge may be required to hear a collision case, it will be found that the decisions have been, as a whole, sound in seamanship as well as in law. The most experienced judges are not infallible, of course, and not infrequently circuit courts of appeal, co-ordinate in rank, will differ on some disputed point of collision law and the issue must be settled by the Supreme Court. The basic rules have existed in substantially their present form for so many years that most doubtful questions have long since been decided by that august tribunal, and the law for the most part may be regarded as pretty well settled. It remains only for the mariner to familiarize himself with the gist of the important ruling decisions, many of which are set forth in textbooks dealing with the subject.

The international character of the Collision Regulations requires that they should be understood by the seamen of different nations in the same sense. It is therefore of importance that the construction placed on them by courts of different countries should be uniform. This has been distinctly recognized in the United States:

> The paramount importance of having international rules, which are intended to become part of the law of nations, understood alike by all maritime powers, is manifest; and the adoption of any reasonable construction of them by the maritime powers . . . affords sufficient ground for the adoption of a similar construction . . . by the courts of this country.[16]

Rules Apply According to Location of Vessel

A fifth principle to be borne in mind by the mariner is that he must always be careful to observe the particular rules that apply in the locality of his vessel during the approaching situation. There are important differences in the rules to be followed on the high seas and those of many local waters throughout the world. The inland waters of the continental United States are governed by a set of special rules laid down in the Inland Navigation Rules Act of 1980. While as consistent as much as possible with the international rules, there are major differences in the special rules applicable to United States domestic waters. The significant point to which attention is drawn here is that obedience to the wrong set of rules, where they are in conflict, constitutes just as serious a breach as does the deliberate disregard of the law altogether. To illustrate: the use of a single

[16]The *Sylvester Hale*, 6 Bened. 523.

short blast of the whistle by a stand-on vessel holding her course and speed when crossing at a distance within half a mile is proper in inland waters, as provided by Rule 34(a)(i), but might make the vessel liable for a collision if she used the same signal on the high seas, where one short blast indicates a change of course to starboard.

History of the Rules of the Road

The International Rules date back to rules introduced in 1863 by England and France, and similar rules adopted by the United States in 1864, which in turn were adopted, with amendments, before 1886 by the United States, England, France, Germany, Belgium, Norway, and Denmark. These early rules were modified at a conference of representatives of the maritime nations of the world at Washington, D.C., in 1889, and subsequently adopted by the respective nations concerned. In the United States they became effective in 1897. Amended slightly in 1910, unsuccessful attempts were made to amend them further in 1913–1914 and 1929.

In 1948 at London, the International Conference on Safety of Life at Sea proposed a revision of the 1889 International Rules as Annex B to its Final Act. The 82nd Congress, by Public Law 172, approved 11 October 1951, adopted this revision and authorized the president to fix its effective date by presidential proclamation. Similar legislative steps were taken by the other governments participating in the 1948 Safety Conference. By 19 December 1952, thirty-seven maritime nations had agreed to this revision, at which time the United Kingdom, in accordance with the provisions of the Final Act of the 1948 Conference, fixed 1 January 1954 as the date when the revised international rules would be in force and effect. This then made it possible for the individual nations concerned to proclaim that date as the effective date of the 1948 revision of the international rules.

The 1948 rules, unlike the 1889 rules, were of short duration. The 1948 rules were reconsidered and revised again at a similar conference of the same name and scope held in London in 1960. By similar national and international process, the 1960 rules were made effective 1 September 1965.

In the United States, the 1960 rules were enacted into law by the 88th Congress as Public Law 131 on 24 September 1963 and proclaimed effective pursuant to that law on 1 September 1965 by Presidential Proclamation 3632 of 29 December 1964.

In 1961 a joint working group from the British, French, and German Institutes of Navigation devised separation schemes for the Dover Straits. A further working group, with additional representation from other countries, was set up in 1964 to consider similar schemes for other areas of the world. The proposals were accepted by IMCO and recommended for use by mariners in 1967.

The most recent International Conference was held in London during October 1972. The revision was a major one, making the rules more comprehensive—including the provision of mandatory separation schemes—and resulted in a completely new format. The 1972 International Rules became effective July 15, 1977. Executive Order 11964 ratified the rules for the United States.

The international rules do not apply to the inland waters of the United States, the authority to make special rules being preserved in Rule 1(b). Section 6 of the act adopting the inland navigational rules makes the international rules nonapplicable to vessels inshore of the lines of demarcation between the inland waters of the United States and the high seas.

Prior to this act different statutes subdivided U.S. inland waters into three sections, with a distinct set of rules for each, and the statutory rules were supplemented in each case by a corresponding body of pilot rules, formulated and issued by the Commandant, U.S. Coast Guard. The three sections of inland waters referred to were (1) the Great Lakes and connecting and tributary waters as far east as Montreal; (2) the Red River of the North, and certain other rivers and their tributaries whose waters flow into the Gulf of Mexico; (3) all other inland waters of the United States. In each of these sections, the pilot rules, except where they were in conflict with the statutory rules, had coextensive jurisdiction with them, and the mariner had therefore to be equally familiar with both. To add to the complexity, Rules and Regulations of the Corps of Engineers, Department of the Army, supplemented the pilot rules for the "Great Lakes." They, too, when not in conflict with the statutory rules, had coextensive jurisdiction with them, in the same manner as did the pilot rules. There were also special navigational light requirements for small commercial and recreational vessels of not more than sixty-five feet in length found in the 1940 Motorboat Act. While these seven sets of rules were alike in many respects, there were significant differences between them, as well as between the international rules, that tended to generate confusion among mariners.

An attempt to unify the various laws, rules, and regulations that affected the navigation of vessels on the waters of the United States was made during the early 1960s. It was then decided that unification of the inland rules into a single system should be deferred until the international maritime community further developed the international regulations. The latter was completed by the entry in force, in 1977, of the International Regulations for Preventing Collisions at Sea, 1972. Since then, the Coast Guard actively pursued the unification of the inland rules and regulations. To assist in developing the best possible set of rules, the Coast Guard established the Rules of the Road Advisory Committee, whose membership was composed of a cross section of maritime interests that used or were familiar with the various inland waterway systems. After deliberations and discussions over a period of eight years, the Advisory Committee developed a proposal to unify the inland navigation rules. That proposal was the basis of the Inland Navigation Rules Act of 1980.

Simply stated, the act established a unified system of local or special navigation rules applicable to U.S. domestic waters, as consistent as possible with the international system of navigational rules. In certain instances the international rules were considered inappropriate for specific waters of the United States because of local problems, and as a result, some previous domestic rules and practices have been adopted. Thus, reference will still be found in the act to the Western Rivers and the Great Lakes. Until otherwise decided, it can be assumed that some court interpretations of the earlier rules for inland waters will remain valid precedents.

Despite the differences that remain between the inland rules and international regulations, the Act of 1980 is a major advance along the road to revision and unification, a matter that received considerable support in previous editions of this book. It is to be hoped that the power granted to the Secretary of Transportation to issue regulations and establish technical annexes to the act will provide the flexibility to meet changing conditions and prevent the United States rules from becoming unduly out of line with international practice.

In 1972 the Ports and Waterways Safety Act was passed by Congress and amended by the Port and Tanker Safety Act of 1978. The act as amended authorizes the secretary of the department in which the Coast Guard is operating, to establish and operate vessel traffic services for navigable waters of the United States, as well as to exercise control over environmental quality in the same waters. In some aspects the act is complementary to the traffic separation schemes of the international rules, and the systems of both dovetail neatly at the boundaries of the high seas.

SUMMARY

Federal courts, sitting in admiralty, have jurisdiction over collisions occurring on public navigable waters, which have been held to be waters navigable, though not necessarily navigated, in interstate commerce. State courts may have concurrent jurisdiction, when provided by local statute, over certain cases on public navigable waters within a state. Admiralty law permits action to be brought against a vessel *in rem* (against a thing) or against her owners *in personam* (against a person). Whenever both vessels in a collision are at fault, under United States law the liability must be apportioned between them. A vessel may be sold at auction to satisfy a judgment against her, but the right to collect for damage may be modified by the provisions of the Limited Liability, Harter, and Carriage of Goods by Sea acts. As a matter of public policy, naval and other publicly owned vessels cannot be libeled, but the government permits itself to be sued *in personam* for faulty collision damage by a public vessel.

Vessels on the high seas are subject to the international rules; when in specified inland waters, various local rules, which differ in important respects, govern their action in collision situations. The rules are mandatory, must be obeyed in a timely manner, apply alike to all vessels, must be understood in the light of court interpretation, and have application within fixed geographical limits. While uniformity of the rules has been greatly advanced, the mariner must observe the differences in requirements according to the immediate location of his vessel.

12

Lawful Lights

The Function of Running and Riding Lights

The importance of proper running and riding lights on vessels using public navigable waters can scarcely be over-emphasized. During the hours of darkness it is the function of these lights in clear weather to give such timely and effective notice to one vessel of the proximity of another that all doubt as to her character and intentions will be satisfactorily settled before there is any serious risk of collision. Even in restricted visibility, with the mariner's safety in an approaching close-quarters situation dependent upon radar and sound rather than upon sight, it is often the welcome glimmer of these same lights through the haze that finally enables each fog-enshrouded vessel to feel her way safely past the other. To the student of collision law it is significant that twelve of the thirty-eight international regulations and eleven of the thirty-eight inland rules relate wholly or in part to lights, and that both sets of rules are supplemented by annexes that go into great detail as to the technical specifications of lights and also provide additional lights for vessels engaged in fishing. In cases of collision the courts are as certain to hold a vessel at fault for improper lights as for a violation of signal requirements or for failure to maintain a proper lookout.

Lights at Anchor Must Conform

It is evident from the court cases that mere volume of light, even for a vessel at anchor on a clear night, does not constitute the due notice to which approaching vessels are entitled or satisfy the requirement for regulation lights. In an interesting decision affirmed by the circuit court of appeals, the large seagoing tanker *Chester O. Swain* collided with the government cotton carrier *Scantic* on the Mississippi River 600 feet off the docks at New Orleans, and the latter vessel was found equally at fault with the *Swain* because of the irregularity of her anchor lights. It appeared that on the day preceding, fire had spread from a cotton warehouse to the *Scantic* and her cargo, damaging the vessel's running lights before it could be extinguished. The *Scantic* was shifted from her pier to a temporary anchorage across the river from the regular anchorage, so that if the fire in her cargo again broke out, as sometimes happens with cotton, it would not result in damage to other vessels at anchor. The lights were not repaired during the day, as they might have been; according to the testimony, the following lights were shown by the *Scantic* as soon as darkness fell: an oil lantern lashed to the jack staff; an oil lantern on an awning spreader aft; four sets of cargo cluster lights (each containing four large bulbs), one at the center of the bridge, one on the forward part of the boat deck to light No. 3 hatch, one at the after end of the boat deck to light No. 4 and No. 5 hatches, and one on the starboard side near No. 4 hatch to light the pilot ladder that led down to a motorboat tender standing by to render aid in case of the fire's recurring. In charge of a pilot, the *Swain*, proceeding down the river from Baton Rouge with a load of oil, rounded Algiers Point and crossed over to within 600 feet of the left bank, when the lights of the anchored vessel were first sighted against the background of city lights about 1,500 yards ahead. Instead of maneuvering promptly to avoid collision, the *Swain* first continued ahead, then stopped, then backed, then stopped and drifted helplessly into the mass of lights that marked the *Scantic*. The tanker was held at fault for her vacillating conduct and for the lack of vigilance of her lookout, but in also inculpating the *Scantic* the circuit court held that:

> while the lights she [the *Scantic*] installed ought to have been discoverd by the *Swain* in time to prevent the collision, it cannot be said that better and more conventional lights would not have got the attention of the *Swain* sooner than those which the *Scantic* rigged and would not have averted the trouble.[1]

The importance of having anchor lights conform to the specific requirements was brought out in a number of early cases in which incorrect lights, though visible, proved misleading to approaching vessels. In one such case, a 75-foot dredge in New York Harbor exhibited two white lights at anchor instead of the stationary single light and was mistaken by an approaching tug with tow for a tug underway and being overtaken. The tug did not discover her error until within 300 feet of the dredge, when she avoided collision herself by a hard-over helm but was unable to pull her tow clear. Although the tug was held at fault for failure to

[1]The *Chester O. Swain* (CCA NY 1935) 76 F (2d) 890.

discover sooner that the dredge was stationary, nevertheless the dredge shared the damages because of technically improper lights.[2]

Notwithstanding this, it seems that a ship will not necessarily be held at fault for a collision caused by improper lights, if her own regulation lights have recently been destroyed. A steamship at anchor with her masthead light up instead of her proper riding lights, was held free from blame in an English court. Her riding light had been broken shortly before the collision in a previous collision for which she was not at fault.[3]

Lights Underway Must Conform

The tragic collision of the USS *S-51* and the steamship *City of Rome* off Block Island 23 September 1925 affords a striking example of the importance attached by the courts to proper running lights underway. It also indicates the strictness with which the exemptions authorized for vessels of special construction by Rule 1(e), of both international regulations and inland rules, regarding lights, will be construed. This collision, which attracted unusual public interest because of the protracted but futile efforts that were made in the face of Atlantic storms to rescue possible survivors in the sunken submarine, happened on a clear night, and the masthead light of the *S-51* was under continuous observation from the *City of Rome* after being reported by her lookout twenty-two minutes before the collision. It was first made out as a faint white light broad on the starboard bow, and its bearing did not appreciably change until shortly before the collision, when it was observed to be closing in on the steamship's course and to be growing brighter. According to the testimony brought out at the trial, the captain of the *City of Rome* had been watching the light for some twenty minutes without taking any action, although in some doubt as to its character. He concluded at this point that he was overtaking a small tug or fishing vessel and ordered the course changed to the left to give it a wider berth. But a few seconds later the red sidelight of the submarine appeared a little to the right of the white light, indicating for the first time that she was not an overtaken vessel but was crossing the course of the *City of Rome* from right to left, as she had a right to do under the rules. Although the liner's rudder was ordered put the other way immediately and a few seconds later the engines were reversed, it was then too late to avoid the collision. The submarine sank very quickly, and of the three survivors picked up none had been on deck during the approaching situation or could give evidence as to the navigation of the submarine. However, notwithstanding that the *City of Rome* was flagrantly at fault for failure to reduce her speed when in doubt regarding

[2]The *Arthur* (NY 1901) 108 F 557.
[3]The *Kjobenhavn* (1874) 2 Asp. MC 213.

the movements of the other vessel and for failure to signal her change of course to the left by two short blasts as required under the then effective international rules, both the district court and the circuit court of appeals found the *S-51* at fault for improper lights. Referring to Article 2 of the 1889 International Rules, then in force, the court said:

> The *S-51* was 240 feet 6 inches long; her beam 25 feet; her surface displacement upwards of 1,000 tons; her forward white light was not 20 feet above the hull but was only 11 feet 2 inches above the deck; the side lights were fixed in a recess on the chariot bridge 7½ feet above the hull; they were not fitted with inboard screens projecting at least 3 feet forward from the lights. There is also testimony that the red light was so constructed and in such close proximity with the masthead light, only 3½ feet apart, that the visibility of the relatively dim red light was materially reduced by the greater brilliancy of the white masthead light.

To the student of collision law, a very significant feature of this case was the comments of both trial and appellate courts on the government argument that it was not practicable to have *S*-boats comply with the literal provisions in regard to lights, and moreover as a special type of naval vessel such craft were not under compulsion to comply. The district court said:

> I cannot accept the view that submarines running on the surface through traffic lanes are immune from the usual requirements regarding lights. . . . The obvious answer to the contention that the nature of their construction and operation makes it impractical for them to comply with these rules is that if this be so they should confine their operation to waters not being traversed by other ships. The fact that they are more dangerous should not be a reason for their disregarding rules which other ships must observe. . . . The testimony conclusively shows not only that the failure of the *S-51* to show proper side lights might have contributed but that it was a principal cause of the disaster.

In confirming the decision of the lower court to hold the submarine equally at fault with the liner, the circuit court of appeals added:

> There remains only the contention that the submarine was not subject to the ordinary rules of the road. It does not appear that it was impossible for her to comply with these, but it would make no difference if it did. . . . It is apparent that the rules regulating lights were meant to apply to ships of war; Article 13 would be conclusive if the preamble alone were not enough. If unfortunately it is impossible to equip submarines properly, they must take their chances until some provision has been made for them by law. We have no power to dispense with the statute nor indeed has the Navy. As they now sail they are unfortunately a menace to other shipping and to their own crews, as this unhappy collision so tragically illustrates. We cannot say that they are not to be judged by the same standard as private persons. The safety of navigation depends upon uniformity; only so can reliance be placed upon what masters see at night.[4]

[4]Ocean SS. Co. of Savannah v. U.S. (CCA NY 1930) 38 F (2d) 782.

To illustrate the international character of court findings, a similar collision occurred between HM Submarine *Truculent* and a merchantship in the Thames estuary, with equally tragic loss of life in the submarine. The masthead light of the submarine, like that of the *S-51*, was improperly placed according to the existing rules (also the 1899 ones), and was held to have contributed to the collision and damage. Evidence was accepted showing that the difficulties, in the case of submarines, of complying with the requirement for a masthead light were great, if not insuperable, and that the positioning of lights in submarines of all navies was similar but:

> . . . these considerations afford no answer to the charge that the steaming light of HMS *Truculent* constitutes a breach of the regulations.

It was further held that if it was really impossible in the case of a submarine to fit a masthead light that complied with the rules, then there was a duty to issue a warning to mariners of the fact.[5]

Due warning, in the format of Notices to Mariners, was issued in 1953. The next revision of the international rules, effective 1 January 1954, provided the basis for the legal exemptions referred to in the *S-51* case. Subsequent revisions of the rules have maintained the exemptions for naval or military vessels unable to comply, and Rule 1(e), International and Inland Rules, extends them further to any vessel of special construction approved by her government. Exemptions in the United States are issued by the Secretary of the department in which the Coast Guard is operating, for USCG and commercial vessels, and by the Secretary of the Navy for naval vessels, as certificates of alternative compliance. Similar exemptions for foreign warships are to be found in national notices to mariners and in the appropriate sailing directions. All these vessels must be certified as being unable to comply with the requirements regarding lights and *must be in the closest possible compliance* with the literal requirements of the rules. Should a collision such as was cited occur, it undoubtedly would still be indefensible, on the grounds of inadequate compliance. The variation in the number or position or character of the lights required to be shown cannot be misleading. On the contrary, the obvious intention of the authorized exemptions is to recognize the inability of certain vessels to comply with the literal requirements, while requiring that the nature and spirit of the lighting requirements of the rules be maintained. Thus, it may be presumed that the courts will be extremely critical of any unnecessary departure from the literal requirements or failure to give proper notice of the character and position of the lights carried by such vessels of special construction.

[5]The *Truculent*, Admiralty v. SS Devine [1951] 2 Lloyds Rep. 308 w.

Improper lights contributed to the tanker *Frosta* being held solely to blame for a collision with another tanker, the *Fotini Carras*, in the Indian Ocean in December 1968. The *Frosta* was overtaking the *Fotini Carras*, about 500 yards off, when her steering gear jammed. Among other actions, she promptly exhibited her not-under-command lights but failed to extinguish her masthead lights. Mr. Justice Brandon held that the *Fotini Carras* was not at fault because, among other factors, her chief officer, who was on watch

> . . . would at this stage be concerned mainly with judging the heading of the FROSTA and the extent to which she was closing FOTINI CARRAS and that, while doing so, he might well not notice that not-under-command had been switched on above the bridge of FROSTA so long as the masthead lights continued to show. . . . I am not prepared to find the chief officer was at fault in not noticing the not-under-command lights before he took action for other reasons.[6]

Improper Lights May Be a Fault in Naval Vessels

In the United States the courts have held all vessels to a strict observance of the rules even in time of war. Thus, when the armored cruiser *Columbia* sank the British freighter *Foscolia* off Fire Island during the Spanish-American War, the *Columbia* was held solely liable for the loss, the court finding that even a vessel of the navy in time of war cannot be excused for masking her lights, orders of the squadron commander notwithstanding, since there is no statutory authority for such an order.[7] Similarly, in a collision off the East Coast in 1918 between the privately owned cargo steamships *Proteus* and *Cushing*, both vessels, because of danger from hostile submarines, and under authority of the Navy Department, were proceeding without lights, and both were held at fault because they did not turn on their lights in time to avoid collision;[8] and in still another case, less than three weeks later, the U.S. Navy tanker *Hisko*, running without lights, was held solely at fault for sinking the cargo steamship *Almirante*, which was showing proper sidelights, though no masthead light.[9]

During the 1939–45 war, Allied convoy orders had statutory force and, for nations affected, overrode all contrary provisions contained in the international regulations.[10] Vessels remained otherwise bound to comply with the remainder of their obligations, including the duties of good seamanship, which were not affected by the convoy orders.[11]

[6]The *Frosta*, Q.B.(Adm. Ct.) [1973] 2 Lloyd's Rep. 348.
[7]Watts v. U.S. (NY 1903) 123 F 105.
[8]The *Cushing* (CCA NY 1923) 292 F 560.
[9]Almirante SS. Corp. v. U.S. (CCA NY 1929) 34 F (2d) 123.
[10]The *Vernon City* (1942) 70 Ll. L. Rep. 278.
[11]The *Scottish Musician* (1942) 72 Ll. L. Rep. 284.

General Characteristics of Sidelights

The precise nature of the lights required in inland waters was not always clear from the wording of the rules, as indeed had been the case previously on the high seas. Earlier rules dictated some compromise since the sidelights, mounted at the sides of a ship, would have to shine across the projected fore-and-aft line of the vessel at some point in order to be visible from ahead, or else there would be a theoretical "dark lane" ahead that could result in vessels meeting exactly end on being unable to see each others sidelights. The revised international regulations and inland rules cover this eventuality by allowing a practical cut-off outside the prescribed sector. The full details concerning arcs of visibility, contained in Annex I to the 1972 International Rules, are shown below. Similar specifications are contained in Annex I to the Inland Rules and will have to be ultimately met by all vessels tht sail solely on the inland waters of the United States.

> 9. Horizontal Sectors
>
> (a) (i) In the forward direction, sidelights as fitted on the vessel must show the minimum required intensities. The intensities must decrease to reach practical cut-off between 1 degree and 3 degrees outside the prescribed sectors.
>
> (ii) For sternlights and masthead lights and at 22.5 degrees abaft the beam for sidelights, the minimum required intensities shall be maintained over the arc of the horizon up to 5 degrees within the limits of the sectors prescribed in Rule 21. From 5 degrees within the prescribed sectors the intensity may decrease by 50 percent up to the prescribed limits; it shall decrease steadily to reach practical cut-off at not more than 5 degrees outside the prescribed limits.

Previously, vessels that did operate solely on the inland waters of the United States were governed to a certain extent by court interpretations. An analysis of early collision cases involving improper sidelights, mostly concerning sailing craft and steamships around the turn of the century, shows that out of the confusing phraseology of the rules, the courts have reached the following definite conclusions:

> (1) A vessel in collision is liable if any obstruction prevents the side lights from being visible to a vessel closing from ahead.[12]
>
> (2) A vessel in collision is liable if the inboard screens are not of the prescribed length and if the side lights are seen excessively across the bow.[13]
>
> (3) A vessel in collision is liable if both side lights are visible from a point on her own bow, i.e. both can be seen from the stem of the vessel.[14]

In a crossing collision between two steamers off the New Jersey coast on a clear night, the stand-on vessel had her sidelights set in the rigging in

[12]The *Vesper* (NY 1881) 9 F 569; the *Johanne Auguste* (NY 1884) 21 F 134; *Carleton* v. U.S. (1874) 10 Ct. Cl. 485.

[13]The *North Star* (1882) 27 L. Ed. 91.

[14]Clendinin v. the *Alhambra* (NY 1880) 4 F 86.

such a manner that they crossed at her stem, and her green light could be seen by the burdened vessel 3 points across the bow. In finding her at fault for improper lights, the Supreme Court made the following emphatic comment:

> This rule in regard to setting and screening the colored lights cannot be too highly valued, or the importance of its exact observance be overstated. Better far to have no side lights than to have them so set and screened as to be seen across the bow. In that situation they operate as a snare to deceive even the wary into error and danger.[15]

Modern vessels have far less cluttered superstructure than the earlier ships mentioned in the above cases and are, perhaps, less liable to obstruction or poor fittings of sidelights. Nevertheless it is as vital as ever to check that sidelights do show over the correct arcs.

The positioning and spacing of sidelights are also defined in the revised rules. In the past the inland rules somewhat loosely stated that the lights were to be on the port and starboard side, which is to say, apparently, that moving the sidelights inboard would not make a non-seagoing vessel liable for a collision as long as such lights were visible over the prescribed arcs.[16]

Annex I to the 1972 International Regulations, which forms the basis for Annex I of the Inland Rules, is more precise:

> 2. Vertical positioning and spacing of lights
> (g) The sidelights of a power-driven vessel shall be placed at a height above the hull not greater than three quarters of that of the forward masthead light. They shall not be so low as to be interfered with by deck lights.
> (h) The sidelights, if in a combined lantern and carried on a power-driven vessel of less than 20 metres in length, shall be placed no less than 1 metre below the masthead light.
> 3. Horizontal positioning and spacing of lights
> (b) On a vessel of 20 metres or more in length the sidelights shall not be placed in front of the forward masthead lights. They shall be placed at or near the side of the vessel.

The placing of sidelights abaft the forward masthead light should aid other vessels in the visual assessment of a vessel's course. Their position "at or near the side of the vessel" bears out the prophecy of footnote 16. However, for vessels of special construction, particularly smaller warships, the placing of lights abaft the usually single masthead light will not always be practicable. Mariners will constantly have to bear in mind the possibility that a combination of sidelights ahead of a single masthead

[15]The *Santiago de Cuba* 10 Blatch. U.S. 444.

[16]But see *Samuel H. Crawford* (NY 1881) 6 F 906, which although not faulting the vessel in this instance as the port light did show ahead without obstruction, nevertheless intimated that the fitting of sidelights on a deckhouse of a sailing schooner rather than on the rigging "is not to be approved."

light could well indicate the presence of a vessel much larger than 20 meters in length.

The precise positioning and spacing of lights can cause a problem for very large ships. On certain common types of VLCCs in ballast condition and normal trim, which represents some 50 percent of their working life at sea, it is physically impossible to see their sidelights from dead ahead if the observer is in a small vessel up to two miles away. The specifications of paragraph 2(g) or Section 84.03(g) of Annex I result in the inability to carry the sidelights on the bridge wings because they would then be higher than the forward masthead light. They are placed lower, and because of the normal trim of the vessel in ballast, they are cut off between one-half and three-quarters of the length of these very long ships, leaving a blind-zone some two miles dead ahead. A small ship that did not have the height of eye to see "above" the VLCC's bow would, when in the blind-zone, be denied any information from the VLCC's sidelights and might disastrously consider her farther away because only the masthead lights would be visible. Fishing vessels, sail boats, and other small craft engaged in coastal passages would do well to bear this limitation in mind.

Vessels constructed under previous rules enjoy an exemption of nine years from the horizontal repositioning of sidelights, to date from the entry into force of the international regulations and inland rules respectively.

Masthead Lights; Stern Lights

In addition to describing the sidelights for power-driven vessels and sailing vessels, the rules also describe in considerable detail the masthead lights to be carried by an independent power-driven vessel. Essentially there are few significant differences between the masthead and sternlights prescribed by both the international regulations and the inland rules. Both seagoing and non-seagoing vessels carry a 20-point masthead light forward and may, if less than 50 meters, and must, if 50 meters or more in length, carry a second 20-point masthead light higher than and abaft the forward one. Minor deviations are permitted for the positioning of a single masthead light for vessels less than 20 meters in length. All such vessels carry a 12-point sternlight, though in inland waters a power-driven vessel of less than 12 meters in length is allowed to carry an all-round white light in lieu of the masthead and stern lights. It is expected that the international regulations will accept this latter rule when the first amendment to the 1972 Convention comes into force in the early summer of 1983.

A major difference from the international regulations is that on the Great Lakes, a power-driven vessel of any length may carry an all-round white light in lieu of the second masthead light and sternlight. This light is carried in the position of the second masthead light and is visible at the same minimum range. Such a departure from the otherwise uniformity of rules governing United States waters was granted in recognition that the all-round light, being higher on the vessel than a sternlight, provided a greater degree of safety during the frequent low-hanging fogs prevalent on the Great Lakes. This permissive use of an all-round light to power-driven vessels underway is limited to the Great Lakes and is not extended to any other classification of vessels or to any other area.

The international regulations make equally specific requirements for the positioning of masthead lights, as they do for sidelights:

> 2. *Vertical Positioning and Spacing of Lights*
> (a) On a power-driven vessel of 20 metres or more in length the masthead lights shall be placed as follows:
> (i) the forward masthead light, or if only one masthead light is carried, then that light, at a height above the hull of not less than 6 metres, and, if the breadth of the vessel exceeds 6 metres, then at a height above the hull not less than such breadth, so however that the light need not be placed at a greater height above the hull than 12 metres;
> (ii) when two masthead lights are carried the after one shall be at least 4.5 metres vertically higher than the forward one.
> (b) The vertical separation of masthead lights of power-driven vessels shall be such that in all normal conditions of trim the after light will be seen over and separate from the forward light at a distance of 1000 metres from the stem when viewed from sea level.

Vessels are granted a nine-year exemption on repositioning of lights due to paragraph 2(b) above and permanent exemption of any repositioning of lights as a result of conversion from imperial to metric units. Rule 38(d)(v) of the inland rules gives similar exemption for vessels in inland waters.

Importance of Second Masthead Light

While the combined masthead, stern, and side lights are generally adequate to convey a satisfactory indication of one of the above vessels' movements to another, there is at least one situation peculiarly fraught with the danger of misunderstanding, where the arrangement of two white lights in range is invaluable. This is where two power-driven vessels are crossing at a very fine angle on courses differing by as much as 170°, and the give-way vessel, seeing both sidelights of the other a little on the starboard bow, mistakenly assumes that a starboard-to-starboard meeting is proper. The privileged vessel, seeing only the green light of the other slightly to port and expecting her to give way, maintains course and speed, which may result in a stalemate until the two vessels are in dangerous proximity, particularly if the approaching vessel veers somewhat to port.

The obvious advantage of the second masthead light here is that, with vessels as nearly end-on as indicated, the slightest change of course by either is instantly revealed to the other. For this reason, notwithstanding its partly optional nature under both the international regulations and the inland rules, the after masthead light should be carried whenever practical.

Apart from the meeting situation, the presence of the second masthead light can provide valuable visual confirmation of the course of the other ship, as well as give an indication of approximate length. The international regulations require:

> *3. Horizontal Positioning and Spacing of Lights*
> (a) When two masthead lights are prescribed for a power-driven vessel, the horizontal distance between them shall not be less than one half of the length of the vessel but need not be more than 100 metres. The forward light shall be placed not more than one quarter of the length of the vessel from the stem.

The inland rules make similar provision, and under both rules, permanent exemption from repositioning is given to vessels of less than 150 meters in length, with nine years grace for vessels of greater length.

Visibility of Lights

Both the 1972 International Regulations and the 1980 Inland Rules increase the visibility range requirements of lights, and give vessels four years from entry of force of the respective rules to comply.

> Rule 22. Visibility of Lights
> The lights prescribed in these rules shall have an intensity as specified in Section 8 of Annex I to these regulations so as to be visible at the following minimum ranges:
> (a) In vessels of 50 metres or more in length:
> —a masthead light, 6 miles;
> —a sidelight, 3 miles;
> —a sternlight, 3 miles;
> —a towing light, 3 miles;
> —a white, red, green or yellow all-round light, 3 miles.
> (b) In vessels of 12 metres or more in length but less than 50 metres in length:
> —a masthead light, 5 miles; except that where the length of the vessel is less than 20 metres, 3 miles;
> —a sidelight, 2 miles;
> —a sternlight, 2 miles;
> —a towing light, 2 miles;
> —a white, red, green or yellow all-round light, 2 miles.
> (c) In vessels of less than 12 metres in length:
> —a masthead light, 2 miles;
> —a sidelight, 1 mile;
> —a sternlight, 2 miles;
> —a towing light, 2 miles;
> —a white, red, green or yellow all-round light, 2 miles.

The equivalent Rule 22 of the Inland Rules is the same as the above, except for the addition of a special flashing light with a visibility of 2 miles for use in certain categories of tows and a white all-round light with a visibility of 3 miles for partly submerged tows.

Towing lights

Perhaps next in importance to proper running lights for independent vessels are the lights of vessels towing and being towed. Generally speaking, such vessels move slowly, but tugs are hampered by their tows and relatively unable to maneuver in close situations, while vessels and rafts in tow are, of course, almost completely helpless. The lights for a vessel towing or pushing as set forth in Rule 24, International Rules, consist of the usual sidelights, a fixed sternlight, and in place of the masthead light a pair of similar lights in a vertical line at least 2 meters apart (for a vessel 20 meters or more in length); or if towing *and* the tow extends more than 200 meters astern of the tug, then a third similar white light in line with the others, the three at such a height that the lowest is at least 4 meters above the hull. An after masthead light is mandatory if the towing vessel is 50 meters or more in length; the light is optional in the case of smaller vessels. When towing astern, a yellow towing light is required above the sternlight.

Referring to the inland rules, we find that the lights for vessels towing and pushing are essentially the same as required by the international regulations, but with modifications considered necessary to be more consistent with previous United States rules. International Regulation 24(a) has been changed to make it clear that the corresponding Inland Rule 24(a) applies to a vessel towing astern. Instead of two or three masthead lights forward in a vertical line, the inland rules permit the carriage of these lights on either the forward or after mast. If they are carried forward, then a masthead light higher than and abaft the forward white lights must be carried. If they are carried aft then a forward 20-point masthead light must be carried. Behind the difference from the international regulations lies the fact that, in many cases, the requirement to carry two or three white masthead lights on the foremast of small, non-oceangoing tugs would cause glare problems on the bridge. The 1972 International Regulations will make similar provision for tugs on the high seas, however, when the proposed amendments go into effect in the early summer of 1983.

A power-driven vessel pushing ahead or towing alongside in inland waters carries the same two masthead lights as a vessel on the high seas but adds two yellow towing lights in a vertical line. The international regulations do not require such distinctive stern lighting, and the provision for them in the inland rules is to eliminate the problem of an overtaking vessel, seeing only a white light and not appreciating the task in which the overtaken vessel is engaged.

On the Western Rivers, and on any other waters specified by the Secretary, vessels towing alongside or pushing ahead are exempt from the requirement to display any white masthead lights. This continuation of existing practice makes allowance for the limiting vertical clearances of bridges under which the tug will pass. If white lights were fitted, they would be close to the pilothouse and create excessive reflection and backscatter, impairing night vision. Also, little additional warning would be afforded by using white lights because of the frequent bends on the Western Rivers.

Finally, the inland rules provide for the case where in distress situations, the mariner who normally does not engage in towing operations cannot comply with the rules for towing lights when he is attempting to render assistance to another vessel. This provision is also expected to be incorporated into the international regulations on the adoption of the amendments by IMCO in early summer 1983.

Lights for Tows

The international rules make a distinction between vessels towed and vessels being pushed ahead. The latter-type vessels are lighted by a single pair of sidelights at the forward end, whether being pushed singly or in a group. However, there is a modern development in which tugs and barges are capable of being mechanically locked so rigidly that they can operate in the pushing mode as one unit even on the high seas. Between them they show only the lights for a single power-driven vessel.

The lights for vessels being towed are the same as for sailing vessels underway, consisting of the usual sidelights and the fixed 12-point sternlight required of all vessels. All vessels being towed on the high seas must conform to these requirements, whether sailing vessels, steamers, barges, or even a crib of logs or a dracone of fuel oil. The last two examples, however, are probably covered by Rule 24(g), which provides:

> Where from any sufficient cause it is impractical for a vessel or object being towed to exhibit the lights prescribed . . . all possible measures shall be taken to light the vessel or object towed or at least indicate the presence of the unlighted vessel or object.

The inland rules, while retaining the above international regulation, have gone further and prescribed lighting for towed objects to which sidelights and a sternlight cannot be meaningfully affixed, but to which one or more all-round lights could be attached. Inland Rule 24(g) applies to vessels that are partially submerged, such as dracones, which are bags to carry liquids, and which are becoming more common. Because the vast majority of these types of towing situations are done at slow speeds, it is not impracticable to require a float with a white light on it. This rule also applies to logs, log rafts, and similar objects being towed.

One other difference between the international regulations and the inland rules is that the latter prescribe a special flashing light for certain towing operations. This light is additional to the normal sidelights for a vessel, or a group of vessels lighted as one vessel, being pushed ahead. It is a 16- to 20-point yellow light placed as far forward as practicable on the centerline of the tow. The frequency of flashes of this light should clearly distinguish it from the flashing light used by non-displacement craft.

Tug and Tow Jointly Responsible for Proper Lights on Tow

A tug is not only liable for showing improper lights herself that contribute to a collision, but is jointly responsible with the vessel in tow for proper lights on the tow. An interesting case involving in liability both tug and tow occurred some years ago in Norfolk Harbor, when a tug with a covered barge on each side brought one of the barges into collision with a ferry crossing from starboard and therefore having the right-of-way. On the showing that there were no lights on the barges, which were high enough to blanket the tug's sidelights, and that towing lights of the tug, instead of being in a vertical line, were suspended from either end of a horizontal spar across the flagstaff at distances below the spar supposed to differ by three feet, the tug and the tow were found fully liable for the damage to the ferry.[17] Another case confirming the doctrine of the towing vessel's responsibility for the absence of proper lights on her tow was decided in a collision in New York Harbor between a ferryboat which should have kept out of the way and a barge in tow having no light on her bow. In a unique decision, the three vessels involved—ferry, tug, and barge, the last two belonging to the same owner—were found equally liable, with the result that the ferry recovered from the other two an amount equal to two-thirds of her damage.[18]

It should be noted that all of the above cases refer to the towing of barges, presumably with no motive power available, i.e., "dumb" barges. In a situation where the tug is the servant of the tow—i.e., the tug is assisting and responding to the orders from a manned ship, perhaps when maneuvering in confined waters—then it is more probable that it is the tow's responsibility for exhibiting proper lights.[19] In England, the Court of Appeal, in 1953, found the uncompleted aircraft carrier *Albion*, although under tow on the high seas, at fault after a collision in which a

[17]Foster v. Merchants and Miners Transportation Co. (Va 1905) 134 F 964.
[18]The *Socony No. 123* (NY 1935) 10 F Supp 341.
[19]The *Mary Hounsell* (1879) 4 PD 204.

merchant vessel was sunk. The carrier, and not the tugs, was held liable for her defective port sidelight and also for her failure to show not-under-command lights.[20]

Anchor Lights

Reference has already been made to a case proving that the same strictness in regard to lights is applied to vessels at anchor as when underway.[21] The *Scantic* shared the damages in that case because of her inability to satisfy the court that her irregular lights *could not have contributed to the collision.* It was the same relentless test that is put on every infraction of the rules that precedes a collision.

The lights for a vessel at anchor are described in Rule 30 of both International and Inland Rules. The international regulation is as follows:

> Rule 30 Anchored vessels and vessels aground
> (a) A vessel at anchor shall exhibit where it can best be seen:
> (i) in the fore part, an all-round white light or one ball;
> (ii) at or near the stern and at a lower level than the light prescribed in subparagraph (i), an all-round white light.
> (b) A vessel of less than 50 metres in length may exhibit an all-round white light where it can best be seen instead of the lights prescribed in paragraph (a) of this rule.
> (c) A vessel at anchor may, and a vessel of 100 metres and more in length shall, also use the available working or equivalent lights to illuminate her decks.
> (e) A vessel of less than 7 metres in length, when at anchor or aground, not in or near a narrow channel fairway or anchorage, or where other vessels normally navigate, shall not be required to exhibit the lights or shapes prescribed in paragraphs (a), (b) or (d) of this rule.

The inland rule is the same as that of the international regulations with but one change and one addition. The change is an exemption for vessels less than 12 meters in length from showing the two red all-round lights when aground. The addition provides that vessels less than 20 meters in length need not show anchor lights when in a specially designated anchorage area.

Especially noteworthy is the fact that anchor lights are visible all around the horizon, and that where the length of the vessel requires two lights, or a small vessel elects to be lighted as a large vessel, it is the forward light that is the higher, reversing the relative heights of the underway range lights. One practical import of this is that in maneuvering to avoid a large vessel at anchor forward, it should be remembered that any effect of wind or current would be in the direction indicated by a line from the higher

[20]The *Albion*; Thomas Stone (Shipping) Ltd. v. Admiralty [1953] Lloyds Rep. 239.
[21]The *Chester O. Swain* (CCA NY 1935) 76 F(2d) 890.

(bow) light toward the lower. While the courts failed to find a steamship at fault for having her forward anchor light 10 feet higher than her after light instead of the then required 15 feet higher, in a case where the testimony showed the lights were seen by other pilots 5½ miles away and the collision was due to improper lookout on the colliding steamer,[22] nevertheless, the distinction in the requirements due to a vessel's length is likely to be strictly enforced, particularly when the vessel involved is 50 meters in length or longer. Thus, in the case of a 271-foot barge anchored in Chesapeake Bay that was struck by a pilot boat five minutes after her after light had blown out in a puff of wind and while a seaman was overhauling the light, the barge (which was sunk) was held solely at fault for the collision. As said by the court in referring to Article 11 of the then relevant inland rules:

> The provisions of this article, which requires a vessel of 150 feet or upwards in length when at anchor to carry two lights, one at the forward part and the other at or near the stern at a lower height, to indicate the length of the vessel and the direction in which she is pointing, is of great importance and must be strictly observed, and, if violated, a loss caused by a collision resulting must be borne by the vessel so violating it.[23]

There is also a requirement for vessels at anchor to illuminate their decks with such working lights as they may have. It is possible that such lights could be confusing, particularly against background lighting, and that illumination of the stack could be a more effective measure.

Subject to any local rules, a vessel secured to a buoy may be regarded as a vessel at anchor. She is made fast to moorings which are themselves attached to the ground by an anchor or the equivalent of an anchor.[24]

Vessel Made Fast to Another Must Have Own Lights

Of particular interest to the naval service, with the frequent nesting of vessls at anchor, is the requirement of the courts that a vessel made fast to another at anchor must maintain her own proper anchor lights. It was so held in a collision between one of two barges attached to each other and a car float in Norfolk Harbor.[25] It has long been held by the courts that every vessel in an anchor nest in fog must make proper sound signals for a vessel at anchor whether her own anchor is down or not.[26] On the other hand, in a case with the impressive title Emperor of All the Russias v. the *Heipershausen*, the district court of New York held that a steam launch made fast to a boom projecting sixty feet from the side of a man-of-war at

[22]The *John G. McCullough* (CCA Va 1916) 239 F 111.
[23]The *Santiago* (Pa 1908) 160 F 742.
[24]The *Dunhelm* (1884) 9 P.D. 164, 171.
[25]The *Prudence* (Va 1912) 197 F 479.
[26]The *Cohocton* (CCA NY 1924) 299 F 319; the *Southway* (NY 1924) 2 F (2d) 1009.

anchor in New York Harbor need not exhibit any light. It seems that the Russian cruiser *Dimitri Donskoi* was visiting the East Coast during the Columbian Naval Exposition of 1893, and while she was lying in the North River at anchor with proper lights burning, two tugs with a tow of barges passed between her and the piers, so close that one of the barges struck and sank the steam launch, at the same time knocking the boom around against the captain's gig, which was also damaged. It was argued that the steam launch was not lighted, but the court held that inasmuch as the boom did not project over sixty feet, the minimum passing distance at that time prescribed by statute in New York Harbor, no light was necessary on the launch and the tugs were fully liable for the damage.[27]

Lights for a Vessel Alongside Wharf

Strictly speaking, a vessel moored at a wharf is not at anchor, as implied by the distinction in the rules: A vessel is "underway" when she is not at anchor, or made fast to the shore, or aground. From the standpoint of her ability to maneuver to avoid collision, of course, there is little to distinguish the three not-underway situations. The practical necessity of lights to give notice to approaching vessels varies according to circumstances. In several old cases the courts have decided that a vessel moored in the usual way alongside a wharf and not in the way of other boats need not exhibit lights.[28] But a vessel lying moored at the end of a wharf on a dark night in the navigable part of a narrow stream, constantly traversed by different kinds of craft, is required by the special circumstances, in the exercise of common prudence, to carry a light, whether or not it is expressly required by the rules.[29] And, of course, the rule may be further modified by a local harbor regulation, disregard of which is legally as serious as disregard of the inland or international rules. The following extract from the Seattle Harbor ordinance is typical of the rule that prevails in many organized ports:

> Every vessel or obstruction . . . while lying at any pier or other structure in Seattle Harbor between the hours of sunset and sunrise shall display at least one (1) white light at the outer end of the vessel or obstruction, which white light or lights shall show clearly from seaward and be so constructed and of such character as to be visible at least one (1) mile in clear weather.[30]

Lights for barges moored to a bank or a dock are included in the Pilot Rules of Annex V of the Inland Navigational Rules Act of 1980.

[27]The *Dimitri Donskoi* (NY 1894) 60 F 111.
[28]Denty v. the *Martin Dallman* (Va 1895) 70 F 797; City of New York v. the *Express* (NY 1891) 61 F 513.
[29]The *Millville* (NJ 1905) 137 F 974.
[30]Seattle Harbor Ordinance (1921).

Lights for a Vessel Aground

Rule 30, International Regulations, also contains the following provision:

> Rule 30. Anchored vessel and vessel aground
> (d) A vessel aground shall exhibit the lights prescribed in paragraph (a) or (b) of this rule and in addition,where they can best be seen:
> (i) two all-round red lights in a vertical line;
> (ii) three balls in a vertical line.
> (e) A vessel of less than 7 metres in length, when at anchor or aground, not in or near a narrow channel, fairway or anchorage, or where other vessels normally navigate, shall not be required to exhibit the lights or shapes prescribed in paragraphs (a), (b) or (d) of this Rule.

The significance of this rule is, of course, that it provides statutory sanction to the fact that a vessel aground is in the same helpless category as a vessel at anchor, though additional lights must be carried not only to indicate her predicament, but also to warn other vessels.

The inland rules follow the same pattern, but separate the exemption for vessels aground from International Regulation 30(e) and place it in Inland Rule 30(f) with an increase of length from less than 7 meters to less than 12 meters.

Fishing Vessel Lights

Previously, the lights for vessels engaged in fishing in inland waters were slightly different from those pertaining to the high seas. Under the Inland Navigational Rules Act of 1980 both are now the same, eliminating the variance in requirements and extending the application of the lights to all waters of the United States.

Vessels Not Under Command or Restricted in Their Ability to Maneuver

Other than for the Great Lakes, there were no previous rules for vessels of the above category in our internal waters. Now Inland Rule 27 provides for the lighting of vessels not under command or restricted in their ability to maneuver. The rule is essentially the same as International Regulation 27, with some minor modifications. These United States modifications are expected to be adopted by the IMCO Assembly on the occasion of the first amendment to the International Regulations, currently scheduled for early summer 1983.

A vessel not under command is one which through some *exceptional* circumstance is unable to maneuver as required by the rules. She then, by night, displays two all-round red lights in a vertical line, which gives her right-of-way over all other categories of vessels underway that are in sight of her. This important privilege must not be assumed for mere convenience but must be justified by genuine defects that prevent compliance with the rules. A collision occurred in the Dover Straits on 10 November 1969 in good visibility, but with a force-eight wind blowing. One of the ships, the *Djerba,* had been subjected to four days of heavy weather and was carrying not-under-command lights, although she had full use of her engines and steering. It was held in the admiralty court that the *Djerba* was not justified in carrying not-under-command lights and that she was in breach of the collision regulations. Moreover, as a ship that would otherwise be a give-way vessel under the crossing rule, she was not relieved of her duty to give way, although she carried such lights. Furthermore, under the 1960 Rules then in force, she should not have carried a white masthead light as well as the two red not-under-command lights. On apportioning 60 percent of the blame for the collision to the *Djerba,* Mr. Justice Brandon said:

> It is important that ships which are genuinely disabled from manoeuvring adequately should have both the right and the duty to advertise the fact by exhibiting appropriate signals and so to make it clear to other ships that they must take steps to keep clear of them. It is equally important that ships which are not genuinely disabled, although they may be under certain difficulties, should not claim this special right and privilege . . . without proper justification.[31]

This judgment was upheld on appeal, when it was clearly stated that it was not right to say that the *Djerba* was not under command and that:

> even if only the red lights had been shown instead of the white masthead light and red lights, *Djerba* was not excused from complying with rule 19 (duties of a give-way vessel in the crossing situation) unless she was actually not under command.[32]

A vessel restricted in her ability to maneuver is covered in those cases where the nature of her work imposes restriction on the movement of the vessel to such an extent that she cannot readily keep out of the way of another vessel. Rule 3(g) of both sets of rules list identical operations that justify displaying these lights, though the preamble in the inland rules, unlike the international regulations, makes it clear that the list is not exhaustive. It can be expected that the courts are unlikely to find favor with a vessel that claimed the privilege that these lights bring without proper justification. It is not intended that vessels engaged in routine towing operations should declare that they are restricted in their ability to deviate from their course.

Within the category of vessels restricted in their ability to maneuver there are four different kinds of lighting prescribed:

(1) minesweepers show three all-round green lights in addition to the

[31]The *Djerba,* Q.B. (Adm. Ct.) [1976] 1 Lloyd's Rep. 50.
[32]The *Djerba,* C.A. [1976] 2 Lloyd's Rep. 41.

normal lights for a power-driven vessel underway;
(2) dredgers and similar vessels show three all-round lights (red-white-red) in a vertical line in addition to the normal lights for a power-driven vessel underway and, where an obstruction exists, two red lights on the obstructed side and two green lights on the clear side;
(3) tugs towing astern with a difficult tow show the same red-white-red configuration in addition to normal towing lights;
(4) other vessels show the red-white-red configuration in addition to normal lights.

There are minor differences in the lighting requirements between the inland and international rules, e.g., for a dredger operating while at anchor, but again it is expected that the latter regulations will be amended to reflect the former in summer 1983.

Vessels Constrained by Their Draft

Unique to the international regulations is the three all-round red lights in a vertical line for a vessel whose draft in the available depth of water severely restricts her ability to change course. Not only the depth of water, but also the available navigable water width should be used as a factor to determine whether a vessel may consider herself as constrained by her draft. A vessel navigating in an area with a small underkeel clearance but with adequate space to take avoiding action in good time should not be regarded as a vessel constrained by her draft.

This category originated from the very real problems experienced by deep-draft ships on passage through shallow seas, where shallow-water effect should impinge on normal maneuverability. The signal does not give right-of-way to the constrained vessel, but International Regulation 18(d) directs other vessels to avoid impeding her. There have been complaints voiced that large ships are displaying these lights in areas where there are no constraints caused by depth of water or sea room, obviously hoping that other vessels will keep clear of them. In the event of a collision it is certain that the courts would look closely at the reasons why the three red lights were displayed.

Although the lights are optional on the high seas, they are mandatory in some ports of the world. On 11 October 1972, a collision occurred in the Thames Estuary in darkness and in clear weather, between the West German coaster *Rustringen* and the Dutch tanker *Kylix*. Both ships were inward bound, and while overtaking the coaster, the *Kylix* struck and sunk the *Rustringen*. Among many other faults that emerged, the *Kylix* was in breach of a local rule, Bylaw 27, for failing to show three red all-round lights.[33]

[33]The *Kylix*, Q.B. (Adm. Ct.) [1979] 1 Lloyd's Rep. 133.

The Inland Navigational Rules Act of 1980 avoids the subjective nature of the international regulations, pointing out in the accompanying House of Representatives report that it could lead to abuses and result in situations wherein a vessel might claim a right-of-way to which she was not entitled, thereby creating a dangerous situation. Accordingly, there is no provision in U.S. inland waters for lights of a vessel constrained by her draft. Nonetheless, mariners should be aware of the international signal, which could be met on the high seas outside of the demarcation line for inland waters.

Requirements Regarding Lights Should Be Strictly Observed

In conclusion, it may be said that the courts are strict in enforcing the requirements for proper lights—and rightly—but not unreasonably so. In a collision between two sailing vessels approaching so that one of them had her red light toward the other, it was held to be immaterial that her green light was out, since its presence could not have averted, nor its absence contributes to, the collision.[34] It should be remembered that neglect by a vessel to show regulation lights does not relieve another vessel from observing the rules of navigation and using every precaution to avoid collision with her,[35] and that a steamer failing to see lights on a sailing vessel, which were clearly visible from the steamer's position, was not absolved from responsibility merely because the lights were somewhat faultily placed. This case, like many others, became one of mutual fault.[36] In any doubtful situation the courts are very likely to have the attitude that:

> The rule requiring lights may as well be disregarded altogether as to be only partially complied with, and in a way which fails to be of any real service in indicating to other vessels the position and course of the one carrying them.[37]

After all, the only safe rule to follow is to be sure that ones running or riding lights during darkness or restricted visibility conform as closely as possible to the lights specified by the rules *in force where the vessel may happen to be*, remembering that any departure from the rules that is followed by a collision will be subject to the test so often annunciated by the Supreme Court:

> But when a ship at the time of collision is in actual violation of a statutory rule intended to prevent collision, it is no more than a reasonable presumption that the fault, if not the sole cause, was at least a contributory cause of the disaster. In such a case the burden rests upon the ship of showing, not merely that her fault

[34]The *Robert Graham Dun* (CCA NH 1895) 70 F. 270.
[35]Swift v. Brownell (CC Mass 1875) Fed. Cas. No. 13,695.
[36]The *Samuel H. Crawford* (NY 1881) 6 F 906.
[37]The *Titan* (CC NY 1885) 23 F 413.

might not have been one of the causes, or that it probably was not, but that it could not have been.[38]

If lights are lost or extinguished, then this must be detected by the watch and repair or replacement carried out as soon as possible. Emergency lights, whether oil or battery, should comply with the regulations as strictly as normal lights and should be properly maintained and kept ready for use. If severe weather prevents immediate attention, because of danger to personnel, a delay may be justified, but should be recorded in the official deck log.

SUMMARY

The importance of proper running and riding lights is emphasized by the fact that so many of the international and inland rules relate to them, and that in collision cases the courts construe strongly against vessels that disregard their requirements.

The international regulations and the inland rules are now largely in agreement, with clear requirements for the placing of lights. Important differences between the rules are now confined to the range light on the Great Lakes, the special flashing light on the Western Rivers and, for the time being, dredgers working at anchor.

With virtually no major recent court cases concerning the incorrect siting of lights, the lessons to the mariner today must be:

(1) display the correct lights for the situation he is in;

(2) scrupulously avoid a misleading mixture of lights when the situation changes;

(3) do not falsely claim privilege through the display of improper lights; and

(4) be alert for those vessels that cannot comply with the letter of the law and may give an inadvertent impression of their course, size, and proximity.

[38] The *Pennsylvania* (1875) 19 Wall 125.

13

Whistle Signals

Difference in Meaning of Signals

The most important differences to be found in the rules of the road on the high seas and in the inland waters of the United States have to do with the sound signal requirements for vessels approaching one another in clear weather. The purpose of this chapter is to analyze the differences in whistle signals prescribed by the respective rules for vessels meeting, overtaking, or crossing in good visibility.

As a preliminary consideration, it may be pointed out that there is a fundamental difference in the meaning of the conventional one- and two-short-blast signals in the two jurisdictions. Under international rules the signals are purely rudder signals, to be given when, and only when, a change of course is executed. It is therefore unnecessary to specify the use of a signal in a particular situation, a general rule being stated that provides for a signal in every situation when the course is changed. Under the inland rules, on the other hand, the one- and two-short-blast signals are not for the purpose of announcing a change in course, but to indicate the side on which an approaching vessel will pass.

Signals Compulsory Since 1890

It is interesting to note that it was not until the adoption of the 1889 International Rules in 1890—made effective in the United States by presidential proclamation 1 July 1897—that the use of sound signals by power-driven vessels at sea, except in fog, became compulsory. The first international rules, adopted by England and France in 1863, and the revised rules, adopted in 1885 by those countries and the United States, Germany, Belgium, Japan, Norway, and Denmark, authorized certain whistle signals but made their use discretionary with the navigator. At the

International Convention of 1889, however, the delegates, after some debate on the subject, decided that they would no longer leave to the discretion of the mariner the question of whether, in a particular case, there was less risk of using the whistle and being misunderstood or in not using it at all. Accordingly, the former Article 28, International Rules, was passed to read:

> Art. 28. The words "short blast" used in this article shall mean a blast of about one second's duration.
> When vessels are in sight of one another, a steam vessel under way, in taking any course authorized or required by these rules, shall indicate that course by the following signals on her whistle or siren, namely:
> One short blast to mean, "I am directing my course to starboard."
> Two short blasts to mean, "I am directing my course to port."
> Three short blasts to mean, "My engines are going at full speed astern."

When the 1889 rules were revised in 1948, the International Conference reiterated the substance of the former Article 28, modifying the three-blast signal's meaning to correspond to the interpretation of the courts, namely, that the backing signal is proper when the engines are going astern at any speed. A limited in-doubt signal was provided and additional special whistle signals were authorized. In 1960 when further revision of the rules occurred at the International Conference on Safety of Life at Sea, 1960, a new optional whistle light was authorized. The International Conference in 1972 added a new set of signals to be given and answered by vessels in an overtaking situation in a narrow channel or fairway, and removed the restriction that the optional in-doubt signal be used only by the stand-on vessel, making its use mandatory for all vessels in doubt. The present rule now reads:

> Rule 34. Manoeuvring and warning signals
> (a) When vessels are in sight of one another, a power-driven vessel underway, when manoeuvring as authorized or required by these rules, shall indicate that manoeuvre by the following signals on her whistle:
> —one short blast to mean "I am altering my course to starboard";
> —two short blasts to mean "I am altering my course to port";
> —three short blasts to mean "I am operating astern propulsion."
> (b) Any vessel may supplement the whistle signals prescribed in paragraph (a) of this rule by light signals, repeated as appropriate, whilst the manoeuvre is being carried out.
> (i) these light signals shall have the following significance:
> —one flash to mean "I am altering my course to starboard";
> —two flashes to mean "I am altering my course to port";
> —three flashes to mean "I am operating astern propulsion";
> (ii) the duration of each flash shall be about one second, the interval between flashes shall be about one second, and the interval between successive signals shall be not less than ten seconds;
> (iii) the light used for this signal shall, if fitted, be an all-round white light, visible at a minimum range of 5 miles, and shall comply with the provisions of Annex I to these Regulations.

(c) When in sight of one another in a narrow channel or fairway:

(i) a vessel intending to overtake another shall in compliance with Rule 9(e)(i) indicate her intention by the following signals on her whistle:

—two prolonged blasts followed by one short blast to mean "I intend to overtake you on your starboard side";

—two prolonged blasts followed by two short blasts to mean "I intend to overtake you on your port side".

(ii) the vessel about to be overtaken when acting in accordance with Rule 9(e)(i) shall indicate her agreement by the following signal on her whistle:

—one prolonged, one short, one prolonged and one short blast, in that order.

(d) When vessels in sight of one another and approaching each other and from any cause either vessel fails to understand the intentions or actions of the other, or is in doubt whether sufficient action is being taken by the other to avoid collision, the vessel in doubt shall immediately indicate such doubt by giving at least five short and rapid blasts on the whistle. Such signal may be supplemented by a light signal of at least five short and rapid flashes.

(e) A vessel nearing a bend or an area of a channel or fairway where other vessels may be obscured by an intervening obstruction shall sound one prolonged blast. Such signal shall be answered with a prolonged blast by any approaching vessel that may be within hearing around the bend or behind the intervening obstruction.

(f) If whistles are fitted on a vessel at a distance apart of more than 100 metres, one whistle only shall be used for giving manoeuvring and warning signals.

If we scrutinize the present rule with a little more than ordinary care, five important provisions will become apparent:

(1) The mandatory use of the signals mentioned in Rule 34(a) is evident in the words "shall indicate." Even if it is thought that the signals may not be heard,[1] or the officer on watch considers they would disturb his own ship, especially the master,[2] does not matter—they still must be given.

(2) The one-blast signal indicates a lawful change of course to the right and the two-blast signal indicates a lawful change of course to the left. That is, both the one- and the two-blast signals are rudder signals and therefore should never be used, either for an original signal or a reply, except when the course is changed. Thus, these signals are not required if the rudder is used to counteract the effect of the wind and tidal current,[3] or if using the rudder to check the swing of the ship when backing,[4] and even, apparently, if a vessel rounds the bend of a river with her rudder amidships.[5]

(3) If a change of course is made or the engines are reversed in accordance with the rules, the one-, two-, or three-blast signal must be given by a power-driven vessel whenever another vessel is in sight. The

[1]The *Haugland* (1921) 15 Aspinall MC 318.
[2]The *Fremona* (1907) Deane J.
[3]The *Gulf of Suez* (1921) 7 Ll. L. Rep. 159.
[4]The *Aberdonian* (1910) 11 Aspinall MC 393.
[5]The *Heranger* (1937) 58 Ll. L. Rep. 377.

obligation of signaling is plain, whether the other vessel is ahead, abeam, or astern, and whether she is a power-driven vessel with a whistle or a sailing vessel without one.[6] Thus, a tug in collision with a sailing vessel was found at fault for failure to signify by two short blasts her alteration of course to port.[7] And conversely, the use of any of these signals under conditions of visibility so low that neither the other vessel nor her lights can be seen is prohibited.[8] However, a vessel is unlikely to be exonerated for not sounding signals through failure to sight another vessel because of a poor lookout[9] or because the other vessel was momentarily out of sight.[10]

(4) The signal of five or more short blasts is required of any vessel in doubt and, like the blind bend and overtaking signals, is not restricted to power-driven vessels. It can, however, be used only by vessels in sight of one another.

(5) The one-, two-, three- and five-blast signals may be supplemented by light signals. The latter need not be synchronized with the sound signals and may be repeated at intervals, without the sound signal, while the immediate maneuver is in progress. If a further alteration is made shortly after the initial maneuver is completed, it is still necessary to make a sound signal for the second maneuver even if it is identical to the first. The light signal alone will not suffice. As the reception of sound signals in other vessels can never be certain, especially in diesel and gas-turbine ships with high noise levels, the visual light signals are an important additional indication of action taken or for the reinforcement of the in-doubt or wake-up signal of five short blasts.

Signals under International Rules

Let us consider the application of the sound signals in Rule 34 when two vessels encounter each other outside the inland waters of the United States.

(1) Vessels meeting under international rules. Rule 14(a), governing the head-on or meeting situation, states: "when two power-driven vessels are meeting on reciprocal or nearly reciprocal courses so as to involve risk of collision each shall alter her course to starboard so that each will pass on the port side of the other." It is clear that when risk of collision exists in this situation, each vessel must turn to the right, at least enough to render safe a port-to-port passing, and when the necessary change in course is made, Rule 34 requires the one-blast signal. It is equally clear, in the language of the rule, that the mandate to change course does not apply if

[6]The *Comus* (CCA NY 1927) 19 F(2d) 774.
[7]The *Kathy K* (Can. Ct.) [1972] 2 Lloyd's Rep. 36.
[8]The *Parthian* (CCA 1893) 55 F 426.
[9]The *Lucille Bloomfield* [1966] 2 Lloyd's Rep. 245.
[10]The *Heire* (1935) 51 Ll. L. Rep. 325.

the present course of the two vessels will carry them well clear of each other. If the course is not changed, no whistle signal is authorized or permitted, the one-blast signal being a mandatory rudder signal. Should one vessel be in doubt as to whether holding on will give sufficient clearance and accordingly execute right rudder and give the one-blast signal, the other vessel then finding the clearance even more ample without change of course on her part, may lawfully hold her course. But if she does, she cannot use her whistle, since to do so would indicate an action that she is not taking. The practical risk of this apparent ignoring of the other vessel's signal is obvious, and may be avoided by the very simple expedient of keeping within the law by changing course to the right and blowing one blast. The point to remember here is that there is legal fault if you use your whistle without at least a slight change in course.

Conversely, you cannot, in the meeting situation under international rules, lawfully change course at any time without using the one-blast signal. In the collision of the *Anselm* and the *Cyril*, the two vessels met end on in the estuary of the Amazon River, and when they were about two miles apart the *Anselm* changed course slightly to the right. A moment later, seeing that the *Cyril* was apparently changing to port, she again executed right rudder, and this time gave the required one-blast signal. In the collision that followed the *Anselm* was held at fault by the British court of appeals for failure to blow one blast the first time she altered her course.[11] Under very similar circumstances the *Malin Head* met the *Corinthian* in the St. Lawrence River. The *Malin Head* changed course to the right and blew one blast, but did not repeat the signal when the failure of the *Corinthian* to go to starboard compelled her again to change course, after steadying for about two minutes, this time with hardover helm. As in the other case, the court of appeals held the offending vessel accountable for the omission, notwithstanding that the lower court exonerated her on the grounds that testimony showed the failure did not contribute to the collision.[12]

In a harbor collision at Harwich, England, the coaster *Thuroklint* was held mainly to blame when, among other things, she failed to indicate a series of alterations of course to starboard by sounding signals of one short blast. An inward-bound vessel, she had changed her pilots on the port side of the approach channel and then crossed ahead of the outcoming ferry *Koningin Juliana* in an attempt to gain her correct side. The *Koningin Juliana* was also castigated for using an incorrect whistle signal, namely two short blasts, when she stopped her alteration to starboard, having turned nearly 30 degrees, and attempted to pass ahead of the

[11]The *Anselm*, 10 Aspinall M.C. (N.S.) 438.
[12]The *Corinthian*, 11 Aspinall M.C. (N.S.) 264 (Admiralty).

Thuroklint on a steady course across the channel. The signal was unlawful because she "was not directing her course to port at the time when she sounded them."[13]

In another meeting case in a narrow channel, the *Joaquin Ponte Naya* was proceeding up the River Parana on the wrong side of the channel when she met the *Martin Fierro* coming down river. In attempting to cross to her starboard side the *Joaquin Ponte Naya* failed to sound one short blast on altering course, a fault that contributed to her receiving 85 percent of the blame for the subsequent collision.[14]

An interesting interpretation of the use of the one-short-blast signal arose from the court findings of a collision on the River Maas. The *City of Capetown* was inbound and collided with the outward bound *Adolf Leonhardt*. The latter vessel failed to get to her own starboard side of the channel promptly enough after leaving her dock and was found to be two-thirds to blame. However, she was found not to be at fault for failing to sound one short blast when first altering to starboard some five minutes before the collision, because she was not then taking action to avoid a collision situation. Although this finding was based on the local collision regulations, which supplemented the international regulations, it was further said that even if the latter had required her to sound such a signal, the failure to do so was not causative to the collision.[15] This, perhaps, highlights the point that maneuvering and warning signals under the international regulations are required only when "manoeuvring as authorized by these Rules," which in this meeting case did not apply until the second alteration of course to starboard. However, it is unlikely that a vessel would be faulted for sounding the *correct* whistle signals when *maneuvering* on the high seas in sight of another vessel, even if risk of collision did not exist at that moment.

As indicated in the wording of the rule, vessels meeting end on or nearly end on are bound to pass port to port. It is only when the vessels are so far to starboard of each other as not to be considered as meeting head-on, that they escape the requirements to change course to starboard; and in that case no change of course, and therefore no whistle signal, is required. It is difficult to picture a meeting situation at sea where the two-blast signal would be valid, for in any borderline case where there is doubt as to whether the two vessels are meeting end on or are already heading clear for a starboard-to-starboard passing, then the vessel(s) in doubt must assume it is a head-on situation in accordance with Rule 14(c).

(2) *Overtaking and overtaken vessels under International Rules.* When any vessel is overtaking another vessel at sea, her action is governed primarily

[13]The *Koningin Juliana* Q.B.(Adm. Ct.) [1973] 2 Lloyd's Rep. 308.
[14]The *Martin Fierro* Q.B.(Adm. Ct.) [1974] 2 Lloyd's Rep. 203.
[15]The *Adolf Leonhardt* Q.B.(Adm. Ct.) [1973] 2 Lloyd's Rep. 324.

by Rule 13, International Rules. This rule, which applies not only when the overtaking vessel is coming from well aft but when she is so near the dividing bearing between an overtaking and a crossing vessel as to be in doubt as to her status, provides in part:

> Notwithstanding anything contained in the Rules of this Section any vessel overtaking any other shall keep out of the way of the vessel being overtaken.

The overtaken vessel, thus having the right-of-way, is governed by Rule 17, International Rules, which provides in part:

> Where one of two vessels is to keep out of the way the other shall keep her course and speed.

Applying Rule 34 to this situation, it is evident that the overtaking vessel, unless she slows down, must change her course to one side or the other if necessary to clear, announcing any change in course with an appropriate whistle signal. The overtaken vessel, on the other hand, being under compulsion to maintain her course and speed, cannot answer the whistle of the other, though she may, if in doubt as to whether the overtaking vessel is taking sufficient action to avert collision, sound the signal of five or more short and rapid blasts to call the attention of the overtaking vessel to its duty to keep clear. Because of the provisions of Rule 34, the use of a rudder or reversing whistle signal here, too, is unlawful when there is no change in course or speed. When the overtaking vessel has passed well clear of the overtaken vessel and changes course a second time in order to return to her original heading, she must again use the proper signal, which will, of course, if the first signal was one blast, be two blasts. This signal, likewise, must remain unanswered.

In narrow channels or fairways, the overtaken vessel is often required to assist in the maneuver, and the international rules provide for an *exchange* of whistle signals "when overtaking can take place only if the vessel to be overtaken has to take action to permit safe passing." The vessel intending to overtake shall indicate her intention by sounding one of the following signals on her whistle:

—two prolonged blasts followed by one short blast to mean "I intend to overtake you on your starboard side";

—two prolonged blasts followed by two short blasts to mean "I intend to overtake you on your port side."

The vessel to be overtaken shall, if in agreement, sound the following signal on her whistle: one prolonged, one short, one prolonged, and one short, in that order (International Code group "Charlie" meaning "affirmative"). The overtaken vessel shall then take steps to permit safe passing. If the overtaken vessel is not in agreement, she may sound instead the doubt signal of five or more short blasts. The overtaken vessel should not attempt passing until an agreement is reached, nor does agreement relieve her of her obligation to keep out of the way until well past and clear.

The tanker *Kylix*, which was mentioned earlier in chapter 12 for not showing three red lights, was found to be at fault for failure to indicate that she was about to overtake the coaster *Rustringen* in the Thames Estuary. Local Bylaw 42 then made similar provision to the international regulation described above and the court found:

> if such a signal had been sounded at a proper time, it would have alerted the master of *Rustringen* to what *Kylix* was doing appreciably earlier . . . which might well have prevented, or at any rate mitigated, the collision.[16]

The *Rustringen* compounded matters by altering course to starboard in front of the overtaking vessel, without making a signal of one short blast. The collision occurred one and one-half minutes later, with the *Rustringen*'s unsignaled alteration being held contrary to the international regulations and one of several causative factors.

(3) Crossing vessels under International Rules. When two power-driven vessels are crossing so as to involve risk of collision, Rule 15 requires the vessel that has the other on her own starboard side to keep out of the way of the other. The stand-on vessel in the crossing situation, like the overtaken vessel, must hold her course and speed and consequently is forbidden to use the one-, two- or three-blast signals of Rule 34. The give-way vessel, on the other hand, is required to keep out of the way, to take positive early action toward this end, to avoid, if the circumstances admit, crossing ahead of the other, and on approaching her, if necessary, to slacken her speed or stop or reverse. If she turns to starboard to go under the stand-on vessel's stern, Rule 34 requires her to sound one blast; if she avoids crossing by sheering to port, a questionable maneuver, two blasts; and if she reverses her engines, three blasts; and all of these signals, so far as the stand-on vessel is concerned, must remain unanswered for the reason already given, that the latter can change neither course nor speed.

In a crossing case off Cape St. Vincent, a tanker, the *Statue of Liberty*, was the give-way vessel but took tardy and indecisive action to pass astern of the motor-vessel *Andulo*, which was the stand-on vessel. When about 1,600 yards apart, the *Andulo* first made a small alteration to port (mainly to get her on course for Casablanca), during which time the *Statue of Liberty*

[16]The *Kylix* Q.B. (Adm. Ct.) [1979] 1 Lloyd's Rep. 141–142.

altered farther to starboard. Both ships increased their turn but a collision occurred. With regard to sound signals, neither ship signaled their first alteration nor did the *Statue of Liberty* indicate her second. Both were found at fault for these omissions, but only that of the *Statue of Liberty* was considered causative to the collision, which, compounded with other faults, resulted in her receiving 85 percent of the blame.[17]

A vessel doubting the intentions or actions of the other vessel, whether stand-on or give-way, is required to give the signal prescribed in Rule 34(d). However, the sounding of this signal was judged as no excuse for a give-way vessel's failure to take appropriate action in a crossing situation in the Dover Straits. The Liberian ship *Genimar* was navigating contrary to the flow of a traffic separation scheme and approached the Greek bulk carrier *Larry L* on the latter's starboard bow. *Larry L* was navigating correctly in accordance with the traffic separation scheme and her master's attitude appears to have been that since *Genimar* was proceeding in the wrong direction, it was for her to keep out of the way of *Larry L.*

> This attitude is exemplified by his two peremptory signals of five short blasts and his failure to take any action of any kind . . . In adopting this attitude the master of *Larry L* was plainly wrong.[18]

The *Larry L* was found two-thirds responsible for the subsequent collision.

In addition, 1972 Rule 17(a)(ii), subject to the qualification in Rule 17(c), now permits the stand-on vessel to "take action to avoid collision by her manoeuvre alone, as soon as it becomes apparent to her that the vessel required to keep out of the way is not taking appropriate action in compliance with these Rules." When taking action under this rule, which applies equally to any stand-on vessel, not just the one in the crossing situation, the appropriate whistle signal in Rule 34 (a) must be given. When, from any cause, the stand-on vessel "finds herself so close that collision cannot be avoided by the action of the give-way vessel alone, she *shall* take such action as will best aid to avoid collision." Even in extremis, she may not be excused for failure to indicate such action by the proper whistle if at that point she sheers to starboard or to port, or reverses her engines.[19]

(4) *Nearing a bend in a channel under the International Rules.* Rule 34(e) prescribes a signal for a vessel approaching a bend or similar obstruction:

> A vessel nearing a bend or an area of a channel or fairway where other vessels may be obscured by an intervening obstruction shall sound one prolonged

[17]The *Statue of Liberty*, (H.L.) [1971] 2 Lloyd's Rep. 277.
[18]The *Genimar*, Q.B.(Adm. Ct.) [1977] 2 Lloyd's Rep. 25.
[19]The *Comus* (CCA NY 1927) 19 F(2d) 774.

blast. Such signal shall be answered with a prolonged blast by any approaching vessel that may be within hearing around the bend or behind the intervening obstruction.

A prolonged blast is defined in Rule 32(c) to be a blast of four to six seconds duration. A vessel around the bend hearing the signal must answer with a like signal. Henceforth, no other signals are given by either vessel until and unless one of the vessels changes course or backs down.

Signals Under Inland Rules

We will next consider the action required of two power-driven vessels approaching each other so as to involve risk of collision in the inland waters of the United States, with particular reference to sound signals when meeting, crossing, or overtaking. Except that a somewhat larger angle on the bow is included by the courts to take care of vessels proceeding in opposite directions through winding channels, the three situations in the inland rules are similarly defined. In addition, the secretary of the department in which the Coast Guard is operating is authorized to develop local pilot rules as necessary to implement and interpret the inland rules.

The first important difference to note in the rules for inland waters is that the one- and two-short blasts are signals of intent and not of action. This will be discussed in the meeting and crossing situations sections later. The three-short-blast signal remains, however, as in the international regulations, a signal of action meaning "I am operating astern propulsion." Finally, there are different signals used in inland waters for the overtaking situation, also to be discussed later.

The In-Doubt or Danger Signal

The definition and use of the in-doubt signal of at least five short and rapid blasts are the same for both sets of rules, although the inland rules also refer to it as the danger signal. It should be noted that this signal is in no sense optional but must be used whenever any vessel is in doubt as to the intentions or actions of another, whether the vessels are meeting, crossing, overtaking, or being overtaken. A vessel was found at fault for failure to use it in clear weather when in doubt as to the course or intention of another vessel whose lights were obscured by a deck load of lumber but whose whistle signal was heard somewhere ahead.[20] More recently, following a collision on the Mississippi River in 1974 when a tanker crossed from her starboard to her port side of the river ahead of a tug and tow on a reciprocal course, the tug was found partially at fault for not "immediately sounding what is commonly called a danger signal"

[20]The *Virginia* (CCA 1916) 238 F156.

when uncertainty as to the tanker's intentions arose.[21] Although heard under Canadian jurisdiction, a similar opinion was voiced following an incident on the St. Lawrence Seaway where a motor vessel, forced into collision with a dock because another motor vessel failed to give sufficient room to pass, was to a minor extent at fault "for not giving the danger signal."[22]

The in-doubt signal is clearly intended by Inland Rule 34(d) to be used by vessels in sight of one another. If the vessels are not in sight of one another due to restricted visibility, only the signals prescribed in Rule 35 are required.

Signal for Blind Bend and Power-Driven Vessel Leaving Her Berth

The signal of one prolonged blast to be used by *all* vessels nearing a blind bend is the same under both sets of rules. Unlike the international regulations, this same signal is required in inland waters for a power-driven vessel leaving a dock or berth in clear weather. If she should back out of the berth, then the prolonged blast must be followed by three short blasts as soon as she comes in sight of another vessel.

Bridge-to-Bridge Radiotelephone and Whistle Signals

A feature unique to the inland rules is that vessels reaching agreement on how to pass each other by use of bridge-to-bridge radiotelephone need not sound whistle signals. Mariners should not allow this practice to affect their use of whistle signals on waters not covered by the inland rules. For example, although no collision occurred, an ore carrier, *The Ore Chief*, was found totally to blame for forcing the tanker *Olympic Torch* aground when overtaking her in the River Schelde. Apart from condemning the unseamanlike behavior of the *Ore Chief*, it was held that

> the misunderstanding between the pilots as to the intentions of *Ore Chief* illustrated the undesirability of substituting imprecise signals on VHF for precise exchanges by whistle signals. . . . The provisions . . . were mandatory and it was not open to navigators to disregard them by agreement, whether overt or tacit. . . .[23]

Clearly, this last does not apply to waters governed by the inland rules. Moreover, mariners on inland waters should not neglect to use the bridge-to-bridge radiotelephone. In the case mentioned in the earlier discussion of danger signals, the tanker *Anco Princess*'s

[21]Gulf Coast Transit Co. v. M.T. *Anco Princess* et al. E.D. La (1977), [1978] 1 Lloyd's Rep. 299.

[22]Liquilassie Shipping Ltd. v. M.V. *Nipigon Bay* (Canada Fed. Ct. 1975), [1975] 2 Lloyd's Rep. 286.

[23]The *Ore Chief*, Q.B.(Adm. Ct.) [1974] 2 Lloyd's Rep. 433.

> failure to attempt to contact the *Libby Black* via bridge-to-bridge radio was contrary to the purpose, if not the terms, of Vessel Bridge-to-Bridge Communications Act, 33 U.S.C. ss 1201 etc. . . . The mere presence of a radio on the bridge of a vessel obviously is meaningless if it is not put to use.[24]

Thus, on the inland waters of the United States, the radiotelephone should be used—at a minimum—to communicate intentions and thus supplement whistle signals. Experience has shown that it is often possible to arrange a passing agreement on the radiotelephone prior to exchanging whistle signals, in which case the latter may be dispensed with. However, a timely exchange of whistle signals is required if for any reason a positive agreement cannot be reached.

One- and Two-Short-Blast Signals, Inland Waters

As mentioned earlier, there is a major difference between the international regulations and the inland rules over the use of one- and two-short-blast signals by power-driven vessels in sight of each other. Inland rule 34 retains the "signals of intent and reply" that are imbedded in the maritime custom of the United States. They were considered to be safer for use in confined inland waters than the international regulations' "signals of action" that are used on the high seas. Disregarding for the time being the overtaking situation, the one- and two-short-blast signals have identical meaning in both the meeting and crossing situation. One short blast means "I intend to leave you on my port side" and two short blasts means "I intend to leave you on my starboard side." The signals may be supplemented by one-second flashes on an all-round white or yellow light, synchronized with the whistle.

Inland Rule 34(a) uses the expression "when maneuvering as authorized or required by these Rules." This is not interpreted as meaning the short blasts should be used only when a course alteration is anticipated, for the preamble to Rule 34(a) also contains the qualification when "meeting or crossing at a distance within half a mile of each other." This is a retention of a former pilot rule, often referred to as the half-mile rule. Effectively it means that signals of intention and agreement by power-driven vessels should be exchanged if the closest point of approach will be within half a mile, whether or not a course change is required. Previously, the half-mile rule only applied to meeting vessels but it is now extended to cover crossing vessels. Clearly vessels should not wait until within or reaching half a mile of each other before exchanging signals, but should initiate the proper signals in a timely manner, taking full advantage of the bridge-to-bridge radiotelephone.

[24] Gulf Coast Transit Co. v. M.T. *Anco Princess* et al. E.D. La (1977), [1978] 1 Lloyd's Rep. 299.

> If signals are to be of any value, they must be given with an allowance of a sufficient time to exchange signals and agree on a passing, taking into consideration the speed, power and apparent agility of the vessels.[25]

In the crossing and meeting situations one- or two-short-blast signals may be initiated by either vessel and must, if in agreement, be acknowledged by the other vessel repeating the same signal. If, however, the other vessel is in any doubt whatsoever as to the safety of the proposal, she must sound the danger signal of at least five short and rapid blasts on her whistle. She should not respond to the signal that worried her by sounding an alternative proposal, e.g., to respond to a two-short-blast signal with a one-short-blast signal. Despite an earlier history of acceptance, such "cross signals" have been invalid since 1940. In a crossing situation between the steamships *Eastern Glade* and *El Isleo* in a channel approaching Baltimore harbor, in which the privileged *El Isleo* answered the two-blast proposal of the *Eastern Glade* to yield her right-of-way with the danger signal *followed by one blast,* she was exonerated by both the lower courts, but the Supreme Court reversed the decree and remanded the case to the circuit court of appeals.[26] Upon rehearing of this case, the circuit court of appeals duly reversed its original decision and held both vessels at fault. It seems highly probable, therefore, that in future cases in inland waters, vessels will be held at fault for using cross signals. Having sounded the danger signal, each vessel takes "appropriate precautionary action" *before* concluding a safe passing agreement.

Meeting Vessels, Inland Waters

Inland Rule 14, describing action to be taken in the head-on or meeting situation is virtually identical to that of the international regulations. It is clearly the intent of the inland rules to have vessels meeting head-on in inland waters pass in the same way as required by the corresponding international regulations. If there is any doubt that vessels are meeting on reciprocal or nearly reciprocal courses, then the vessel(s) must assume the situation is a head-on one and act accordingly.

While the actions required to maneuver to avoid each other are the same, namely both power-driven vessels alter course to starboard and pass port-to-port, the signals differ in meaning. In inland waters one short blast is initiated by either vessel stating her intention to leave the other on her port side, which must then be similarly answered by the other vessel, if in agreement. Both vessels are committed to take the steps necessary to effect a safe port-to-port passage, though the alteration of course, and perhaps speed, may take place after the exchange of signals.

[25]River Terminals Corp. v. U.S., DC La 1954, 121 F Supp 98.

[26]Postal S.S. Corp. v. *El Isleo* (1940) 308 U.S. 378, 84 L. Ed. 335, reversing (CCA NY 1939) 101 F(2d)4: rehearing (CCA NY 1940) 112 F(2d)297.

If no alteration of course is necessary to achieve a safe passing, signals are still required if the vessels will pass within half a mile of each other. Should the vessels be passing port-to-port, they exchange the one-short-blast signal. If the vessels anticipate passing starboard-to-starboard then two short blasts are required. It is not considered that the latter case would be a frequent occurrence on most waters: a vessel proceeding along a channel is bound by Inland Rule 9(a)(i) to keep over to her starboard side, thus facilitating a port-to-port passage. However, Inland Rule 9(a)(ii) provides priority for a downward bound vessel, with a following current, in narrow channels or fairways of the Great Lakes, Western Rivers, or other waters specified by the secretary. This right-of-way supersedes the normal meeting rules of Inland Rule 14 and could well lead to exchanges of two short blasts, initiated by the downbound vessel.

In summary, vessels in a meeting situation have a compulsory duty to signal their intentions by means of a whistle signal,[27] unless agreement has been reached by bridge-to-bridge radiotelephone. In the meeting case twice mentioned earlier in this chapter, the tanker *Anco Princess*'s faults were compounded by her failure to signal intentions to the *Libby Black* either by whistle signal or by radio communication.[28]

Crossing Vessels, Inland Waters

Inland Rule 15(a) describes the crossing situation and is the same as in the international regulations. The whistle signals to be made are identical to those required for meeting vessels, being contained in the same Inland Rule, 34(a), though, again, these are signals of intent and assent rather than action signals. It is the intent of the rules that the give-way vessel should avoid crossing ahead of the stand-on vessel, if the circumstances of the case permit. Thus, the normal situation will be for the two vessels to leave each other on their port sides, requiring an exchange of signals of one short blast. Having achieved agreement, the stand-on vessel maintains her course and speed while the give-way vessel can either alter course to starboard, slow, stop, or reverse her engines. Only the last action requires a further signal—one of three short blasts.

If the circumstances do not permit a port-to-port passing, then an agreement for the give-way vessel to cross ahead of the stand-on vessel will have to be negotiated by an exchange of two short blasts. This should be an unusual occurrence and one best avoided whenever possible. However, Inland Rule 9(d) prescribes that a vessel should not cross a narrow channel or fairway if she will impede the passage of a vessel that is

[27]Chotin, Inc. v. S.S. *Gulfknight*, E.D. La (1966) 266 F Supp 859 aff'd, 5 Cir.(1968) 402 F(2d)293.

[28]Gulf Coast Transit Co. v. M.T. *Anco Princess* et al. E.D. La (1977), [1978] 1 Lloyd's Rep. 299.

confined to the channel, and in some circumstances, this could lead to the use of two-short-blast signals. If the vessel navigating along the channel is uncertain of the crossing vessel's intention, then she is required to sound the danger signal.

Different from the international regulations is Inland Rule 15(b), which requires a vessel crossing a river to keep out of the way of a power-driven vessel ascending or descending on the Western Rivers and the Great Lakes and can be extended by the secretary to include other waters. Unlike Rule 9(d), this rule is not limited to narrow channels, nor does it depend on the maneuverability of the ascending or descending vessel. Thus, if the crossing vessel should be a power-driven one, an exchange of two-short blasts on these waters could be as common as that of one short blast.

Overtaking and Overtaken Vessels, Inland Waters

In the overtaking situation signals of one- or two-short blasts have a different meaning from those used in the meeting or crossing situations. They are still signals of proposal and require consent before overtaking can commence and are applicable only to power-driven vessels. If the overtaking vessel desires to overtake on the starboard side of the vessel ahead, she sounds one short blast, and two short blasts if she desires to pass on the port side. The overtaken vessel answers with the same signal if in agreement or she must answer promptly with the danger signal if in disagreement.

The Inland Navigational Rules Act of 1980 retained the previous signals for overtaking in inland waters, as they were considered superior to the corresponding international regulations: they reduce the risk of collision by applying to *all* overtaking situations of power-driven vessels, are firmly implanted in U.S. maritime practice, and have worked well. One might add that the simpler inland rules avoid the cacophony of noise possible under the international regulations.

Meaning of Maneuvering Signals a Major Difference Between Rules

Probably the most significant difference between the international regulations and the inland rules can be found in Rule 34 concerning maneuvering and warning signals for power-driven vessels in sight of each other. The arguments over the relative merits of the international system, which restricts the use of one- and two-short-blast signals to indicate change of course, and the inland system of proposal and agreement are finely balanced. Certainly the hodge-podge of previous inland signals has been clarified, and the signals brought closer to the international ones by the passing of the Inland Navigational Rules Act of 1980. The retention in this act of "signals of intent" seems appropriate for the more restrictive waters inside the demarcation lines. Regrettably, however, cases continue to occur, outside of U.S. inland waters where there are no local rules to authorize such use, of vessels improperly using whistle signals to express intentions rather than rudder action.[29] Thus the mariner must continue to use sound signals with the most careful regard to the geographical location of his vessel in any particular situation, and to fall back on the exercise of good seamanship when he gets into a collision approach where two vessels are on opposite sides of the demarcation line and neither set of rules governs both vessels.

SUMMARY

Under international regulations, one- and two-short-blast signals, as provided in Rule 34, are rudder signals that must be given whenever, and only when, a vessel in sight of another changes course, irrespective of the kind of approach. Under inland rules, one- and two-short-blast signals are prescribed to indicate the manner of passing and to be given *and answered* regardless of change in course. At least five short and rapid blasts is the in-doubt signal for vessels in sight of each other in both sets of rules, although the inland rules additionally refer to it as the danger signal. The mariner must be aware of the important differences in the signal requirements for all situations on the high seas and in inland waters.

In the meeting situation, vessels under international regulations sound one short blast concurrent with a change of course to the right; under inland rules both vessels sound one short blast for a port-to-port passage within half a mile, whether or not either changes course.

In the crossing situation under inland rules, both vessels again exchange one short blast to express their intentions to leave each other to port. Under international regulations the stand-on vessel may not announce her intention of maintaining course and speed with one short blast. She can only question the actions of the give-way vessel, when in doubt that that vessel will keep clear, by sounding the in-doubt signal.

In the overtaking situation on the high seas, the overtaking vessel sounds one or two short blasts as she makes course changes to avoid the overtaken vessel. In a narrow channel or fairway subject to the international regulations, or in inland waters, the overtaking vessel must first give a signal indicating the side on which she intends to pass, and receive an answer before carrying out the maneuver. To propose passing on the starboard side, the signal is two prolonged blasts followed by one short blast under the international regulations, or one short blast in inland waters. To propose passing on the port side, the signal is two prolonged

[29]The *Shell Spirit* [1962] 2 Lloyd's Rep. 252; *see also* the *Friston* [1963] 1 Lloyd's Rep. 74; the *Centurity* [1963] 1 Lloyd's Rep. 99; the *Koningin Juliana* [1973] 2 Lloyd's Rep. 308.

blasts followed by two short blasts under the international regulations, or two short blasts in inland waters. In inland waters agreement is signaled by the overtaken vessel's answering with the same signal, while under the international regulations agreement is signaled by one prolonged, one short, one prolonged, one short. Under both sets of rules, disagreement with the proposal is indicated by use of the in-doubt or danger signal.

Unique to the inland rules is that whistle signals may be dispensed with if agreement for passing is reached by use of the bridge-to-bridge radiotelephone. This is not permitted on the high seas.

The bend signal of one prolonged blast is identical under both sets of rules, but is also used in inland waters for a power-driven vessel leaving her moorings.

14

Head-On Encounters

The Meeting Situation

Rule 14 of the inland rules and the international regulations are identical. They describe an approach situation by two power-driven vessels in sight of each other meeting in a head-on situation, the conditions being that (1) they are approaching on reciprocal or nearly reciprocal courses ahead or nearly ahead of each other so as to involve risk of collision, (2) by night one vessel sees the masthead lights of the other in line, or nearly in line, and/or both sidelights, and (3) by day one vessel observes the corresponding aspect of the other vessel.

Both sets of rules require both vessels to alter course to starboard and pass port-to-port. If there is any doubt as to whether a vessel is in a meeting or crossing situation, she must assume it is the former and act accordingly.

Clearly the description contained in the rules only applies to the initial sighting and assessment of the situation. Two vessels on exactly opposite courses at night, following tracks that will take them clear of each other, port-to-port by, say 500 yards, will, in the earlier stage of their approach, each see both sidelights of the other, almost, although not exactly ahead. It is obvious that sooner or later the green light of each vessel will be shut out to the other, leaving vessels "red-to-red," but this certainly does not remove the application of the head-on rule.

Again, if two vessels are approaching end-on, as soon as one of them, in obedience to the rules, makes a marked change of course to starboard, she will at night exhibit one sidelight ahead as seen by the other vessel, and by day will be seen ahead crossing the course of the other.

Still another discrepancy between the literal phraseology and its practical application may be expected because of the physical imperfections of sidelights, which have a certain amount of leakage outside the designated arcs they cover. The specifications in the technical annexes to both sets of rules provides for sidelights to be screened so as to show between one and three degrees across the bow, thus avoiding a theoretical dark lane immediately ahead of the ship. This overlap, plus the effect of any yawing, may make appreciation of whether a situation is a head-on or a crossing one somewhat uncertain. In such circumstances Rule 14(c) applies, both on the high seas and on inland waters, and requires the vessel in doubt to assume a meeting situation.

If, in a marginal case, a vessel considers herself privileged in a crossing situation, the latitude granted in Rule 17(a)(ii) of both sets of rules allows her to take action early rather than having to stand on until in extremis. Such action should be in accordance with Rule 17(c), applicable in both jurisdictions, by avoiding an alteration to port. Thus, there is less need to draw a fine distinction between the boundary line dividing the two situations.

From the wording of Rule 14(b) it is clear that the course referred to in Rule 14(a) is the direction of the ship's head and not the course being made good over the ground. This could be significant where strong cross winds or tidal currents are experienced, transferring what would be, in still waters, a head-on situation into more of a crossing one. The temptation to alter course to port in such circumstances must be resisted, as witnessed by a collision in a swept channel to the approaches of Belfast in 1945. A strong tidal current was sweeping across and two approaching ships were necessarily compensating for it, with the result that their headings were different from their tracks along the channel. The *British Engineer* was found at fault for altering course to port to avoid a green light seen almost dead ahead.[1] *Karanan*, the other ship, altered to starboard as was her duty as give-way vessel.

In a case of a collision in the open sea, where two vessels approached each other so that their courses were within ¾ point (about 8 degrees) of opposite, the court held them to be under the crossing rule because of the lights that could be seen:

> These vessels were on crossing courses: as respects the *Knight* because she saw only the other's green light; as respect the *Gulf Stream* because the two colored lights were seen not ahead but from a half a point to a point and a half on her own starboard bow for a considerable time before any risk of collision commenced, so that she showed to the *Knight* only her green light.[2]

[1] The *British Engineer* (1945) 78 Ll. L. Rep. 31.

[2] The *Gulf Stream* (NY 1890) 43 F 895. *See also* the *Comus* (2 CCA) 1927 AMC 860, in which two vessels at sea intersecting courses 1° from head and head until within 2 miles of each other were held to be crossing vessels.

In an even earlier case in the open waters of Lake Erie, when a schooner and a bark collided after approaching each other within half a point (about 5½ degrees) on opposite courses, the Supreme Court held that:

> the variation, in any view of the evidence, did not exceed half a point by the compass, which is clearly insufficient to take the case out of the operation of that (end-on) article.[3]

It would therefore seem that while the head-on rule must always be construed in the strictest accordance with the explanatory words of the rule, construction was a matter for the court.[4] In other words, every case should be tested against the explanatory words of the rule itself, rather than solely against previous cases or an arbitrary cut-off point based on the differences between the courses of the vessels involved. Nevertheless, the tendency of decisions appears to have been for a difference between courses (- 180 degrees) of one point or more to be treated as crossing situations, and a difference of half a point or a little more to be treated as head-on situations.[5] We shall see shortly how this construction has varied for that hazardous passing encounter between two ships meeting in a narrow channel.

It is clearly the intent of Inland Rule 14 that vessels meeting head-on in inland waters should each alter to starboard to achieve the same port-to-port passage that is required under the international regulations. Yet the whistle signals of Inland Rule 34(a), which were discussed in chapter 13, equally clearly provide the means for agreeing to a starboard-to-starboard passage. The retention of the half-mile rule in the preamble to Inland Rule 34(a) effectively means that an exchange of signals will be made when two vessels expect to close to within that distance when passing on opposite courses. If the vessel were sufficiently apart that they could pass starboard to starboard *without altering course* to avoid meeting head-on, then such a passing is probably acceptable. Apart from special circumstances in some rivers, such occasions should be rare. A starboard-to-starboard passage within half a mile of another vessel, under any rules, can give one an uncomfortable feeling and is best avoided.

Crossing Vessels May Be Meeting

A modification in court interpretation of the old head-and-head rules relates to the arc of approach included under the rule where vessels meet in a river or winding harbor channel. In a daylight collision between an inbound steamship and an outbound trawler in the New York ship channel it was held that:

[3]The *Nichols* (1869) 19 L Ed. 157.
[4]The *Kaituna* (1933) 46 Ll. L. Rep. 200; also the *Gitano* (1940) 67 Ll. L. Rep. 339.
[5]Marsden, *Collisions at Sea*, p. 551.

Approaching vessels whose courses diverge not more than one or two points are meeting end on or nearly so within Art. 18 of the Inland Rules and are required to pass port to port.[6]

In a collision between two ocean steamships on the Delaware River the court similarly ruled that:

If they are approaching as much as 1½ to 2 points, they are to be considered as head and head and bound to pass port to port.[7]

Remembering that 2 points are equal to 22½°, it will be seen that the head-on rule has been made to include, under some circumstances, cases that might have the first appearance of crossing at a considerable angle. Indeed, as pointed out in the case of the *Milwaukee*:

In determining how vessels are approaching each other, in narrow tortuous channels like the one here in question, their general course in the channel must alone be considered, and not the course they may be on by the compass at any particular time while pursuing the windings and turnings of the channel.[8]

From this it may be seen that vessels approaching a sharp bend in a river from opposite directions might even be moving at right angles when first mutually sighted, yet still be under the head-on rule. One might logically raise the question as to whether vessels are ever crossing in the legal sense, when they approach in the ship channel of a river. The best discussion found on this point, together with a definite answer to the question in the affirmative, was in a case of collision between two steamships on the Whangpoo River, in which the learned justice of the Supreme (admiralty) Court of China and Japan thus commented on the rules:

The cases of the *Velocity*, the *Ranger*, and the *Oceano* have explained and illustrated the distinction which exists in the effect of the crossing rule as regards vessels navigating the open sea and those passing along the winding channels of rivers. The crossing referred to is "crossing so as to involve a risk of collision" and it is obvious that while two vessels in certain positions and at certain distances in regard to each other in the open sea may be crossing so as to involve risk of collision it would be completely mistaken to take the same view of two vessels in the same positions and distances in the reaches of a winding river. The reason, of course, is that the vessels must follow, and must be known to intend to follow the curves of the river bank. But vessels may, no doubt, be crossing vessels in a river. It depends on their presumable courses. If at any time, two vessels, not end on, are seen keeping the courses to be expected with regard to them respectively, to be likely to arrive at the same point at or nearly at the same moment, they are vessels crossing so as to involve risk of collision; but they are not so crossing if the course which is reasonably to be attributed to

[6]The *Amolco* (CCA Mass 1922) 283 F 890.
[7]The *Sabine Sun* (Pa 1927) 21 F (2d) 121.
[8]The *Milwaukee* (1871) Fed. Cas. No. 9,626.

either vesssel would keep her clear of the other. The question therefore always turns on the reasonable inference to be drawn as to a vessel's future course from her position at a particular moment, and this greatly depends on the nature of the locality where she is at the moment.[9]

It was in the fairly wide Irako channel connecting the Japanese port of Nagoya to the open sea, that a British ship, the *Glenfalloch* was held 20 percent liable for a collision that occurred after she maintained course and speed until in extremis, having assessed she was in a crossing situation. The Pakistani vessel *Moenjodaro* was outbound down the channel and strayed over towards her port side, thus presenting a green sidelight to the inbound *Glenfalloch* and

> . . . although *Moenjodaro* was steering a course of about 10 degrees to port of a direct course along the channel she was a ship proceeding along the channel in a contrary direction to that of *Glenfalloch*, rather than a ship crossing more or less directly from one side of the channel to the other . . . the Collision Regulations did not apply so as to oblige *Glenfalloch* to keep her course and speed until the two ships were so close to each other that a collision could not be avoided by the action of *Moenjodaro* alone; on the contrary *Glenfalloch* was free to alter either her course or her speed. . . .[10]

Narrow Channels

Aside from the doctrine above discussed at some length, it may be pointed out that vessels proceeding up or down a river are usually under Rule 9(a) and Rule 9(a)(i), international regulations and inland rules, respectively, with an obligation to keep to the side of the channel that would naturally produce a port-to-port passing. It should be noted that Rule 9, unlike Rule 14 for the head-on situation, is contained, in both sets of rules, in the section concerning action under any condition of visibility and applies to all vessels, not just power-driven ones. This rule places a burden on any vessel that is on the port side of a channel to have a need for being there or to establish agreement for a starboard-to-starboard passage.

However, in narrow channels or fairways on the Great Lakes, Western Rivers, or any other water specified by the secretary, a power-driven vessel proceeding *downbound with a following current* has right-of-way over an upbound vessel and the responsibility for proposing the manner and place of passage. Inland Rule 9(a)(ii) makes it crystal clear that this right-of-way supersedes the normal requirement to keep to the starboard side of the channel (Rule 9(a)(i) and the normal head-on situation of having to pass port-to-port (Rule 14(a)). Found in similar waters throughout the world, this rule recognizes the limited maneuverability of a down-

[9]The *Pekin* (1897) AC 532, Supreme Court for China and Japan.
[10]The *Glenfalloch* Q.B.(Adm. Ct.) [1979] 1 Lloyd's Rep. 247.

bound vessel and the *occasional* need to deviate from the standard port-to-port passing as a result of river current patterns when rounding a bend in twisting, narrow channels and fairways. Giving the right-of-way and choice in method of passing to such vessels in these designated waters is essential for safety of navigation. It is clearly incumbent on the down-bound vessel to initiate in good time an exchange of one- or two-short-blast signals so that the upbound vessel may adjust course and/or speed as necessary. The latter vessel may well often have to slow or hold her position against the current in order to allow the downbound vessel to negotiate a bend.

Apart from the unequivocal right-of-way for downbound vessels in certain waters, it is necessary for all vessels less than 20 meters in length, all sailing vessels, and all fishing vessels engaged in fishing, in any channel governed by either the international regulations or the inland rules, to navigate in such a manner so as not to impede the passage of a vessel that can safely navigate only within the narrow channel or fairway. The use of the word "impede" in Rules 9(b) and (c) of both sets of rules does not confer right-of-way status on the vessel confined to the channel. These rules are precautionary ones and advise all mariners, operating large as well as small craft, that vessels navigating a narrow channel or fairway can be at a disadvantage and that the navigational situation between them and less restricted vessels should be considered so as not to impede their safe transit within the channel. In some channels the small craft or fisherman might be able to clear the channel, or hold up at a convenient passing place, but if this should not be possible then the normal strictures of keeping to starboard and of the head-on situation apply, except, of course, in those waters covered by Inland Rule 9(a)(ii).

Collisions have sometimes led to later dispute in court over the meaning of the term "narrow channel," and it might be worthwhile considering some past interpretations. In the *Stelling* and the *Ferranti* case,[11] it was held that the word "fairway" meant the whole area of navigable water between lines joining the buoys on either side and that "mid-channel" meant the center line of that area. In the *Crackshot*,[12] it was held that "narrow channel" meant the dredged channel marked by the pecked lines on the chart and that mid-channel meant the centerline of that dredged channel. In the *American Jurist* and the *Claycarrier* case,[13] the potential contradiction between the two previous decisions was noted, but no further ruling emerged until the *Koningin Juliana* case, already mentioned earlier in this text. Then, the judge

[11]The *Stelling* and the *Ferranti* (1942) 72 Ll. L. Rep. 177.
[12]The *Crackshot* (1949) 82 Ll. L. Rep. 594.
[13]The *American Jurist* and the *Claycarrier* [1958] 1 Lloyd's Rep. 423.

> did not think that it would be right to hold that, in relation to these waters, the expression "narrow channel" means the dredged channel and no more. A great many vessels using the harbour can and do navigate outside the dredged channel; and so to hold would mean that such vessels were not subject to rule 25(a)[14] at all. . . . It does not follow, however, because the expression "narrow channel" means the whole of the navigable water, that the word "mid-channel" means the centre line of that area without regard to the existence and position within that area of the dredged channel . . . as a matter of common sense . . . the centre of the dredged channel must be regarded as mid-channel, not only in relation to the dredged channel itself but also in relation to the wider navigable area as a whole.[15]

In amplification of his phrase "as a matter of common sense," the judge stated that he used it because it was essential that the keep to starboard rule should be applied uniformly to all ships navigating up and down the river, irrespective of their draft and whatever the state of the tide might be. Common sense indeed; and, apart from the special circumstances of some tricky rivers, it would seem equally pertinent to most narrow channels.

But keeping to starboard is not always sufficient to avoid collision in a close port-to-port passage. Careful consideration must be given to the positioning and speed of a vessel in relation to the depth of water, the width of the channel, and the proximity of the channel's edge. When two vessels close each other in relatively shallow and confined channels, interaction can be expected not only between the ships but also between the banks of the channel. Fortunately, the complicated interactions that are set up are of less magnitude in the more common meeting situation than they are in the less frequent overtaking situations. Nevertheless, they do exist, with the magnitude of pressures around a ship varying approximately with the square of the speed of the ship. There is a critical speed when the combination of circumstances produces a sheer that will take the ship away from the near bank and that will be rapid and uncontrollable. Such was the experience of the East German ship *Schwarzburg* when she attempted to pass port-to-port with the Liberian steamship *Sagittarius* in a dredged channel in the River Plate. Both ships had sighted each other at some ten miles distance and, when about half to one mile apart, had each sounded the one-short-blast signal as they adjusted their course along the channel. Shortly after this the *Sagittarius* heard three short blasts from the *Schwarzburg* and put her wheel to starboard and stopped engines. Simultaneously, the *Schwarzburg* appeared to swing rapidly to port. Later it

[14]1960 International Regulations, Rule 25(a) provided that: In a narrow channel every power-driven vessel when proceeding along the course of the channel shall, when it is safe and practicable, keep to that side of the fairway or mid-channel which lies on the starboard side of such vessel.

[15]The *Koningin Juliana*, Q.B.(Adm. Ct.) [1973] 2 Lloyd's Rep. 308; (C.A.) [1974] 2 Lloyd's Rep. 353; (H.L.) [1975] 2 Lloyd's Rep. 111.

transpired that this was due to an involuntary sheer. Her engines were put full astern but some two or three minutes later she had crossed the channel, struck the *Sagittarius* at an angle of about 60 degrees, killed four people, and caused severe damage. In apportioning two-thirds of the blame to the *Schwarzburg*, the court held that she:

> would not have got into the difficulty, and would accordingly not have sheered uncontrollably to port across the channel, if she had not been going so fast; the dangers of interaction between a ship and the bottom and sides of a channel . . . at anything but carefully controlled speeds should have been well known . . . the speed of *Schwarzburg* was excessive while approaching a passing with *Sagittarius*.[16]

The *Schwarzburg*'s speed initially had been 15½ knots, with a reduction to about 12 knots shortly before the sheer began—with the result that, despite putting her engines full astern, she struck the *Sagittarius*, whose own headway was down to four knots, at about eight to nine knots. Expert testimony was that a speed of eight to nine knots was required for a safe passing, provided that the *Schwarzburg* had got down to that speed well before meeting the other ship.

Although not criticized for placing her engines full astern, there is little doubt that such action reduced the effectiveness of the *Schwarzburg*'s rudder, increased the suction of the stern towards the near bank, and aggravated the sheer of the bow across the channel. When proceeding in narrow channels it is a sound practice to do so at a speed well below the maximum available, thus maintaining a ready reserve of revolutions to improve rudder effect by going full ahead for a short time. A brief burst of high revolutions improves water flow past the rudder without appreciable gain in speed and is far more effective in correcting a sheer than going astern. Also, the use of an anchor can be of great assistance, and this valuable brake should always be imediately available when in confined waters.

Physical Characteristics of Head-on Situation

The head-on situation has certain definite characteristics that it would be well to bear in mind before proceeding to a consideration of further illustrative cases. Vessels approach each other in this situation at a rate equal to the sum of their speeds, whereas in the overtaking situation the rate of approach is, of course, the difference of the two speeds, and a varying, but intermediate, rate of approach marks the crossing collision. When vessels strike full on, even at slow speed, the result, in accordance with the well-known mathematical rule expressing the respective kinetic

[16]The *Schwarzburg*, Q.B.(Adm. Ct.) [1976] 1 Lloyd's Rep. 39.

energies, is extremely destructive. The rule is $E = W \times V^2 \div 2g$; that is, the energy of each vessel (in foot tons) equals the weight (in tons) times the velocity squared (in feet per second) divided by 64.4. There was such a collision on the Columbia River at night between two large, full-powered ocean ships, the steamship Edward Luckenbach and the Italian motorship, *Feltre*, in which the latter was said to have been sunk with three distinct gashes in her hull, two of which must have been inflicted after a double recoil by the crumpled bow of the *Edward Luckenbach*. In this situation the least time is allowed for proper action to avoid collision—hence the primary importance of taking such action in a timely manner. To offset the superior hazard of this situation having the most rapid approach, it is true that the target is the smallest; that any change of course by one vessel—through changed alignment of mast by day or of masthead lights by night—should be instantly apparent to the other; and that the smallest change of course will be much more effective than in either the crossing or the overtaking situation, with a strong possibility of a glancing blow if the collision cannot be avoided altogether. It is also true, however, that many so-called head-on collisions become physically, though not legally, crossing collisions, as one vessel makes a last desperate attempt to swing clear of the other and succeeds only in attacking or being attacked at a more vulnerable angle.

Legal Characteristics

Legally, the head-on situation resembles the crossing, and differs from the overtaking, situation in that the manner of passing is not optional but prescribed; and it is unlike both of these in that neither vessel has the right-of-way, and both must therefore take definite and positive action to avoid collision. Under international and inland rules alike, vessels meeting head-on *must* pass port to port, and each *must* alter her course to starboard enough to make that maneuver safe, with the concomitant obligation of a proper whistle signal.

Signal Differences

At this point attention is again called to the fundamental difference in the meaning of one- and two-short-blast whistle signals under international and inland rules, discussed in the preceding chapter. Under international regulations, these signals are described in Rule 34(a) and apply alike to all situations in clear weather. Under Inland Rule 34(a), the one- and two-short-blast whistle signals are provided to indicate the manner of passing without reference to changes in course. The signals shall be given and answered by pilots in compliance with these rules, not only when meeting head-on or nearly so, but at all times when the power-driven vessels are in sight of each other, when passing or meeting at a distance within half a mile of each other, and whether passing to the starboard or port. As explained in chapter 13, in the meeting situation the practical import of these differences is that:

(1) *Outside the inland waters of the United States*, the one-blast signal is a rudder signal indicating a change of course to the right and two blasts are a similar signal indicating a change of course to the left. Hence, the one-, two-, or three-blast signal *must* be given by a power-driven vessel in sight of another vessel whenever she changes course, or reverses her engines, the obligation being plain whether the other vessel is ahead, abeam, or astern, and whether she is a power-driven vessel with a whistle or a sailing vessel without one. Additional changes in course require repetition of the signal. Conversely, the one- or two-blast signal can never be used, either for an original signal or a reply, except when the course is changed, nor can either be used under conditions of visibility so low that neither the other vessel nor her lights can be seen.

(2) *In the inland waters of the United States*, the one-blast signal is required of either vessel as an announcement to the other of a port-to-port passing, whether a change in course is made or not. The two-blast signal is required of either vessel as an announcement to the other of a starboard-to-starboard passing, whether a change in course is made or not; this signal is proper only when the courses of the two vessels are so far on the starboard of each other as not to be considered as meeting head-on. It will be noted that the rule requires a prompt reply in each case. An exchange of signals is required when the predicted distance of passing each other in clear weather is within half a mile.

Port-to-Port Required

It will be seen from the rules that when two vessels meet head-on or nearly so, whether at sea or in inland waters, they are not merely permitted, but are *required*, to pass port to port, and to alter course to starboard *as may be necessary to make such a passage safe*. In this connection it should be noted that the change in course should be both timely and substantial and that *a change of 20° or more is far more effective than one of only 2°*. A sound rule, where circumstances permit, is that the amount of alteration should be more than sufficient to clear the other ship, even if she fails to alter course to starboard. If the vessels are exactly end-on, so that both have to change course to the right, then both the maneuver and the whistle signal of each vessel will be the same under international rules as under inland rules, always bearing in mind that the whistle is blown by each vessel in the first case *because* she is changing course, and in the second case *because* she is proposing or accepting a port-to-port passage. In a case with both vessels so far to port of each other as to pass port to port without a change in course by either, the one-blast signal required in inland waters would be omitted at sea. Similarly, with both vessels so far to starboard of each other as to pass starboard to starboard without a change in course by either, the two-blast signal required in inland waters would be omitted at sea.

The first point to be emphasized here is that the obligations to alter course and to signal are mutual in their application, and therefore neither vessel, after a head-on collision, can ever sustain the plea that she was waiting for the other vessel to change course or to signal *first*. The second point is that the injunction to turn to the right or to pass port to port given in all these sets of rules is meant to be obeyed, and should be disregarded for only one purpose—to avoid immediate danger.

It is a common error of experienced mariners to act on the assumption that since neither vessel in the meeting situation has the right-of-way or its attendant obligation to hold course and speed, the vessel getting out her signal first properly determines whether the passage shall be port to port or starboard to starboard. Certainly the widespread disregard of the port-to-port requirement is not due to any ambiguity in the rules. The international rules contain no reference whatever to a starboard-to-starboard passage. In the inland rules, the two-short-blasts signal is available to propose a starboard-to-starboard passage but, apart from special circumstances, this should not be the norm, as Rule 14(a) mandates an alteration to starboard. The rules are therefore unanimous in requiring a port-to-port passage in every genuine head-on situation. Moreover, it is interesting to note that numerous court decisions have agreed in upholding this basic provision of the rules. A two-blast signal has been held to be merely a proposal to depart from the rules, not binding upon the other vessel unless and until she assents by a similar signal,[17] and a vessel initiating the two-blast agreement assumes all risks of the attempt, including a misunderstanding of signals.[18] In the light of these decisions it would seem to be the part of common sense to adhere scrupulously to the rules, and at least never to *propose* a departure unless special circumstances necessitate.

Keynote Is Caution

Referring again to the fact that in the head-on situation neither vessel has the right-of-way over the other, in the broader sense even a proper port-to-port signal may well be regarded by the vessel giving it as merely a proposal until it is accepted by the whistle of the other. It is true that the other vessel is legally bound to accept it. But the proper keynote of the meeting situation is caution, and the degree of caution required goes

[17]Southern Pacific Co. v. U.S. (NY 1929) 7 F Supp 473.
[18]The *St. Johns* (1872) 20 L Ed. 645.

much further than a perfunctory observance of the rule. Indeed it is just about the time of giving the first signal that the real necessity for caution may be said to begin. The other vessel may misunderstand or fail to hear the signal, she may ignore it, she may deliberately disregard it, or she may make a simultaneous counterproposal of two blasts. Unfortunately the first vessel is far from being in a position to say, "We have changed course to starboard and blown one blast. The rest is up to you." The moment the signaling vessel discovers definite evidence of the other's failure to obey the rules, however flagrant the fault, she must take immediate steps to avert collision. The first and most important step, in the eyes of the courts, is to reduce headway to a point where she is under perfect control; if she fails to do this she is practically certain, in the event of collision, to be held guilty of contributory fault. Vessels have been so held for failure to stop or reverse as soon as there was any uncertainty of the other vessel's course,[19] or an apparent misunderstanding of signals,[20] or where the other vessel was seen to be using left rudder in the face of a proper one-blast proposal.[21]

It must be remembered that the signal requirements of both the international regulations and inland rules include the obligation to use the in-doubt signal, five or more short blasts, whenever there is doubt as to the course or intention of another vessel. In a head-on approach in inland waters, such doubt should be deemed to exist and the danger signal used whenever a signal is ignored, after one repetition, or disputed; whenever the vessel is slowed or stopped as a precautionary measure; and whenever the other vessel proposes, or is seen attempting to execute, a dangerous maneuver.[22] If the engines are reversed, the three-blast signal in accordance with Rule 34(a) must be given. In a collision below Owl's Head Buoy in New York Harbor between two steamships that sighted each other exactly end on when about a mile apart, it was held that the inbound vessel that blew one whistle, righted her rudder, reversed her engines when the other swung to port, gave a second signal of one whistle, and then blew three whistles, had done all that could be expected of her in efficient endeavor to avoid collision, and the outbound vessel was held solely liable.[23]

If the situation is under international rules, the one- and two-blast signals denote change in course rather than proposal-agreement on the method of passing. However, if either vessel fails to understand the intentions or actions of the other, she is required to sound the interna-

[19]The *Munaires* (CCA NY 1924) 1 F (2d) 13.
[20]The *Transfer No. 9* (CCA NY 1909) 170 F 944; the *Teutonia* (La 1874) 23 L Ed. 44.
[21]The *Albert Dumois* (La 1900) 44 L Ed. 751.
[22]The *Commercial Mariner* (2 CCA) 1933 AMC 489.
[23]The *Bilbster* (CCA 1925) 6 F (2d) 954.

tional in-doubt signal of five or more short blasts. In addition she should immediately consider revising her safe speed (Rule 6), which would possibly, in a head-on situation, mean reducing or stopping her way in accordance with Rule 8(e).

The Two-Blast Proposal

Whether to assent when a vessel approaching from ahead proposes a two-blast agreement is a question that must be answered very often in practice. The decision of the prudent navigator will be on the basis of what he considers the lesser risk under the circumstances. Many mariners are of the opinion that to dispute a proposal that can safely be accepted is bad practice regardless of the technical invalidity of the proposal. On the other hand, the view is also widely held that the Rules of the Road are so nearly collision-proof that it is always safer to obey them and to refuse to become a part of their nonobservance. When we remember that more than 99 percent of all collisions follow infractions of these rules by one or both vessels, there is some force to this argument. Our first consideration in all cases should be to avoid collision and our second to avoid liability when the wrongful act of another vessel threatens to force collision upon us. The immediate effect of assenting to a two-blast proposal in the meeting situation is to put both vessels under the rule of special circumstances.[24] As said by the circuit court of appeals:

> A two-whistle agreement varying what would otherwise be the normal method of navigation creates a situation of special circumstances. If the proposal is made when there is reasonable chance of success the other vessel is justified in assenting.[25]

This places an equal burden on both vessels to navigate with caution and to take necessary steps to avoid collision, as agreed, by altering course to port. In effect, the assenting vessel puts the stamp of approval on meeting contrary to law, and thereby assumes an equal responsibility with the proposing vessel to carry out the maneuver in safety. Such assent should therefore never be given when, because of current conditions, bends in the channel, speed, proximity of the land, low visibility, or other unfavorable circumstances a starboard-to-starboard passage seems to be hazardous. It will then be much the better procedure to reply promptly to the two-blast signal with the danger signal and take appropriate precautionary action, such as a marked reduction in speed, until a safe passing agreement has been reached.

In a head-on, daylight collision in New York Harbor between two ocean

[24]The *Newburgh* (CCA NY 1921) 273 F 436.
[25]The *Transfer No. 15—Lexington* (2 CCA) 1935 A.M.C. 1163.

steamships, the vessel assenting to a two-blast proposal in a tight situation was found equally at fault with the vessel initiating it. The weather was foggy, with a visibility of 500 yards, and the *El Sol*, going up the wrong side of the channel against the tide at 2 knots sighted the *Sac City* almost ahead coming down at 7 knots and immediately blew two blasts. The vessels were then about two minutes and four ship lengths apart. The *Sac City* replied with two blasts, but was unable to swing left fast enough and after one more exchange of two-blast signals, both vessels reversed, but too late to prevent collision. Although the *Sac City* dropped her anchor in a final effort to stop, she struck the *El Sol* in the starboard quarter with sufficient force to penetrate for eighteen feet, sinking her almost at once. The *El Sol* was, of course, primarily at fault for being on the wrong side of the channel and for proposing a starboard-to-starboard passage without justification, but the *Sac City*, which failed to stop her engine when first hearing the *El Sol*'s fog signals and was guilty of excessive speed, was also specifically held at fault:

> for assenting to the two-blast signal of *El Sol* when, having regard to distance between the vessels, she must have known it was risky to try to pass starboard-to-starboard when there was so little time left to swing, and for not stopping, reversing, and blowing alarm signals as soon as she heard *El Sol*'s two-blast signal.[26]

When Starboard-to-Starboard Passage Is Proper

The reader should not infer from the foregoing discussion that a starboard-to-starboard passage is never legitimate. If such were the case, there would be nothing for meeting vessels to do when approaching each other starboard to starboard within two points but follow the old merchant marine precept (meaning right rudder) of, "port your helm and show your red." Indeed, on the high seas under international rules, with the vessels visible to each other through several miles of approach, there is much to justify such a procedure, at least in cases where it is doubtful if the vessels can pass a safe distance off without some change in course. A critical study of the statutory language raises a doubt whether a two-blast signal is ever strictly proper at sea with the vessels on approximately opposite courses. That is to say, if they are so far to starboard of each other as to be able to clear, and actually do clear, without a change of course, no whistle signal is permissible; and if they are somewhat to starboard of each other, but not enough for safe clearance, then they are within the purview of the head-on rule and bound to pass port to port. Several collisions have been caused by one vessel altering to port to increase the passing distance and the other vessel turning to starboard. If

[26]The *El Sol* (NY 1930) 45 F (2d) 852.

it is thought necessary to increase the distance of passing starboard to starboard, then the implication is that risk of collision exists and the situation should, in a timely manner, be treated as a head-on one. In such a case, the first vessel to alter course to starboard must cross the course of the other, and hence it is increasingly important that both the visible action and the audible notice of it be timely.

The situation under inland rules is somewhat different. The two-short-blast signal is sanctioned for use in the head-on situation and thus provides for the possibility of a starboard-to-starboard passage. The signal is required both as an original and a reply, regardless of whether there is a change of course. We have seen earlier in this chapter that starboard-to-starboard passages can be particularly envisaged on certain rivers where passing in such a manner might be the most seamanlike procedure for a vessel navigating with the current. Outside of those designated waters, a starboard-to-starboard passing should be rare, but not totally extinct. Any sound signal defining a proposed maneuver, particularly such an unexpected one as two-short blasts, should be given whenever possible in ample time to allow for disagreement before the vessels are *in extremis*, or as a federal judge aptly expressed it, in time to maneuver out of a misunderstanding.

Prior to the Inland Navigational Rules Act of 1980 there had been specific recognition, albeit in cumbersome language, of the starboard-to-starboard passage. Consequently, when conditions were appropriate, the courts were inclined to enforce that part of the former rules almost as strictly as the port-to-port provision. An examination of these past decisions discloses that at least in inland waters a vessel in a proper position for a starboard-to-starboard passage was at fault for a collision that occurred because she insisted on meeting port to port.

There have been precious few comparable decisions under international regulations, and it may be that the courts accepted the absence of any reference in the rules to a starboard-to-starboard passage as an indication that it is not an enforceable procedure. There was, however, one case following a disastrous collision where the possibility of a safe starboard-to-starboard passing on the open seas was implicitly recognized. Due to the deaths of so many involved and the lack of a lookout, it was difficult to reconstruct what happened, or why, in this attempted conversion of a starboard-to-starboard passage to a port-to-port passage.

The Brazilian motor tanker *Horta Barbosa*, of 62,619 tons gross, was on voyage from Rio de Janeiro in ballast and met the South Korean motor tanker *Sea Star*, of 63,988 tons gross, which was outward bound in the Gulf of Oman for Rio de Janeiro fully laden with crude oil. The visibility was good and the night clear. Both ships detected each other on radar at about 14 to 16 miles and could see each other's masthead lights at 8 miles.

Their courses were nearly reciprocal, and both were proceeding at 16 knots. For some six minutes or so the bridge of the *Horta Barbosa* was unmanned while the lookout and cadet called their reliefs and the officer on watch was in the charthouse obtaining a radar fix. According to this officer, the *Sea Star* was about three to four miles away bearing 30 degrees on the starboard bow when he had left the bridge. Just before 0400 the lookout returned to the bridge, saw the *Sea Star* on a crossing course and called the officer. The engines were ordered full astern, but collision took place almost at once and was followed by an explosion. The *Sea Star* was destroyed by fire and sank; eleven of her crew lost their lives including four of the five who were then on the bridge—the cadet, who was in the charthouse, alone surviving. The *Horta Barbosa* had to be abandoned, though she was later towed, gutted by fire, to harbor.

It was claimed in court for the *Horta Barbosa* that both ships would have safely passed starboard-to-starboard at a distance of one mile if the *Sea Star* had not altered course. In defense it was alleged by *Sea Star* that the two ships were end-on or nearly so and that both ships should have altered to starboard. The judge was not impressed by the evidence presented by either side, finding the witnesses from the *Horta Barbosa* "lacking in a proper sense of responsibility" and the cadet survivor from *Sea Star* too young and inexperienced to appreciate all the implications of what had happened. Fortunately, there was a third ship some distance off, the *Amoco Baltimore*, whose evidence was crucial in the court's acceptance of the *Horta Barbosa*'s broad case that the two ships were green to green, with each on the starboard bow of each other before:

> . . . the *Sea Star* made an apparently inexplicable alteration to starboard which changed a relatively safe situation into one of acute danger. There must have been reasons why such an alteration was made, even though they were bad reasons. Because the officer or officers responsible are dead, however, it is not possible to ascertain what those reasons were.[27]

The blame for the collision was divided; the *Sea Star* 75 percent because the situation of danger was created by her fault, and the *Horta Barbosa* 25 percent because of her failure to react properly to the situation so created.

Considerably more legal precedent exists for inland waters. In a collision in New York Harbor where two steamers approached each other with ample clearance for meeting starboard to starboard, the one that proposed a port-to-port passage and then made a wide sheer to starboard was held at fault by the circuit court of appeals.[28] In another case a passenger vessel was held negligent for leaving her course and turning to

[27]The *Sea Star*, Q.B.(Adm. Ct.), [1976] 2 Lloyd's Rep. 123.

[28]The North and East River Steamboat Co. v. Jay Street Terminal (CCA NY 1931) 47 F (2d) 474.

starboard into the path of an oncoming steam lighter without getting an answer to her one-blast signal.[29] In another case, a steamship and a ferry exchanged two-blast signals and then the steamship, confused by the ferry's whistle to another vessel, suddenly changed course in an attempt to pass port to port, for which she was held solely liable for the collision.[30]

Similarly, in an old case in Long Island Sound at night between the side-wheeler *Rhode Island* and the steam propeller *Alhambra*, the court found after an analysis of unusually conflicting testimony that the vessels were in a proper position to pass starboard to starboard, and that the *Rhode Island*'s red light had already shut in and her green light was widening on the *Alhambra*'s bow, when she suddenly blew one blast and swung to starboard. The *Alhambra* reversed, but a collision followed for which the *Rhode Island* was solely liable.[31] When two tugs with tows met in the Delaware River in a starboard-to-starboard approach, one of them made a one-blast proposal when they were only 700 feet apart and the other accepted it. Both were at fault. As said by the court:

> The *Crawford*'s fault was in giving the wrong signal, offering to pass on the wrong side, and in attempting to carry out a dangerous maneuver. The *American* was at fault in accepting an obviously improper proposal and in taking part in the effort to carry it out.[32]

In a more recent case, two steamships met above the Delaware River Bridge, and the *Manchester Merchant*, bound down the river, had just overtaken a tug and houseboat close aboard to starboard, when in answer to a two-blast signal of the *Margaret*, bound upward, she sounded one blast and went to starboard. Again both were held at fault for the collision, the *Manchester Merchant* for attempting to force a port-to-port passing, and the *Margaret* for not sounding the danger signal and for not stopping when the *Manchester Merchant* refused to accept her starboard-to-starboard signals.[33] From these decisions it may be reasonably inferred that the rule is as pointed out many years ago by the circuit court of appeals:

> When two meeting vessels by keeping their courses would pass to the left of each other in safety, one of them, which insists on the naked right of passing to the right, and changes the course when it is attended with danger, is in fault for a collision which results.[34]

No doubt this finding will continue to be applicable to open stretches of water where ships in their normal course of business happen to meet

[29]The *Kookaburra* (NY 1932) 60 F (2d) 174.

[30]The *General Putnam* (CCA NY 1914) 213 F 613. *See also* Kiernan v. Stafford (CC NJ 1890) 43 F 542.

[31]The *Alhambra* (CC NY 1887) 33 F 73.

[32]The *American* (Pa 1912) 194 F 899.

[33]The *Margaret* (Pa 1927) 22 F (2d) 709.

[34]The *City of Macon* (CCA Pa 1899) 92 F 207.

broad on each other's starboard bow while on opposite, or nearly opposite, courses and can safely pass starboard to starboard. However, in narrow channels, the injunction for vessels to keep to their starboard side will place an onus on a vessel provoking a starboard-to-starboard passage to justify her presence on her port side of the channel.

The Usual Cause of Head-On Collisions

In conformity with the foregoing discussion it may be said that a striking similarity appears in the causes of most head-on collisions. It is preeminently a situation where the surest way to produce a collision is for one vessel to obey, and the other to disobey, the rule. The end-on rule, when obeyed by both, is so nearly collision proof that it will be no surprise to the reader to be told that the very large majority of collisions of this kind occur simply because *one of the two vessels turns to the left*. While various reasons may be assigned for this disregard of the rule—a misunderstanding of signals, a mistaken notion that the other vessel is intending to turn left, a misapprehension of the requirement to pass port to port, a deliberate usurpation of what is considered the favorable side of a channel—they all point to the same result, and the vessel wrongfully swinging left collides with the vessel properly swinging right. If this obvious cause of disaster could be impressed on the consciousness of every navigator, cases of head-on collision at sea would become extremely rare, being confined largely to those border-line cases at night where there is doubt as to whether the situation is meeting or crossing, and to the even less frequent case where a vessel makes a mistaken attempt to convert a normal starboard-to-starboard passage into one that is port to port.

Indeed, it is not on the high seas that the majority of clear-weather head-on collisions occur, but in restricted channels, such as rivers and fairways. The open spaces of the high seas, with the mutual sighting of ships at relatively long ranges and the ample room to maneuver, lend themselves to the bold, simple action necessary to turn an uncertain meeting situation into a safe and timely passing. In restricted waters there is greater navigational risk, not just from collision but from grounding, shipwreck, interaction, tides, reduced warning time, and an often confusing backdrop of lights, that adds to the dangers of a meeting situation. In addition, the existence of local rules and the ubiquitous presence of a pilot can sometimes produce uncertainty in what, on the high seas, was an efficient bridge organization.

Illustrative Cases

To illustrate these situations, the following recent cases have been selected.

Last minute attempt to change side of channel. Firstly, there is the case in inland waters where two vessels, one a tug with a tow, meeting end on in the Mississippi River collided after the upbound ship, a tanker, abandoned her position on her starboard side of the river. Severe damage was caused to the tanker, the *Anco Princess*, and to the barge, the *Barbara Vaught*, which was under tow by the tug *Libby Black*. The *Anco Princess* entered the river, took a pilot aboard, and proceeded up river under the direction of the pilot at full speed and at about 400 feet from the right-hand bank. The *Libby Black* and her tow were making their way down river. It was afternoon and visibility was excellent. The pilot on the tanker first sighted the tug when the two vessels were some five miles apart. He made no attempt to contact her by radio or by signal. The pilot, in the belief that the *Libby Black* was too close to the east bank to permit a port-to-port passage, ordered the *Anco Princess* to change course so that the vessels would pass starboard-to-starboard. In doing so, the *Anco Princess* cut across the course of the *Libby Black*, which rammed her broadside on.

The Eastern District Court of Louisiana concluded that the tug was navigating approximately along the mid-channel and that she was on the tanker's port bow just prior to the *Anco Princess*'s change of course to port. This alteration of course by the pilot of *Anco Princess*, made while the master was absent from the bridge making a radiotelepone call to the ship's agents, "amounted to gross fault and was the major cause of the collision." This fault was compounded by the failure of the pilot to signal his intentions to the *Libby Black*, either by whistle signal or by radio communication, as was mentioned in chapter 13. In court, the operators of the tanker pleaded that, after the *Anco Princess*'s course alteration, the courses of the vessels were "so far apart on the starboard of each other as not to be considered meeting head and head," and therefore a starboard-to-starboard passing was required. This plea received short shrift from the judge, who said that a meeting situation existed before the alteration of course, and that being so, "a vessel cannot change what is an end-on situation to one which requires a starboard-to-starboard passage." The *Anco Princess* was held 85 percent at fault, with the *Libby Black* attracting 15 percent for her failure to signal in time her doubt as to the intentions of the *Anco Princess* when the latter did not reply to either her one-short-blast whistle signal or to calls on the radiotelephone.[35]

Failure to give other vessel sufficient room to pass. Secondly, a case in clear weather in the St. Lawrence Seaway, where vessels did not fully consider the effects of interaction. The motor vessel *Liquilassie* proceeded up river from the lock and, one and one-half miles farther, she had to pass the

[35]Gulf Coast Transit Co. v. M.T. *Anco Princess* et al., E.D. La (1977), [1978] 1 Lloyd's Rep. 293.

motor vessel *Nipigon Bay*. The latter had been coming down river but was now stationary. There was little room to pass, and creeping along at about 2 knots, the *Liquilassie* attempted a starboard-to-starboard passage, as the greatest water existed on that side. During the course of this maneuver she came near the stern of the *Nipigon Bay*. The latter then gave a kick to her engines to straighten herself out, the force from which was sufficient to cause the *Liquilassie* to take a sheer and collide with the dock, despite taking bold and drastic actions with her engines and dropping an anchor. The Federal Court of Canada held that the bank suction that forced the *Liquilassie*'s critical sheer resulted not so much from any errors of judgment on her master's part in going too close to the bank or too fast, as from the fault of the *Nipigon Bay* in not allowing the *Liquilassie* the full half of the navigable space to which she was entitled. The *Nipigon Bay* was, if not actually on the wrong side of the channel, at least in the dead center of it. The collision was caused by the kick delivered to *Nipigon Bay*'s engines when the *Liquilassie* was so close to her stern and to the bank.

> The *Nipigon Bay* should have known that a kick of this magnitude and this duration delivered when the *Liquilassie* was so close to the stern of the *Nipigon Bay*, and was so much lighter, would have had the effect which it did on her.[36]

The *Nipigon Bay* received 80 percent of the blame, and as discussed in chapter 13, the *Liquilassie* attracted 20 percent for her failure to sound a danger signal and for not coming to a stop while waiting for the stationary ship to move.

Failure to regain own side of channel. Thirdly, a case where a vessel proceeding down stream in the River Maas failed to get over to her correct side and struck the upbound motor vessel *City of Capetown*. It was a fine, clear night with the tidal stream ebbing at 2 knots. The *City of Capetown* was proceeding on her starboard side of the channel with engines at dead slow ahead. Her pilot attempted to contact the downbound vessel, *Adolf Leonhardt*, on the VHF radio, without success. The *Adolf Leonhardt*, which was by then distant about half a mile, five degrees on the *City of Capetown*'s starboard bow, opened her red sidelight and closed her green sidelight. The *City of Capetown* then stopped her engines, later putting them to full speed astern. On board the *Adolf Leonhardt*, due to the bend of the river, only the green sidelight and masthead lights of the *City of Capetown* could be seen, and it was thought that the latter was attempting to cross the channel. The *Adolf Leonhardt* attempted a port-to-port passage, but with her engines first at slow ahead and then at emergency full astern, she drifted downstream into the other vessel.

[36]Liquilassie Shipping Ltd. v. M.V. *Nipigon Bay*, Fed. Ct. Canada, [1975] 2 Lloyd's Rep. 279.

The admiralty court held that the *Adolf Leonhardt* was navigating on the wrong side of the channel. She had found herself there initially because of an earlier passing situation, but had failed to cross back to her correct side as quickly as she reasonably could and to remain there. Sufficient time and space existed for the *Adolf Leonhardt* to do this but:

> . . . the burden of the case against her is that because those on board did not appreciate how far to the south she was, they did not take such action in time.[37]

Not knowing where they were laterally in the channel, the *Adolf Leonhardt* misconstrued the meaning of the other vessel's green sidelight. She received two-thirds of the blame, with the *City of Capetown* one-third for failure to reduce speed earlier and, even though on her correct side of the channel, for not edging farther to starboard where sufficient sea room existed.

Failure to keep to own side of channel. Finally, a case where a ship allowed the elements to carry her to the wrong side of the channel. The *City of Leeds* was inward bound along a dredged channel on the River Mersey. She passed one ship successfully port to port, and knowing that another outward bound ship, the *Anco Duchess*, was due to leave her dock, she slowed and eventually stopped to allow passing at a suitable place in the channel. The weather was fine and clear, and it was still substantially dark although dawn was beginning to break. The wind was west-south-west force 4 and pushed the *City of Leeds* over to the port side of the channel. To clear a buoy close on the port side the engines were put ahead for two minutes and once again stopped, with the ship still well over to the port side of the channel. To prevent further leeway the engines were later put to slow ahead, and the ship crept slowly up the port side of the channel with her head angled about 20 degrees to starboard of an up-channel course so as to counteract the very considerable effect of the wind on her starboard side. Thus, her red sidelight was presented to the *Anco Duchess*, who had just cleared her dock and entered the channel. Because of the bend, the *Anco Duchess* was showing the *City of Leeds* her green sidelight only. The latter decided it was time to get over to her correct side of the channel, and she sounded one short blast, but she had left it too late. As she crossed the channel, the *Anco Princess* sounded two short blasts and altered to port. Collision took place one and one-half minutes later.

The *City of Leeds* was found more culpable and apportioned 70 percent of the blame. The *Anco Duchess* was also criticized for not appreciating that the upbound vessel would attempt to pass port to port, for not proceeding at a much slower speed, and for not hearing the *City of Leeds*'s whistle

[37]The *Adolf Leonhardt*, Q.B.(Adm. Ct.) [1973] 2 Lloyd's Rep. 318.

signal—thus receiving the remaining 30 percent apportionment of fault.[38]

SUMMARY

A head-on approach is where two vessels approach on substantially opposite courses, and although in confined waters, because of winding channels such vessels may first sight each other almost at right angles. The manner of passing is prescribed as port to port in both international and inland waters. The international regulations make no mention of a starboard-to-starboard passage, but the inland rules provide a two-short-blast signal for the meeting situation that assumes the possibility of a starboard-to-starboard passage for ships passing within half a mile. However, apart from the special cases in designated waters, such a passing should be uncommon, particularly as both sets of rules require vessels to keep to the right in a narrow channel or fairway.

The meeting situation is also characterized by maximum speed of approach, maximum effect of a small change in course by either vessel in avoiding collision, and maximum damage in case of actual collision. In as much as neither vessel has the right-of-way, neither is under obligation to hold course and speed. On the contrary, both are charged with a positive duty to avoid collision, and hence the keynote of this situation should be extreme caution from the moment the approach becomes evident until collision between the two vessels is no longer possible. By the rules each vessel is bound to sound the in-doubt signal and otherwise bring herself under control whenever it is apparent that the other vessel's maneuver/proposal is a cause for concern. It is important to remember that when making a proposal in inland waters, signals to be effective must be given in plenty of time, and that nearly all head-on collisions occur in narrow channels where one vessel fails to keep to the correct side or turns to the left in violation of the rules.

[38]The *City of Leeds*, Q.B.(Adm. Ct.), [1978] 2 Lloyd's Rep. 346.

15

The Overtaking Situation

Definition and Characteristics

Rule 13(b) of both the international regulations and the inland rules describes the overtaking situation in the following clear-cut terms:

> (b) A vessel shall be deemed to be overtaking when coming up with another vessel from a direction more than 22.5 degrees abaft her beam, that is, in such a position with reference to the vessel she is overtaking, that at night she would be able to see only the sternlight of that vessel but neither of her sidelights.

It will be noted that the rule attempts to define the overtaking approach at night in terms of the visibility of sidelights, and the tolerances allowed for the arcs of these lights will, of course, cause some fluctuation in the actual limits of the overtaking approach in terms of arc of the compass, even if the overtaken vessel is proceeding on a perfectly steady course. Moreover, the exact angle of approach cannot always be determined even in daylight, as frankly conceded by the rules in the admonition to regard all doubtful cases as subject to the overtaking rules:

> (c) When a vessel is in any doubt as to whether she is overtaking another, she shall assume that this is the case and act accordingly.

Rule 13 of both sets of rules differs from the rules describing the meeting and crossing situations in that the words "so as to involve risk of collision" are not included—the stricture being that "any vessel overtaking any other shall keep out of the way of the vessel being overtaken." Therefore, apart from being applicable to all kinds of vessels, not just power-driven ones, it is open to interpretation as to whether risk of collision is a necessary condition for the overtaking rule to be in force, a point we shall return to shortly. Before then, however, it is necessary to point out one difference between the two jurisdictions' overtaking rules.

Rule 13(a) of the international regulations currently applies only to vessels in sight of one another. The applicability of Inland Rule 13(a), however, extends not only to vessels in sight of each other but also to vessels in any condition of visibility, although it does stop short of including Rule 19 (conduct of vessels in restricted visibility) in its embrace. This is undoubtedly right, for the overtaking situation, as well as the crossing and meeting situations, cannot be properly determined in restricted visibility. Nevertheless, the implication of Inland Rules 13(a) is that the overtaking vessel should have less of a problem in keeping clear and avoiding collision than the vessel being overtaken, regardless of visibility. It is expected that the international regulations will be amended to agree with Inland Rule 13(a) on the occasion of the first amendment to the former.

Taking only the case where vessels are in sight of each other and risk of collision does exist, there are really two characteristics necessary to any overtaking situation: first, the overtaking vessel must be proceeding in the same general direction as the other, that is, within six points of the same course; and second, her speed must be greater. Naval vessels in column are neither overtaking nor overtaken in the legal sense, as long as their speed remains uniform, though, of course, a vessel joining such a column is an overtaking vessel until she takes up her position. From the nature of the case, it will be seen that the speed of approach in this situation tends toward the difference in speed of the two vessels, and will be exactly that difference when the vessels approach on identical courses. This fact gives more time for decisive action to avoid collision and usually lessens the force of the blow if a collision finally occurs. An ocean vessel has been sunk, however, by being overtaken and struck on the quarter by another vessel making only 3 or 4 knots more speed.

Every vessel overtaking any other is obliged by Rule 13(a) to keep clear of the vessel to be overtaken, even if the usual test for risk of collision—the bearing of the other vessel—is changing appreciably. If a vessel comes up relatively close to another vessel from any direction more than 22.5 degrees abaft the latter's starboard beam, draws ahead, and subsequently turns to port to come onto a crossing course, she is not relieved of her duty to keep clear. Both sets of rules lay down in Rule 13(d):

> (d) Any subsequent alteration of the bearing between the two vessels shall not make the overtaking vessel a crossing vessel within the meaning of these rules or relieve her of the duty of keeping clear of the overtaken vessel until she is finally past and clear.

If she had passed at a considerable distance away from the slower vessel, however, the overtaking rule would not have applied and the other vessel could be obliged to keep clear if a subsequent crossing situation brought risk of collision between the same two vessels. This eventuality was first formulated in the judgment on the collision between the *Baine Hawkins* and the *Moliere*, where, although the latter was found guilty of improperly overtaking, it was said;

> . . . It may, on the other hand, be that, when there is no risk of collision at the time (of overhauling)—if, for example the vessel comes within sight of a sidelight at a considerable distance—the crossing rule comes into force; . . .[1]

If the faster vessel is in any doubt, she should assume the obligation to keep out of the way (Rule 13(c)) if it becomes necessary to turn onto a crossing course that brings about risk of collision. In such a case, she should bear in mind the admonition in Rule 16 to take early action to keep well clear and, even if she should consider herself the stand-on vessel in a crossing situation, she should avoid an alteration of course to port (Rule 17(c)).

The actual separation between vessels when such a judgment comes into force probably depends not only on whether risk of collision originally existed, but also on the relative speed of approach once the crossing situation comes into being. For very slow vessels proceeding on similar courses with hardly any difference in their speeds it might be as little as a mile. In one case, two vessels proceeding in approximately the same direction collided after the leading ship, which was two points on the port bow of the second ship, altered course to starboard, flying the correct international flag hoist for adjusting compasses. The ships were between two to three miles apart, and it was held that, up to the time of the alteration, the regulations did not apply as there had been no risk of collision and therefore no overtaking situation. After the alteration the vessels were considered to be in a crossing situation before colliding.[2]

A further case hinging on whether the crossing rules or overtaking rules were applicable arose from a collision between the Cypriot ship *Olympian* and the Polish ship *Nowy Sacz* off Cape St Vincent. Both ships were proceeding on about parallel courses in a northerly direction, the *Olympian* at 14½ knots and the *Nowy Sacz* at 12½ knots. The night was clear and the visibility good. At about 0245 the *Olympian* bore about 25 to 30 degrees abaft the starboard beam of the *Nowy Sacz* at a distance of some three miles. By 0300 the *Olympian* no longer was more than two points abaft the beam of *Nowy Sacz* and could see the masthead lights and green sidelight of the latter. At 0330 the *Olympian* was on the *Nowy Sacz*'s starboard beam and it appeared to her that the *Nowy Sacz* was closing on a crossing course from port to starboard at an angle of 25 to 30 degrees. At

[1]The *Molière* (1893) 7 Aspinall MC 364.
[2]The *Manchester Regiment* (1938) 60 Ll.L. Rep. 279.

0350 the *Nowy Sacz* made five short flashes on the signaling lamp and, receiving no response, slowed and stopped her engines. The *Olympian*, then 300 to 400 yards away, sounded one short blast and altered hard to starboard. The *Nowy Sacz* put her engines full astern, sounding three short blasts and also altered hard to starboard. Shortly afterwards, at 0357, a collision occurred between the stem of the *Nowy Sacz* and the port quarter of the *Olympian* at an angle of about 10 degrees.

The Admiralty Court held that the overtaking rules were only applicable if, before 0300 when the *Olympian* was still more than two points abaft the beam of the *Nowy Sacz*, two conditions were fulfilled—first, that the two ships were in sight of each other, and second, the risk of collision between them had by then already arisen. The court found the first condition fulfilled, but not the second:

> . . . such risk did not arise until much later when the distance between the ships had decreased to something like a mile. That was about 0330 by which time . . . the *Olympian* was bearing about abeam.[3]

Accordingly, the situation was deemed a crossing and not an overtaking one and the *Nowy Sacz* was found three-quarters to blame with the *Olympian* one-quarter for taking tardy emergency action.

This decision was overturned by the court of appeals, who pointed out that the overtaking rule does not contain the qualification "so as to involve risk of collision" but does contain the words "coming up with another vessel." Furthermore, the latter part of the rule contemplated that there may be alterations of bearing after the rule had begun to apply:

> Since a constant bearing between two ships points to a risk of collision and an alteration of bearing against such a risk, this part of the rule suggests that the rule applies before there is a risk of collision.[4]

The *Olympian* was found to be a vessel "coming up with another vessel." Although risk of collision had not yet arisen the words "coming up with another vessel" were deemed to incorporate the concept of proximity in space, which the *Olympian*, passing *Nowy Sacz* at about 400 yards, clearly met. Accordingly the *Nowy Sacz* was the stand-on ship and the *Olympian* the give-way vessel. The lower court's apportionment of blame was reversed: three-quarters to the *Olympian* and one-quarter to the *Nowy Sacz*.

In considering the above case, it should be borne in mind that neither vessel altered course or speed as the close quarters situation developed until a very late stage. Where a faster ship comes upon a slower vessel heading in the same general direction and, while some distance from her,

[3]The *Nowy Sacz*, Q.B.(Adm. Ct.) [1976] 2 Lloyd's Rep. 695.
[4]The *Nowy Sacz*, (C.A.) [1977] 2 Lloyd's Rep. 95.

alters course, an overtaking situation may not exist. A collision occurred off the west coast of Spain between two vessels, the Spanish ship *Manuel Campos* and the Italian ship *Auriga*, both proceeding in a southerly direction at speeds, respectively, of about 12 and 14 knots. The *Manuel Campos* saw, in good visibility, the two white masthead lights and red sidelight of the *Auriga* bearing about 10 degrees off the starboard quarter, distant about three miles, and assumed the *Auriga* was overtaking. The *Auriga* was proceeding at full speed, and the courses of the two vessels were diverging at an angle of 7 degrees. Later, the *Auriga*, when almost abeam of *Manuel Campos* at a range of 2¼ miles altered course to port. The two vessels were then converging at an angle of 24 degrees, with a risk of collision if the courses were maintained. They were—and some twenty minutes later the collision occurred, causing substantial damage to both vessels. Apart from castigating the very bad lookout kept by both vessels, the admiralty court again reiterated its *Nowy Sacz* finding that the overtaking rules only applied when risk of collision existed. An overtaking situation did not exist during the period when the *Auriga* was bearing more than two points abaft the beam of the *Manuel Campos*, because there never was at any time a risk of collision between them since the courses of the two vessels were diverging and the distance between them too great for risk of collision to exist. Accordingly, the crossing rules applied when the two vessels were subsequently on convergent courses.[5] However, the *Manuel Campos*, although therefore the give-way vessel, was only apportioned 40 percent of the blame since the risk of collision coming into being was entirely the fault of the *Auriga* for altering her course at an improper time.

It will be seen from the rules that an overtaking vessel has the option of passing on either side of the overtaken vessel, subject only to the modification that in a narrow channel the overtaken vessel should be on the right-hand side of the channel and it would ordinarily be better seamanship in such waters for the overtaking vessel to pass on the left. The overtaking situation thus becomes the only one of the three approaches where any discretion is given as to the lawful manner of passing. Meeting vessels must pass port to port, and in the crossing situation the vessel having the other to starboard must keep out of the way. But while the rule does not prescribe any specific maneuver, this situation is, to a greater degree than any other, one of privilege and burden, and the obligation is put upon the overtaking vessel to keep well clear not only throughout the approach and during the actual passing, but long enough afterward so that she is in the most literal sense "finally past and clear."

It should be noted that International Rule 13, the overtaking rule, takes

[5]The *Auriga*, Q.B.(Adm. Ct.) [1977] 1 Lloyd's Rep. 386.

precedence over all other rules in Part B, Section II, Conduct of Vessels in Sight of One Another and that Inland Rule 13 takes precedence over Inland Rules 4 through 18. Therefore, sailing vessels and all other vessels given priority in Rule 18, even those hampered in some way by their activity, size, or casualty, must consider themselves bound by Rule 13 and keep out of the way of a vessel they are overtaking. Normally this should be possible by alteration of course or speed; however, some activities do not readily permit an alteration, such as an aircraft carrier recovering aircraft or a minesweeper with gear fully streamed. The special circumstances of the case may then deem it prudent for a vessel being overtaken by the like to get out of the way.

Although International and Inland Rule 13 does override other rules, it would still be good sense for small vessels, sailing vessels, or vessels engaged in fishing, to avoid impeding the passage of any vessel overtaking them in a narrow channel or following a traffic separation scheme. Such vessels should conduct themselves so as to keep clear of the deeper part of channels to allow passage for those vessels restricted to the channel. It should be borne in mind by small vessels that many large ships can only transit a channel at certain states of the tide and that they have a limited time to achieve the passage safely. Nevertheless, in the event of coming upon smaller vessels in a channel, the burden of keeping clear still rests with the overtaking vessel, in accordance with Rule 9(e)(ii).

Signals

As pointed out in chapter 13, one must bear in mind the fundamental difference in the meaning of clear weather sound signals in the overtaking situation as in the meeting and crossing approaches. On the high seas, *A*, if a power-driven vessel, overtaking *B* and desiring to pass her, does not signal unless the approach will be sufficiently close to make advisable a change of course to clear her. The *A* blows one blast if changing to the right and two blasts if changing to the left, in accordance with Rule 34(a). *B* is required by Rule 17(a)(i) to hold course and speed, and cannot properly give any whistle signal except the in-doubt signal. When *A* has reached a position well ahead of *B*, so as to be past and clear within the meaning of the rule, she may return to her course, using the proper signal to indicate the direction she now turns and once again receiving no answering signal from *B*. In a narrow channel or fairway under the jurisdiction of the international regulations, where the overtaken vessel must take action to permit a safe passage, the overtaking vessel, whether a power-driven one or not, sounds two prolonged blasts followed by one short blast or two short blasts, indicating a desire to pass on the starboard or port side, respectively. The overtaken vessel signals agreement by sounding a prolonged-short-prolonged-short signal. The overtaken vessel may answer with the in-doubt signal if not in agreement.

In inland waters the overtaking vessel, if a power-driven one, must initiate an exchange of signals with another power-driven vessel she intends to overtake, regardless of the need to maneuver. The requirement to exchange signals is not restricted to use in narrow channels and fairways but is for use in *all* inland waters, and unlike the signals for the crossing and meeting situations, there is no reference to the half-mile rule. Under Inland Rule 34(c), *A* signifies a desire to pass to starboard with one blast, or to port with two blasts, and *B* is required to answer promptly, returning *A*'s original signal as assent to the proposed maneuver or blowing the danger signal as dissent to it. In the latter case, *A* should not press her attempt to pass, though she could propose subsequently to overtake on the other side. When *A* is finally past and clear she may come back to her original course, but no whistle signal, such as is required by the international regulations, is provided for announcing the maneuver in inland waters. Perhaps the most important difference to remember in the signal requirements of this situation on the high seas and in inland waters is that under international regulations *B* cannot properly use her whistle except to warn *A* of her duty to keep clear, while under inland rules she must whistle, and has repeatedly been held at fault for failure to answer *A*'s proposal by signaling one way or the other.[6]

Rule Applies in All Waters

The overtaking rule has long been recognized by the courts as applying not only in restricted waters but wherever an attempt to pass might mean risk of collision. An early case in point was a collision on a winter night in the wide open waters of Narragansett Bay in which the tug *M. E. Luckenbach*, with a tow 2,500 feet long consisting of two coal-laden barges, overtook and passed the tug *Cora L. Staples* with a similar tow of three barges. The *Staples*, with the heavier tow, was making about 4 knots, and the *Luckenbach*, at 6 or 7 knots, overtook and passed the other on the starboard hand without any signal and then crowded to port in such a manner that the hawser to her barge fouled the wheelhouse of the *Staples*. Despite frantic danger signals from the latter, the *Luckenbach* held on while her heavy hawser ripped off the wheelhouse roof, smokestack, and mast of the *Staples*, seriously injured the master, and knocked the mate overboard. The mate was picked up by a passing vessel after clinging to the wreckage of the wheelhouse for an hour and a half. Both he and the master libeled the *Luckenbach* and the first barge in her tow. The vessels had been on nearly parallel courses, and in holding the *Luckenbach* at fault for close-shaving, for changing course toward the other when passing, for not signaling as required by the inland rules, and for attempting to pass

[6]The *Mesaba* (NY 1901) 111F 215; Ocean SS. Co. v. U.S. (CCA NY 1930) 39 F (2d) 553.

without the consent of the *Staples*, and the barge for contributory fault in sheering to port as she was passing and thereby causing the hawser to foul the *Staples*, the court answered the plea of the *Luckenbach* that the overtaking rule was meant for narrow channels by saying:

> I do not think that the rule was designed for narrow waters only but for any waters where an attempt to pass would involve danger of collision.[7]

Similarly, the overtaking rule was invoked against a steamer that overtook a sailing schooner on the open ocean 60 miles northeast of Cape Hatteras on a dark, overcast night, and collided with her after she failed to show a torch, as at that time was required. While the schooner was unquestionably at fault for this omission, the steamer was held equally at fault when it developed that the mate sighted the schooner less than half a point on the bow, knew from the invisibility of sidelights that she was on the port tack, and could easily have cleared her by executing left rudder. The court said:

> Though the overtaken sailing vessel failed to exhibit a torch, as required, the burden is still upon the overtaking vessel to show that she used all reasonable diligence to avoid the other.[8]

Important Significance of Exchanging Signals

There is an important legal significance in this requirement as interpreted by the courts. In one sense it is doubtful whether *B*'s dissenting signal really adds anything to the obligation of *A*. It must be remembered that if *B* maintains her course and speed as required under the rule, all the burden of avoiding collision by keeping sufficiently clear is put upon *A*—whether she has *B*'s permission to pass or not—and that in the event of collision under such circumstances, *A* would ordinarily be solely liable anyway. To be sure, she would violate good seamanship by passing in the face of *B*'s danger signal, but if guilty already on the charge of passing too close, this could not, under American admiralty law, add to *A*'s liability. In another sense, however, in inland waters and in narrow channels and fairways under jurisdiction of the international regulations the requirement that *A* obtain permission by signal to pass the overtaken *B* has a most important bearing on the duty of each vessel to the other. It means that while on the high seas, an overtaken vessel incurs the obligation of a privileged vessel to keep her course and speed from the moment the overtaking vessel approaches until the latter is finally past and clear, while within the jurisdiction of inland rules, or international regulations in a narrow channel or fairway, the overtaken vessel is under no obligation

[7]The *M.E. Luckenbach* (NY 1908) 163 F 755.
[8]The *City of Merida* (NY 1885) 24 F 229.

unless and until she assents to the overtaking vessel's proposal to pass. If the overtaking vessel neglects her duty to signal, the overtaken vessel is free to reduce speed or alter course, however abruptly, provided, of course, she does not delay such action until the other vessel is so close that it would make collision inevitable, which would be a violation of good seamanship. But as held by the Supreme Court in a very early case on the Mississippi above New Orleans, so long as the overtaking vessel can avoid collision by reversing or sheering out, she must pay for her silent approach with full liability for a collision resulting from an unexpected change in course or speed by the vessel ahead.[9]

A sound discussion on this point was made by Judge Learned Hand of the circuit court of appeals in New York in the case between the steam lighter *Industry* and the tug *Viking*, which were in a daylight collision in Kill Van Kull off Myers Wharf at Port Richmond, Staten Island, New York. The *Industry* was overhauling the *Viking* on a parallel course 50 to 60 feet to starboard of her and was about 300 feet astern when the *Viking* swung across her bow to go into her wharf. Although the *Industry* had not signaled, the *Viking* saw her and announced her intention of crossing by a two-blast whistle, a technically improper signal. The *Industry*, instead of stopping, held her speed and likewise executed right rudder, following the *Viking* around until she was boxed in by the wharf alongside which the tug was attempting to land. She reversed too late to avoid collision and struck the tug on the quarter. There was no question of the fault of the *Industry*, but her owners appealed from the decision of the district court on the grounds that the *Viking* was at fault for her change of course while being overtaken. In denying the appeal and holding the *Industry* solely at fault, despite the improper signal of the *Viking*, which was found not to have contributed to the collision, the court said:

> Article 24[10] puts the burden upon the overtaking vessel to keep out of the way till she is past and clear and Article 21[11] in general provides that, whenever one vessel must keep out of the way, the other shall keep her course and speed. Therefore, taken without recourse to the other rules it might be inferred that, whenever one vessel is in fact overtaking another, the vessel ahead must hold her course and speed. However, Article 18, Rule VIII,[12] further modifies the relative duties when each vessel is a steamer. If she would pass, an overtaking steamer must signal, and before passing must get the consent of the steamer ahead. The rule concludes by saying that the steamer ahead shall in no case attempt to cross the bows of the overtaking steamer or to crowd upon her course. It is perhaps possible to read this language as referring to the period before the exchange of signals as well as thereafter, but it seems to us unreasonable to do so. In the first place, so construed, it would add nothing to the general

[9]Thompson v. the *Great Republic* (1874) 23 L Ed. 55.
[10]Former overtaking rule, equivalent to current Rule 13.
[11]Former rule governing duties of privileged vessel, equivalent to current Rule 17.
[12]Former rule governing whistle signals, equivalent to current Rule 34(c).

duty prescribed by Article 21. Rather, we think it intended to make clear that it is only after the exchange of signals that the duty of the vessel ahead begins at all. This is besides the only reasonable construction. The vessel ahead is usually overtaken because she has less speed and cannot avoid it; the overtaking vessel may always slow down and keep astern. The rules provide for no signal by which the overtaken vessel may declare her purpose to change her course and speed, and, if she is bound to keep both, she is, as it were, frozen in her navigation from the moment that risk of collision begins, merely because the overtaking vessel begins to overhaul her, something which it is not always easy to ascertain. Thus, she may be compelled to abandon her intended destination, or to pass her berth, merely because the overtaking vessel insists upon passing. It is therefore fair to suppose that her consent is required, not alone because she may think the passing dangerous in any case, but also because she may need to change her course for her own purposes, to which the interests of the overtaking vessel can hardly be considered equal. The latter should therefore be obliged to hold herself in check against unexpected changes of course, and be prepared to meet them, until by the consent of the vessel ahead, she gets assurance that it is convenient for her to hold on.

And after pointing out that this right of the overtaken vessel to change course or speed in the absence of a signal from the overtaking vessel is questionable only when the latter has crept up so close that any change by the former would instantly create a situation *in extremis*, the court found that:

while it remains possible for the overtaking vessel by proper navigation to accommodate herself to a change in course or speed of the vessel ahead, that vessel is not held to any duty, but may execute her purpose, regardless of the overtaking vessel and in reliance upon her duty to keep out of her way. That is the situation at bar, for the *Industry*, by starboarding or backing, could, if alert, have avoided collision as soon as the *Viking* began to port.[13]

On the high seas, where no exchange of signals is necessary, similar judgments have been made that the stand-on vessel need not rigidly adhere to its original course and speed. The requirement under international regulations, now Rule 17(a)(i), for the privileged vessel to maintain course and speed has been interpreted as follows: "Course and speed"—in Rule 17(a)(i)—"mean course and speed in following the nautical manoeuvre in which, to the knowledge of the other vessel, the vessel is at that time engaged."[14] This judgment was given in a case where two steamships were each making to pick up a pilot, and it was held that the burdened vessel was justified in slowing down, as the other vessel should have been aware of her intention to pick up a pilot. This principle has been consistently applied in all situations, whether they be overtaking or crossing.[15]

[13]The *Industry* (CCA NY 1928) 29 F (2d) 29; certiorari denied (1929) 73 L Ed. 985.
[14]The *Roanoke* (1908) 11 Aspinall MC 253.
[15]The *Echo* (1917) 86 L.J.P. 121; the *Taunton* (1929) 31 Ll.L. Rep. 119; the *Manchester Regiment* (1938) 60 Ll.L. Rep. 279; the *Statue of Liberty* [1971] 2 Lloyd's Rep. 277.

Equally, in numerous decisions applied to collisions in inland waters, the doctrine outlined in the *Industry-Viking* case, above, has been followed. In several cases where the overtaking vessel was held solely at fault, the circuit court of appeals has gone so far as to excuse the overtaken vessel from not discovering the presence of the other vessel before the collision. Thus, in a case off Pier 4 in New York Harbor, in which a rapidly overhauling steamer that had not announced her approach rammed and sank a small tug that was deflected from her course somewhat by the action of the tide, the appellate court held that:

> An overtaken vessel receiving no signal from an overtaking vessel is not required to look behind before she changes course however abruptly. . . . If the overtaking vessel without signal comes so close to the overtaken vessel that a sudden change of course by the latter may bring about a collision, the fault is that of the overtaking vessel.[16]

However, this approval of what under some circumstances would be an improper lookout aft should not be accepted too literally. It really amounts to little more than a refusal to inculpate a vessel that *is* flagrantly run down by another vessel that has no right to overtake her without a signal and is still able by smart maneuvering to avoid the collision. It merely says that with respect to a particular offending vessel approaching from more than two points abaft the beam, there is no obligation to sight her before hearing her required whistle. It should not be construed as a blanket provision removing the obligation of a proper lookout astern with eyes as well as ears. On the contrary, it is distinctly poor seamanship to change course or reduce speed materially without first checking the situation all around the compass. There may be an overtaking vessel under sail, which is not under the signal requirement of Inland Rule 34(c) and has no means provided of requesting passage. (Similarly, the rule does not apply to a power-driven vessel overtaking a sailing vessel in inland waters because the sailing vessel has no way of assenting or dissenting.) As already pointed out, one power-driven vessel overtaking another, whose whistle has not been blown, or whose signal has not been heard, may be so close to the stern of the other that a sudden change in the action of the vessel ahead may make collision inevitable, in which case the latter will be liable for at least part of the damages and may not be able to recover from the other vessel at all.[17] An overtaking vessel following another in a narrow channel and for the time being having no intention to pass is ordinarily under no obligation to signal.[18] The point is that in as

[16]The *M. J. Rudolph* (CCA NY 1923) 292 F 740; see also the *Holly Park* (CCA NY 1930) 39 F (2d) 572.

[17]Long Id. R.R. Co. v. Killien (CCA NY 1895) 67 F 365; the *Pleiades* (CCA NY 1926) 9 F (2d) 804.

[18]The *Pleiades*, supra.

much as we are bound to observe a vessel astern under some circumstances, the only safe procedure is to look astern under all circumstances where a change in maneuver can be followed by an overtaking collision. As long ago held by the Supreme Court in a case where a steamer was overtaking a schooner in Delaware Bay and sank her when she tacked unexpectedly across the steamer's bow:

> While a man stationed at the stern as a lookout is not at all times necessary no vessel should change her course materially without having first made such an observation in all directions as will enable her to know how what she is about to do will affect others in her immediate vicinity.[19]

The schooner was solely liable for the collision.

If the overtaking vessel in inland waters signals one or two blasts and the vessel ahead does not answer, it is her duty to repeat the signal. In a very early case it was held that this was necessary, and that even when the signal is repeated and unanswered, the overtaken vessel's silence may not be regarded as acquiescence.[20] If the overtaking vessel persists in passing without receiving the necessary response, both vessels are breaking the rule and will divide the damage.[21] There is, thus, a requirement so rigid with respect to the overtaking vessel in inland waters that unless she signals and receives permission to pass, the only question usually left the courts to decide after a resulting collision is whether she shall pay all the damages or only some proportionate share.

Cases Where Rule Applied

The overtaking rule has been held to apply in a number of cases having certain peculiar points of interest outside the usual run of one vessel trying to pass too close to another on the same course. As early as 1886 a state court in New York brought out the fact that a ship is an overtaking, and not a crossing, vessel within the meaning of the terms used in the navigation rules although there is a difference of three points in the courses of the two vessels.[22] (Of course, under the rule the vessels might differ by any amount less than six points.) In the *George W. Elder*, an ocean steamer that approached on the quarter of a tug engaged in picking up a tow of barges on the Columbia River at night and sank the tug, was solely at fault for violation of the overtaking rule, although the tug, which was properly lighted, was stationary at the moment.[23] In an earlier case it was held that a steam vessel coming up with another from a direction more than two points abaft the beam did not cease to be an overtaking vessel

[19]The *Illinois* (1881) 26 L. Ed. 562; *see also* the *Philadelphian* (CCA NY 1894) 61 F 862.
[20]Erwin v. Neversink Steamboat Co. (1882) 88 NY 184.
[21]The *Mesaba*; the *Pleiades* (CCA NY 1926) 9 F (2d) 804.
[22]Aldrich v. Clausen (NY 1886) 42 Hun 473.
[23]The *George W. Elder* (CCA Ore 1918) 249 F 956.

merely because during the approach the overtaken vessel took off all her headway and was at the time of collision actually going astern. The overtaking vessel was liable for not keeping clear.[24]

Similarly, a schooner under sail that ran into the hawser of a barge tow half a mile long in Chesapeake Bay, cutting the hawser in two, was at fault for not keeping clear as an overtaking vessel and liable for the salvage of two of the barges, although during the approach the tow was either motionless, or, because of a 5-knot adverse tide, possibly making sternboard.[25] A steamship that overtook another and attempted to pass her while rounding Corlears Hook near New York, without obtaining her consent—in violation of the overtaking rules and the state law limiting speed at this point—was held liable in part for a collision between the overtaken vessel and a tug, which was caused when the steamship's wrongful acts crowded the overtaken steamer too close to the shore and compelled her to reverse into the following tug. The overtaken steamer was also at fault for failure to recognize the critical situation and take preventive action sooner, as required by the general prudential rule.[26] In a daylight collision on the East River, an overtaking tug with tow, which was warned by the exchange of two-blast signals of an overtaken tug with tow and a large steamer coming the other way, was held solely at fault when she failed to conform her movements to the announced change in course by the tug ahead. As a result the two tows were brought into collision.[27] A ferryboat starting out of her slip with her bow pointing astern of a steamer, and then swinging across the steamer's bow in a circling course was held to be an overtaking vessel.[28] However, in an old case in the East River, when a steamer had a large steamship to starboard in the act of turning around in such a way as to cross her course, it was held that the larger vessel was entitled to have room to maneuver, and the steamer was in fault when she got across her bows and then claimed to be an overtaken vessel.[29]

Cases Where Rule Did Not Apply

The overtaking rule was held not to apply in a case where it was doubtful whether two tugs were crossing or overtaking, each insisted on the right-of-way, and both proceeded without regard to the danger of collision. When they collided, both vessels were found at fault.[30] In another case, a steamship and a steam pilot boat started almost together,

[24]The *Sicilian Prince* (CCA NY 1905) 144 F 951.
[25]The *Charles C. Lister* (CCA NY 1910) 182 F 988.
[26]The *Plymouth* (CCA NY 1921) 271 F 461.
[27]O'Brien v. the *White Ash* and the *Winnie* (NY 1894) 64 F 893.
[28]The *Venetian* (Mass 1886) 29 F 460.
[29]The *State of Texas* (NY 1884) 20 F 254.
[30]The Steam Tug No. 15 (NY 1907) 157 F 142.

and after first one and then the other had forged ahead, they entered a narrow pass abreast without exchanging signals and were brought into collision when a sudden current sheered the steamship into the pilot boat.[31] Here again, the overtaking rule was found not to apply, and both vessels were held at fault. In a third case, where two steamers were originally on meeting courses but one of them rounded up to make a landing, and at the time of collision had no definite course, her maneuver did not create an overtaking situation, but like the two preceding cases, rather one of special circumstances.[32]

Initial Signal by Overtaken Vessel

An overtaken vessel is under no obligation to make an initial signal to a vessel coming up astern if there appears to be ample room for her to pass in safety. It was so held when two tugs with tows in Baltimore Harbor on slightly converging courses, but with one a little in the lead, gradually drew together in broad daylight and finally collided. It developed that the master of the overtaking tug was partially blind in one eye and did not see the other vessel until too late to prevent collision; his tug was, of course, fully liable for the damage.[33] But it is clearly the duty of the overtaken vessel, both in inland waters and on the high seas, to sound the proper danger/in-doubt signal if she actually sees that collision is possible. In a somewhat unusual case the steamship *Howard*, on her regular run from Baltimore to Providence, overtook the seagoing tug *Charles F. Mayer* with two laden coal barges under tow on a clear night near Point Judith and sank one of the barges, drowning three of the crew. The watch of the *Howard* had just been changed, and the relieving watch officer had his attention suddenly attracted to port by sighting the red light of a steamer, on opposite course altogether too close for comfort. He immediately changed course sharply to the right, which cleared the other steamer, but brought him into dangerous proximity to the barges of the *Mayer*, which he had not noticed. The barges were in tandem on a very long towline, with about a thousand feet between them, and the mate of the *Mayer*, seeing by the *Howard*'s lights that she was heading for the barges, called the captain instead of taking preventive action himself. The captain rushed to the bridge and, instead of blowing the danger signal to warn the *Howard*, immediately slowed his engines in the hope that the towline between the barges would sag and that the *Howard* would pass clear above it. Of course, this was a violation of his duty, knowing the other vessel's presence, to keep course and speed; and on the showing that the mate had

[31]The *Joseph Voccaro* (La 1910) 180 F 272.
[32]The *John Englis* (CCA NY 1910) 176 F 723.
[33]The *Albemarle* (Md 1927) 22 F (2d) 840.

deferred proper action until the captain had taken charge, this could not be excused as an error *in extremis*. The *Howard*'s fault was obvious, but the *Mayer* was also held in fault, not only for her untimely reduction in speed but for her failure to sound the inland danger signal as soon as the danger became apparent, and for failure to have an officer on watch who was qualified instantly to take the decisive action required.[34]

Legal Effects of Assent to Passing

The international regulation requiring an exchange of signals in a narrow channel or fairway assumes that the overtaken vessel must assist in the maneuver. The legal effect of her assent, therefore, is an agreement to maneuver in such a way as to give the overtaking vessel more room, if necessary. While the overtaken vessel is required to help, the overtaking vessel is still responsible for the safety of the maneuver.

When an overtaken vessel in inland waters assents to the proposal of an overtaking vessel to pass, she neither yields her right-of-way in the slightest degree nor assumes responsibility for the safety of the maneuver, and the fear that either of these results will follow is not an excuse for failure to answer. As already pointed out, in inland waters she *must* answer, and promptly. She must examine the situation ahead as thoroughly as conditions permit, and immediately express the assent or dissent provided by law. As said by the circuit court of appeals:

> The passing signal from an overtaking vessel is not solely a request for permission to pass. It also asks for information which the overtaking vessel is entitled to have. When the overtaken vessel knows of conditions which may make the passing unsafe it has no right to refuse to inform the overtaking vessel of such conditions, and if it does refuse it cannot throw the entire blame for an accident upon the other vessel.[35]

This should not be taken to mean that the mere act of assenting to the proposal ensures a safe passage, however, and inculpates the overtaken vessel in an action following collision. The general rule is that it does not inculpate her at all. In narrow channels, where these collisions usually take place, she can be charged with fault for her assent only when it is given in the face of conditions that ought to make it apparent that a passing is fraught with serious peril and is almost certain to cause a collision. In other words, she cannot deliberately lead the following vessel into a trap and escape liability. But if, in her judgment, the overtaking vessel can, with the exercise of a high degree of skill, successfully make her way past, then she is legally justified in giving her signaled consent to

[34]The *Howard* (CCA Md 1919) 256 F 987.
[35]The *M. P. Howlett* (CCA Pa 1932) 58 F (2d) 923.

the attempt. The burden of clearing her is left almost entirely to the overtaking vessel. As said in a very old case:

> The approaching vessel, when she has command of her movement, takes upon herself the peril of determining whether a safe passage remains for her beside the one preceding her, and must bear the consequences of misjudgment in that respect.[36]

In a typical case illustrating the application of this doctrine by the Supreme Court, the *Gulf Trade*, a single-screw steamship 429 feet long, overtook and attempted to pass on the starboard side of the tug *Taurus*, which was towing in tandem, i.e., in column, four scows, two of them loaded, just as the *Taurus* swung her long tow out of the Delaware River into the quieter waters of the Schuylkill. The *Gulf Trade*, which had previously proposed twice to pass the *Taurus* and had received an assent each time, made the actual attempt after the third exchange of signals, in time to collide with two of the scows, which were swung somewhat in her path by the action of the flood tide as the tug moved out of the Delaware. The district court exonerated the *Taurus* and the circuit court of appeals found both steamers liable, but the Supreme Court held the *Gulf Trade* solely at fault for the collision, reiterating the precedent established by that tribunal many years before. The high court said:

> We cannot conclude that the *Taurus* was in fault. She was prudently navigated in plain view of the *Gulf Trade* who knew the relevant facts; and by assenting that the latter might pass she certainly did not assume responsibility for the maneuver. At most the *Taurus* obligated herself to hold her course and speed so far as practicable, to do nothing to thwart the overtaking vessel, and that she knew of no circumstances not open to the observation of the *Gulf Trade* which would prevent the latter from going safely by, if prudently navigated. Of course no ship must ever lead another into a trap. There was ample room for the *Gulf Trade* to pass. But if not she should have slowed down and kept at a safe distance. Her fault was the direct and sole cause of the collision. Under these regulations the duty of the *Gulf Trade* was clear. She should have anticipated the effect of the flood tide in the Delaware upon the flotillas as they rounded into the still water of the Schuylkill and kept herself out of the zone of evident danger.[37]

If an overtaking vessel proposes by signal to pass and the overtaken vessel assents under conditions of great and obvious danger, and a collision follows, then both vessels will be at fault. Such a case occurred on the Neches River in Texas. At the time of collision the Neches was a narrow and tortuous stream, at the point where the collision occurred about 600 feet in width, reduced for navigation purposes to a dredged channel 30

[36]The *Rhode Island* (NY 1847) Fed. Cas. No. 11,745.
[37]Charles Warner Co. v. Independent Pier Co. (1928) 73 L Ed. 195.

feet deep, steep-to on one side, and because of a shelving slope on the other side having a bottom width of only 150 feet. In this narrow pass a twin-screw Norwegian steamship, the *Varanger*, assisted by two tugs, one on either side, was overtaken by the *Dora Weems*, a single-screw lake-type steamer. The *Varanger* was 489 feet long, 60 feet wide, and at the time drawing nearly 28 feet. The *Dora Weems*, the faster vessel, was 261 feet long, 40 feet wide, and had a maximum draft of nearly 17 feet. She followed the *Varanger* for several miles, and when the channel reached a straight stretch about a mile in length, speeded up and blew two blasts to the larger vessel for a port-side passage. The *Varanger* promptly responded with two blasts, and the *Dora Weems* at full speed attempted the narrow opening between the shelving bank and the *Varanger*'s port-side tug. The double suction produced by the bank and the deep-draft *Varanger* caused the *Dora Weems* to sheer to starboard and although her rudder was put hard left and her engines kept at full speed, the break of her forecastle head struck the port side of the *Varanger* well aft of amidships, and serious damage was done to both vessels. The tug on the collision side of the *Varanger* let go to save herself, the *Varanger* backed her port engine, and by the time the *Dora Weems* scraped clear the *Varanger* had swung into the bank and stranded.

Both vessels in the case were in the charge of veteran river pilots, and the district court and the circuit court of appeals agreed in finding them guilty of negligent navigation, the one in proposing what was, in the opinion of various witnesses, a very dangerous maneuver, and the other in sanctioning the proposal. In holding the *Varanger* equally at fault with the *Dora Weems*, the appellate court said:

> Neither a master nor a pilot, whether in charge of a favored or a burdened vessel should acquiesce in a maneuver so inherently dangerous that it cannot likely be accomplished with safety. Both overtaking and overtaken vessels in a river are under duty to avoid collision if possible. The pilot of an overtaken vessel in a narrow channel is charged with knowledge that the force of double suction from a river bank and his vessel will come into play when an overtaking vessel attempts to pass. While the overtaken vessel may acquiesce if the proposed maneuver can, in its judgment, be accomplished with safety, although requiring an unusually high degree of skill, it must not permit the passing if it knows, or has reasonable cause to believe, that passing is fraught with positive danger.[38]

That the overtaken vessel that consents to being passed in a narrow channel is still entitled to use mid channel if necessary was brought out in an early case in Hell Gate, and reiterated in the *Varanger* decision. The overtaken vessel may ease to make more room for the other, without

[38]The *Varanger* (CCA Md 1931) 50 F (2d) 724.

being chargeable with changing her course, but is not obliged to do so if it involves danger for herself. As said in the earlier case referred to, the rule implies:

> that the overtaking vessel will, in passing, fulfill her statutory duty of keeping out of the way of the overtaken vessel, and that the latter will keep her course so far as practicable, consistent with the knowledge that the overtaking vessel is to pass her to port. The overtaken vessel has the right to keep in mid channel so long as there is sufficient room on the port side for the overtaking vessel.[39]

We may sum up the import of these decisions on required signals in the overtaking situation in inland waters by pointing out that while the overtaken vessel will very rarely be held to account for consenting to a proposed passage, she will never be held to account for holding up a following vessel with the prescribed danger signal, and this option should always be used in any situation where a reasonable doubt exists as to the safety of a proposed passage.

Duty of Overtaking Vessel to Keep Clear

The courts have not been more specific than the rule itself in regard to the duration of the overtaking vessel's duty to keep clear. The rule lays that obligation upon her until "she is finally past and clear." Many years ago the point was settled that the rule applies not merely until the overtaken vessel is abeam, but until she has completely passed the other.[40] In a later case, when an overtaking steamship, just after passing a tug with barges, lost headway in order to come to anchor and was struck by one of the barges, the steamship was held not to have fulfilled her duty to keep clear.[41] In a New York Harbor collision between a ferry and a steam lighter, in which the ferry had overtaken and passed the lighter some three-quarters of a mile before the point where they came together, and had been obliged to stop her engines to avoid another vessel cutting across her bow, and had herself changed course across the lighter when still from 300 to 500 feet ahead of the lighter, the court held that the ferry was no longer an overtaking vesssel, but was, within the meaning of the rule, finally past and clear. The steam lighter, temporarily in charge of a deck hand while the master was at dinner, negligently rammed the ferry and was held at fault, although what the courts considered an unnecessary change in course by the ferry resulted in her having to pay half the damages.[42] Perhaps the best policy in deciding this point in practice is to fall back on the literal provision of the overtaking rule, and consider that

[39]The *Dentz* (CC NY 1886) 29 F 525.
[40]Kennedy v. American Steamboat Co. (1878) 12 R. I. 23.
[41]Brady v. the *Bendo* and the *Sampson* (Va 1890) 44 F 439.
[42]The New York Central No. 28 (CCA NY 1919) 258 F 553.

the overtaking vessel must keep clear until far enough ahead so that her maneuver cannot embarrass the vessel she has overtaken, as long as the latter holds course and speed. Of course, if she has to stop in the path of the vessel she has left astern there is an added reason for the prompt use of the danger signal.

Under both sets of rules, the action by a give-way vessel is covered by Rule 16 and enjoins her to pass "well clear," an injunction that does not prevent her from passing ahead of the stand-on vessel. Thus, an overtaking vessel may cross ahead, but the stricture of Rule 8(d) makes it imperative that it shall be at a "safe distance," with the maneuver not cut so fine as to result in danger. Where vessels have a large advantage in speed, such as a surface effect ship overtaking a more pedestrian vessel, then such a maneuver is probably safe. However, the constant risk of mechanical breakdown should always be borne in mind in assessing what is a "safe distance" ahead. For more conventional vessels the more prudent action would be to alter course or reduce speed to pass under the stern of the vessel being overtaken. Such action must be taken in good time, for there is a risk that a vessel being overtaken may become sufficiently concerned to take action that might be conflicting under Rule 17(a)(ii). Early action is particularly important in a marginal case, where a vessel to starboard of another may assume herself the give-way vessel in an overtaking situation and turn to port to duck under the stern, whereas the vessel to port thinks she is the give-way vessel in a crossing case and alters to starboard. Perhaps the best solution is for the vessel to starboard, if she is relatively close to the other, to turn to a parallel course and pass ahead, or slow and wait for enough room to pass safely under the other vessel's stern, keeping a sharp lookout for any maneuver by the other.

The Usual Cause of Overtaking Collisions

In concluding this discussion, it may be said that the duties of the respective vessels in the overtaking situation are clear, and it is only by disregarding them that collision is likely to occur. The burden of keeping a safe distance away is placed on the overtaking vessel. The overtaken vessel has the simple obligation of maintaining course and speed as far as practicable, and in inland waters, or in a narrow channel or fairway under international rules, the additional duty of answering the other vessel's signals. In an overwhelming majority of overtaking collisions, the cause is the impatience or negligence of the faster vessel, which in an ill-advised effort to save minutes or seconds, crowds the slower vessel too closely, or attempts to pass her when the time and place are not safe. The *Frosta*, already discussed in chapter 12, was an example of the former when she chose to overtake the *Fontini Carras* at a distance apart of only 500 yards, despite having the open expanse of the Indian Ocean available. The sudden failure of the *Frosta*'s steering gear resulted in that narrow gap between the two vessels being closed in less than three minutes. The *Frosta* was found totally liable for the collision, in part for overtaking too close when:

> both sides accepted that a proper overtaking distance for ships of this size in the open seas was not less than half a mile.[43]

An example of the latter type of case occurred in the Manchester Ship Canal when the motor-vessel *Saint William* got underway from her berth just before the outbound motor-vessel *Jan Laurenz* was due to pass. The *Jan Laurenz* started to overtake on the *Saint William*'s starboard side shortly after, when both vessels were approaching some narrows where a viaduct crossed the canal. Collision occurred as a result of interaction between the vessels in such a confined space, but in describing the "race for the viaduct" the court regarded the case as one of

> two bull-headed navigators, each determined not to give way to the other, and as a result, taking or persisting in action which was almost certain to end up in collision.[44]

The *Jan Laurenz* was found at fault for attempting to overtake at an unsuitable time and place, though the *Saint William* was also seriously at fault for leaving her berth when she did. Blame was apportioned equally.

There are many cases in the books where vessels have ignored the well-known danger of suction and disaster has resulted. A collision due to suction is nearly always chargeable to the overtaking vessel because it is prima facie evidence that she tried to pass too close. While the effect of suction is undoubtedly strongest, and therefore most properly to be anticipated in shallow channels,[45] it has been alleged to have caused a collision in the deep waters of the Hudson River with vessels more than 200 feet apart.[46]

Suction, or more properly interaction, between two vessels is present whenever they are relatively close aboard each other. When the vessels are moving in the same direction with little difference of speed between them, and especially if they are of dissimilar size, the risk from interaction is greatest. Even in deep water interaction effects may be experienced by fast-moving vessels overtaking at close distances. The collision between the *Queen Mary*, overtaking at 28½ knots, and the cruiser HMS *Curacao*, proceeding at 25 knots, was considered partly due to interaction even

[43]The *Frosta*, Q.B.(Adm. Ct.) [1973] 2 Lloyd's Rep. 355.

[44]The *Jan Laurenz*, Q.B.(Adm. Ct.) [1972] 1 Lloyd's Rep. 404; (C.A.) [1973] 1 Lloyd's Rep. 329.

[45]The *Sif* (Pa 1910) 181 F 412; the *Aureole* (CCA Pa 1902) 113 F 224.

[46]The *Cedarhurst* (CCA NY 1930) 42 F (2d) 139; certiorari denied (1930) 75 L Ed. 767.

though the charted depth was approximately 120 meters. Hence, it is not prudent to attempt to pass close in open waters when there is plenty of sea room available. As was said in the *Queen Mary-Curacao* case:

> . . . the vessels should never have been allowed to approach so closely as to bring the forces of interaction into existence.[47]

In shallow waters, where the flow of water beneath the keel is restricted, the effects of interaction between vessels are enhanced. In addition, without another ship being in close proximity, proceeding at a speed too great for the amount of water depth available will cause the rudder to become "sloppy" and, in extreme cases, result in a loss of directional stability; i.e., the ship can take an uncontrollable sheer. When the shallow water is confined to a channel, there is also interaction between a vessel and the channel bank—sometimes called "canal effect"—which can also lead to a sheer to the far bank. For very large vessels with small keel clearances, there is also the prospect of "squat" or bodily sinkage that might result in touching the bottom and producing an appreciable trim by the bow, with consequent impairment to shiphandling abilities. The sum effects of all types of interaction are much exaggerated when vessels are overtaking in shallow channels, as the process takes longer than with meeting vessels. Before overtaking another vessel in such waters full consideration should be given to the risk of collision with other marine traffic; to the effects of interaction between vessels; to canal effect causing a possible sheer towards the far bank; to sinkage effect for large vessels and, for all vessels, to the possible loss of control due to proceeding at too high a speed for the water depth available. The first three of these points were specifically mentioned in a recent case dealing with a collision, while overtaking, on the River Schelde, Belgium.[48]

Unlike open waters, where sea room is available, overtaking in shallow channels requires vessels to pass fairly close to each other. The risks involved in overtaking should be carefully weighed against the benefits of getting past the other vessel, especially if she is moving at a high speed, albeit slower than yours, and is of a dissimilar size. Above all, ships should bear in mind that they are less maneuverable in shallow waters, particularly when overtaking or when passing from "deep" to "shallow" water.

In all cases under the rule the keynote is caution, bearing in mind that as the stand-on vessel the one ahead has a very strong presumption in her favor. In the eyes of the courts she has an indisputable right to use the public navigable waters in which she is navigating, and so far as that particular stage of the voyage is concerned, she was there first and no one

[47]The *Queen Mary* (1949) 82 Ll.L. Rep. 303.
[48]The *Ore Chief* [1974] 2 Lloyd's Rep. 427.

passes but by her leave. Perhaps the whole matter has not been more succinctly stated than by the Supreme Court as to a collision at night more than 100 years ago, when one sailing schooner overtook and sank another in the open waters of Chesapeake Bay:

> The vessel astern, as a general rule, is bound to give way, or to adopt the necessary precautions to avoid a collision. That rule rests upon the principle that the vessel ahead, on that state of facts, has the seaway before her, and is entitled to hold her position; and consequently the vessel coming up must keep out of the way.[49]

SUMMARY

An overtaking vessel is one that approaches another from a direction more than 22.5 degrees abaft her beam. The overtaking rule, which requires the overtaking vessel to keep out of the way and thus binds the overtaken vessel to hold course and speed, applies in open waters as well as in narrow channels. In the overtaking situation, the vessels are proceeding in the same general direction and the vessel astern is faster than the vessel ahead. The overtaking vessel may pass on either side, subject to the modification in narrow channels that if the leading vessel is keeping to the right in accordance with Rule 9(a), it may be better seamanship to pass her on her port side.

Under international regulations in open water, the overtaking vessel signals course changes during passing with one or two short blasts. In a narrow channel or fairway under international regulations, where the overtaken vessel has to take action to permit safe passing, a proposal to pass to starboard is signaled by two prolonged and one short blasts, and two prolonged and two short blasts for passage to port. The same proposals in inland waters are made without the prolonged blasts, one short to starboard and two short blasts for passage to port. Agreement is signified by answering with the same signal in inland waters, or by a prolonged-short-prolonged-short signal in international waters. Under both sets of rules, disagreement with a proposal must be made known with the in-doubt/danger signal. In international waters only, agreement signifies that the overtaken vessel will assist in the maneuver.

In inland waters, an overtaken vessel has been found equally at fault with the overtaking vessel when she failed to answer the latter's signal and her silence was mistaken for acquiescence. When the overtaken vessel in inland waters assents, she does not assume responsibility for the safety of the maneuver but merely agrees not to thwart the attempt of the other vessel to pass; however, in the face of apparent danger, it is her duty to prohibit the passage by sounding the prescribed danger signal, and if she

[49]Whitridge v. Dill (1860) 16 L Ed. 581.

assents instead, she will also be held at fault. Even after a proper assent, the overtaking vessel is bound to pass a safe distance off, and will be liable for a collision brought about by passing too closely, provided the other vessel maintains course and speed. Most overtaking collisions are due to this cause, and particular care must be given to pass far enough off to avoid the effect of suction, especially in shallow water. The obligation of the overtaking vessel to keep clear holds until her maneuvers can no longer embarrass the overtaken vessel, or until she is "finally past and clear."

16

The Crossing Situation

Crossing Situation Most Hazardous

There is no other approach of vessels at sea or in inland waters so trying to the souls of seamen as that of two vessels on a near-collision course in the crossing situation. What navigator of a stand-on vessel, about to cross another, has not experienced certain tense moments when it appeared doubtful if the give-way vessel was going to do her duty and give way? Moments when the impulse to reverse or sheer out and yield the right-of-way was barely balanced by a realization of the risk, legal and physical, of changing course or speed? Or what navigator of a give-way vessel in the same situation has not at some time convinced himself by a pair of bearings that he had plenty of time to cross ahead, and then experienced a harrowing interval wondering whether he was going to make it, conscious that if the race ended in a tie he was hanged higher than a kite? According to the case books, about 40 percent of faulty collisions are crossing cases. When we consider the nervous psychology nearly always present on one or both vessels in these cases, the wonder is that the collision rate is not higher. One is almost inclined to the belief that the crossing situation should have a law against it. It was to reduce, perhaps, this tenseness that the 1972 conference introduced the most revolutionary change in recent years to the crossing situation: namely, the stand-on vessel may now maneuver before reaching in *extremis*. This change is now also reflected in the rules for preventing collisions on the inland waters of the United States.

In the past it was unfortunate that under American law an approach that is naturally fraught with a certain degree of mental hazard should have been made legally so complex that some of its intricacies were puzzling even to admiralty lawyers and were completely baffling to the average mariner, whose duty is not to argue fine distinctions of law but to avoid collision. The Inland Navigational Rules Act of 1980 was a bold attempt to clarify the *ad hoc* collection of rules that governed the crossing situation as well as attempting to achieve unity, not only within inland waters, but with the international regulations. Some differences, however, remain and these will be discussed later in this chapter. Of more importance is what will happen to the wealth of legal precedents concerning the crossing situation that has been built up in United States courts. Some former judgments will undoubtedly be cast aside by the new inland rules, but there is much that was founded upon common sense that will remain relevant. The new rules have yet to be tested in court, and therefore the illustrative examples provided in this chapter are, in the opinion of the authors, those whose decisions seem to remain valid.

The Same Maneuvers Are Required by Both Sets of Rules

Despite the differences that remain between the separate jurisdictions it must be stressed that both sets of rules employ, in open water, the same principle for maneuvering to avoid collision. International Rule 15, which is identical to Inland Rule 15(a), clearly spells out the status of ships in a crossing situation:

> When two power-driven vessels are crossing so as to involve risk of collision, the vessel which has the other on her own starboard side shall keep out of the way and shall, if the circumstances of the case admit, avoid crossing ahead of the other vessel.

Therefore, any power-driven vessel that is involved in risk of collision with another in sight on her starboard side—that is, the arc from dead ahead up to 22.5 degrees abaft the starboard beam—and the situation being neither a meeting nor an overtaking one, is a give-way vessel. She is enjoined not to cross ahead of the other by Rule 15, bound to take early and substantial action in accordance with Rule 16, as well as being required by Rule 8 to see that the action is positive, made in ample time, and acceptable to the practice of good seamanship. Thus, the timely options available to the give-way vessel are to keep out of the way by altering course to starboard, to slacken speed or stop or reverse, or to alter to port sufficiently to avoid passing ahead. Whatever action is taken it must be bold enough to be clearly understood by the other vessel. The most common maneuver is for the give-way vessel to alter sufficiently to starboard to place the stand-on vessel well on the port bow. Nibbling alterations to starboard do not constitute bold action to keep clear, as was illustrated in the case mentioned in chapter 13, when the *Statue of Liberty*'s two belated and minimal alterations were adjudged inadequate in the circumstances.[1]

The action to be taken by the stand-on vessel, in any situation, is governed by Rule 17(a)(i) of both sets of rules:

> Where one of two vessels is to keep out of the way the other shall keep her course and speed.

This is a clear enough injunction that satisfactorily covers the majority of crossing situations in open waters. However, the stand-on vessel is not required to maintain her course and speed blindly until collision occurs. If the other vessel fails to give way and approaches so close that any avoiding measure by her alone cannot avoid collision, then the stand-on vessel must take whatever action, under Rule 17(b), she thinks best will avert collision. This traditional maintenance of course and speed by the stand-on vessel until reaching *in extremis* is now further modified by Rule 17(a)(ii) where she

> . . . may, however, take action to avoid collision by her maneuver alone, as soon as it becomes apparent to her that the vessel required to keep out of the way is not taking appropriate action in compliance with these Rules.

Further, and unique to the crossing situation, Rule 17(c) goes on to stipulate that, in taking the permissive avoiding action, the stand-on vessel

> . . . shall, if the circumstances of the case permit, not alter course to port for a vessel on her own port side.

The theory, long enshrined in the rules and upheld by courts, that collision is less likely to occur if one vessel is directed to keep clear of another vessel that is required to stand-on, is not invalidated in the face of Rule 17(a)(ii). Only when it has become apparent that the give-way vessel has failed to assume her obligation to keep clear is the stand-on vessel permitted to deviate from faithful adherence to course and speed. The danger of the theory breaking down lies in the occurrence of the delayed compliance of the give-way vessel to maneuver to avoid collision, coincident with the decision of the stand-on vessel that the time had come to exercise her prerogative to take avoiding action. Now awakened to danger, the give-way vessel can no longer be sure that the stand-on vessel is in fact standing-on. In order to avoid this situation, it becomes even more important for the give-way vessel in a crossing situation to adhere strictly to the requirement of Rule 16—that is, to take early action to keep well clear.

[1]The *Statue of Liberty* (H.L.) [1971] 2 Lloyd's Rep. 277.

And the stand-on vessel? When is she advised to proceed under Rule 17(a)(ii)? Certainly not until risk of collision has been established. Probably not until she has sounded the signal of doubt/danger as a wake-up signal (supplemented by flashes of a light if possible), possibly after failure to achieve VHF contact, but definitely prior to *in extremis*. The distance at which to take action under Rule 17(a)(ii) will vary considerably. It will be much greater for high-speed vessels involved in a fine crossing situation than for low-speed vessels in an abaft-the-beam crossing situation, which is not to say that the latter is the least risky situation! For a crossing case in open waters it is suggested that the maximum range for most ships to take permissive action would be in the order of about 2 to 3 miles. However, this figure will clearly vary with the capabilities and characteristics of individual ships, with a very large ship having a "domain" in excess. No court decisions are yet on hand to offer guidance, though it is unlikely the courts would ever establish a norm, but always consider each case on its merits: the ships involved, the proximity of other dangers, the diligence of those on watch, etc., etc. The ordinary practice of good seamanship should serve to guide the mariner in his decision to take permissive action.

The important point is to realize that the stand-on vessel in a crossing situation has not been granted license to maneuver at will. She is not permitted to alter course or change speed under Rule 17(a)(ii) solely because she thinks it might help the situation or once she observes the other vessel maneuvering to meet her obligations as the give-way vessel.

In considering permissive action, it is well for the stand-on vessel to bear in mind that the give-way vessel may belatedly wake up to her responsibilities, probably by altering course to starboard, and to adhere to the provision of Rule 17(c)—that is, not to alter course to port for a vessel on her own port side. Additionally, a reduction of speed might make it more difficult for the give-way vessel to pass astern. A major alteration to starboard may well be the best course of action, not forgetting to sound the proper sound signal so that the other vessel might speedily understand the intent of the maneuver.

Action to avoid collision becomes compulsory for the stand-on vessel when she finds herself so close that collision cannot be avoided by the give-way vessel alone. The distance between the two vessels when this moment arrives will vary with the direction and speed of approach of the other vessel, an estimate of her maneuvering characteristics, and a thorough knowledge of one's own vessel's capabilities and limitations. The precise point at which to cease maintaining course and speed is difficult to determine, and some latitude has been allowed in court cases. When it is shown that the other vessel had been carefully watched and that the stand-on vessel had endeavored to do her best to act at the correct moment, she will not be held to blame, though subsequent analysis shows she waited too long or acted too soon.

> The conduct of a prudent seaman in such circumstances is not to be tried by mathematical calculations subsequently made.[2]

An alteration to starboard to avoid a vessel close on the port bow could be most dangerous, as it will take the stand-on vessel across the other ship's bow. Left too late and she might be struck amidships, the most vulnerable spot. Rule 17(c) therefore does not apply when in extremis is reached, as a vessel is permitted to take any action that might best avert collision. Turning towards the other vessel may well be the best action to take at close quarters, particularly if she is likely to strike abaft the stand-on vessel's beam.

When collision, despite all efforts, appears to be inevitable, the aim should be to reduce the effect to a minimum. A glancing blow is normally better than a direct impact, though with tankers and the like it is unlikely to reduce the risk of fire. If a glancing blow is impossible, it is probably best to take the impact on the bow, forward of the collision bulkhead. An alteration away that exposes the vulnerable ship's side might well be the most damaging course of action to take.

To make totally clear to a give-way vessel that the permissive action open to a stand-on vessel in no way excuses the former from her responsibilities, Rule 17, under both jurisdictions, concludes:

> (d) This Rule does not relieve the give-way vessel of her obligation to keep out of the way.

Thus, the give-way vessel must not hang on in the hope that the stand-on vessel will opt out of the situation early, nor is she released from her duty to ensure that a safe passing distance is achieved even if the stand-on vessel does maneuver.

Special Circumstances Not Substitute at Will for Crossing Rule

An examination of past cases disclosed among navigators a common misconception of the law in regard to the crossing situation. This was that the application of the crossing rule was modified by special circumstances to such an extent that whenever the stand-on vessel recognized any risk of collision whatever, she immediately had complete discretion as to whether or not she would attempt to pass ahead. This is not so, for any departure from the rules is only permitted by Rule 2(b) when there are special circumstances *and* there is immediate danger. Both conditions must apply.

There are many navigators who openly express the belief that, even when they have the right-of-way, in a close situation the safest plan is to

[2]Compagnie des Forges d'Homecourt (1920) S.C. 247; 2 LL. L. Rep. 186.

yield it at once upon the slightest evidence of dispute. The trouble with this in practice is that oftentimes it may lead us into the very collision we are seeking to avoid, with the stigma of legal liability added to physical injury. The road hog is a notorious bluffer, and there is always the danger that his intention to usurp the right-of-way may weaken, and he may sheer suddenly to starboard in conformity with the requirement just at the time the stand-on vessel slows down or sheers to port to avoid him, with incriminating consequences to the latter if collision ensues. The moral culpability of the stand-on vessel in such a case may be slight, but legally she will be held as guilty as the other, and may even be found solely at fault.

With the law on this point as specific as it is the courts could hardly decide otherwise. The crossing rule represents the second of two distinct methods of procedure prescribed in the rules when two vessels approach each other so as to involve risk of collision. The first requires each vessel to take some positive action that has the effect of changing her *status quo*. It is illustrated by the rules that govern when vessels meet end-on. Both must alter course to starboard, both are equally bound to signal, and neither can claim that the other should have whistled first. The second method is based on the assumption that one vessel shall maintain her *status quo*, and that any positive action to avoid collision shall be taken by the other. It is illustrated by the rules when one vessel is overtaking another, or a power-driven vessel meets a sailing vessel, or a sailing vessel with the wind on the port side meets another with the wind on the starboard side, or two power-driven vessels meet on crossing courses. The whole theory of such rules is that collision is less likely to occur if one vessel is directed to avoid the other and the other is then required to continue exactly what she is doing. As discussed earlier, the theory breaks down the moment that the give-way vessel cannot rely with certainty on the correct action by the stand-on vessel.

Thus, the permissive action now available to the stand-on vessel makes it imperative that the give-way vessel takes her avoiding action well before the former gets sufficiently concerned to depart from her course and speed. Neither should the stand-on vessel be precipitate.

Action should not be taken by the stand-on vessel without first determining that risk of collision does in fact exist. Compass bearings should be observed accurately, and the radar should be used to measure the range of the approaching vessel. The earliest moment for permitted action will obviously be related to the range and rate of change of range. Without such measurement the stand-on vessel might have difficulty in later justifying her action:

> Mere apprehension of danger of collision will not justify change of course in a

> vessel whose duty under the rules is to keep her course. A change should only be made where there is actual danger.[3]

One such case, where special circumstances were deemed to override an apparent crossing case, arose from the collision between the Spanish motorship *Alonso de Ojeda* and the Italian motorship *Sestriere* in the River Plate. Both ships were approaching the drop-off point for disembarking their pilots. The *Alonso* arrived first and altered course to a northerly heading at about 3 knots to drop her pilot. The *Sestriere* was on *Alonso*'s port bow, heading 110 degrees, and reducing speed to about 1 to 2 knots. Both ships recognized the risk of collision and both sounded the in-doubt signal. The *Alonso* went hard to starboard, but collision was not averted. In the subsequent court case it was alleged that the *Sestriere* was to blame for failing to keep out of the way of the other ship, who was on her starboard side in a crossing situation. It was held by the admiralty court that for the crossing rule to apply it was necessary that the stand-on ship should be on a clearly defined course apparent to the other ship.

> The Court was not satisfied that *Alonso* ever settled on such a course, for she was free to adjust her heading as necessary while performing the operation of dropping her pilot . . . thus she was not on a sufficiently settled course to bring the crossing rule into operation; nevertheless . . . it was the duty of *Sestriere* . . . in the special circumstances of the case . . . to take timely action to keep clear of *Alonso* which had arrived on the scene first.[4]

The *Sestriere* was blamed for not taking her way off but was expressly *not* found at fault for not altering course to starboard. The *Alonso* was blamed for unseamanlike action in not putting her engines astern, and both ships shared liability equally between them.

Meaning and Use of Signals Differ on High Seas and Inland Waters

Despite the uniformity between the two sets of rules on the maneuvers to be taken in the crossing situation, there is a major difference between them on the use of sound signals. The meaning and use of three short blasts is, however, identical in both rules: namely that the vessel making the signal is operating astern propulsion.

As mentioned in chapter 13, whistle signals under the international regulations are signals of action. Thus, in the crossing situation the give-way vessel sounds one short blast if she alters to starboard, two short blasts if she alters to port and three short blasts if reversing. She sounds no signal if she slows or stops.

[3]The *General U.S. Grant* (NY 1873) Fed. Cas. No. 5.320.

[4]The *Sestriere* Q.B.(Adm. Ct.) [1976] 1 Lloyd's Rep. 130; *see also* the *Avance* Q.B.(Adm. Ct.) [1979] 1 Lloyd's Rep. 143.

The stand-on vessel on the high seas does not answer these signals if she maintains her course and speed. Only if she takes permissive action under Rules 17(a)(ii) and 17(c) would she sound a whistle signal. This is likely to be one short blast for the preferred alteration to starboard. If she stood on until *in extremis* she must also sound the requisite one, two- or three-short blasts when taking the action that best avoids collision. Should she at any time stop or slow there is no signal to signify the reduction in speed.

Long before reaching *in extremis*, and preferably before taking permissive action, the stand-on vessel should have been so concerned by the action or non-action of the give-way vessel to have sounded the in-doubt signal of five or more short blasts, ideally supplementing it with five short flashes of a light. Hopefully, hearing or seeing this signal, the give-way vessel will be reminded of her responsibility. Equally the give-way vessel can use the in-doubt signal if she should fail to understand the intentions or actions of the stand-on vessel. Such a situation might arise if the stand-on vessel should, inadvisably, take permissive action by altering to port.

In inland waters a totally different principle is employed. Here, one and two short blasts are signals of intent that demand a reply from the other vessel. Such signals are required when two power-driven vessels in sight will cross within a predicted distance of half a mile of each other. Either vessel may initiate the exchange, which common sense decrees should not be delayed until the vessels close to half a mile.

One short blast means "I intend to leave you on my port side." This should be the normal signal in most crossing situations. If initiated by the give-way vessel it indicates an intention that can be, if necessary, subsequently executed by altering course to starboard when no further signal is made, or by reduction of speed. On hearing the signal the other vessel, in this example the stand-on vessel, must, if in agreement, immediately respond with one short blast even though she will be making no change to her course and speed.

The provisions of Inland Rule 34(a) also allow for a signal of two short blasts, meaning "I intend to pass you on my starboard side." Such a signal, apart from the special case of narrow channels yet to be discussed, should not be normal in the crossing situation. Nevertheless, the rule provides for such an eventuality. Given the stricture of the half-mile clause and the constraints of confined waters, it is feasible to recognize that occasions can occur when a two-short-blast signal is appropriate within the dictates of good seamanship. If the proposal is made by one ship and the other agrees, she must reply immediately with two short blasts herself. Thus, the give-way vessel is granted acquiescence to pass ahead of the stand-on vessel, a situation about which any mariner should feel distinctly uneasy.

It must be clearly stated that "cross" signals are not provided for—that is, the answering of one short blast with two, or two short blasts with one. Until 1940 the lower courts were agreed that it was always proper to "cross" an improper signal and that the stand-on vessel, unless agreeing to a two-short-blast signal, was bound to hold course and speed until literally in the jaws of collision. The proper procedure had been to respond with the danger signal and then a single short blast until *in extremis* was reached.

Such was the decision of both the district court and circuit court of appeals in New York in 1939 in the case of the Baltimore harbor collision of the steamships *Eastern Glade* and *El Isleo.* In this case the two vessels sighted each other at night over a mile apart in converging channels, and the *Eastern Glade,* which was the give-way vessel, blew two blasts. The *El Isleo* interpreted this as an announcement of intention to cross her bow and immediately responded with the danger signal, followed by one blast, and continued to hold course and speed until the two vessels were *in extremis,* when she sheered out to starboard in a futile effort to avoid collision and was rammed nearly amidships at a point about 200 yards outside the buoyed channel. The fault of the *Eastern Glade* was glaring, and in accordance with the long line of its own previous decisions, the circuit court of appeals upheld the district court in exonerating the *El Isleo* and then finding the *Eastern Glade* solely liable for the collision. The court took the opportunity to express dissatisfaction with the line of reasoning underlying these decisions but felt powerless to change what had become established procedure without an opinion from the Supreme Court. The case was appealed, and the highest tribunal reversed the lower court and held that:

> . . . when two steamships are on crossing courses, the privileged vessel has no absolute right to keep her course and speed, regardless of danger involved; her right to maintain her privilege ends when there is danger of collision; and in the presence of that danger both vessels must be stopped and backed, if necessary, until signals for passing with safety have been made and understood. . . .[5]

The case was remanded to the circuit court of appeals, second circuit, which reconsidered its former verdict and held both vessels at fault. In its new decision 3 June 1940, the circuit court said:

> There can be no doubt that the Supreme Court meant to hold in a crossing case, when the holding-on vessel gets two blasts from the giving way vessel, which are unacceptable to her, she must neither cross the signal, nor keep her speed, but must at least stop her engines, and if necessary back "until signals for passing with safety are made and understood."

Effectively, Inland Rule 34(a)(ii) now places this historic interpretation of

[5]Postal SS.Corp. v. *El Isleo* (1940) 84 L Ed. 335, reversing (CCA NY 1939) 101 F (2d) 4.

the Supreme Court into the body of the rules. In effect it means that the stand-on vessel is deprived of her right and relieved of her duty to hold course and speed the moment there is an unacceptable or disputed signal that must be challenged by the use of the in-doubt signal. In practice, the right referred to has so often proved unenforceable and the duty so difficult that the mariner is, on the whole, much better off under the new requirement.

In summary, although one and two short blasts have different constructions between international regulations and inland rules, the basic intent is the same: the give-way vessel should avoid passing ahead of the stand-on vessel. When doubt exists as to the sufficiency of action/intention, both sets of rules provide for a signal of five short blasts. Both rules allow the stand-on vessel to take early permissive action to resolve the issue, though only the inland rules clearly specify that the sounding or receipt of the danger signal requires precautionary action by both ships. It should be noted, however, that International Regulation 8(e), as indeed does Inland Rule 8(e), requires *any* vessel to "slacken her speed or take all way off" if necessary to avoid collision or to give more time to size up the situation.

Two Blasts by Give-Way Vessel Do Not Give Her Right-of-Way

In inland waters in the past there was a belief that an initial two-short-blasts signal by the give-way vessel conferred upon her the right-of-way. Apart from those waters specified in Inland Rule 15(b), where there might be some justification, such a belief is a misconception of the rules, despite its frequent use in the past. Whatever may be the local custom, it is never safe to assume that it can displace a positive statutory provision. As held in a very old case:

> If there is a custom which permits Sound steamers to claim exemption from the operation of this article when approaching the ferries in the East River on the ebb tide, such custom is opposed to law, and cannot prevail.[6]

Moreover, notwithstanding local custom, the practice is at least questionable, and every navigator who makes use of it should fully understand its legal and judicial significance. It is true that there was at one time a pilot rule that permitted the vessel having the other on her starboard hand to proceed if it could be done without risk of collision, and provided a two-blast signal therefor; but even this was regarded as contrary to the spirit of the crossing rule, and was repealed after being invalidated by the federal court stating that the rule

[6]The *Pequot* (1887) 30 F 839; the *Mohegan* (CCA NY 1900) 105 F 1003.

> . . . which permits the vessel having the other on her starboard hand to cross the bows of the other if it can be done without risk of collision, is invalid, as repugnant to the starboard hand rule.[7]

In the second place, if the give-way vessel, assuming that she has ample time to cross ahead, initiates the two-blast signal, which in the absence of any statutory provision can have no sanction but that of local custom, she thereby expresses an admission that risk of collision exists, and lays herself open to the charge of proposing a violation of the law. Recognizing this, the lower courts have repeatedly ruled that such action is at best no more than a proposal that the other vessel is under no obligation to accept; that unless and until the proposal is agreed to by the other, as indicated by a reply of two blasts, the stand-on vessel must hold on; and that even after such agreement the give-way vessel assumes any risk of carrying out the maneuver.

The following decisions, although antedating the present rules for inland waters, are ample evidence of the attitude of the courts on this question:

> If a burdened steamer, by her signals, invites a departure from the ordinary rules of navigation she takes the risk both of her own whistles being heard, and, in turn, of hearing the response, if a response is made, and of the success of the maneuver.[8]
>
> Two whistles given in reply to a signal of two whistles from a steamer bound to keep out of the way mean only assent to the latter's course at her own risk, and an agreement to do nothing to thwart her. It does not relieve the latter of her statutory duty to keep out of the way; but when collision becomes imminent, both are bound to do all they can to avoid it whether the previous signals were of two whistles or one. If imminent risk of collision is involved in the maneuver assented to, and the maneuver was unnecessary, both are responsible for agreeing on a hazardous attempt.[9]
>
> When the boat having the right of way fails to respond to the signal of the boat whose duty it is to keep out of the way, the latter has no right to assume, because of such silence, that the former abandons her right of way.[10]
>
> The failure of the privileged vessel to assent to a signal contrary to the rule is equivalent to a dissent which holds the burdened vessel bound to observe the starboard rule.[11]
>
> The privileged vessel is entitled to assure that, although the burdened vessel may at first propose to exchange rights of way, it will, if such a proposal be rejected, conform to the rules of navigation.[12]

In the light of these decisions the risk in the use of an initial two-blast signal by the give-way vessel is apparent, and should generally be avoided.

[7]The *Pawnee* (NY 1909) 168 F 371.
[8]Hamilton v. the *John King* (1891) 49 F 469; the *Admiral*, 39 F 574.
[9]The *Nereus* (NY 1885) 23 F 448.
[10]The *Pavonia* (NY 1885) 26 F 106.
[11]The *Eldorado* (NY 1896) 32 CCA 464, 89 F 1015.
[12]L. Boyers' Sons Co. v. U.S. (NY 1912) 195 F 490.

Certainly it can never be legally justified when the vessels are approaching each other so as to involve risk of collision; and when it is possible for the vessel having the other to starboard to pass so far ahead of the other that risk of collision may be deemed not to exist, whistle signals are not only unnecessary, because of the half-mile clause, but are better omitted.

Departure from Rules by Agreement

If, as sometimes happens, the stand-on vessel offers to yield the right-of-way by blowing two blasts first, the situation with regard to the give-way vessel is somewhat different. Once the give-way vessel assents to the arrangement by answering with two whistles and the desired agreement is thus established by the interchange of signals, she becomes, in a limited sense, a privileged vessel, though not under the same legal obligation as a stand-on vessel to hold her course and speed. In a crossing collision between a tug and a ferryboat in New York Harbor in 1903, where the stand-on tug gave two blasts and then failed to go astern of the ferry, the courses being almost at right angles, the circuit court of appeals held that:

> An agreement by signal, initiated by the privileged vessel, by which she was to pass under the other's stern, justified the latter in keeping her course and speed.[13]

However, in another crossing collision in the North River, decided fourteen years later, in which the vessels were crossing at a finer angle, the circuit court of appeals held both vessels at fault, involving the give-way vessel for not cooperating with the stand-on vessel to avoid collision by also altering course. The court found that:

> When a privileged vessel proposes that the burdened vessel cross her bows and gets an assent to such proposals, she assumes the risk of the proposal . . . the case is one of special circumstances, and the burdened vessel is not rigidly bound to keep her course and speed.[14]

An interesting sidelight on the legal complexities that are introduced when the stand-on vessel initiates the two-blast signal, the signal is accepted, and a collision follows, is revealed in the deliberations of the circuit judges in the case cited:

> It is good law that when the burdened vessel decides to keep out of the way by crossing the bows of the privileged vessel, though she gets an assent to such a proposal, she assumes the risks involved in choosing that method. The duty of the privileged vessel in such cases is to cooperate and she need not keep her course. The situation, at least in this circuit, after the agreement, is one of special circumstances. But such an agreement initiated by the privileged and

[13]The *Edwin J. Berwind* (NY 1906) 144 F 664.
[14]The *Newburgh* (NY 1921) 273 F 436.

> assented to by the burdened vessel, might be regarded as creating other duties. It could be considered as a proposal that the duties of the vessels should be reversed, and that the burdened (now the privileged) vessel hold her course and speed, so that the privileged (now the burdened) vessel might be able to forecast her positions at future moments precisely as the rule requires when no agreement has been made.
>
> We have been unable to find much in the books that touches on this precise point. In the *Susquehanna*, 35 F 320, the burdened vessel was exonerated because she did not "thwart" the proposal, having apparently kept her course. On the other hand, in the *Columbia*, 29 F 716, Judge Brown thought it a matter of indifference which vessel proposed the change; the burden always remaining upon the vessel originally burdened. In Stetson v. the *Gladiator*, 41 F 927, Judge Nelson said that the exchange justified the burdened vessel in keeping her course.
>
> In none of these cases was the originally burdened vessel held to any duty to keep her course. On the whole we are disposed to think that any agreement to change the usual rules should be treated as creating thereafter a position of special circumstances. If so, we think that, although the proposal emanates from the privileged vessel, and should be taken as meaning that she will undertake actively to keep out of the way, it need not absolve the burdened vessel from her similar and original duty also to keep out of the way, nor will it impose on her a rigid duty to hold her course and speed. It is true that that duty is imposed by the rule generally as a correlative to the duty to keep out of the way, but only in cases where no agreement has been reached. Some convention is essential when neither knows the other's purposes, but where both have agreed upon a maneuver by an exchange of signals their accord should be left for execution by movements adapted to the circumstances. For example, if the angle of crossing is wide, it will usually be best for the originally burdened vessel to hold her course and speed; but if it be narrow, it is safest for both to starboard and pass at a greater distance. No doubt the proposal involves the proposer in a duty to give a wide enough margin for safety, even though the assenting vessel does not starboard.

The foregoing discussion shows that even the seasoned admiralty judges of the circuit court of appeals are sometimes compelled to struggle with the intricacies of the law of crossing. If a single useful fact emerges from the involved discussion in this decision, it is that changing the lawful signal by both vessels has the important effect of destroying the right-of-way of one of them, and making her share with the vessel contemplated by the rules the burden of avoiding collision. This in itself is an excellent argument, in the crossing situation, for sticking to the procedure in practice provided by law.

Two Blasts by Stand-On Vessel

From the viewpoint of the stand-on vessel, for her to initiate the two-short-blast signal must appear as a very foolish act. For she is either suggesting that in her opinion the vessel to port already has ample clearance, a questionable opinion as the other vessel must be passing ahead within half a mile for the need to signal to arise, or else she is giving notice to the give-way vessel that she (the stand-on vessel) proposes to waive her privilege and act contrary to the law, in which case, if the other assents by answering with two short blasts, she is bound to do her part in carrying out the maneuver.

Except in certain narrow channels and rivers, not only should the stand-on vessel scrupulously avoid proposing a two-short-blast signal but she should be somewhat chary about assenting when it is proposed by the give-way vessel. It would be unwise to lay down an arbitrary rule here, and the individual case must be decided on its merits, remembering that avoidance of collision is always the prime desideratum. The navigator must determine, in the particular instance, which action involves the least risk: agreeing to the proposal with two answering blasts or dissenting by sounding five short blasts, taking way off and proceeding only after passing signals are satisfactorily exchanged. This is the procedure approved by the Supreme Court.[15] However, when adopting the first action it should be borne in mind that while a stand-on vessel is not in fault, according to the ciruit court of appeals, for holding her course even though she fails to receive a response to her first signal of one whistle,[16] if she assents to the crossing of her bows by the other vessel she waives her privilege absolutely.[17] When she assents to this by repeating the two whistles, it is her duty at once to assist the maneuver, a point which the United States Supreme Court, on appeal, declined to review.[18] However, this point is now incorporated in Inland Rule 34(a)(ii), whereby a vessel if in agreement with any proposal in the crossing, and indeed meeting, situation "shall . . . take the steps necessary to effect a safe passing."

Finally, a vessel that assents by signal that another shall cross her bows cannot urge the attempted maneuver as a fault, though it results in a collision.[19] These decisions make it evident that even an undisputed acceptance of an irregular proposal carries with it certain risks and that the second procedure, compelling both vessels to stop, often has decided merit.

Narrow Channels

As for the overtaking and meeting situations, the crossing situation can be modified by the rules for narrow channels and fairways. Both international regulations and inland rules have common Rules 9(b) and (c):

[15]The *Transfer No. 15* (CCA NY 1906) 145 F 503; Yamashita Kisen Kabushiki (CCA Cal 1927) 20 F (2d) 25; the *Norfolk* (Md 1924) 297 F 251; Postal SS. Corp. v. *El Isleo* (1940) 84 L Ed. 335.

[16]The *E. H. Coffin*, Fed. Case No. 4,310.

[17]The *Sammie* (NY 1889) 37 F 907; the *Albatross* (1910) 184 F 363.

[18]The *Boston* (NY 1922) 258 U.S. 622, 66 L Ed. 796.

[19]The *Arthur M. Palmer* (NY 1902) 115 F 417.

(b) A vessel of less than 20 meters in length or a sailing vessel shall not impede the passage of a vessel that can safely navigate only within a narrow channel or fairway.

(c) A vessel engaged in fishing shall not impede the passage of any other vessel navigating within a narrow channel or fairway.

Thus, a power-driven vessel navigating in such channels can reasonably expect those engaged in fishing, under sail, or less than 20 meters in length to avoid creating a situation that places her in a give-way role. Specifically, for a vessel that wishes to cross a narrow channel or fairway, Rule 9(d) of both jurisdictions goes on to forbid any such attempt if it will impede the passage of vessels, confined for navigational reasons, to such channels and fairways. This rule has wide application in harbor channels and river areas. It recognizes the problems experienced by mariners in narrow channels—currents, congestion, restricted maneuverability, squat, interaction, as well as time and tidal constraints. However, the words "avoid impeding" do not have the same unequivocal meaning as do those of "give-way." When, because of unseen circumstances, a crossing situation occurs in a narrow channel, who then has right-of-way: the power-driven vessel confined to the channel? or the power-driven vessel (not fishing and over 20 meters in length, of course) on the former's starboard bow? Guidance proposed by IMCO[20] is that a vessel required not to impede the passage of another shall, as far as practicable, navigate in such a way as to avoid the development of risk of collision. If, however, as postulated above, a situation develops so as to involve risk of collision, the relevant steering and sailing rules are to be complied with. For the crossing situation this means, depending on location, International Regulation 15 or Inland Rule 15(a) and, in our scenario, the vessel confined to the channel is the give-way vessel. Apart from possible minor adjustment of course to starboard, the only option open to her is to slow, stop, or reverse engines, all maneuvers not always navigationally prudent in narrow channels. If these are likely to be inadequate, or if there is doubt to the movements of a crossing vessel, then the vessel proceeding along the channel *may*, under the international regime, sound the in-doubt signal or, in inland waters, *must* sound it. This signal should help to clarify the crossing vessels intentions and alert her to the anxiety of the other. It is important for the crossing vessel, who is well able to navigate outside the channel, to ease the anxiety on the bridge of the vessel confined to the channel. Early exchange of signals and/or unambiguous alterations of course and speed are essential in order to avoid forcing risky give-way

[20]IMCO, Sub-Committee of Safety of Navigation—25th Session, Annex 3 of NAV XXV/4 dated 10 Dec 80.

action on the part of the vessel in the channel. However, it must be reiterated, that both International and Inland Rule 9(d) do not automatically give the right-of-way to larger vessels over smaller craft in narrow channels should the latter fail to avoid impeding the former, who may therefore have to yield to the inconsiderate behavior of a vessel crossing on her starboard bow.

In many ports of the world, local by-laws, but not the international regulations, clearly give right-of-way to vessels in the channel and relegate crossing vessels, such as ferries, to a permanent give-way status. Similarly, in inland waters, the normal crossing situation of Inland Rule 15(a) is abrogated, and the precautionary nature of Inland Rule 9(d) made mandatory, by Inland Rule 15(b):

> Notwithstanding paragraph (a), on the Great Lakes, Western Rivers, or water specified by the Secretary, a vessel crossing a river shall keep out of the way of a power-driven vessel ascending or descending the river.

The meaning of the rule is unambiguous. Unlike Rule 9(d), it is not limited to narrow channels or fairways, nor does it depend on the maneuverability of the ascending or descending vessel. However, clear delineation of when a river ceases to be a river would seem essential to avoid uncertainty of application at river mouths.

Adequate and Up-To-Date Charts Must Be Carried

Not only must delineation of rivers, as well as narrow channels and fairways, by unambiguous, but this must be known to ships. In a collision off Nagoya, Japan, the British motorship *Troll River*, outbound for New Orleans, was struck by the inbound Israeli motorvessel *Shavit*. It transpired that the *Shavit* was misled about the extent of the fairway by having an out-of-date chart and formed an erroneous view of what the other ship was likely to do, thinking that the *Troll River* was a give-way vessel in a crossing situation who failed to alter course to starboard sufficiently early. The *Shavit* was found alone to blame.[21]

In 1979 the Greek cargo ship *Aeolian Sky* was involved in a collision with the West German cargo ship *Anna Knuppel* off the Channel Isles. The *Aeolian Sky* subsequently sank after being holed in the port bow, and although no case has come to court, speculation has been made that she was not carrying the latest edition of the chart depicting the traffic separation scheme because her track had followed an obsolete scheme.[22]

In the summer of 1979, an Anglo-French survey of ships' outfits revealed a significant minority (15 percent) held charts displaying out-of-

[21]The *Troll River*, Q.B.(Adm. Ct.) [1974] 2 Lloyd's Rep. 181.

[22]*Seaways*, April 1980 p. 13.

date information.[23] Hopefully, the 1974 SOLAS convention, which came into force in May 1980 and requires ships to carry the latest editions of charts, has rectified this situation. Only time will tell.

Traffic Separation Schemes

Even with up-to-date charts there remain instances of ships proceeding contrary to the direction of traffic flow laid down for traffic separation schemes. Where collisions have occurred, the courts have been consistent in finding that, despite the rogue vessel's contravention of International Regulation 10(b)(i), the other rules of the collision regulations applied in all respects. Even though the *Larry L* was navigating in accordance with the traffic separation scheme, and the other vessel, the *Genimar*, was not, both vessels were found at fault for not taking appropriate action in a crossing situation, with the *Larry L* attracting two-thirds of the blame for failing to give way.[24]

Inland Rule 10 is somewhat different from its international equivalent. It is considerably less detailed and leaves specifics to be promulgated by regulation. The inland rules' parallel to traffic separation schemes (TSS) are vessel traffic services (VTS).

When Crossing Rule Begins to Apply

The navigator is sometimes puzzled to know just how close a vessel approaching from starboard must be to make the law of crossing apply. It will be noted that the same crossing rule is applicable to both inland and outside waters *when vessels are crossing so as to involve risk of collision.* In another part of the rules we are told that risk of collision can, when circumstances permit, be ascertained by carefully watching the compass bearing of an approaching vessel, and that if the bearing does not appreciably change such risk should be deemed to exist. This should not be taken to imply that obtaining a series of bearings of an approaching vessel is an absolute test of risk of collision. It is undoubtedly a valid test to the extent that if the bearings observed are constant, collision will occur if neither vessel changes course or speed, and consequently, vessels have been held at fault for failure to take such precautionary bearings.[25] But it does not follow that in all cases where the bearing is changing, no risk of collision is involved. As said by the circuit court of appeals:

> This section is not a rule of navigation, but merely a suggestion of one circumstance which denotes that there is danger of collision; and a steamer is not

[22] *Seaways*, April 1980 p. 13.
[23] *Seaways*, April 1980 p. 26.
[24] The *Genimar*, Q.B.(Adm. Ct.) [1977] 2 Lloyd's Rep. 25; *see also* The *Estrella*, Q.B.(Adm. Ct.) [1977] 1 Lloyd's Rep. 534.
[25] The *President Lincoln* (1911) 12 Asp MC 41.

> justified in assuming that there is no risk because there is an appreciable change in the compass bearing of the lights of a sailing vessel seen at night, which would manifestly be an unwarranted assumption under some circumstances.[26]

Thus, in the crossing situation, the bearings of a vessel on the starboard bow may be constant, indicating that the two vessels will reach the point of intersection at the same time; they may be drawing ahead, indicating the stand-on vessel will reach the point of intersection first; or they may be drawing aft, indicating the give-way vessel will reach that point first. The rapidity of change in bearing depends on the distance apart and the relative speeds of the two vessels. It would be a mistake to assume that only in the first case is there risk of collision. It is true that if the bearings of the stand-on vessel draw ahead with a certain degree of rapidity, there is a presumption that she will have clearance across the bow of the burdened vessel; but this does not relieve the latter of her obligation to watch the stand-on vessel closely, and to slow down or stop or reverse, if necessary, before coming into dangerous proximity. Conversely, if the bearings of the stand-on vessel draw aft with sufficient rapidity, there is a presumption that the burdened vessel might cross with safety; but wide indeed must be the margin in that case before she is legally justified in making the attempt. As a matter of law, it must be wide enough so that no risk of collision is involved; as a matter of practice and of common prudence, it should be wide enough so that no collision can occur no matter what the other vessel does.

This is a sweeping statement, but its validity is established by court decisions in both the United States and in England. As early as 1869 the Supreme Court held:

> Rules of navigation such as have been mentioned (as to the duties of two vessels approaching each other) are obligatory upon such vessels when approaching each other from the time the necessity for precaution begins; and they continue to be applicable as the vessels advance so long as the means and opportunity to avoid the danger remain. They do not apply to a vessel required to keep her course after the approach is so near that collision is inevitable, and are equally inapplicable to vessels of every description while they are yet so distant from each other that measures of precaution have not become necessary to avoid collision.[27]

This decision was referred to by the district court of Michigan in a Great Lakes case a short time later, and amplified in the following unmistakable language:

> Risk of collision begins the very moment when the two vessels have approached so near each other and upon such courses that by departure from the rules of

[26]Wilders SS. Co. v. Low (Hawaii 1911) 112 F 161.
[27]The *Winona* (1873) 19 Wall. 41.

> navigation, whether from want of good seamanship, accident, mistake, misapprehension of signals, or otherwise, a collision might be brought about. It is true that prima facie each man has a right to assume that the other will obey the law. But this does not justify either in shutting his eyes to what the other may actually do or in omitting to do what he can to avoid an accident made imminent by the acts of the other. I say the right above spoken of is prima facie merely, because it is well known that departure from the law not only may, but does, take place, and often. Risk of collision may be said to begin the moment the two vessels have approached each other so near that a collision might be brought about by any such departure and continues up to the moment when they have so far progressed that no such result can ensue. But independently of this, the idea that there was no risk of collision is fully exploded by the fact that there was a collision.[28]

Similarly, in the case of the *Philadelphia*, the district court said:

> The term "risk of collision" has a different meaning from the phrase "immediate danger" and means "chance," "peril," "hazard," or "danger of collision"; and there is risk of collision whenever it is not clearly safe to go on.[29]

Two statements from decisions of Dr. Lushington, famous admiralty jurist of the mid-nineteenth century, will suffice to show the English parallel of this doctrine. In a case of 1851 he said:

> This chance of collision is not to be scanned by a point or two. We have held over and over again that if there be a reasonable chance of collision it is quite sufficient.

In another case he said:

> The whole evidence shows that it was the duty of the *Colonia* . . . to have made certain of avoiding the *Susan*. She did not do so, but kept her course till she was at so short a distance of a cable and a half's length [1,000 feet] in the hope the vessels might pass each other. Now it can never be allowed to a vessel to enter into nice calculations of this kind, which must be attended with some risk, whilst it has the power to adopt, long before the collision, measures which would render it impossible.[30]

It was such an interpretation of the term *risk of collision* that caused the circuit court of appeals to reverse the lower court in a New York Harbor collision between the tug *Ashley* and the tug *Volunteer*. The district court absolved the *Volunteer* on the theory that the starboard hand rule did not apply to her because it appeared that there was time for her to get across before the courses would intersect; but the circuit court of appeals, in reversing the decision said:

> The starboard hand rule is intended to avoid just such speculations. When the courses as being steered are crossing courses they involve risk of collision and

[28]The *Milwaukee* (1871) Fed. Case No. 9,626.
[29]The *Philadelphia* (Pa 1912) 199 F 299.
[30]Marsden's *Collisions at Sea*, 11th ed., p. 462.

> the burdened vessel is required to keep out of the way and the privileged vessel to hold her course and speed. The account given by the master of the *Volunteer* brings the situation precisely within this article.[31]

Assumptions about another vessel's intentions are equally no cause for departing from the crossing rules. Two vessels were approaching a junction of three channels in the Thames Estuary. The inward bound *Homer*, in one channel, mistakenly formed the impression that the outward bound *Elisa F*, in another channel then on her port bow, was navigating to pass into the third channel. However, on reaching the junction, the *Elisa F* turned to starboard to enter the *Homer*'s channel, sounding one short blast as she did so. The *Homer* never heard the blast and altered course early to port, without signal, to enter the *Elisa F*'s former channel. Belatedly, VHF messages were exchanged, mutually conflicting orders were passed to helmsmen, and the two vessels swung together with resultant collision. Apart from bad aural and visual lookout and failure to sound two short blasts, the *Homer* was at fault because;

> the situation was one of crossing vessels and *Homer* should have kept her course and speed.[32]

The *Elisa F* was not at fault and the *Homer* was alone to blame. An appeal by the *Homer* that *Elisa F*, in view of the lower court's finding, should have altered earlier to starboard was dismissed.

Earlier in this chapter, the case of the *Sestriere* was discussed where a crossing situation was found not to exist because the other ship, the *Alonso*, never settled on a clearly defined course.[33] Clearly then, this must be a test as to whether a crossing situation is applicable, as in the case arising from the collision between the Dutch tanker *Forest Hill* and the Greek tanker *Savina* in the roadstead of Ras Tanura. The *Forest Hill* was proceeding from an anchorage in the southern part of the roadstead on a northerly course. The *Savina* was proceeding to an anchorage in the northern part of the roadstead on an easterly course. The stem of the *Savina* struck the port side aft on the *Forest Hill*.

The *Forest Hill* had altered course to starboard as she got underway, eventually some eight minutes before the collision, steadying on a course of 350 degrees. From this time, the court held that a crossing situation existed but not at any earlier time:

> because, until C − 8, *Forest Hill* was not on a definite course at all.

Both ships were castigated for keeping a poor lookout; the *Savina* for

[31]The *Ashley* (NY 1915) 221 F 423.
[32]The *Homer*, Q.B.(Adm. Ct.) [1972] 1 Lloyd's Rep. 429; C.A. [1973] 1 Lloyd's Rep. 501.
[33]The *Sestriere*, Q.B.(Adm. Ct.) [1976] 1 Lloyd's Rep. 125.

failing to stop as give-way vessel and the *Forest Hill* for increasing speed when stand-on vessel. Blame was apportioned 40 percent to *Savina* and 60 percent to *Forest Hill*, the latter attracting major blame because her increase of speed was calculated to force drastic avoiding action by the *Savina*. This interpretation of the *Forest Hill*'s increase of speed was not accepted by the court of appeal, which reversed the apportionment of blame, as the *Savina*

> having failed to take the necessary avoiding action was the more to blame of the two.

However, on further appeal, the House of Lords decided that there was evidence that entitled the admiralty court to hold that the *Forest Hill* did intend to force her way across the head of the *Savina* and that therefore the original apportionment would be respected.[34] Among these vicissitudes of apportionment, the salient point for the mariner to hoist in is that all three courts upheld that the crossing situation came into effect when the *Forest Hill* steadied on her course of 350 degrees at collison minus eight minutes (C − 8).

No special rights accrue to a vessel lying stopped on the high seas, unless she is in one of the categories of Rule 27. She must keep out of the way of a vessel, with whom risk of collision exists, which approaches from her starboard bow. However, a recent decision of the Scottish Court of Session[35] caused a flurry of alarm about this principle. Contrary to an earlier decision by the English Admiralty Court (The *Broomfield* 1905) and of courts in the United States, the Scottish court held that a power-driven vessel that was underway but lying stopped, was not a give-way vessel in relation to another power-driven vessel approaching from her starboard side. The collision arose between two fishing vessels—neither at the time actually engaged in fishing—one of whom, the *Mayflower*, was sunk. The case was, regrettably, not the subject of appeal. However, international representations were made to IMCO, who provided clarification[36] stating that the word "underway" in Rule 3(i) also applied to a vessel that was lying stopped in the water. Hopefully this will take care of the aberration of the Scottish court decision.

Rule 15, the crossing situation on the high seas, is abrogated if one of the two vessels is not under command, restricted in her ability to maneuver, or engaged in fishing. Rule 18(a) applies in such circumstances and a vessel that normally would be stand-on in a crossing situation is required to keep out of the way. Passing ahead is not ruled out, though an altera-

[34]The *Savina*, Q.B.(Adm. Ct.) [1974] 2 Lloyd's Rep. 323; C.A. [1975] 2 Lloyd's Rep. 148; H.L. [1976] 2 Lloyd's Rep. 132.
[35]The *Devotion II*, Sc. Ct. [1979] 1 Lloyd's Rep. 509.
[36]IMCO, Sub-Committee on Safety of Navigation, Annex 3, Nav XXV/4 dated 10 Dec 80.

tion to starboard may prove most prudent. A tug towing is not automatically conferred any rights, unless she shows the shapes or lights authorized by Rule 27(b), but it behooves a stand-on vessel to take into account the hampered movements of a tug and tow when contemplating action under Rules 17(a)(ii) or 17(b).

Under International Rule 18(d), all vessels are enjoined to avoid impeding the safe passage of a vessel constrained by her draft. As mentioned earlier in this chapter, the expression to "avoid impeding" is different from "to keep out of the way," being far less imperative. If a power-driven vessel is unable to avoid impeding the passage of a vessel constrained by her draft, then the rules for the crossing situation, or indeed the meeting or overtaking situations, apply.

There is no equivalent rule in inland waters, as the inland rules do not provide for the category of vessels constrained by their draft. This was a deliberate exclusion for inland waters, as it was felt that such a subjective rule might lead to abuses and result in situations where a vessel constrained by her draft might claim a right-of-way to which she was not entitled, thereby creating a dangerous situation. There are indications that some ships on the high seas do indeed display the appropriate lights or shapes for a vessel constrained by her draft when the circumstances do not justify such use.

Crossing Rules Equally Binding on Both Vessels

Perhaps in the very nature of the case common sense would permit no other interpretation. For in the final analysis, it is as logical to place an absolute obligation upon one vessel not to cross ahead as upon the other vessel to maintain course and speed. And when a collision occurs because a give-way vessel that thought she had time to get across is hit by a stand-on vessel that sheered to the right in an ill-timed attempt to clear her, the courts cannot consistently find that one was under compulsion to hold course and speed if they do not find that the other was equally under compulsion to avoid crossing her bow.

If we exclude all those cases that frequently arise in crowded harbors where the presence of a third, or even other additional vessels, creates special circumstances that modify the rules, then we may draw a very practical lesson from the foregoing decisions—namely, the manifest danger of crossing a vessel to starboard unless she is so far away that it would be impossible for her to bring about a collision. If she is that far away, the navigator need not worry, for there is no risk of collision.

SUMMARY

In accordance with the theory of stand-on and give-way the crossing situation requires, both in inland waters and on the high seas, that the vessel having the other to port maintain course and speed, and that the vessel having the other to starboard keep out of the way, avoid crossing ahead, and if necessary slacken speed or stop or reverse. This arrangement creates a serious hazard when, as frequently happens, either vessel fails to do her duty.

On the high seas and in inland waters, the stand-on vessel is required to maneuver when *in extremis*. Under both jurisdictions, the stand-on vessel *may* maneuver earlier when it becomes apparent to her that the give-way vessel is not taking appropriate action.

The give-way vessel under international regulations must signal if she alters course or reverses engines, and the same applies *in extremis* and in the exercise of Rule 17(a)(ii) to the stand-on vessel. In inland waters the use of one or two short blasts is a declaration of intent that requires an immediate response from the other vessel. If doubt or disagreement arises, the danger signal must be sounded and both vessels must take appropriate precautionary measures until agreement is reached. Should engines be reversed, then the signal is the same as the international one of three short blasts. The signal of two short blasts, except in specified waters where the vessel in the river has right-of-way over a crossing vessel, should never be proposed by the stand-on vessel, and should be accepted when proposed by the give-way vessel only when the maneuver indicated can be done with a high degree of safety. The effect of such assent is to take the right-of-way from the stand-on vessel without conferring it on the give-way vessel, and thus to put vessels under special circumstances, with the mutual duty of taking any positive action to avoid collision.

17

The Law in Fog

Restricted Visibility

Despite the concentration of law relating directly to the navigation of vessels in restricted visibility in both the inland rules and international regulations, and the remarkable technological advances in the ability to detect and track vessels by radar, fog remains a major factor in many marine collisions and a prolific source of litigation arising out of those collisions.

Prior to the advent of radar, the regulations depended entirely upon the fog signal: vessels in fog sounded fog signals and when other vessels heard them, they stopped. The widespread use of radar has meant that this approach has had to be supplemented, for the means of navigation in fog has improved substantially. Firstly, in 1960, a radar annex was appended to the international regulations and, in 1972, incorporated into the body of the rules—a move followed by equivalent action for U.S. inland waters in 1980. Indicative of the changes was the discarding of the word "fog" in official phraseology and its replacement by the term "restricted visibility." Fog, while never defined in either the rules or the courts, has a meteorological association with visibility of less than half a mile. With today's high-speed shipping this is too short a range to invoke "fog rules," whereas restricted visibility is a sufficiently elastic phrase to encompass varying states of visibility and varying circumstances that require action at greater ranges.

In the meantime, considerable legal comment has been passed on those collision cases reaching the courts, particularly on the use of radar as an anticollision device. Despite this focus by the courts, it must be emphasized that radar is but one element in the solution to the problem of avoiding collision when visibility is impaired. It does not, for example, supplant the need for fog signals.

Uniformity of Fog Signals

With the entry into force of the Inland Navigational Rules Act of 1980, the differences between sound signals to be used in restricted visibility on the high seas and those for use in inland waters have been largely eradicated. The international reference to a sound signal for "vessels constrained by their draught," however, has been deleted from the inland rules in keeping with the U.S. decision not to legislate for this category of vessel. Also the inland rules allow a waiver for small vessels anchored in designated special anchorage areas. Apart from these two aspects, the only remaining difference is for vessels restricted in their ability to maneuver or vessels engaged in fishing *while at anchor.* Inland Rule 35(c) provides for such vessels to sound the same signal that they would have done if underway. The international regulations are silent on this point but are expected to come into line on the occasion of their first amendment.

When Fog Signals Are Required

Exactly when the weather is thick enough to require fog signals to be sounded can be difficult to determine. In the absence of positive court authority, previous editions of this book have cited the very practical suggestion of La Boyteaux that the prescribed minimum visibility of sidelights (then 2 miles) indicated that notice of approach should be given at least at that distance and that therefore when proper lights cannot be seen by vessels 2 miles apart, fog signals were in order.[1] Now that the required visibility of sidelights for the larger vessels is 3 miles, this guideline is perhaps less relevant, particularly as the audibility range of the whistle on even the largest of ships is, theoretically, only 2 miles,[2] but the underlying rationale remains valid. In particular, the implication that signals should be sounded not only when actually in fog, but when steaming toward, away from, or near, a fog bank is important.

Both inland rules and international regulations specifically cover this point by requiring fog signals when navigating not only in but near an area of restricted visibility. How close one has to be is undefined, but there would seem little merit in sounding signals if the visibility is greater than the audibility range of the appliance concerned. However, care must be taken against using the audibility ranges quoted in the annexes to the rules as an infallible yardstick, for the equipment might be heard at greater distances. Moreover, if there is uncertainty as to the true extent of visibility, the prudent mariner should err on the side of caution and make the appropriate signal at a visibility limit greater than the distance his particular appliance can be heard. In *The Elwick,* failure to sound fog

[1]La Boyteaux, *Rules of the Road at Sea,* p. 67.
[2]International Regulations, Annex III, 1(c).

signals in visibility between one-half to three-quarters of a mile was held to be a fault.[3] However, it would be misleading to try and set any precise limit, for each case will undoubtedly be judged on the circumstances prevailing at the time.

The exigencies of war are no excuse in the eyes of the law for failure to sound fog signals, even when dispensation has been given by duly authorized naval orders.[4] While these decisions do not mean that in time of hostilities vessels have to go about giving legal due notice of their presence to an enemy, it does mean that damage that results from any errors of omission or commission must be paid for out of the public treasury. However, the facts of modern naval warfare make it unlikely that there is anything to be gained by not sounding fog signals.

Time Interval Between Signals

Under both sets of rules, vessels underway make the appropriate sound signal on their whistle at intervals not exceeding two minutes, whereas bell and gong signals of vessels not underway have an interval of not more than one minute.

The two-minute maximum interval for whistles was standardized because temporary deafness can be caused by too frequent a sounding, and there must be sufficient listening time between one's own signals to be able to hear those of an approaching vessel. This does not mean that signals should not be made at lesser intervals than those specified, particularly when other vessels are close and bearing in mind they may not have operational radar. Indeed there are occasions where good seamanship requires signals to be given much oftener. Vessels at anchor have been found at fault for not striking their bell more frequently than the specified interval,[5] while a vessel turning into an occupied anchorage was equally at fault for not increasing the frequency of her whistle to warn vessels already at anchor.[6] And so when vessels are feeling their way past each other in dense fog, there can be little doubt that legal obligation, as well as good seamanship, may require signals to be at greatly reduced intervals until both vessels are past and clear. However, it should be borne in mind that in a crowded harbor area the continuous sounding of fog signals could tend to confuse rather than add to safety.

Two Prolonged Blasts

This signal, applicable both on the high seas and in inland waters,

[3] *The Elwick* (1923), Shipping Law, 131.
[4] Watts v. U.S. (NY 1903) 123 F 105; Thurlow v. U.S. (1924) 295 F 905.
[5] *The Chancellor* (1870) Fed. Cas. No. 2,589; Brush v. the *Plainfield* (1879) Fed. Cas. No 2,058.
[6] The *Quevilly* (1918) 253 F 415.

conveys valuable information to a vessel encountering another in fog. It signifies the presence of a power-driven vessel underway but who is stopped in the water, and its use is only lawful when those conditions are met. Great care must be exercised, particularly at night, to ascertain that the vessel is dead in the water before changing from one blast to two. The courts have been quick to find a vessel guilty when wrongfully using this signal:

> You are not to blow this signal until you are stopped and you must be quite certain that you really are.[7]

Other vessels are entitled to assume that the two-blast signal will only be given when the signaling vessel is stopped.[8] However, that is all they may assume, for:

> The signal of two long [*sic*] blasts . . . advertised only the fact that the other vessel was at that moment stopped in the water. It contained no guarantee that the other vessel would remain stopped in the water. . . .[9]

When a vessel starts making way again after having stopped, she should change her signal to one prolonged blast "immediately she begins to gather headway."[10] However, this is not necessarily applicable when a vessel uses her rudder and engines merely to prevent her bow from falling off.[11]

Hampered Vessels

Vessels that are in some way constrained in their activities are provided with special sound signals in Rule 35(c) and (d). The minor differences between international regulations and inland rules for these vessels have already been mentioned. While the rules for vessels in sight of each other give a pecking order for give-way status among hampered vessels, there is no such explicit status in restricted visibility. Despite the provision of unique signals for hampered vessels, Rule 19—the conduct of vessels in restricted visibility—affords them no specific rights. Strictly, they must behave themselves the same as any other vessel, but clearly the distinctive signals for them have the obvious purpose of causing ordinary vessels to approach them with greater caution and to give them as wide a berth as circumstances permit.

[7]The *Lifland* (1934) 49 Ll. L. Rep. 285; *see also* the *Ansaldo Savoia* (1921) 276 F 719; the *Haliotis* (1932) 44 Ll. L. Rep. 288; the *British Confidence* [1951] 1 Lloyd's Rep. 447; the *Almizar* [1971] 2 Lloyd's Rep. 290.

[8]The *Matiana* (1908) 25 T.L.R. 51; the *Kaiser Wilhelm II* (1915) 85 L.J.P. 26; the *Marcel* (1920) 2 Ll. L. Rep. 52.

[9]The *Cornelis B* [1956] 2 Lloyd's Rep. 540.

[10]The *Dimitrios Chardris* (1944) 77 Ll. L. Rep. 489.

[11]The *Canada* (1939) 63 Ll. L. Rep. 112.

Sailing vessels underway now give the same signal as hampered vessels. With the revised requirement for sound signal appliances it was considered that the previous one-, two-, or three-short-blast signals might be confused with maneuvering signals and were, anyway, of limited value to other vessels.

In-Doubt/Danger Signal in Restricted Visibility

The international in-doubt signal of five or more short blasts is restricted to use when vessels are in sight of each other. The inland danger signal, now Inland Rule 34(d), is equally confined to occasions when vessels are in sight. This used not to be the case, for the danger signal was held to be required when vessels did not understand the course or intention of an approaching vessel in fog,[12] or when a vessel was aground in fog.[13] It has been mooted that the danger signal, while in no way a substitute for fog signals, could be used in restricted visibility by a vessel detecting an immediate situation on radar or by other means that could result in a collision. In such a case, it is argued, the responsibility requirement of Rule 2 would permit the vessel to sound the danger signal if it thought such a signal would help to avoid immediate danger. However, this suggested, permissive use of the danger signal is very different from the positive requirement just cited in the cases above.

Hopefully, this sort of aberration is now obsolescent in the new-found near-uniformity between international and inland fog signals. Certainly it should prevent repetition of the unfortunate circumstances where, in a collision in Juan de Fuca Strait, both navigators acted under the conviction that they were in inland waters and therefore subject to a different regime, whereas the court concluded that they were 100 yards into the international regime.[14]

Some Rules Are Applicable to Any Condition of Visibility

The general requirements of both the inland rules and the international regulations as to lookout (Rule 5), safe speed (Rule 6), risk of collision (Rule 7), action to avoid collision (Rule 8), narrow channels (Rule 9), and vessel traffic services/traffic separation schemes (Rule 10) apply equally to vessels in restricted visibility. These rules, common to whether a vessel is in sight or not, are important in understanding what is required of the mariner should he find himself approaching a close-quarters situation with another ship regardless of the extent of visibility. The great increase in maritime traffic density and in the size and speed of ships, with

[12]The *Virginia* (1916) 238 F 156; the *Celtic Monarch* (1910) 175 F 1006.
[13]The *Leviathan* (1922) 286 F 745.
[14]Border Line Transportation Co. v. Canadian Pacific Ry. Co. (Wash 1919) 262 F 989.

a consequent decrease in the time available to determine how the rules should be applied as vessels approach each other, has led, as previously mentioned, to the need to take action at greater ranges than hitherto. The rules of Part I of both jurisdictions therefore require careful consideration, for they form the bedrock of precautions to take in all kinds of visibility and, with reference to this chapter, the necessary preliminary actions to be taken before nearing restricted visibility. A ship who conscientiously applies the rules of Part I is unlikely to be caught short by the sudden advent of poorer visibility. As in all well-run ships, the time to take precautions, whether they be for seamanship, navigation, or rules of the road reasons, is when there is still time to complete those precautions before the danger arises. Failure to do so, with particular reference to the latest rules of the road, are:

> regarded as serious faults because they are breaches of the regulations committed at a time when there was or should have been plenty of time to consider carefully the correct course of action to be taken.[15]

The subject of maintaining a lookout will be discussed in chapter 20, though many relevant facets of the use of radar as a lookout will be covered later in this chapter, as well as their impact on safe speed. Suffice it to say for now that restricted visibility imposes a greater responsibility on mariners in use all means available in maintaining a proper lookout, not least so that a continuous appraisal can be made of what is a safe speed for the circumstances and conditions prevailing.

Safe Speed Not the Same as Moderate Speed

Previous international and inland rules required a vessel to "go at a moderate speed" in restricted visibility, but there were no strictures on speed in clearer weather. With the advent of very large vessels that carry their way for some considerable time after engines are stopped, thus traveling well over a mile before coming to rest, there is a need to relate speed to all conditions and circumstances prevailing and not just when fog *per se* was present. Thus "safe speed" replaced "moderate speed," with an extended applicability to all conditions of visibility. It is, of course, in restricted visibility that the need for a safe speed particularly applies. Hence, the state of visibility is first in the list of factors that Rule 6 states shall be taken into account in determining safe speed.

Clearly the very term *safe speed* does not preclude the setting of a high speed in appropriate circumstances. Furthermore, even if a collision should occur, it does not necessarily follow that a ship was proceeding at an unsafe speed. Poor lookout or incorrect avoiding action might well be

[15]"The *Eleni V* v. *Roseline*," *Seaways*, August 1981, p. 6.

to blame, rather than a high initial speed. However, for a high speed to be accepted as safe, it must be shown that circumstances and conditions were continuously monitored and speed was adjusted when fresh information came to hand.

> . . . such a speed can only be justified so long as it is safe to proceed and provided that timely action is taken to reduce it or take off all the way in the light of the information supplied. . . .[16]

Too high a speed might render a ship unable to take such "timely action" because of insufficient time to do so after sighting or detecting a vessel at relatively short range. This has often been the case in collisions occurring in fog. More rarely, on the other hand, too low a speed might be unsafe if the ship loses steerageway.

> . . . it was the duty of those on the bridge of the *Ring* to appreciate that they had lost steerage-way and were going off course and it was their duty to correct it by appropriate engine and helm movement.[17]

Excepting this unusual case of stopping engines in the path of a ship following close astern, a prime requirement of safe speed is that a ship can "be stopped within a distance appropriate to the prevailing circumstances and conditions." This might seem to be a mere rewording of the former rules for restricted visibility specifying "a moderate speed, having careful regard to the existing circumstances and conditions." But it is not: in part because safe speed applies to all conditions of visibility, but more crucially because the concept of safe speed is different from the more restrictive term moderate speed. What constituted moderate speed has had a wide variety of construction by the courts, all of which may be characterized as much less liberal than the interpretation put on the term in practice by the most careful of navigators. The original rules' reference to existing circumstances may have intended to leave some discretion on this point to the mariner; but certainly very little discretion has been left by decades of court decisions. Perhaps the arrival of safe speed will restore some of that discretion.

Early United States Supreme Court decisions established what may perhaps be termed the general rules of what was moderate speed. Firstly, moderate speed was defined as bare steerageway,[18] followed a few years later by a second definition that implied the right to navigate a vessel only at such speed as would enable her to stop in half the distance of visibility then existing.[19] It is apparent that the two definitions were at times

[16]The *Kurt Alt* [1972] 1 Lloyd's Rep. 31.
[17]The *Ring* [1964] 2 Lloyd's Rep. 177.
[18]The *Martello* (1894) 153 U.S. 64.
[19]The *Umbria* (1897) 166 U.S. 404.

contradictory; that is, there could be conditions when to maintain even steerageway, a speed would be necessary that would make it impossible to stop in time after first sighting an approaching vessel, or even an anchored vessel. In inland waters in such cases, in order to obey the law, it was therefore necessary sometimes to come to anchor[20] or to delay leaving a dock.[21] These are sensible measures when they can be taken, but are of no value to a ship in deep water who cannot anchor and who would have to drift until visibility improved, an arguably greater danger in crowded or close waters, particularly for large ships with noxious cargoes capable of causing widespread pollution.

The acceptance of the moderate speed rule into the half-distance rule spread world-wide,[22] but received its first check in British courts where it was held that it was not a rule of law[23] and that each case must be judged in the context of the original wording of "existing circumstances and conditions." Several United States courts have espoused a similar approach,[24] with the Supreme Court ruling in one case that

> Implicit in that portion . . . of the Inland Rules of Navigation that directs a moderate speed for vessels proceeding in foggy weather, and in the concomitant half-distance rule, is the assumption that vessels can reasonably be expected to be travelling on intersecting courses. On the facts of this case, it was totally unrealistic to anticipate the possibility that the vessels were on intersecting courses and the rule was not applicable.[25]

The Supreme Court decision did not do away with the half-distance rule, merely modifying its applicability to vessels in the vicinity of a fog bank.

The moderate speed requirement and its associated half-distance rule ignored the realities of today's shipping industry. Ships must often of necessity navigate in reduced visibility and by so doing they undoubtedly proceeded at times at "immoderate speeds." This is not necessarily reckless driving on the part of masters, but is caused by financial realities, which dictate a need to *reasonably* maintain schedules. Rigid application of the half-distance rule took no account of this or of the greatly increased capability of ships to navigate safely in restricted visibility. The rule was developed in an earlier day when speeding blindly through fog was sheer folly. Today, the rule might be appropriate for a vessel without radar, but a ship that is making proper use of radar in, say, the open ocean cannot be

[20]The *Southway* (1924) 2 F(2d) 1009; the *Lambs* (1926) 17 F(2d) 1010, affirming 14 F(2d) 444.

[21]The *Georgia* (1913) 208 F 635.

[22]HMS *Glorious* (1933) 44 Ll. L. Rep. 321; Silver Line Ltd v. U.S. (1937) 94 F(2d) 754.

[23]Morris v. Luton Corporation (1946) K.B. 114 (C.A.).

[24]Hess Shipping Corp. v. the SS *Charles Lykes* (1969) 417 F(2d) 346; Polarus S.S. Co. v. T/S *Sandefjord* (1956) 236 F(2d) 270, 272.

[25]Union Oil Co. v the *San Jacinto* (1972) 401 U.S. 145.

realistically expected to take all way off when the fog becomes so dense that it is not possible to see beyond her bows. Further, too strict a compliance with the half-distance rule might lead a ship, although well capable of pulling up in half the visibility, to proceed at an unsafe speed when other factors should have been taken into account. This is, of course, what Rule 6 and Rule 19(b) now demand of mariners by providing a list of factors, not necessarily exhaustive, for their guidance. It would seem appropriate that the actions of those unfortunate enough to be involved in collisions should be considered in the light of those factors rather than on too strict an interpretation of the half-distance rule. Of course, the handling of the vessel under the circumstances must also be taken into account and should be in accord with the dictates of good seamanship and recognized practices of navigation.

Factors to Be Taken into Account

Safe speed is a relative term. It cannot be defined so as to apply to all cases; it depends on the circumstances of each case. The factors in Rule 6 provide a valuable check-off list of points the courts have considered over the years. Visibility tops the list. Earlier in this chapter there was discussion on when visibility was sufficiently restricted to require sound signals to be made. A similar process must be gone through when deciding visibility is such as to affect safe speed. This is without doubt likely to be at a greater limit than that used to determine when to start fog signals. Visibility of less than five miles should cause sufficient concern about safe speed to, at least, place the engine spaces on alert, even though it would not warrant the sounding of signals. The prudent mariner, however, would do well to consider himself in or near an area of restricted visibility for the purpose of Rule 19 when he cannot visually distinguish the type and aspect of a closing vessel within the distance he would normally expect to do so, a distance for many ships that should be measured in miles rather than in a few hundred yards.

When the visibility is so restricted, other important factors will obviously include the size and maneuverability of a ship, particularly her stopping power. The ability to take way off is relevant in considering a ship's speed in fog:

> It would be absurd, to take an extreme case, to suggest that a speed of ten knots in a destroyer would be as excessive as ten knots in an old collier.[26]

However, not too much weight should be given to this factor, for

> . . . it is not enough to say that because a vessel has remarkable pulling-up power she is therefore justified in proceeding at high speed in fog. Regard must be had

[26]The *Munster* (1939) 63 Ll. L. Rep. 165.

to the chance other ships have of receiving her fog signals. If two vessels are approaching each other at very high speed, it must be quite obvious that their chance of hearing each other's fog signals are very much reduced.[27]

Neither will high speed give much opportunity for assessing what action should be taken when a vessel is sighted or first detected at short range.

Traffic in the vicinity will also be an important factor. In the open sea, with little or no traffic around, a relatively high speed may be appropriate provided a proper radar watch is being kept and the engines are ready for immediate maneuver. However, even a vessel with good stopping power using an advanced collision avoidance system would not be justified in proceeding at high speed in dense fog through congested waters or areas where fishing vessels or other small craft are likely to be encountered.

When visibility is restricted, Rule 19(b) requires a vessel to have her engines ready for immediate maneuver as well as to proceed at a safe speed. This applies in open as well as restricted waters. Because it can take some time to round up the necessary personnel and prepare the engines for instant use, forehandedness should be shown in giving the engineers as much warning as possible. For ships with bridge control of engines, there is, of course, less need for advance notice of a speed change. Such a change is envisaged by Rule 19(e), which lays down occasions when speed is to be reduced to steerageway and, if necessary, for all way to be taken off. Clearly, under the circumstances of Rule 19(e), the term "safe speed" has much in common with the old "moderate speed" (but not necessarily with the half-distance rule), and many previous court interpretations of "moderate" will still be relevant in those circumstances.

In restricted visibility a vessel making proper use of radar may often be justified in going at a higher speed than that which would be acceptable for a vessel not so equipped, but not usually at the speed that would have been considered safe for clear weather. A recent example of what one court considered these speeds should be, in the circumstances it was examining, were: 6 to 7 knots for a non-radar-fitted ship in a busy area at night with one mile visibility; and 8 to 9 knots for a radar-fitted ship in the same vicinity at the same time but with her visibility about 1,400 yards and a caveat that further reduction was necessary on seeing a close-quarters situation developing.[28] Clear weather speeds for both ships were 17 and 13½ knots, respectively. While not too much importance should be attached to the specific speeds quoted, it is instructive to note that the speed accorded the radar-fitted ship was, in those circumstances, not

[27]The *Arnold Bratt* [1955] 1 Lloyd's Rep. 16, 24.

[28]The *Hagen* [1973] 1 Lloyd's Rep. 257; *see also* the *Zaglebie Dabrowskie* [1978] 1 Lloyd's Rep. 570, where a proper speed in ¾ to 1½ mile visibility would have been no more than 8 to 9 knots.

much higher than that for the other vessel. Of course, radar is only one factor to be considered, and the existence of an operational set on board requires several further factors to be taken into account when determining safe speed. These will be covered in the discussion on radar later in this chapter.

Excessive Speed

Any speed that is not safe within the judicial construction of the word is excessive; and an examination of reported cases reveals the fact that vessels colliding in fog have, in an extremely wide range of existing circumstances, been convicted of excessive speed from barely making way to 22 knots, the first named being a case of giving a brief kick ahead on the engine to maintain course.[29] On the other hand, also in a wide variety of visibility and traffic conditions, speeds ranging from 3.5 knots to a half speed of 8 to 9 knots have been specifically approved as moderate, the last named being an instance concerning a rogue vessel in the approaches of a traffic separation scheme in visibility between 50 to 200 meters.[30] There would appear to be a slight upward trend, but not by more than a knot or two, in the more recent cases of what is a safe speed where a proper radar watch was kept. However, it must be stressed yet again that such a safe speed was only applicable for the circumstances prevailing and is invariably accompanied by the qualification that further reductions were necessary should visibility further decrease, or if there was detection of a potential close-quarters situation. It is obvious that the use of radar does enable vessels to travel safely at a somewhat higher speed than would be safe without radar—if the circumstances are right—for the installation of radar is not a license to speed if the conditions are not right. Thus, while the courts may approve a slightly higher safe speed for radar-fitted ships than hitherto in less severely restricted visibility, they are likely to continue to be as strict in their former judgments of moderate speed as they are of safe speed in dense fog when other shipping has been detected. As was said in the *Kurt Alt* case, previously cited

> radar . . . while, if properly used and can be relied upon to indicate all potential dangers in ample time to safely avoid them, it may give some justification for a speed in restricted visibility, which would otherwise be immoderate, such a speed can only be justified so long as it is safe so to proceed and provided that timely action is taken to reduce it or take off all the way in the light of the information supplied or to be inferred from radar.[31]

[29]Afran Transport Co. v. the *Bergechief* (1960) 170 F Supp 893 (SDNY), aff'd 274 F(2d) 469.

[30]The *Zaglebie Dabrowskie* [1978] 1 Lloyd's Rep. 570.

[31]The *Kurt Alt* [1972] 1 Lloyd's Rep. 31.

Many have been the excuses offered by mariners for excessive speed in collision cases, and scant indeed has been their judicial consideration. Some of the arguments tried and found wanting were: (1) that full speed was the safest speed, as it enabled the vessel to get through the fog sooner; (2) that the speed was customary for liners; (3) that the vessel was a passenger steamer and obliged to maintain schedule; (4) that the vessel was carrying passengers to whom it was important to get into port; (5) that the vessel was under contract to carry United States mail; (6) that in the opinion of her officers the vessel could not be properly controlled at a lower speed; (7) that the slowest speed of the engines would not drive the vessel at the moderate speed demanded; (8) that the other vessel was more seriously at fault; (9) that the vessel was a ferry; (10) that the regular speed, or at least a speed faster than that allowed, was necessary to keep track of the vessel's position; (11) that a vessel in convoy had to maintain the speed of her naval escort; (12) that an aircraft carrier in peacetime had planes in the air and that to go more slowly meant danger to the lives of those in them. The judges have listened to these excuses and many more, and have then decided the cases on the coldly practical and unanswerable basis as stated by the circuit court of appeals:

> Speed in a fog is always excessive in a vessel that cannot reverse her engines and come to a standstill before she collides with a vessel that she ought to have seen, having regard to fog density.[32]

Pressure to "catch the tide" or otherwise maintain their schedule sometimes makes masters reluctant to come to a safe speed in poor visibility. The owners and operating authorities have a responsibility in ensuring they exert no such pressure on the master. In a recent case, where a ship underway struck another at anchor and the owners sought to limit their liability, it was held that the owners were guilty of actual fault because their ship had proceeded at excessive speed in fog. It came out in evidence that the master had habitually navigated his vessel in fog at excessive speed over several voyages, and that the marine superintendent, to whom the ship's logs were regularly submitted, failed to check the records, which would have revealed this fact, and thus failed to bring it to the master's attention. Consequently, the owners were unable to limit their liability and the master's certificate was suspended. "Excessive speed in fog is a grave breach of duty, and shipowners should use all their influence to prevent it."[33]

Radar—A Statutory Requirement

The vast majority of ships these days are equipped with at least some of

[32] The *John F. Bresnahan* (1933) 62 F(2d) 1077.
[33] The *Lady Gwendolen* [1965] 1 Lloyd's Rep. 335.

the wide range of modern navigational devices that can remove much of the uncertainty as to a vessel's position. Of these aids to navigation, as was said earlier, radar has received the greatest attention of the courts, primarily in its use as an anticollision device. While acknowledging this supremacy of radar it should not be forgotten that the other equipment can also be of value in the anticollision role—either as an important input to radar systems, e.g., gyrocompass and log; as means of information, e.g., VHF radio; or as an accurate positional device when approaching traffic separation schemes, e.g., hyperbolic fixing systems, inertial navigational systems, and, of course, radar itself. Neglect of these ancillary equipments, and indeed the lack of relevant publications, may not cause collision by themselves, but have sometimes been contributory factors to the scenario that led up to collision.

Until recently there was no statutory requirement for vessels to carry radar. In the United States, the Ports and Waterways Safety Act of 1972 provided the first basis for vessels to carry "electronic devices necessary for the use of" a vessel traffic service, with an amendment in 1977 requiring every self-propelled vessel of 1,600 or more gross registered tonnage (grt) operating in U.S. waters to have on board a marine radar system. In the United Kingdom from 1976, vessels of more than 1,600 grt had to carry specified radar equipment, though without any obligation to use such equipment. Long before that, however, the courts had consistently faulted those who failed:

> to take full advantage of any equipment with which a vessel is equipped . . . this equipment [*radar*] is supplied to be used and used intelligently. . . .[34]

Spurred on by pollution disasters, the cause for mandatory carriage of radar was taken to the international forum. Under the auspices of IMCO, the 1978 Protocol to the International Convention of Safety of Life at Sea 1974, (SOLAS), entered into force on 1 May 1981. Among the many features dealing with maritime safety, the protocol required all ships between 1,600 grt and 10,000 grt to be fitted with at least one radar. All ships of 10,000 grt and above were to be fitted with at least two radars, each capable of operating independently of the other.

Malfunction of Radar

If the required radar should fail, then in U.S. waters the local USCG authority must be informed, and there are some similar requirements in other parts of the world. Courts have held that in the face of serious defects the master of a vessel is not, except in extenuating circumstances, at liberty to postpone attempts to make corrective action. He is obliged to

[34]The *Chusan* [1955] 2 Lloyd's Rep. 685.

ensure that an effort is made, employing the resources available to him, to correct faulty equipment that has a potential influence upon safe vessel operation. A decision not to attempt repairs to such equipment could lead to a presumption of negligence and consequent liability upon a vessel involved in a collision in which the faulty equipment was or should have been an ingredient. The majority of U.S. owners of U.S. flag merchant ships have agreed "repairs shall not be delayed until reaching port if the necessary spare parts are available and the Radio Officer or Radio Electronics Officer has a Radar Endorsement, if such equipment is involved."

It is, of course, a function of a vessel's owner to exercise due diligence to make a vessel seaworthy, which increasingly includes radar:

> It is the owner's duty to see to it that a vessel's equipment is in safe working order.[35]

Even if the radar is only partially defective, and the master attempts to exercise good judgment in his reliance upon it, the owner may not escape liability for

> . . . the navigation of a ship defectively equipped by a crew aware of her condition does not relieve the owner of his responsibility or transform unseaworthiness into bad seamanship. . . .[36]

However, since then IMCO has provided an escape clause that states that

> . . . malfunction of the radar equipment . . . shall not be considered as making the ship unseaworthy or as a reason for delaying the ship in ports where repair facilities are not readily available.[37]

Moreover, if the radar had failed after departure from port, provided he had exercised due diligence in clearly providing for routine repairs and maintenance, the owner probably would escape liability on that score. Of course, if radar does fail, safe speed in restricted visibility must be kept low. The *John C Pappas*, whose radar was out of order, was found 70 percent to blame for, *inter alia*, failing

> to reduce speed substantially on entering the fog (i.e., she should have stopped her engines).[38]

Additional Safe Speed Factors for Vessels with Operational Radar

To be operational in the context of the rules, a radar must not only be technically functioning correctly, but must also be operated by personnel

[35]Greater New Orleans Expressway Commission v. tug *Claribel* (1963) E. D. La 222 F Supp 521, 382 US 974 (1965).

[36]The *Maria* (1937) 91 F (2d) 819.

[37]Regulation 12(e), Chapter V of the 1974 International Conference on Safety of Life at Sea.

[38]The *Almizar* [1971] 2 Lloyd's Rep. 290.

who understand the *characteristics, efficiency, and limitations* of the equipment. A large variety of equipment is available at varying degrees of cost and sophistication, each with its own peculiarities. A watchkeeper must be familiar with the one on his ship before standing his first watch, apart from being well educated in general radar theory. He must know of any blind arcs, and therefore be aware of the need to vary course from time to time to clear those arcs, of how to adjust controls to obtain optimum performance, and of how to monitor that performance. Judging from the inquiry into the collision between the tankers *Atlantic Empress* and *Aegean Captain* off Tobago in 1979, the certificated watchkeeper in the latter ship apparently did not successfully manipulate the rain-clutter control and gain control when a rain squall created a white area on the radar screen. Initially, the other ship, the *Atlantic Empress*, was beyond the clutter, which was obscuring part of the *Aegean Captain*'s radar coverage, but was not detected, probably due to a large reduction in gain. In the *Atlantic Empress*, the watchkeeper, using a 3-centimeter radar, did not pick out the *Aegean Captain* from the rain squall that enveloped her. However, the use of the 10-centimeter radar set, with which she was also fitted, almost certainly stood a better chance of penetrating the rain area and detecting such a large contact.[39] In addition, the watchkeeper in the *Aegean Captain* had left had left the bridge two or three times before the collision.

Interrupted observation of a radar screen is not proper use that can justify a high speed in, or near, restricted visibility, for

> . . . when reliance is placed on the radar, it cannot be too strongly emphasized that a continuous watch should be kept by one person experienced in its use. . . .[40]

Furthermore, that experienced person must know how long it will take him or her to extract further information about the other vessel's track and closest point of approach (CPA), apart from the warning of another vessel's presence that a radar supplies. With manual plotting this is likely to be a minimum of six minutes, a factor that must bear heavily on the decision of what is a safe speed.[41]

Any *constraints imposed by the radar scale* in use will vary, as each scale has its disadvantages. The longer the range the less discrimination is possible, the greater the effect of bearing error, and the less likely it is that a small contact at close range will be spotted. Conversely, with short radar range scales no early warning is achieved and no overall traffic pattern can be discerned. The ideal is to have two radar displays each on a different

[39] "Report on Inquiry, Liberian Bureau of Maritime Affairs in London," November 1980 *Seaways*; also Cockcroft, "The S- and X-band Debate," April 1981 *Seaways*.

[40] The *Fina Canada* [1962] 2 Lloyd's Rep. 113.

[41] Afran Transport Co. v. The *Bergechief* (1960) 70 F Supp 893 (SDNY).

scale, one to give the close-in picture and the other for longer range warning. This avoids changing scales and temporarily disrupting the picture, especially when in a close-quarters situation. If only one display is available, it must not be kept continuously on one scale:

> If the master of the *Nassau* was relying upon radar to justify his speed in reduced visibility it was not good seamanship to have kept his radar permanently on the short range . . . They should have extended the range periodically at intervals appropriate to the circumstances to inform themselves of the general situation and, in particular, of the probable effect of the approach of otherwise invisible vessels upon the action of the vessel known to be, and seen to be, ahead of them . . .[42]

The effect on radar detection of the *sea state, weather, and other sources of interference* can sometimes be so severe that, as was seen in the earlier mentioned collision between two tankers off Tobago, even large ship contacts may be obscured by precipitation. Wave clutter relatively close in to the center of the radar display can seriously degrade the picture, and although adjustment of the controls may enable sizable contacts to be held, there is likely to be suppression of small echoes. Indeed, small craft may only give a radar response when they are on the crest of a wave, leading to an intermittent echo that could be taken for yet another part of the wave clutter. Closer inshore, side echoes from land can confuse the picture and contribute toward collision.[43]

The possibility that *small vessels may not be detected* by radar at an adequate range is ever present, despite the increased use of radar reflectors. The loss of the *Jane*, in the Straits of Juan de Fuca, and the *Powhatan* off the eastern seaboard—both wooden-hulled fishing vessels and neither detected on radar—illustrates the dangers of proceeding too fast in restricted visibility in areas where small craft may be expected. Both of the fishing vessels were visually sighted, but neither merchant ship was able to pull up or avoid.[44]

The number, location, and movement of *vessels detected by radar* clearly have impact on what is a safe speed. The denser the traffic the longer it takes to assess risk of collision and the more complex is the effect of evasive maneuvers. Compared to raw radar displays, automated radar plotting aids (ARPA) will speed this process up, but it still necessary for the mariner to absorb the wealth of information that ARPA provides, which must be allowed for in the chosen speed. Where the presence of other shipping may hinder early avoiding action, a drop in speed is

[42]The *Nassau* [1964] 2 Lloyd's Rep. 514; *see also* the *Bovenkerk* [1973] 1 Lloyd's Rep. 62; "The *Eleni V*," August 1981 *Seaways*.

[43]The *Bovenkerk* [1973] 1 Lloyd's Rep. 62.

[44]SS *Mormacpine* (Jane) 1959 USCGI; the *South African Pioneer* (Powhatan) 1961 USCGI.

essential. The *Larry L*, cited in an earlier chapter, detected the *Genimar* on radar some seven miles distant. Visibility was variable from three to four miles. Because of the presence of another vessel to starboard, the *Larry L* made a succession of small alterations of course to port before she sighted the *Geminar*. The court commented that *Larry L*:

> should have made a substantial alteration of course to starboard at an early stage, or, if she was inhibited from doing so by the presence of the *Pearl Creek* on her starboard side she should have made a substantial reduction of speed.[45]

As the courts have been consistent in faulting ships for not slowing down more in restricted visibility when they detect but one ship with a CPA that would result in a close quarters situation, it can be expected that they will insist on low safe speeds when more than one ship is detected.[46]

Of course, it is radar that makes possible the *more exact assessment of the visibility* by observing the ranges at which other ships or navigational marks either first become visible or are lost from sight. A deterioration in visibility is not always possible to detect with certainty, especially at night, and watchstanders should use the radar to aid the lookout's eye. Particularly they should be alert to the presence of restricted visibility when they fail to sight a vessel detected on radar, or a navigational mark that should have been within the expected visual range.

> . . . the *Gunnar Knudsen* failed to appreciate the presence of this fog bank as soon as they ought to. She was provided with radar . . . and . . . it ought . . . to have been brought into operation . . . when buoy No. 9 was not seen.[47]

Automatic Radar Plotting Aids Also a Statutory Requirement

It can be seen from earlier discussion that radar cannot be used casually to justify a high speed in restricted visibility. To draw the maximum benefit from it requires constant, knowledgeable appraisal by at least one person dedicated to the task and not distracted by other duties. Some mariners, lulled by a radar-induced sense of security, felt that they could rationalize increased speeds without extra vigilance. Added to this overconfidence was sometimes a reluctance to undertake the tedious nature of manual plotting. The result was that those ships did not make proper use of radar.

Partly because of this, since 1974 the United States has been urging that large ships of over 10,000 grt, carrying bulk cargoes of oil and hazardous materials, be fitted with an electronic relative motion analyzer (ERMA),

[45]The *Genimar* [1977] 2 Lloyd's Rep. 25; *see also* the *Nassau* [1964] 2 Lloyd's Rep. 509; the *Sanshin Victory* [1980] 2 Lloyd's Rep. 359.

[46]The *Bovenkerk* [1973] 1 Lloyd's Rep. 62.

[47]The *Gunnar Knudsen* [1961] 2 Lloyd's Rep. 440; *see also* the *Fina Canada* [1962] 2 Lloyd's Rep. 117; the *Almizar* [1971] 2 Lloyd's Rep. 290.

sometimes also called collision avoidance aid or collision avoidance system (CAS). Although IMCO examined this proposal, other nations were unwilling to proceed apace with an equipment considered to be still under development. The IMCO solution was to schedule a gradual introduction of what is now internationally known as an automated radar plotting aid (ARPA), into *all* ships of over 10,000 grt, commencing in 1984 and completing in 1989. This was insufficient to satisfy U.S. domestic concerns, however, particularly after the "winter of the *Argo Merchant*," and an impatient Congress passed the Port and Tanker Safety Act of 1978. This act requires bulk carriers of oil and hazardous materials of over 10,000 grt to be equipped with an ERMA by 1 July 1982. Thus, all such ships calling at U.S. ports must have an electronic system that analyzes their radar picture, providing auto-tracking and vector resolution of up to 20 echoes (depending on whether contact acquisition is automatic or manual) that would otherwise require manual plotting. For ships not calling at U.S. ports, it is expected that the IMCO schedule will be followed.

This use of modern technology is laudable, for as a general rule ARPA, with multiple echoes, should perform as well as or better than a trained radar observer using a radar display with a single echo. There is also provision for an alarm system to alert the bridge watchkeeper, but while the United States requires both audible and visual alarms, IMCO allows the option of either.

ARPA's primary function is to reduce the workload for the bridge personnel, and it can be expected to provide fuller information with greater accuracy and less delay than can be obtained by manual methods. The rapid increase in the installation of ARPAs to ships in the 1980s is likely to compare with the rate of fitting early radars to ships in the 1940s. Proponents of ARPA feel that its installation will result in an early and vast improvement in navigational safety. Radar was also expected to have such an effect, but the anticipated reduction in the incidence of collisions was not achieved. Indeed, there arose an era of "radar-assisted collisions," which only gave signs of improvement when thorough radar training was established and the limitations of this new aid were better understood. The potential benefits of ARPA are considerable, but there is doubt that enough adequate ARPA training will be provided to cope with the rapid fitting of ARPA in ships. More sophisticated equipment by itself is unlikely to bring about an instant improvement in navigational safety unless it is operated by trained personnel who fully appreciate its limitations and who are not lulled into a false sense of security by expecting too much from it. The fitting of ARPA could lead to a number of "ARPA-assisted collisions," similar to the radar ones of the 1950s, unless adequate training is undertaken.

> It is on men that safety at sea depends and they cannot make a greater mistake than to suppose that machines can do all their work for them.[48]

The Use of Radar in Determining Risk of Collision

Rule 7(b) strengthens the many court findings that radar, if fitted and operational, must be properly used to obtain early warning of risk of collision, an admonition reinforced by Rule 19(d) when restricted visibility exists. The first U.S. court case involving a collision of a radar-equipped vessel found the failure of the *Barry* to use her radar a most serious fault when visibility was restricted.[49] The earlier discussion on what factors to take into account when determining safe speed provides an intimation of what the courts will consider constitutes the proper use of radar. An efficient radar watch, with controls at optimum settings, must be maintained with long-range scanning at regular intervals.

In waters where safe navigation of the ship requires accurate fixing of her position and frequent alterations of course, there should be a bridge team of at least two competent officers, one to concentrate on plotting the position of the ship while the other keeps the radar watch.[50] This watch must be continuous and cannot be casual. The choice of display can be important, one that is gyro-stabilized to be preferred where possible. True-motion displays can be of value in congested waters, though they can be subject to errors from log input. Again, the ideal would be two radar displays, one for true motion and the second for relative motion, but where only one is available the choice, in pilotage waters, should be made by the master and not solely by the pilot.[51] Notwithstanding what preference of mode of display is made, the requirement is for early warning of risk of collision, and this can only be obtained by systematic plotting of echoes, either manually or with ARPA.

A radar presentation will show only the bearing and range of an echo. Only plotting will ascertain closest point of approach (CPA), a major determinant of risk of collision, as well as course and speed—all of which are needed to make proper use of radar. Systematic plotting in a busy situation is a labor-intensive and time-consuming procedure, and many vessels have failed to make the effort, but the courts "have been uniform in their condemnation of such failures."[52] No doubt the advent of ARPA will ease the burden of plotting, provided proper use is made of it. However, not all ships will be fitted with ARPA—indeed, it could mal-

[48] The *Fogo* [1967] 2 Lloyd's Rep. 208.
[49] *Barry-Medford* (1946) 65 F Supp 622; *see also Burgan-Bergechief* (1960) 274 F 2d 469.
[50] "The *Eleni V*," *Seaways*, August 1981.
[51] The *Atys*—The *Siena* (1963) Netherlands Inquiry.
[52] Orient Steam Navigation Co. v. United States (1964) 231 F Supp 474 (SD Cal); *Australia Star-Hindoo* (1947) AMC 1630.

function—and the plotting of bearings will be essential if reliable inferences are to be drawn.[53] The well-known use of the cursor and grease pencil marks on the screen will not suffice in restricted visibility,[54] though it must be accepted that in heavy traffic it is often impractical to make and evaluate a comprehensive manual plot. In such circumstances, provided the radar is being carefully and continuously observed, it should be possible to carry out some form of "threat reduction" and discard contacts that are clearly passing well clear, concentrating only on those that are not.

Unless some form of systematic plotting is undertaken, there is a grave risk that action may be taken on insufficient evidence. Rule 7(c) is crystal clear on this aspect that assumptions shall not be made on the basis of scanty radar information. Plotting errors will occur whether computerized or manual methods are used, particularly at longer ranges where inherent bearing inaccuracies of radar may well give an incorrect impression of CPA in the early stage of an encounter. Observation must therefore be based on several successive readings taken over a realistic period of time. Far too often, ships approaching each other in fog, sometimes even in excellent visibility, have mis-assessed the radar CPA and taken inadequate or conflicting action with disastrous results.[55]

Detection by Radar Alone Requires Early Action

If proper use of radar has shown that risk of collision exists and/or that continuing on the present course and speed will lead to a close-quarters situation, then in restricted visibility, ships must take action in accordance with Rule 19(d). This first level of action in restricted visibility applies when the determination of risk has been made by radar alone, without sighting the other vessel or hearing her fog signal. Ships must not delay in taking the action couched in the somewhat awkward terms of Rule 19(d)(i) or (ii) to avoid the impending close-quarters situation. Rule 8(a), equally applicable in any condition of visibility, also requires action to be taken in ample time. The purpose of the requirement to avoid turning to port for a vessel forward of the beam is to reduce the possibility of conflicting action being taken by vessels on opposite courses. Hanging on in the expectation of sighting the other vessel and changing the situation from a restricted visibility to a visual one is to court disaster. Ships have often detected each other on radar at considerable ranges but for various

[53]The *Prins Alexander* [1955] 2 Lloyd's Rep. 8.

[54]Skibs A/S Siljestad v. SS *Matthew Luckenbach* (1963) 215 F Supp 680 (SDNY), aff'd 324 F 2d 563.

[55]The *Evje* (1950) 84 Ll. L. Rep. 20; the *Andrea Doria—Stockholm* (1956) JRIN Vol. 30 No 2, 238; the *Santa Rosa*—the *Valchem* (1959) USCGI; the *Niceto de Larrinaga* [1963] 1 Lloyd's Rep. 205; the *British Aviator* [1965] 1 Lloyd's Rep. 271; the *Toni* [1973] 1 Lloyd's Rep. 79.

reasons have failed to take bold action until far too late.[56] A predicted CPA of several hundred yards is too small in fog on the open seas, and vessels should endeavor to achieve a more substantial distance apart, one measurable, as has been said before, in miles rather than in yards.[57] Two to three miles would seem a seamanlike distance, as maneuvering to achieve this safe distance should be detectable on the other ship's radar, should make due allowance for any radar plotting errors especially at long range, should ensure that the echo is not lost at a later stage in any sea clutter near the center of the display, and should reduce the risk of a conflicting maneuver by the other ship. However, in congested waters when safe speed should be lower, lesser distances may be sufficient. A major reduction of speed, particularly effective for a vessel closing from abeam, is, of course, an option available—as well as, and perhaps in conjunction with—an alteration of course. A change of speed, however, for a vessel closing on a near-reciprocal course will little effect the CPA, but it will give more time for assessment and, should it be necessary, more time to hear her fog signals. Above all, the aim is to avoid getting into a close-quarters situation, and radar, properly used, can help.

If Early Action Is Not Possible or If a Fog Signal Is Heard

Hearing a fog signal apparently forward of the beam or when a close-quarters situation with a vessel ahead cannot be avoided invokes the second level of action required of a vessel navigating in restricted visibility. This portion of Rule 19 is closest in meaning to the former fog rule that demanded vessels to "stop engines and navigate with extreme caution," and it can be expected that previous court rulings will have particular relevance.

Firstly, Rule 19(e) contains the proviso "except where it has been *determined* that a *risk of collision* does *not* exist (emphasis added). This undoubtedly means that, where it is not certain what the position, course, speed, and CPA of the other vessel may be—or to use former terminology "a vessel the position of which is not ascertained"—then the remainder of the rule applies. It was held before the invention of radar that the position of a vessel could only be regarded as ascertained when fog signals placed the matter beyond doubt, but it was rare for the courts to be satisfied that fog signals alone resolved doubt,[58] particularly as the direction and distance of sound signals can be misleading in fog. With the advent of radar, the question of whether the position of a vessel whose fog signal had been heard apparently forward of the beam was "ascertained" when she was

[56]The *Sanshin Victory* [1980] 2 Lloyd's Rep. 359; "*Eleni V*," *Seaways*, August 1981.
[57]The *Verena* [1961] 2 Lloyd's Rep. 133.
[58]*See* generally Marsden's *Collisions at Sea*, pp 536-39.

visible on the radar screen was soon considered.[59] Many navigators thought that one good radar range and bearing, or certainly a series of ranges and bearings, constituted "ascertainment" of the other vessel's position. However,

> many of them whose misfortune had brought them into Court were disabused of this idea by the learned judges; evidently a great more was needed.[60]

Apart from distance and bearing, the minimum amount of knowledge required is the other vessel's course and speed[61] and the probable closest point of approach. On deciding on the latter point the court also commented on the likelihood of expecting routine alterations of course at navigational focal areas and that:

> It must be remembered that radar does not foretell intention, and all these are circumstances which must be considered.[62]

Thus, a mariner must be very certain indeed that risk of collision does not exist when he finds himself in an unavoidable close-quarters situation with another vessel before he can decide not to come down in speed to bare steerageway. Even with a good radar plot, the possibility that a fog signal he heard may come from a different vessel than the one he is tracking must always be taken into account. The fishing vessels, whose sinkings were mentioned earlier in this chapter, were never seen on radar, although their fog signals were heard.

When the strict proviso of Rule 19(e) cannot be met, then a vessel is required to reduce her speed to the minimum at which she can be kept on her course. This is for either a fog signal heard or for a developing close-quarters situation with another vessel forward of her beam. Such reduction in speed must be made in ample time, without waiting for the close-quarters situation to be reached. The greater the initial safe speed, the earlier, and therefore the greater range, at which speed should be reduced:

> . . . it could then be seen from radar that a close-quarters situation with the *Hagen* was developing, and . . . those on board the *Boulgaria* should at this stage have reduced speed further. . . .[63]

Not too rigid an interpretation should be placed on the word "beam," for determining the direction of fog signals is unreliable at best. It was once

[59]The *Prins Alexander* [1955] 2 Lloyd's Rep. 1.

[60] Wylie, *Radar at Sea*, 22 JRIN 35 (1969).

[61]The *Gunnar Knudsen* [1961] 2 Lloyd's Rep. 437; the *British Aviator* [1964] 2 Lloyd's Rep. 403.

[62]The *Sitala* [1963] 1 Lloyd's Rep. 212.

[63]The *Hagen* [1973] 1 Lloyd's Rep. 265; *see also* the *Nora* [1965] 1 Lloyd's Rep. 625; the *Zaglebie Dabrowski* [1978] 1 Lloyd's Rep. 570.

held that in a case where a whistle was heard abaft the beam twice on the same bearing, the vessel should have stopped.[64] Fog signals heard from astern are not covered by the rule, but in one occasion where a vessel stopped her engines because of running into dense fog, she was faulted for not maintaining steerageway and falling off her course across the stem of an overtaking vessel.[65]

The person in charge of a vessel should not delay in reducing speed because he himself has not heard the fog signal, but must act on the report of his lookouts:

> I see no excuse for the failure of the master and pilot to act upon the report made to them by the third officer, when he informed them that he had heard the whistle of a vessel ahead. It seems to me that it is no excuse on the part of either pilot or master to say he did not hear it himself.[66]

Previously, on both the high seas and in inland waters, the rules required a power-driven vessel to stop her engines when she heard, apparently forward of her beam, a fog signal of a vessel whose position had not been ascertained or when there was a vessel detected ahead of the beam with which she could not avoid a close-quarters situation. This action is no longer mandatory in the first instance, but having slowed down to bare steerageway in a timely manner, it may still be prudent so to do at a later stage. Indeed, Rule 19(e) continues that "if necessary all head-way shall be taken off." The hearing of a fog signal close aboard for the first time, or the sighting of a vessel of uncertain course looming out of the fog are examples when such action in urgently necessary, particularly if the radar is suspect. The phrase "if necessary" permits no exception, and while the circumstances might permit way to be run off just by stopping engines, the fastest way to take off all headway is to back engines full.

Some discretion, however, should be used in putting engines in the astern mode unnecessarily:

> . . . when there is any question of listening for signals one is creating the worst possible conditions for hearing them by working the engines at full speed astern.[67]

In addition, going astern may cause the ship's head to fall off and present her beam to a vessel approaching from ahead. By keeping the bow end-on, or nearly so, to such a vessel, a smaller target is presented, which may assist avoiding action by the other vessel, as well as increase the likelihood of taking any impact forward of the collision bulkhead.

[64]The *Bremen* (1931) 47 T.L.R. 505.
[65]The *Ring* [1964] 2 Lloyd's Rep. 177.
[66]The *Chuson* [1953] 2 Lloyd's Rep. 151.
[67]The *Monarch* [1953] 2 Lloyd's Rep. 151.

Alterations of course are usually best avoided if there is a close-quarters situation developing or if a fog signal is heard forward of the beam, unless both the position and track of the other vessel are positively known. Long experience has led to the stricture in Rule 7(c) that assumptions shall not be made on the basis of scanty information. Late alterations of course, in any direction, merely on the faith of a fog signal,[68] or catching an incomplete glimpse of a vessel first looming out of the fog,[69] or as a blind reaction to a close radar contact that was not properly tracked,[70] have been roundly condemned by the courts. A tendency in recent cases has been particularly to fault last-minute alterations of course to port for being contrary to Rule 19(d)(i):

> . . . the action taken by *C.K. Apollo* in putting her wheel hard-to-port was too late and in the wrong direction and was contrary to the express terms of rule 19.[71]

An alteration of course is not always found to be negligent for, as ever, the court will judge the case on the circumstances prevailing. In the *Sedgepool* case two vessels were approaching each other in fog on approximately opposite courses in the Ambrose Channel, New York. One ship held the other on radar on her port bow. Both ships, on hearing the other's fog signals altered to starboard:

> This was a collision which . . . occurred in a narrow channel, and in those circumstances, I should be very slow to blame a ship which hearing a fog signal from another vessel, altered her course to starboard in an attempt to get more over to her proper side.[72]

Vessels Must Navigate with Caution

"And in any event navigate with extreme caution," concludes the restricted visibility rule, "until danger of collision is over." There is subtle irony in these closing words to the navigator, already chafing with impatience at his reduced speed and enforced stops, who pushes on too eagerly with his voyage. For the hardest thing to prove in any court after a collision has occurred is that the danger of collision no longer existed. It behooves the mariner who has conformed to the law up to this point to practice the most scrupulous care in feeling his way past the other vessel. An occasional kick ahead, enough to keep his vessel on her course, with a

[68]*Miguel de Larrinaga* [1956] 2 Lloyd's Rep. 538.

[69]Crown SS Co v. Eastern Navigation Co. Ltd. (1918) S.C. 303; the *Koningin Luise* (1922) 12 Ll. L. Rep. 477; the *Devention* (1922) 12 Ll. L. Rep. 484; the *Wear* (1925) 22 Ll. L. Rep. 59; The *Bharatkhand* [1952] 1 Lloyd's Rep. 470.

[70]The *Anna Salem* [1954] 1 Lloyd's Rep. 475; the *Achille Lauro* [1956] 2 Lloyd's Rep. 540; the *British Aviator* [1964] 2 Lloyd's Rep. 403; the *Linde* [1969] 2 Lloyd's Rep. 568.

[71]The *Sanshin Victory* [1980] 2 Lloyd's Rep. 359; *see also* "The *Eleni V*," *Seaways*, August 1981.

[72]The *Sedgepool* [1956] 2 Lloyd's Rep. 668.

change in that course only justified when he is reasonably certain of the other's position, course, speed, and CPA, and with a prompt and vigorous reversal of engines if the latter decreases or the other's fog signals are steady on the bow, is the action most likely to prevent collision and to win court approval if the worst happens.

Whether in inland waters or on the high seas, it should be remembered that three short blasts must accompany any reversal of the engines if the other vessel is sighted; otherwise, the signal must not be used. Equally, other maneuvering signals are only valid after sighting the vessel that one is feeling one's way past. And then, of course, the regular rules govern, but never when one vessel is not in sight of another.

Rules in Restricted Visibility Are for Safety and Should Be Obeyed

The new regulations have come a long way in recognizing the practical need of maritime trade to run to a reasonable timetable, which, effectively, means that ships underway need to make way in restricted visibility. The strictures of moderate speed, and hopefully of the half-distance rule, have been exchanged for the more flexible, though more extensive in application, guidelines for safe speed. The rules acknowledge that well-equipped and properly manned ships can avoid close-quarters situations without reducing below safe speed. Further, instead of having no option in certain circumstances but to stop engines, ships may now proceed at steerageway, albeit minimum steerageway. It is, of course, the rapid and major technological advances in ships' equipment, particularly radar, that has allowed this less strait-jacketed approach.

However, with this increased capability to "see" and freedom to act in restricted visibility, there comes increased responsibility to use, and use properly, the greatly expanded amount of information on the safe handling of vessels.

> For unto whomsover much is given, of him shall much be required.[73]

Greater attention is required of mariners, particularly in mastering the new aids, than some have shown hitherto. While there will always be a premium upon the established principles of good seamanship, no longer acceptable are the standards of bygone days.[74] However, not too much trust should be placed in the latest marvels of science, and the reader is reminded of the quotation from the *Fogo* case cited earlier (see note 48).

While the law in restricted visibility now makes greater allowance for the fact that shipping exists for the timely and profitable transport of goods, it is nevertheless apparent that obedience to the rules will occasion

[73]The *Nora* [1956] 1 Lloyd's Rep. 625.
[74]Jacobsen, *Technology and Liability* (1977) 51 T.L.R. 1132.

some delay if the weather clamps down. In the past, many mariners had a mistaken form of professional pride in getting their ship in on time regardless of weather conditions, and were influenced also by the fear that their owners would not accept fog as an excuse for any considerable prolongation of the voyage. In some cases this fear was by no means groundless, but it should be less so when responsible owners recognize the risks of liability, not only from collision but also arising from pollution. Thus, the law in restricted visibility does not merely suggest but demands that voyages must often be delayed in the interest of safety of life, of property, and the environment. The navigator who knows the law and deliberately sets himself above it is likely sooner or later to come to grief and find that his skill in dodging innumerable vessels while he proceeds at excessive speed does not excuse him from liability for a final collision. Although written over a century ago, the following remains valid:

> It is unquestionably the duty of every master of a ship, whether in intense fog or great darkness, to exercise the utmost vigilance, and to put his vessel under command, so as to secure the best chance of avoiding all accidents, *even though such precautions may occasion some delay in the prosecution of the voyage.*[75]

SUMMARY

This chapter has examined four major areas relating to the law in restricted visibility: sound signals, safe speed, radar, and close-quarters navigation. All four are inextricably interwoven, and it is fortunate that since near-uniformity of rules has been achieved, there are few distinctions between the rules for the high seas and inland waters. Neither set of rules places a finite value on the meaning of the term "restricted visibility," but the key words are "not in sight of each other." Thus, a radar detection at long range comes under this definition, even though visibility might be, say, over 5 miles.

The few minor differences to be found are mostly in the sounding of fog signals, affecting only a small category of vessels. While the rules do not specify the minimum visibility under which fog signals are mandatory, they should be used when visibility first diminishes the effectiveness of sidelights. The time between signals is a maximum interval that can, and should, be decreased when circumstances warrant. Great care should be exercised that the signal of two prolonged blasts is used only when a ship is truly dead in the water. It should be clearly understood that the signal does not necessarily indicate intention of remaining stopped. The use of the inland danger signal is, by the terms of Rule 34(d), intended for use by vessels in sight of each other, though it has been suggested that in

[75]The *Itinerant* (1844) 2 W. Rob. Adm. 236.

certain circumstances the responsibility requirements in Rule 2 would permit a vessel to sound the danger signal in restricted visibility.

Safe speed has a wide application to all conditions and circumstances, but it is in restricted visibility that it comes closest in meaning to the former "moderate speed." However, the previous concomitance of the "half-distance rule" with moderate speed is less valid—a matter no doubt the courts will decide on when and if necessary. A visibility of 5 miles or less should lead to a reconsideration of present safe speed, with, at least, the engines being placed ready for immediate maneuver. As visibility decreases, so should safe speed, the aim being to have time to detect, plot, evaluate, and avoid at long range a potential close-quarters situation. If avoidance is not possible, then safe speed is again reduced, early, to bare steerageway. On hearing a fog signal ahead, unless absolutely certain of the movements of the ship originating it, bare steerageway is again required with, if necessary, all way being taken off.

Radar is now a mandatory requirement for most ships, but it is not a license for excessive speed in fog. Courts may accord an extra knot or so to the safe speed of a ship properly using radar, but always with the caveat that further reductions are necessary when radar information reveals that the circumstances have changed. Both the rules and the courts provide firm guidance as to the proper use of radar. Systematic plotting is essential, a task in which automated aids have the potential, if used by trained personnel, to be of great assistance.

There are two tiers of action for close-quarters navigation: first, the mariner should avoid the situation at long range by following the express terms of Rule 19(d); and second, if unable to avoid, he should slow right down, taking all way off if necessary and resisting precipitate course alterations based on scanty information.

Collisions in restricted visibility continue to provide a major source of court cases. The rules recognize that ships have to navigate in fog, and the law has made provision that the means to do so are on board. It is up to the mariner to use them intelligently so that he can proceed with his voyage in a timely manner and always in accordance with the rules.

18

Departure from the Rules

Responsibility

Rule 2 of both the international regulations and the inland rules has the appropriate heading of *Responsibility*—responsibility, that is, to use the rules sensibly. In effect, it warns against too rigid an interpretation of the rules. It states that it is the responsibility of mariners to take all seamanlike precautions in all situations, including those of "special circumstances," and, when necessary to avoid danger, to take additional measures not included in the rules, or in some cases, to depart from the rules altogether. For the rules offer practical advice and instructions to mariners faced with particular problems; they are not strict rules to be applied literally come what may. Moreover, the courts have always taken a liberal approach to the rules, construing them in the light of the prevailing circumstances, whether they be special or not. Special circumstances may be such that the rules do not cover them or that to obey the rules creates an even greater hazard than disobeying them.

When Departure from the Rules May Be Necessary

This chapter will confine itself to the second part of the Responsibility Rule: that of lawful departure from the rules, known in the past as the General Prudential Rule.

> Rule 2(b). In construing and complying with these rules due regard shall be had to all dangers of navigation and collision and to any special circumstances, including the limitations of the vessels involved, which may make a departure from these rules necessary to avoid immediate danger.

Such criteria might seem to justify a wholesale abandonment of the rules every time danger threatened. However, common sense, and not selfish interest, clearly requires that safety is attained by the uniform and exact observance of the rules, a situation that the vast majority of encounters permits. Thus, the mariner should never depart from the regulations except when absolutely necessary, even though the rules were never envisaged as a strait-jacket. The whole point of the rules is to avoid collision, indeed risk of collision, and therefore the regulations should not be so applied as to bring this about. The duty to avoid collision is higher than slavish compliance to the rules, and for that purpose, departure may be made. Neither is a ship bound to take a course of action that, if she obeyed the rules, would place her into greater danger.

> Where there was one and only one chance of escape from collision, a seaman was justified in taking the benefit of that chance, although it necessitated a departure from the regulations.[1]

Naturally, to justify a departure from the regulations there must be clear proof that adherence to them would have caused an immediate danger and that the departure was adopted to avoid such danger.

Departure Only Lawful When There Is Immediate Danger

Rule 2(b) does not give the right to take action contrary to the regulations whenever it is considered to be advantageous to do so. Mere convenience is not a justification for departing from the rules, even if collision could be avoided with safety equal to that of compliance with the rules. Neither is departure excused when the rules require both ships to take action but one does not because the action of the other alone was thought sufficient to avoid collision. Nor is a vessel justified in departing from the rules just because she fears the other ship will not comply with them. The view that whenever a perceptible risk of collision exists, the rules, except Rule 2(b), at once cease to apply and from that moment on it is each man for himself, is equally invalid. A moment's reflection will convince anyone of the folly of having rules that would not hold up in a reasonably close situation. The real key to the matter, as brought out in decision after decision of the courts, is in the words *immediate danger*. As said by the federal court of the Virginia district in an early case:

> The rules of navigation must be observed, and the courts have no option but to enforce them unless in cases coming clearly under this rule where it is necessary to avoid immediate danger.[2]

As said by the circuit court of appeals of Maryland:

> Where two courses are open to a vessel, one to follow prescribed rules and the other to depart from them, duty is imperative to observe rules and to assume

[1]The *Benares* (1883) 9 P.D. 16.
[2]The *R.R. Kirkland* (Va 1880) 48 F 760.

> that an approaching vessel will do likewise until after danger has become so manifest as to show there is no proper choice of judgment other than that of departure from the rules. . . . Departure from navigation rules because of special circumstances is only permitted where it is necessary in order to avoid immediate danger, and then only to the extent required to accomplish that object.[3]

And as said by the Supreme Court in still another case:

> Exceptions to the International Rules, provided for by this rule, should be admitted with great caution, and only when imperatively required by the particular circumstances. Therefore under all ordinary circumstances, a vessel discharges her full duty and obligations to another by a faithful and literal observance of these rules.[4]

The danger that justifies a stand-on vessel in a collision situation to alter the course and speed she is normally required to hold, must be very close indeed. Altering too early to avoid ice,[5] or slowing down in a crossing situation because of the presence of a third party some distance beyond the intersection point,[6] emphasize that *immediacy* of danger must be present. Notwithstanding this, the rules do not require a ship to take a measure that puts her in imminent peril. A vessel is not bound to obey a rule if by so doing she will incur serious damage by running aground[7] or by striking a third vessel.[8] Because of the danger to her, she is excused non-compliance,[9] and if in order to avoid collision it is necessary for the other vessel to also depart from the rules, it is her duty to do so:

> . . . due regard shall be had to any special circumstances which may render a departure from them necessary in order to avoid immediate danger. As soon then it was . . . obvious that to keep his course would involve immediate danger it was no longer the duty of the master of the *Tasmania* to adhere. . . . He was not only justified in departing from it but bound to do so. . . .[10]

In fog, near a notorious sand bank with strong tidal currents, a vessel in a crowded traffic route was excused for not stopping engines on hearing repeated fog signals forward of her beam, because it would have been dangerous for her to do so.[11] However, one vessel in the horns of a dilemma was blamed for not choosing the risk of going aground in

[3]The *Piankatank* (CCA Md 1937) 87 F (2d) 806.
[4]The *Oregon* (1895) 39 L Ed. 943.
[5]Joseph Golding v. the *Illinois* (1881) 26 L Ed. 562.
[6]The *Norfolk* (Md 1924) 297 F 251.
[7]The *Lucia Jantina* v. The *Mexican* (1864) Holt 130.
[8]The *Concordia* (1866) Ll. L. Rep. 1 A & E 93.
[9]The *Hazelmere* (1911) P. 69.
[10]The *Tasmania* (1890) 14 P.D. 53.
[11]The *Mount Athos* [1962] 1 Lloyd's Rep. 97.

preference to a risk of collision with its more serious probable consequences.[12]

Cases Held Not to Be Special Circumstances

It may serve to clear up in the reader's mind what the courts recognize as special circumstances if we first consider a number of situations that the courts have held are *not* special circumstances within the meaning of the rule or to a degree entitling a vessel to disregard the ordinary requirements. We have pointed out that there is no such special circumstance if an impending danger is too distant to be considered immediate. A further illustration of this point was the recent case in the River Plate estuary where an outbound vessel, the *Sagittarius*, attempted to excuse her failure, in contravention of the local rules, to hold back at a bend by pleading that she would have been unable to comply with them in relation to the other ships following the first inbound vessel. The court ruled that the international regulations, which were co-applicable with the local ones:

> authorize departures from the other rules only when these are necessary in order to avoid immediate danger. The potential difficulties in relation to other ships did not . . . give rise to immediate danger within the meaning of these rules. On the contrary the immediate danger was that of passing the *Schwarzburg* at a prohibited place.[13]

In an early New York Harbor collision between the tows of two tugs that met off the Battery in the crossing situation, the give-way vessel was held at fault for failure to comply with the former inspectors' rule requiring her to go under the stand-on vessel's stern. The validity of her argument that a special circumstance was created by a strong adverse tide, which would have set her far down the river and thus materially delayed her if she had executed right rudder instead of trying to keep out of the way by going left, was denied by the circuit court of appeals. In thus deciding, the court, in effect, found that the matter of convenience or inconvenience does not carry weight in determining special circumstances.[14] Again, in another crossing collision in the same harbor between two ferryboats, both were held at fault, and the plea of the give-way ferry that its well-known schedule in connection with railroad trains made a special circumstance entitling it to cross ahead of the stand-on vessel was refused.[15]

In a case of head-on collision in 1869 between a brigantine and a schooner in Long Island Sound, conditions of visibility were such that

[12]The *Durhambrook* [1962] 1 Lloyd's Rep. 104; *see also* the *Whitby Abbey* [1962] 1 Lloyd's Rep. 110.

[13]The *Schwarzburg* [1967] 1 Lloyd's Rep. 38.

[14]Scully v. New Jersey Lighterage Co. (CCA NY 1891) 58 F 251; *see also* the *Hopper D* (1920) 4 Ll. L. Rep. 43.

[15]The *Garden City* (NY 1884) 19 F 529.

neither lookout, although properly stationed, saw the lights of the other vessel until collision was imminent. The brigantine was running close-hauled on the starboard tack, and the schooner, on the port tack, had the wind a little free, and under the rule effective at that time, each vessel was bound to turn to the right. The brig, however, went to the left, and in the resulting collision sank the schooner; the Supreme Court denied her plea that the imminence of collision at the moment of discovering the schooner's lights created a special circumstance excusing her violation of the meeting rule.[16]

In a crossing case in 1865, the side-wheeler *America* was in collision off the Battery in New York with the steamship *Corsica*. In this case, the *Corsica*, the stand-on vessel, instead of holding her course down the river, swung left under the mistaken assumption that the *America* intended to hold on across her bow; and the *America*, backing down to keep out of the way in conformity with the rule, was actually making sternway when struck. At that time the rule requiring the stand-on vessel to hold on contained the stipulation "subject to the next article," which was then equivalent to Rule 2(b), and the *Corsica*'s counsel contended that special circumstances were created by fear that the *America* would not give way. In denying this plea, the Supreme Court thus very early settled the obligation of a stand-on vessel to hold course (and under present rules, speed) as long as it is still possible for the give-way vessel to carry out her own obligation and give way.[17]

In another New York Harbor case the *Red Ash*, a give-way tug with a car float on each side, collided with the *Hale*, a stand-on tug without tow. The fact that the *Red Ash* was with tow alongside did not excuse her for failure to back down at once when she sighted the *Hale* 500 yards distant on her starboard bow, nor allow her to invoke special circumstances as an excuse for such failure, when as a matter of fact she did back down after an interval, but not soon enough to prevent collision.[18]

The steamer *Dimock* collided with the steam yacht *Alva*, which was at anchor in the narrow and tortuous channel known as Pollock Rip Slue on Nantucket Shoals, in a dense fog. As she struck the anchored vessel, the *Dimock* was, of course, self-convicted of excessive speed in accordance with the well-known rule of the Supreme Court that if she was going at such a rate as made it dangerous to any craft she ought to have seen, and might have seen, she had no business to go at that rate.[19] However, she sought to invoke the special circumstance rule on the grounds that run-

[16]The *Annie Lindsley* v. Brown (1881) 26 L Ed. 716.

[17]The *Corsica* (1870) 19 L. Ed. 804. *See also* Postal SS. Co. v. *El Isleo* (1940) 84 L Ed. 335.

[18]Thames Towboat Co. v. Central R. R. of NJ (Conn 1894) 61 F 117; *see also* the *Warrior* (1871) 1 Ll. L. Rep. 3 A&E 553; the *American* and the *Syris* (1874) Ll. L. Rep. 4 A&E 226.

[19]The *Nacoochee* 137 U.S. 330 (1890) 34 L Ed. 686.

ning with a swift tide in a crooked channel compelled her to make about 8 knots over the ground in order to have steerageway. While the court admitted that the argument might have had some force had the *Dimock* been compelled to navigate the channel, on a showing that she deliberately entered after the fog set in and kept going instead of anchoring when it failed to abate, it held that special circumstance did not apply.[20]

In still another New York Harbor case, the *Transfer No. 10* was held at fault for a head-on collision with the *Mary J.* because she was navigating up the Manhattan side of the East River, in violation of both inland and harbor rules. The court denied that the local custom of keeping on the left-hand side in an ebb tide to make better speed could create a special circumstance as contemplated by the rule.[21]

The tug *Mohawk* collided with the tug *Howard Carroll* in a dense fog while the latter was moored at the end of an East River pier. Her plea was that she had a defective compass and supposed she was navigating in the middle of the river with proper caution, at 3 knots. However, on a showing that the compass was known to be out of order before entering the fog at Brooklyn Bridge, the court declined to find her predicament a special circumstance, and held her solely liable for the collision.[22]

In a Boston Harbor case one passenger vessel collided with another in a thick fog shortly after an earlier collision between one of them and a third vessel. The court failed to accept the confusion that prevailed on board the vessel that had already been through a collision as a special circumstance excusing her for failure to note the other's fog signals, and ruled that she should not have again got underway until everything was shipshape and the officers had regained their composure.[23]

In a collision on the Delaware River at night the sloop yacht *Venture* was sunk by a barge in tow of the ocean tug *International*, while drifting in a very light wind of insufficient strength to give her steerageway, and while not keeping a proper lookout. The tug was obviously at fault for failing to keep clear of the sailing vessel, but the court held the yacht also at fault, refusing to excuse her situation on the grounds of special circumstance and holding that she should have anchored near the shore instead of allowing herself to drift into mid-stream and into the regular path of moving vessels.[24]

A meeting place of several navigational channels is not necessarily an area where special circumstances allow departure from the rules. In the *Homer* case, it was held that the normal, in this case crossing, rules applied

[20]The *H.F. Dimock* (CCA 1896) 77 F 226.
[21]The *Transfer No. 10* (NY 1904) 137 F 666.
[22]The *Mohawk* (NY 1890) 42 F 189.
[23]The *Stamford* (Mass 1886) 27 F 227.
[24]The *International* (Pa 1906) 143 F 468, 50 L Ed. 1172.

at the junction of the channels when two ships met with no other vessels involved.[25]

Rule 2(b) Not Substitute at Will for Other Rules

The foregoing decisions make it very plain to the mariner that Rule 2(b) is far from being a mere substitute at will for the requirements of the other rules. The United States Supreme Court has explicitly limited the application of the special circumstance rule in three well-known decisions:

> It applies only where there is some special cause rendering a departure necessary to avoid immediate danger such as the nearness of shallow water, or a concealed rock, the approach of a third vessel, or something of that kind.[26]

> Nevertheless it is true that there may be extreme cases where departure from their requirements is rendered necessary to avoid impending peril, but only to the extent that such danger demands.[27]

> Exceptions to these rules, through provided for . . . should be admitted with great caution, and only when imperatively required by the special circumstances of the case.[28]

Referring to the above opinions of the court of last resort in the *H. F. Dimock*, previously cited, the circuit court of appeals remarks that, taking it altogether, these expressions go little, if any, beyond applying the rule of *in extremis*.[29]

Five Kinds of Cases Where Rule 2(b) Applies

With this discussion as a background, we may now consider a number of decisions where special circumstances have been held to justify a departure from the ordinary rules, and may therefore, under similar conditions, be regarded as a basis of action in a collision situation or of defense after a collision has actually occurred. These may be said to fall into five groups: (1) where the situation is *in extremis*; (2) where other apparent physical conditions make obedience to the ordinary rules impracticable; (3) where the ordinary rules must be modified because of the presence of a third or more vessels; (4) where the situation is not specifically covered by the rules; (5) where one of two vessels proposes a departure from the rules and the other assents.

[25]The *Homer* [1972] 1 Lloyd's Rep. 429.
[26]The *Maggie J. Smith* (1887) 123 U.S., 349, 31 L Ed. 175.
[27]Belden v. Chase (1893) 150 U.S. 674, 37 L Ed. 1218.
[28]The *Oregon* (1895) 158 U.S. 186, 39 L Ed. 943.
[29]The *H.F. Dimock* (CCA 1896) 77 F 226.

Situations in Extremis

Whenever two moving vessels approach each other so closely that collision is inevitable unless action is taken by both vessels to prevent it, the situation is *in extremis*. Except in thick weather without radar, or in confined waters, obedience to the rules will generally prevent vessels from coming into dangerous proximity, it being the intent of the rules to prevent not only collision itself but risk of collision. Hence, it will be found almost invariably that when two vessels reach a situation where collision is imminent, one or both of them has violated the rules. This may be illustrated in the crossing situation. If the give-way vessel fails to give way and both hold on long enough, collision will inevitably occur. It has never been the intent of the rules that the stand-on vessel, which is under a specific requirement to maintain course and speed, should hold that course and speed right through the other vessel. On the contrary, as soon as the vessels reach a position where collision is so imminent that it cannot be avoided by the give-way vessel alone, it immediately becomes not only the right but the expressed duty of the stand-on vessel to take such action as will, in the judgment of her commanding officer, best aid to avert collision. Rule 17(b) is a statutory provision to that effect,[30] applying not only in the crossing situation but in every situation where one vessel is privileged and the other is burdened. As stated by the circuit court of appeals in a collision between two tugs at Charleston, South Carolina:

> There is no right of way on which a vessel is entitled to insist when it is obvious that it will result in danger of collision.[31]

And as held by the circuit court of appeals in a collision of two ferryboats in New York Harbor, where the stand-on vessel maintained course and speed after it was manifest that departure therefrom could alone prevent collision:

> When a collision is imminent, each vessel must do all in her power to avert it, no matter what may have been the previous faults, or which may have the right of way.[32]

A similar opinion was stated by the district court in a later case in which a sailing vessel in tow of two tugs collided with an ocean steamship on Puget Sound in foggy weather:

> Even improper navigation of another vessel does not excuse adherence to a

[30]Rule 17(b) When, from any cause, the vessel required to keep her course and speed finds herself so close that collision cannot be avoided by the action of the give-way vessel alone, she shall take such action as will best aid to avoid collision.

[31]The *Hercules* (SC 1892) 51 F 452.

[32]The *Mauch Chunk* (NY 1907) 154 F 182.

definite rule, when such adherence plainly invites collision, and stubborn adherence to rule is sometimes culpable fault.[33]

It was probably this line of reasoning that influenced the Supreme Court in the case of the *Eastern Glade* and *El Isleo* referred to in chapter 16. In the view of the high court the blowing of a wrongful two-blast signal by the give-way vessel in a crossing situation becomes not merely a proposal, but a positive declaration of intent to depart from the rule and thereby create imminent danger of collision; and the rules, considered in conjunction with the general prudential rule, were seen as an effective means of preventing whistle arguments at high speed, and of getting the two endangered vessels under immediate safe control.[34]

Until the implementation of the current rules, the mariner on a stand-on vessel was given a nice question to decide whenever he was brought into the close proximity with a give-way vessel. For on the one hand, he was, and to some extent still is, required by law to hold to course and speed as long as it was possible for the other vessel to conform to the rules in time to escape collision;[35] on the other hand, he was forbidden to hold on the moment that the persistence of the other created an imminence of collision so great as to constitute a special circumstance. There are two things a seaman will do well to remember. One is that the navigator of a stand-on vessel, who has not taken earlier permissive action, should make no change in course and speed until he is prepared to testify that in his judgment the give-way vessel had made collision inevitable without such action. The other is that if he does not change before this, then *most action that he takes in good faith to aid in avoiding collision will be upheld by the courts.* Most action except no action—that is, continuing on into a collision without change.

Under the current rules, both on the high seas and in inland waters, the stand-on vessel *may* "take action to avoid collision by her maneuver alone, as soon as it becomes apparent to her that the vessel required to keep out of the way is not taking appropriate action in compliance with these rules." In Rule 17(c) the stand-on vessel in a crossing situation is warned not to take advantage of this option by turning left with the give-way vessel on her port hand. Thus, the stand-on vessel is no longer held to stand on into the jaws of collision if she can determine the other vessel is not giving way at an early stage. However, if this permissive action is not taken at a relatively long range, then the stand-on vessel continues to be bound to maintain course and speed until *in extremis*.

It is always difficult for the navigator of a stand-on ship to choose the

[33]The *Kaga Maru* (Wash 1927) 18 F (2d) 295.
[34]Postal SS. Corp. v. *El Isleo* (1940) 84 L Ed. 335.
[35]The *Southern* (Md 1915) 224 F 210.

right moment for taking action *in extremis*. In many cases he is in a dilemma in that, if he acts too early, he may frustrate the belated action by the give-way ship to keep out of the way, while if he acts too late, the action taken may be ineffective. Because of these difficulties the courts are slow to criticize a navigator merely because it can be shown, after the event, that the action that he took was either too early or too late. He is only criticized if, in taking action at the time he did, he did not exercise the ordinary skill and care that could reasonably be expected from a mariner faced with a situation of that kind.[36]

Two Kinds of Situations in Extremis

The courts make a distinction here in favor of the vessel that is brought into a situation *in extremis* solely through the fault of another vessel. It is true that she cannot invoke special circumstance to excuse a violation or an improper action unless she comes into court with clean hands. Thus, a merchant vessel may not be excused for an error *in extremis* when at the time she was not under command of a man with a master's license;[37] nor may a vessel navigating without a proper lookout;[38] but almost time without number the courts have agreed with an opinion expressed by the circuit court of appeals more than eighty years ago:

> Where the master of a vessel, who is a navigator of experience and good judgment, is confronted with a sudden peril, caused by the action of another vessel, so that he is justified in believing that collision is inevitable, and he exercises his best judgment in the emergency, his action, even though unwise, cannot be imputed to his vessel as a fault.[39]

A long line of decisions, many quoted in earlier editions of this book, have found vessels excused for action taken in haste when faced with an emergency situation caused by another vessel. As pointed out by both district court and circuit court of appeal in two decisions some sixty years ago:

> If one vessel places another in a position of extreme danger through wrongful navigation, the other is not to be held in fault if she is not navigated with perfect skill and presence of mind.[40]

> The master of a vessel acting *in extremis* is not held to an exercise of that cool and deliberate judgment which facts later developed show would have been a better course.[41]

[36]The *Martin Fierro* [1974] 2 Lloyd's Rep. 209.
[37]The *City of Baltimore* (CCA 1922) 282 F 490.
[38]The *James A. Lawrence* (NY 1902) 117 F 228.
[39]The *Queen Elizabeth* (CCA 1903) 122 F 406.
[40]The *Lafayette* (CCA NY 1920) 269 F 917.
[41]Sullivan v. Pittsburgh SS. Co. (1925) 230 Mich 414, 203 NW 126.

It is indeed not "perfect skill and presence of mind" or "cool and deliberate judgment" that the courts are looking for, but, as said earlier, "ordinary skill and care." In a case of a ship in a channel moving suddenly across the bows of an outbound ship, it was said of the latter:

> In judging the conduct of the master of the *Troll River*, it is necessary to bear in mind that he was placed in an extremely difficult position by the negligence of the *Shavit*, and the Court should not be too astute to find faults in the action he took in good faith to try and cope with the emergency so created.[42]

Equally, where a ship, on meeting another who was on the wrong side of a channel, took action that was aimed to mitigate the seriousness of the collision, she was not criticized for:

> putting her wheel to port, or not immediately reversing her engines, when she saw what the *City of Leeds* was doing, for a serious collision by then was inevitable and different action might have made the damage resulting from it greater rather than less.[43]

Even if there should be a delay, albeit slight, in the reaction of a ship confronted with a collision situation, she will not necessarily be found negligent and may escape liability for the collision:

> . . . the master, in waiting about one minute before taking avoiding action, made what can now in retrospect be seen to have been an error of judgement, but that he was not in all the circumstances negligent . . .[44]

Moreover, even when the action was thought to be wrong, as suggested in the above quote, it is often regarded as less contributory towards collision and may significantly reduce the apportionment of blame:

> . . . this fault was therefore causative but as it was taken when the vessels were at very close quarters and perhaps too hastily it was less blameworthy than a wrong decision taken with ample time for consideration.[45]

However, it must not be considered that any action *in extremis* is necessarily going to be excused by the courts, particularly if it is taken far too late:

> While making all reasonable allowances for the difficulty facing the master of the *Olympian* . . . he was at fault in that he waited too long before taking starboard wheel action . . . and the effort to avoid collision did not succeed.[46]

[42]The *Troll River* [1974] 2 Lloyd's Rep. 188.
[43]The *City of Leeds* [1978] 2 Lloyd's Rep. 356.
[44]The *Avance* [1979] 1 Lloyd's Rep. 153.
[45]The *Sanshin Victory* [1980] 2 Lloyd's Rep. 359; *see also* the *Estrella* [1977] 1 Lloyd's Rep. 526.
[46]The *Nowy Sacz* [1976] 2 Lloyd's Rep. 696; *see also* the *Auriga* [1977] 1 Lloyd's Rep. 395.

Neither, as was said earlier in the restricted visibility chapter, is precipitate action excused:

> The *Fierro* was only at fault in that, faced with a situation of danger entirely of the *Nayas* making, she did not take the right action in time to avoid or mitigate a collision . . . for putting her wheel to hard to port in the face of an up-coming ship without sufficient evidence of her intentions, does not involve too high a standard of skill and care.[47]

To conclude this point, it may be said that special circumstances exist and vessels are *in extremis* regardless of the cause, whenever the situation becomes one in which, because of the proximity of the vessels, adherence to the normal rules is reasonably certain to cause collision. While the courts adopt an understanding attitude to action departing from the rules in these circumstances, they still must be convinced that such action was that of a mariner of ordinary care and was instigated to avert or minimize a collision.

When Apparent Physical Conditions Prevent Compliance with Rules

A tug with her engines working full speed astern struck a pier on the East River with such force that the master was knocked unconscious, and then backed out in a semicircle with no one in control, until it struck another tug, with tow, coming up the river. Four minutes elapsed between the collision with the wharf and the collision with the tow. While the fault of the first tug in miscalculating her speed and striking the pier was not questioned, the circuit court of appeals found the second tug also liable for not sooner recognizing the erratic action of the other, both in her course and in her failure to answer signals, as special circumstances, and for not reversing more promptly to avoid the collision. this is admittedly a border-line case, with only a two-to-one decision by the circuit judges, the dissenting opinion agreeing with the lower court that the second tug was not at fault.[48]

A vessel completely disabled is clearly unable to comply with ordinary meeting and passing rules. She is, however, under a corresponding obligation to apprise other vessels that may approach her of her plight. Should she break down, her condition should be advertised by the use of the required two black balls or shapes in daytime, the two red lights at night, and the whistle signals of one prolonged and two short blasts when underway in fog.

A wholly disabled steamer being brought into her slip by two tugs damaged a vessel already moored at a pier. So far as the steamer was

[47]The *Martin Fierro* [1974] 2 Lloyd's Rep. 209.
[48]The *Transfer No. 19* (CCA NY 1912) 194 F 77.

concerned, this was a case of special circumstances, and the liability for the damage was attached by the circuit court of appeals to the owner of the tugs.[49]

Presence of More Than Two Vessels

It frequently happens, of course, in crowded harbors, in straits, and off headlands, that more than two vessels are involved in an approaching situation. In all such cases, special circumstances may be deemed to exist the moment any of the vessels are prevented from obeying the usual rules. Thus, if vessel *A*, heading north, is meeting vessel *B*, heading south, while vessel *C* is approaching from eastward to cross them, a complex situation arises; for *A* and *B* with respect to each other should alter course to the right and sound one blast, but with respect to *C*, *B* should maintain course and speed. *C*, on the other hand, is required simultaneously to maintain course and speed with respect to *A* and to give way with respect to *B*, and a one-blast signal by her would indicate both maneuvers, a physical impossibility. Similarly, if *A* is overtaking *B* and *C* is crossing from starboard, *B* would be bound to hold course and speed with respect to *A* and to yield with respect to *C*. In such cases, a timely and judicious use of whistle and radio signals will frequently solve the dilemma with a minimum delay to any of the vessels, although great care must be taken to guard against collision resulting from the acceptance by one vessel of a signal intended for another. Thus, in a situation in inland waters where one vessel is heading north to pass between two vessels proceeding south, but far enough apart so that the maneuver is practicable without a change in course by any of them, a signal of one blast will usually be exchanged between the single vessel and the one to be passed to port, and two blasts between the single vessel and the one to be passed to starboard; under International Rules, with the conditions as stated, no signals would be used.

A point to remember is that because special circumstances exist every vessel must, at the first evidence of confusion, be prompt to reduce her headway or to take any other steps necessary to avoid collision. It is a situation where the unpardonable sin is to maintain a dangerous rate of speed on the theory of a preconceived right-of-way that would apply were there only two vessels involved. As a precaution on the other side, the situation is not one of special circumstances if the relative distances apart and speeds are such that obedience to the ordinary rules will cause the vessels to encounter each other two at a time; in that case, these rules must be followed. Thus, where there were other vessels in the vicinity that were alleged to have hampered the movement of the stand-on vessel, but they

[49]The *Ascutney* (CCA NY 1921) 227 F 243.

were not close enough to prevent her compliance with the steering rules, there was not a case of special circumstances.[50] And in a very early decision it was held that embarrassment by proximity to vessels at anchor was no excuse for the failure of a give-way crossing vessel to keep out of the way of a stand-on vessel where there was no justification for her being so close to the anchored vessels.[51]

A number of illustrative decisions will serve to show the treatment by the courts of this type of special circumstance:

> The fact that a meeting vessel is in danger from a third which was in full view of the pilot of the other meeting vessel is a "special circumstance," which required the latter to slacken speed or to stop and reverse.[52]
>
> A tug was proceeding up the Delaware River, and a steamer was coming down on an opposite course, so that both were bound to change course to starboard. At this time a schooner was towed out from a pier and ran across the channel. As neither the tug nor the steamer could safely turn across the schooner's bow both turned to cross under her stern as closely as possible and collided. Neither having attempted to stop, both were at fault.[53]
>
> A collision between the tows of two meeting tugs in the East River was held due solely to the fault of the up-bound tug in attempting to pass through the narrow space between two descending tugs instead of passing on the port side of both.[54]
>
> Where a sloop and a lighter were sailing close-hauled on the same tack, on courses varying by only 1½ points, the sloop being the leeward vessel and overtaking the lighter, and a tow lay directly across their course, the lighter was bound to tack in time to keep out of the way of the necessary tack by the leeward vessel regardless of which was privileged.[55]
>
> A sheer made suddenly by an overtaking vessel to avoid the one ahead, which caused her to collide with a third vessel coming in the opposite direction before she could recover her course, was a fault.[56]

The situation of three vessels may be further complicated if additional vessels are involved, and of course the greater the number of vessels the greater the necessity of caution by each one. In general the same principle applies: that special circumstances must be deemed to exist until the regular rules can be obeyed with safety. A collision occurred on the East River when the side-wheel passenger steamer *Plymouth,* crowded too close to the shore by the overtaking steamer *Northland* when the latter passed her without an assenting signal, reversed full speed to avoid hitting the

[50]The *Morristown* (CCA NY 1922) 278 F 714.
[51]The *Hansa* (CCA NY 1870) Fed. Cas. No. 6,038.
[52]The *C. R. Hoyt* (NJ 1905) 136 F 671.
[53]The *Reading* and the *David Smith* (Pa 1888) 38 F 269.
[54]The *Volunteer* (CCA NY 1917) 242 F 921.
[55]The *Commodore Jones* (NY 1885) 25 F 506.
[56]The *Alaska* (NY 1887) 33 F 527.

Brooklyn ferry slips and was herself hit by a following tug. A half dozen other vessels were in the immediate vicinity, and the circuit court of appeals, while condemning the *Northland*, also found the *Plymouth*, the overtaken vessel, at fault for not reversing sooner when she saw the *Northland* attempting to pass without signal and knew the traffic ahead made the attempt dangerous.[57]

Situations Not Specifically Covered by the Rules

When any situation arises that is not specifically covered by the steering and sailing rules, the rule of special circumstances governs. For example, there is nothing in the rules about maneuvering around a wharf, except the requirement for the bend signal. A typical decision of the New York Circuit Court of Appeals has held that:

> Where a vessel is entering or leaving a slip and has not yet begun to navigate on a steady course, and a tow is going up or down the river the ordinary steering and sailing rules and signals made for vessels navigating on definite courses do not apply, but each vessel must proceed with due regard to all dangers of navigation and collision.[58]

Where a vessel is coming out of a dock or harbor into the channel, she must undock at a proper time, having regard to any vessels navigating outside.

> A ship which is coming out of a dock, or any side channel, into the main stream must, it is clear, do so at the proper time, and in a careful manner, having regard to traffic that may be passing up or down the main channel. The burden is on her not to cause embarrassment to any up-coming ship. That does not, of course, mean that the up-coming ship has anything in the nature of a right of way, because, as has been frequently laid down, there must be some give and take between vessels. What is wrong is for the vessel entering the main channel from the side to do so at such a time and in such a manner as to require the up-coming ship to take drastic action.[59]

The other vessel should be maneuvered with consideration for the difficulties of the emerging vessel, that is to say, special circumstances apply.

The circuit court of appeals has held in several cases that the starboard-hand rule does not apply to a steamer backing out of a slip before she gets on her definite course, but the special circumstance rule applies to steamers maneuvering to get on their course.[60] Again, where a tug with tow had to pass astern of a steamer backing out from a pier in a narrow channel, the special circumstance rule required the tug, which under the star-

[57]The *Plymouth* (CCA NY 1921) 271 F 461.
[58]The *Transfer No. 18* (CCA NY 1934) 74 F (2d) 256.
[59]The *Adellen* [1954] 1 Lloyd's Rep. 138; *see also* the *Jan Laurenz* [1972] 1 Lloyd's Rep. 329.
[60]The *M. Moran* (CCA NY 1918) 254 F 766.

board-hand rule would have been privileged, to give the steamer a wider berth, and the court found her at fault for not so doing.[61] In such cases the special circumstance rule applies to both vessels until the maneuvering vessel has proceeded far enough definitely to indicate her course.[62] Similarly, where a vessel is navigating near pier ends while a tug is bringing boats from a near-by slip to make up its tow, the case is one of special circumstance,[63] and where a tug is maneuvering with her tow in harbor waters, the situation is likewise one of special circumstances, governed by Rule 2(b) of the inland rules.[64] In the *Daniel McAllister*, the court held that the *Transfer No. 9*, a tug trying to rescue a drifting barge that had been knocked from her moorings by the *McAllister*, was not chargeable with a collision between the scow and a third vessel, which occurred notwithstanding her efforts. From the standpoint of the *Transfer No. 9* this was a case of special circumstances.[65]

It frequently happens in a harbor that when vessels are maneuvering to change their berths, one vessel approaches another while going astern, and occasionally a collision has occurred where both vessels approached each other stern first. The only reference in the rules to signals by a vessel backing is the requirement of Rule 34(a), international regulations and inland rules, that three short blasts be sounded if another vessel is in sight, and this signal must, of course, be given before any other maneuvering signal. The rules are silent regarding meeting and passing signals of backing vessels.

While it is the practice of seamen to regard the stern of a vessel as her bow when she is actually proceeding stern first, numerous court decisions such as those already cited justify the opinion that it is the special circumstance rule that properly governs such a situation. It is true that the stern may be regarded as the bow to the extent that it enables the mariner to determine what passing whistle in inland waters to propose to the other vessel. Thus a power-driven vessel backing west, desiring to back across a vessel to the southward proceeding north, should sound one blast as a proposal to the other vessel.

That the stern of a backing vessel is regarded as the bow only in this limited sense, and that the special circumstance rule replaces the regular meeting and passing rules when one or both vessels approach on a collision course stern first, is logical when we remember that vessels maneuver with much less certainty of control when backing and that no mariner can be deceived into mistaking the stern of a vessel for her bow.

[61]NY Central *Tug No. 27* (NY 1924) 298 F 959.
[62]The *Edouard Alfred* (NY 1919) 261 F 680.
[63]The *William A. Jamison* (CCA NY 1917) 241 F 950.
[64]The *John Rugge* (NY 1916) 234 F 861.
[65]The *Daniel McAllister* (NY 1917) 245 F 183.

From time to time single vessels attempt to pass through, or close ahead, of a squadron of warships or merchant ships in convoy. This can be most dangerous, and single vessels are advised to take early measures to keep out of the way. Mariners are expected to take note of the cautions and recommendations given in various national, official publications, details of which can usually be found in Sailing Directions or Notices to Mariners. Action taken at long range, before risk of collision, to avoid a fleet or convoy on the port bow would not be a departure from the Rules. If, however, a vessel in a formation or convoy is approached close enough for risk of collision to exist, then the Steering and Sailing Rules apply equally to both.

Cautions are usually printed on nautical charts warning mariners of the existence of submarine exercise areas. A vessel should give a wide berth to a warship flying the international code hoist "HP" or "OIY" denoting the presence of submarines submerged in the vicinity, particularly if she is of great draft and in relatively shallow waters.

Every vessel is liable for any damage caused by its swells, either to property along the shore or to passing vessels and their tows. This liability is not excused by the plea that the swells causing the damage were not as large as might have been produced by a high wind, or that the speed was customary for vessels of her class,[66] or that other vessels passed were not injured, or that the vessel injured could have escaped damage by taking unusual precautions. It may be said to apply whenever such speed is used as to cause injury to another vessel of a kind properly in the waters she is navigating in a proper manner.[67] The *Maid of Kent*, at about 19 knots, passed half a mile clear of the *Dunedin Star*, which was embarking a pilot from a launch alongside. The combination of sea state and wash crushed the pilot between the superstructure of the launch and the ship's side. The *Maid of Kent* was found liable.[68]

It has long been a doctrine of the rules that when vessels are approaching so as to involve risk of collision, a subsequent change of course by one of them cannot change a give-way vessel into a stand-on vessel. For example, a vessel overtaking another and passing her on her starboard hand cannot then swing across her bow and claim the right-of-way as a stand-on crossing vessel. The situation, which is partially covered by the rules, is one of special circumstances, and the overtaking vessel crosses at her peril.[69]

The rule requiring stand-on vessels to hold course and speed is mod-

[66]Nelson v. the *Majestic* (NY 1891) 48 F 730.
[67]The *Maid of Kent* [1974] 1 Lloyd's Rep. 435.
[68]The *Asbury Park* (NY 1905) 144 F 553.
[69]The *Horatio Hall* (NY 1904) 127 F 620.

ified whenever required by the approach of the stand-on vessel to pier ends, the windings of the channel,[70] or the necessity of stopping at a guard ship or a pilot ship within plain sight of the give-way vessel to report or to pick up a pilot.[71] The action of the stand-on vessel in slowing down or stopping is then justified under what is sometimes called the doctrine of presumable course and speed.[72]

Departure from Rules by Agreement

In two of the three possible approaching situations between power-driven vessels, the manner of passing is prescribed by the rules. An overtaking vessel may choose the side on which to pass, but a meeting vessel is required to go to starboard and a crossing vessel must comply with the rules of stand-on and give-way. The dangers and the occasional advisability of being a party to a departure from the usual procedure in the crossing situation have been discussed in chapter 16. In a crossing collision in New York Harbor the stand-on vessel proposed a two-blast signal, the give-way vessel assented with two blasts, and a collision followed. In finding both vessels at fault the circuit court of appeals pointed out that:

> The situation in this circuit, after the agreement, is one of special circumstances.[73]

The same arguments apply when two vessels meet head-on and one of them proposes a starboard-to-starboard passing, contrary to the statute both in inland waters and on the high seas. As a concluding statement, these arguments may be summarized as follows:

(a) A proposal to proceed contrary to law is not binding upon the other vessel.

(b) Unless and until such proposal is assented to by the other, both vessels must proceed in accordance with the rules.

(c) When such proposal is assented to by the other, neither vessel thereafter has the right-of-way, but both are equally bound to proceed with caution under the rule of special circumstances.

SUMMARY

Rule 2(b) international regulations and inland rules, provides for a departure from the ordinary rules when special circumstances make this necessary to avoid immediate danger. Rule 2(a) suggests that special

[70]The *Interstate* (NY 1922) 280 F 446.
[71]The *Roanoke* 11 Aspinall M.C. (NS) 253.
[72]La Boyteaux. *The Rules of the Road at Sea*, p. 127.
[73]The *Newburgh* (CCA NY 1921) 273 F 436.

circumstances may require action in addition to a full observance of the ordinary rules.

That the ordinary rules do not govern close situations is a popular fallacy among mariners. The courts have repeatedly held that these rules do hold and must be obeyed as long as it is reasonably possible for them to prevent collision. They have also held that rules may not be disregarded on the plea of special circumstances if an alleged danger is too distant, or it is suspected that a stand-on vessel is not going to perform her duty, or the wrong action is taken because there is not time for protracted deliberation, or the compass is defective, or a vessel unnecessarily enters a narrow channel in foggy weather and is forced by current conditions to proceed at excessive speed.

The Supreme Court has said that the rule of special circumstances "applies only where there is some special cause rendering a departure necessary to avoid immediate danger such as the nearness of shallow water, or a concealed rock, the approach of a third vessel, or something of that kind." It is characteristic of the special circumstance rule that when it is properly invoked, *neither vessel thereafter has the right-of-way and both are required to navigate with extreme caution.*

The rule of special circumstances applies: (1) whenever an approaching situation reaches the condition *in extremis*; (2) when physical conditions that should be apparent to both vessels prevent compliance with the ordinary rules; (3) when an approaching situation simultaneously involves more than two vessels; (4) when the situation is not specifically covered by the rules; and (5) when action contrary to the rules is proposed by a signal of one vessel and accepted by a signal of the other. In regard to the last, it should be remembered that such a proposal by one vessel is not binding on the other; that unless and until such proposal is assented to, both vessels are bound to proceed in accordance with the rules; and that after such assent both vessels are burdened, under the rule of special circumstances, to the extent that they must then proceed with caution.

19

Good Seamanship

Good Seamanship Defined in Rules

Seamanship, according to the dictionary, is the skill of a good seaman. The working definition of good seamanship in Rule 2(a) of both the international regulations and the inland rules is simply this: *any precaution which may be required by the ordinary practice of seaman, or by the special circumstances of the case.* It is one of those things that the mariner may not neglect with impunity. What is required of seamen is ordinary skill and ordinary intelligence; they are not expected to foresee and provide against *every* eventuality. Seamanship does not authorize departure from the rules, but provides:

> a solemn warning that compliance [with the rules] does not terminate the ever present duty of using reasonable skill and care.[1]

The rule says in effect that nothing in the rules shall exonerate any vessel or her owner or master or crew from the consequences of any neglect of good seamanship. What is good seamanship is a question of fact, to be decided on after a consideration of all relevant circumstances.[2]

To the careful student of the rules of the road it is apparent that the lawmakers who formulated the rules—and they included the leading professional seamen of their day—were at great pains to make them definite, specific, and comprehensive. Every possible situation was considered, and what was in the opinion of the delegates the most effective course of action to prevent collision in each case was prescribed. Thus certain crossing, overtaken, and other vessels were designated as having

[1]The *Queen Mary* (1949) 82 Ll. L. Rep. 341.
[2]The *Heranger* (1939) A.C. 101; the *Queen Mary* (1949) 82 Ll. L. Rep. 334; the *F.J. Wolfe* (1945) P.97.

the right-of-way, and vessels encountering them were directed to take all the action necessary to avoid them, with the understanding that such action should be based on the assurance that the stand-on vessel would maintain course and speed. Two vessels meeting end on were both specifically directed to avoid collision by turning to the right. Vessels were required to go at safe speed, to sound fog signals when required, and to reduce to minimum speed on hearing fog signals forward of the beam. It follows that obedience to these rules, which represent the lawmakers' ideas of what is the proper procedure under given circumstances, constitutes the first test of good seamanship; and conversely, disregard of the rules is generally prima-facie evidence of bad seamanship. However, though the regulations do represent good seamanship, they do not cover more than a small part of seamanlike practice.

In the preceding chapter, it was pointed out that the responsibility rule allowed departure from the other rules when special circumstances and immediate danger existed. Superior to this dispensation, however, is the responsibility to use good seamanship *in all cases*; i.e., an obligation is imposed in all circumstances to act in accordance with the recognized practice of skilled seamen. Such skills are not static, but constantly evolving as technology provides increased capabilities. The obligation might refer to the mariner's conduct leading up to, and perhaps even bringing about, the actual collision situation, or to his conduct in avoiding a collision thrust upon him by the fault of the other vessel, or to his conduct in maneuvering to lessen or to aggravate the damage of a collision after it had become inevitable. In considering the mariner's responsibility in such matters, it must be stressed again that the collision regulations are but one part, and a small part, of the principles of good seamanship. A separate book could be written on the all-encompassing needs of good seamanship, but this chapter will concentrate on those related to collisions.

Collision Illustrating Good and Bad Seamanship

A collision that occurred between two high-powered ferries in the channel approaching Puget Sound Navy Yard offers a good illustration of both bad and excellent seamanship in the sense contemplated by the rule. The *Chippewa*, on a course westward through Rich's Passage, was heading to cut inside the turning buoy at Orchard Rocks, while the larger, streamlined ferry *Kalakala*, having rounded Glover Point, was steering an approximate mid-channel course between Orchard Rocks Buoy and Middle Point, with Orchard Point light a little on her starboard bow. The courses of the two vessels were perhaps three points less than opposite, though the fact that they were following the windings of the channel made them technically meeting, rather than crossing, vessels. The *Chippewa*, in charge of her first officer, elected to hold on across the bow of the

Kalakala, and signified her intention by a two-blast signal. This maneuver was, of course, wrong, whether she regarded herself as a give-way crossing vessel or as a meeting vessel, and was made even less excusable by the fact that she was passing between the buoy and the rocks and could have started the turn before getting past the buoy into the channel. The justification urged by the officer on watch was that he was afraid to try to veer toward the other vessel because to do so he would be turning against a strong ebb tide. With a handy, full-powered vessel and only a moderate tide, this fear was probably groundless, but even if it were not, putting his vessel in such a situation by cutting inside a buoy would still lay him open to a charge of faulty seamanship. The two-blast signal was misunderstood by the *Kalakala*, also in charge of her first officer, as a one-blast signal, which she would expect, either as a stand-on crossing vessel or as a meeting vessel, to indicate a port-to-port passing. As the two ferries were approaching each other at a combined speed of more than 30 knots, matters developed very rapidly. As soon as the *Chippewa* heard the one blast of the *Kalakala* she blew the danger signal and reversed full speed; the *Kalakala* followed suit, and a few seconds before the collision, both skippers arrived on their respective bridges.

The steps taken almost instantly by the two seasoned veterans of the ferry line, both arriving on the scene in the jaws of a collision which by that time was inevitable, were impressive. Their testimony agreed on one point: that the two vessels, both backing to port and so preserving the angle of attack, would have struck at an angle of 30 to 40 degrees at a speed of not less than 10 knots, with the result that the *Kalakala* would have cut the *Chippewa* in two. The skipper of the *Chippewa*, however, ordered hard right and half speed ahead, and the skipper of the *Kalakala* hard left and half ahead, with the fortunate result that at the moment of impact, when both engines were again reversed, the vessels were crossing at a finer angle, and the high bow of the *Kalakala* merely raked the house of the *Chippewa* for a few feet, with the consequent destruction of three or four automobiles, but no injuries to passengers or hull. This was clearly an instance on the part of both captains of the best practice of seamen under difficult circumstances—a case where seasoned and promptly applied judgment prevented a major casualty.[3]

Forehandedness Is Essential

The scope of good seamanship is wide, but its practice has one underlying quality—that of forehandedness, or of thinking ahead. Sensible

[3]This case was investigated by the local inspectors of the former Bureau of Marine Inspection and Navigation, but as both vessels were owned by the same line, did not reach the courts. Investigations of this type are now conducted by the U.S. Coast Guard.

precautions are essential whether it be, to name but a few, the securing of cargo, a reduction of speed, the placing of the upper deck out of bounds, operating the fathometer, tuning a radar, calling extra hands, clearing away anchors, posting lookouts, or mentally working out what to do if something goes wrong. The skippers in the case just referred to above were sufficiently experienced seamen to have thought out in advance how they might react to certain circumstances. Such preparation is the hallmark of a good seaman.

One of the most obvious requirements of good seamanship in a vessel, whether underway or not, is the maintenance of a lookout appropriate to the circumstances.

> The duty of the lookout is of the highest importance. Upon nothing else does the safety of those concerned so much depend. In the performance of this duty the law requires indefatigable care and sleepless vigilance.[4]

Prior to the present rules, the responsibility of maintaining a proper lookout was included among the duties of good seamanship. It is the awareness provided by such advanced observation, not just in seeing ships but of recognizing environmental conditions, that allows seamanlike forehandedness to be implemented. A proper lookout is so essential, particularly with regard to observance of the collision regulations, that the courts have built up considerable doctrine relating to it. This necessary prerequisite to obedience to the rules will be more fully discussed in the following chapter. So important is the subject regarded that the latest international regulations and inland rules have provided a separate rule to cover it.

Presumption against Moving Vessel

It is a matter of seamanship that any vessel underway keeps clear of a vessel not underway. There is a definite presumption, by the courts, in favor of a vessel moored or at anchor, as against the vessel that collides with her—and quite properly of course, because of the relative helplessness of the fixed vessel to avoid collision. In restricted visibility a vessel striking another at anchor is practically self-convicted of excessive speed, since she has conclusively demonstrated her inability to stop in time. In clear weather, with lights on the anchored vessel at night, the other vessel is equally self-convicted of faulty seamanship, either through improper lookout or bad maneuvering. As frequently held:

> A moving vessel is prima facie in fault for a collision with one which is moored.[5]

[4]The *Ariadne* (1872) 13 Wall. 475.
[5]The *Banner* (Ala 1915) 225 F 433.

> Where a collision occurs between a vessel moored to the wharf and another steamer which is under way and susceptible of control and management, the presumptions sustained are in favor of the moored vessel, and against the one under way.[6]

A recent case to illustrate the outcome of a collision between one ship at anchor and another underway arose from an incident in the Shatt al Arab. With a following current of about 1½ to 2 knots, and a moderate breeze from aft, the *Ayra Rokh* attempted at a very late stage to cross the bows of the anchored *Aghios Gerassimos*. An unusually good anchor watch was being maintained in the latter ship, but as the *Ayra Rokh* had altered course to starboard across the bow within the last minutes before the collision, it was too late for any effective action to be taken on board the *Aghios Gerassimos*. The court found that

> The collision was not caused by any negligent act or omission on board *Aghios Gerassimos* but by the indecision and wrong action on board *Ayra Rokh* who was alone to blame for the collision.[7]

So strong is presumption of fault against the moving vessel that rare indeed is the case where all the liability is put upon the vessel moored or at anchor. Such an exception was the case of the *Jumping Jack* and the *Pinta*, two fishing vessels that collided in the Promised Land Channel near New York, under the following circumstances. The *Jumping Jack*, a sea skiff 32 feet long, with dark-varnished stern, was at anchor in the middle of the channel and her electric riding light had gone out. The moon had set, it was two hours before daylight, and the *Pinta*, a 60-foot oyster schooner, was proceeding down the channel at 6 knots. Her captain was at the wheel, a seaman on lookout in the bow, and a second seaman was outside the wheelhouse to relay signals from the lookout to the captain because of a noisy diesel engine, when the *Jumping Jack* was sighted almost under foot. The *Pinta* reversed full speed, and put her rudder hard over, but was unable to avoid sinking the *Jumping Jack*. Flagrant as was the fault of the latter in obstructing a narrow channel without lights, the district court divided the damages, but the circuit court, on appeal, exonerated the *Pinta*, finding that she infringed none of the rules and took all reasonable precaution to avoid the collision.[8]

In another case, a barge moored to the end of a New York pier was damaged by a Cunard liner attempting to make a landing at an adjacent pier. An hour before the steamship's arrival, the barge was given notice to move out of her dangerous position while the liner landed, and was

[6]Wood v. Harbor Towboat Co. (La 1881) 1 McGloin, 121.
[7]The *Ayra Rokh* [1980] 1 Lloyd's Rep. 68.
[8]The *Jumping Jack* (CCA NY 1932) 55 F (2d) 925.

offered the free services of a tug to aid her in moving out of the way and back again. The court held that she refused to move at her peril and dismissed her libel against the steamship.[9]

These cases are very rare, however, and ordinarily the best that can be hoped for by the vessel unfortunate enough to strike an anchored or moored vessel in clear weather is a division of damages, on one of four grounds: (1) improper position of the anchored vessel; (2) no lights, or improper lights, on the anchored vessel at night; (3) failure of the anchored vessel to maintain anchor watch where circumstances required; (4) failure of the anchored vessel to take proper steps to avoid the collision.

Anchored Vessel Partly Liable for Improper Position

(1) *Improper position of anchored vessel.* In a very early case of some interest, the Supreme Court emphasized the obligation of a moving vessel to keep clear of an anchored vessel regardless of whether or not the latter lay in a proper anchorage. It seems that on 1 August 1870, the salvage tug *Clara Clarita* saw a fire break out on a ferryboat moored on the Jersey side of New York Harbor and promptly set out to her rescue. After vainly trying to extinguish the flames, the tug was engaged by the ferryboat's master to tow the ferry clear of the wharf to prevent the spread of the fire, which she undertook to do with a Manila towline. Shortly after getting underway the flame spread to the towline, the ferry went adrift and struck the schooner *Clara*, which lay at anchor in her path, injuring her by the collision and setting her on fire. The tug extinguished the flames on the schooner and sought to defend the subsequent libel by arguing that the schooner, which had a proper anchor light and a man on deck, was anchored in a wrongful place. While the schooner was able to satisfy the courts that she was not obstructing the channel and was exonerated, the Supreme Court, agreeing with the lower courts, held that:

> Undoubtedly, if a vessel anchors in an improper place, she must take the consequences of her own improper act; but whether she be in an improper place or not, and whether properly or improperly anchored, the other vessel must avoid her if it be reasonably practicable and consistent with her own safety.[10]

In another early case, a 28-ton oyster schooner was improperly anchored in a harbor channel some 500 yards from a wharf and was directly in the path of a sidewheel steamer approaching to make her regular landing, anchorage in this locality being forbidden by a law of the state of Maryland. The evidence showed that the schooner's anchor light

[9]The *Etruria* (NY 1898) 88 F 555; *see also* the *Express* (NY 1892) 49 F 764.
[10]The *Clarita* and the *Clara* (1875) 23 L Ed. 146.

was burning. The steamer was proceeding at about 7 knots and did not discover the schooner until too late, though she attempted to avoid collision by reversing. The damage was slight, but the court held the side-wheeler at fault for excessive speed in a crowded harbor, declaring that:

> Where a steamer collides with a vessel unlawfully anchored in an improper and dangerous place, while negligently maintaining too high a rate of speed, the damages will be equally divided.[11]

In a later case a launch moored outboard of two other launches was struck by a passenger ship attempting to land at the city dock in the harbor of Juneau, Alaska. It was shown that the launch projected at least half her width outside a line drawn from the corner of the pier to a dolphin against which the stern of incoming steamers was expected to swing, but nevertheless the court found the steamship fully liable for the damage, holding that the launch's position was not legally improper and that the steamer must be treated as a moving vessel colliding with a vessel at anchor and without fault.[12]

In the case of the *Westernland*, that steamer was in collision with a schooner in New York Harbor that was anchored in an improper place too close to the wharf where the steamer had been lying. The steamer notified the schooner to move, but did not offer to provide a tug or to assist her. The schooner refused, and instead of calling on the harbor master to enforce the regulations and compel the schooner to move, the steamer attempted to back out of her slip in a strong ebb tide, and was carried against the schooner. Both vessels were held at fault, the schooner for being in an improper place and refusing to move, and the steamer for proceeding into obvious danger, a violation of good seamanship.[13]

Thus, a vessel should not anchor in a fairway unless she has no choice in the matter. If she should anchor in an improper place she must shift as soon as possible.[14]

Anchored Vessel Partly Liable for Improper Lights

(2) *No lights, or improper lights, on an anchored vessel at night.* A vessel at anchor at night without lights is prima facie at fault; nevertheless, there have been numerous decisions inculpating the moving vessel with the anchored vessel, the courts finding that even an unlighted vessel would have been discovered by a vigilant lookout in time to avoid her.[15] If a

[11]Green v. the *Helen* (Md 1880) 1 F 916.
[12]*Haho* v. the *Northwestern* (1920) 6 Alaska 268.
[13]The *Westernland* (NY 1885) 24 F 703.
[14]The *British Holly* (1924) 20 Ll. L. Rep. 237.
[15]The *Cambridge* v. the *Omega* (Md 1866) Fed. Cas. No. 2336; the *Premier* (Wash 1892) 51 F 766.

power-driven vessel maintaining a proper lookout and otherwise navigating properly reverses as soon as she sees the unlighted vessel, she would not be at fault, and in most cases the unlighted vessel at anchor has been held solely liable for the collision.[16] Where the anchored vessel has lights, but they do not conform to the specific requirements for vessels of her class, she will share the liability for a collision unless it can be proved the faulty lights could not have misled the approaching vessel or have been a contributing factor. In one case, a steamship at anchor at New Orleans, following a fire, with makeshift oil lanterns and electric cargo-cluster lights, was struck by an oil tanker coming down the river, and both were at fault.[17] In another case of mutual liability, a 75-foot dredge wrongfully exhibiting two white lights at anchor was mistaken by a tug with tow for a tug being overtaken.[18]

Failure to Maintain Necessary Anchor Watch

(3) *Failure of anchored vessel to maintain an anchor watch where circumstances required.* As said by the court in a very old case:

> A small vessel at anchor in a safe harbor in ordinary weather is not required by any rule or custom of navigation to set an anchor watch.[19]

In the case of the *Clara* and the *Clarita,* already cited, the Supreme Court, in absolving the anchored schooner from fault, made the comment that the statute does not require a watch on a vessel at anchor.[20] A vessel in the naval service is required by regulations to have an anchor watch, and in many harbors such a watch is specified for all vessels at anchor by harbor ordinance, which has the full force of law. However, despite the absence of any specific provision in the international or inland rules, the courts have found that an anchor watch is sometimes required under the rule of good seamanship. Thus, it was held in the Supreme Court that a schooner at anchor inside the Delaware breakwater during a storm, when numerous vessels were seeking shelter, was in fault for not having a watch on deck, and when sunk by another vessel that was properly navigated and on her way to anchor could not recover damages.[21] Again, in a Massachusetts case it was held that when a vessel is at anchor where other vessels are frequently passing, and navigation is difficult and dangerous because of shoals and a channel only 1½ miles wide, special vigilance is required, including not only a watch on deck but

[16]The *Westfield* (NY 1889) 38 F 366.
[17]The *Chester O. Swain* (CCA NY 1935) 76 F (2d) 890.
[18]The *Arthur* (NY 1901) 108 F 557.
[19]The *Fremont* (Cal 1876) Fed. Cas. No. 5,904.
[20]The *Clarita* and the *Clara* (1875) 23 L Ed. 146.
[21]The *Clara* (NY 1880) 26 L Ed. 145.

someone on lookout to warn off an approaching vessel.[22] In foggy weather, there is, of course, a special reason for requiring an anchor watch on a vessel in a busy harbor;[23] but even in clear weather, if the night is dark and the anchored vessel is in the way of traffic, good seamanship demands it. As said by the court when a schooner barge at anchor in the middle of the Elizabeth River below Norfolk, with an anchor light but no anchor watch, was run down and sunk by a steamship bound down the river:

> Anchored where she was on such a night, she was bound to take every precaution to warn approaching vessels of her presence. A vigilant watch on her deck might by shouting and swinging a lantern have attracted the attention of those on the steamboat to her presence in the locality where she lay at anchor in time to have enabled the steamship to have avoided her. . . . Both vessels being found in fault the damages will be apportioned.[24]

Thus, the requirement for an anchor watch will depend on the circumstances, including the type of vessel involved and the number in her crew.[25] It was held that a small coaster, with a crew of twenty-five, who, while at anchor in the River Thames on a clear, fine night, was struck by a vessel underway, was not expected to maintain a full anchor watch and that, in the prevailing circumstances, the employment of a single lookout was sufficient precaution.[26]

Finally, inasmuch as a vessel is liable for any damage she may do to another vessel by dragging, whenever weather, current conditions, or poor holding ground are such to indicate that possibility, a competent anchor watch becomes essential to prevent it.[27] This watch must be experienced and have received the master's instructions.[28]

Failure of Anchored Vessel to Take Mitigating Action

(4) *Failure of anchored vessel to take proper steps to avoid the collision.* While a vessel at anchor is relatively unable to maneuver, there are two acts that are required of her under all circumstances, and omission of either of them may involve her in fault for a collision under the rule of good seamanship. In the first place, she must not anchor too close to another vessel, the legal presumption being that the vessel anchored first has a right to ample swinging room, upon which the later arrival must not infringe. Secondly, she must anchor securely—that is, with sufficient chain out—and if heavy weather or a strong current make it necessary,

[22]The *Henry Warner* (Mass 1886) 29 F 601.
[23]The *Lydia* (CC NY 1873) Fed. Cas. No. 8,615.
[24]The *Guyandotte* (NY 1889) 39 F 575; *see also* the *Lehigh* (NY 1935) 12 F Supp 75.
[25]The *Gerda Toft* [1953] 1 Lloyd's Rep. 257.
[26]The *Cedartrees* [1957] 1 Lloyd's Rep. 57.
[27]The *Forde* (CCA NY 1919) 262 F 127.
[28]The *Sedulity* [1956] 1 Lloyd's Rep. 510.

she must drop a second anchor to prevent dragging into another vessel. Failure to do this has resulted in numerous decisions holding the dragging vessel at fault for collision.[29] In one such case, the steamship *Bragdo*, at anchor off Staten Island in a December gale with 45 fathoms out, dragged across the chain of the steamship *British Isles* and set her adrift. In finding the *Bragdo* at fault, despite the fact that the gale had reached hurricane strength, the circuit court of appeals cited the rule from Knight's *Seamanship*, recommending a length of cable equal to seven times the depth of water for ordinary circumstances with more if weather conditions cause the vessel to put excessive strain on the chain.[30] It is *prima facie* evidence of negligence for a vessel to drag her anchor, and all seamanlike precautions have to be taken to prevent it. In the *Velox*, a ship dragged her anchors in weather of unusual severity and collided with another. The relevant sailing directions contained a warning about the holding ground, and it was found that when the weather became bad the *Velox* should have used her engines to prevent initial dragging and to have taken even more drastic action later to arrest the dragging:

> . . . no seaman can be called upon to exercise more than ordinary care: but . . . when a seaman is called upon to face wholly exceptional conditions, ordinary care of itself necessarily demands that exceptional precautions may have to be taken.[31]

If an anchored vessel finds herself threatened by another vessel, whether the latter be underway or not, there are some actions she can take to avoid or reduce the effect of collision. These measures may often be futile, because the need to take them does not become apparent until too late for them to be effective.[32] However, the courts have occasionally found that such measures should have been taken, particularly the working of chain whether it be pulling up on the cable[33] or the more common exigency of veering more chain. Two old cases will serve to illustrate the latter. In one instance, a tug and helper were going up the Hudson River at night in a flood tide, with a tow 1,600 feet in length consisting of nine tiers of canal boats, when a steamship was discovered half a mile ahead, anchored somewhat outside the prescribed anchorage. The tug and her helper went right somewhat, but not promptly enough to avoid collision between the last tier in the tow and the steamship, and one of the canal boats was sunk. The anchor watch on the steamship saw the flotilla

[29]The *Bertha* (Va 1917) 244 F 319; the *Djerissa* CCA (1920) 267 F 115; the *Jessie* and the *Zaanland* (1917) P 138.

[30]The *British Isles* (CCA 1920) 264 F 318.

[31]The *Velox* [1955] 1 Lloyd's Rep. 376.

[32]The *Ceylon Maru* (Md 1920) 266 F 396; the *Beaverton* (NY 1919) 273 F 539.

[33]The *Prospector* [1958] 2 Lloyd's Rep. 298.

approaching in ample time so that if he had given her chain the tide would have carried his vessel back and out of danger. He failed to do this, and on that point, and not because she was technically in an unlawful anchorage, the circuit court of appeals found the steamship liable for half the damages.[34]

In another somewhat different case of mutual fault, which is of particular interest because of the emphasis laid by the court on good seamanship in its comments to both parties, the three-masted steamer *Cochico*, lying at anchor near the outer entrance to Hampton Roads in a strong ebb tide, was in the act of heaving up her anchor, preparatory to getting underway. The day was fair, and a large fleet of perhaps 150 vessels, which had put in for shelter the day before and anchored from one to seven miles above the *Cochico*, was proceeding to sea *en masse*, helped along by a light following breeze and a 1- or 2-knot tide. The *Cochico* was in the middle of a 2-mile channel, and the fleet was so numerous that the colliding vessel, the *Kelsey*, did not see her until within 250 yards, when an intervening vessel hauled out of line. The *Kelsey* changed course about a point, enough barely to clear the *Cochico*, but the latter sheered slightly and the *Kelsey*, striking her at a fine angle, knocked off her bowspirit and did some other damage. Just before the impact, the master of the *Kelsey* hailed the *Cochico* to starboard her helm and pay out chain, but she did neither. In finding both vessels liable, the *Kelsey* for not avoiding the *Cochico*, which on evidence she might easily have done by prompt and effective measures, and the *Cochico* for her entire lack of prudence, attention, and assistance in avoiding danger while voluntarily remaining as an obstruction in the midst of a fleet of moving vessels, Judge Brown made the following significant observations:

> [To the *Kelsey*] Such a sheer should have been expected: hence her master was at fault for not allowing a sufficient margin of safety, amid the contingencies of navigation, and not taking in time the decisive measures at his easy command. As I must find that the master had sufficient time and space to keep out of the way had he acted with the promptness and decision that reasonable prudence demanded, and as there was no other vessel that prevented his doing so, the *Kelsey* must be found in fault.
>
> [To the *Cochico*] Ordinarily a vessel anchored in a proper place in the daytime and in fair weather, is not expected, or legally required, to be on the watch, and to stand prepared to take measures to avoid vessels under way, and having control of their motions. But under exceptional circumstances, where the vessel under way is subject to special difficulties or embarrassments in her navigation, some care and precautions on the part of the vessel at anchor may become obviously prudent and necessary that would not otherwise be obligatory. Such I think is plainly this case.[35]

[34]Riley v. the *Richmond* and the *E. Heipershausen* (CCA NY 1894) 63 F 1020.
[35]Wells v. Armstrong (NY 1886) 29 F 216; also the *Bacchus* (Va 1920) 267 F 468.

The last case mentioned also points to another action open to a ship at anchor: that of using her rudder to sheer clear of an approaching vessel. If a vessel at anchor can avoid collision by use of her helm, she may be held to blame if, having the opportunity to do so, she did not exercise it.[36] However, she must not take such action until it becomes clear that the ship underway cannot, by her own action, avoid a collision.[37]

In conjunction with working her cable and helm, a ship at anchor with power ready can, and should, use her engines to assist avoiding action. Following a collision in the River Schelde, the *Sabine* was faulted for not paying out more cable and using her engines in order to make more room for the *Ore Prince* to pass.[38]

Standing Too Close to Other Vessels

The same line of reasoning that presumes it bad seamanship to hit a vessel that is moored or at anchor applies when a collision occurs between a vessel with way upon her and one that is lying dead in the water. Thus, when a steamship that had stopped off the quarantine station on the Delaware River for examination, but was not at anchor, was approached so closely by a passing tug that the two heavy scows in her tow both struck and injured the steamship, the court held the tug solely liable for the damage.[39] As was said in another case:

> The obligation on the part of free vessels to avoid risk of collision with those incumbered, or at rest, is imperative, and one that the admiralty courts must enforce, having regard to the perils of navigation and the importance of the rule of the road in respect thereto.[40]

Similarly, off pilot stations where ships may be encountered lying stopped it has been said:

> . . . it was in any event the duty of *Sestriere*, as a matter of good seamanship in the special circumstances of the case . . . to take timely action to keep clear of the *Alonso* who arrived on the scene first. . . .[41]

However, as discussed in an earlier chapter, it was the *Alonso* who took action that, unfortunately for her, was not excused as being taken in the agony of the moment, and she joined the *Sestriere* in being found equally liable for bad seamanship. More recently, in the Gulf of Thailand the

> *Thomaseverett* was negligent in adopting her course . . . so as to pass between *Esso*

[36]Marsden's *Collisions at Sea*, 11th Ed. p. 607.
[37]The *Viper* (1926) P. 37.
[38]The *Sabine* [1974] 1 Lloyd's Rep. 473.
[39]The *John F. Gaynor* (Pa 1902) 115 F 382.
[40]The *Shinsei Maru* (Va 1920) 266 F 548.
[41]The *Sestriere* [1976] 1 Lloyd's Rep. 131.

> *Chittagong*, which had been waiting for a pilot, and the pilot vessel, there being no more than a cable clearance on either side.[42]

However, while it may be good seamanship and manners to avoid a vessel stopped in the water on the high seas, it must be made clear that she does not have the privileges of a ship at anchor. She is underway and where risk of collision exists should comply with the regulations as far as she is able.[43]

When slowing down to embark a pilot, sufficient allowance must be made for leeway. A large steamship in light trim had hardly any steerage-way when, on embarking the pilot, a squall blew her downwind onto a vessel at anchor. The *William Wilberforce*:

> was alone to blame for the collision . . . which resulted from the inadequate margin of safety which she left.[44]

Even when both vessels are making way, it is unseamanlike to approach so close to another that, for example: upon that ship's stopping or altering course slightly a collision cannot be avoided;[45] or where insufficient sea room is left for corrective action to overcome a steering casualty;[46] or where there is a narrow stretch of river ahead;[47] or to attempt to squeeze through a narrow gap when the vessel ahead swings in the river.[48]

Restriction of Speed in Good Visibility

Discussion of safe speed applicable to *any* condition of visibility was given in chapter 17. Rule 6 of both the international regulations and the inland rules contains a list of factors that must be taken into account, regardless of how good the visibility is, in determining what is a safe speed.

In most canals and in many rivers and harbors, a specific speed limit is fixed by local statute or ordinance, and such a regulation unquestionably has the force of law. When a speed regulation exists, it invariably means speed over the ground, and allowance must therefore be made when the rate is accelerated by a favorable current.[49] One of the important applications of the rule of good seamanship as interpreted by the courts is in the restriction of speed. An examination of the cases shows numerous vessels at fault for excessive speed in inland waters in the absence of any specific

[42]The *Thomaseverett* [1979] 2 Lloyd's Rep. 402.
[43]Marsden's *Collisions at Sea*, 11th Ed. p. 612: but *see also* discussion in chapter 16 re the *Devotion II* [1979] 1 Lloyd's Rep. 509.
[44]The *William Wilberforce* (1934) 49 Ll. L. Rep. 219.
[45]The *Kate* (1933) 46 Ll. L. Rep. 348.
[46]The *Frosta* [1973] 2 Lloyd's Rep. 348.
[47]The *Ore Chief* [1974] 2 Lloyd's Rep. 427.
[48]The *Sabine* [1974] 1 Lloyd's Rep. 472.
[49]The *Plymouth* (CCA NY 1921) 271 F 461.

speed limit, these speeds ranging all the way from 4 to 17 knots. It may be stated as a general rule that any speed in a harbor or narrow channel is excessive (1) if it causes damage to other property by the vessel's swell, or (2) if it renders the vessel herself unmanageable in maneuvering to avoid collision.

(1) *Damage caused by vessel's swell.* In one of the earliest cases decided on this point, the Circuit Court of New York declined to find a steamer at fault for damage to a number of canal boats in a tow caused by the steamer's swells when she passed the tow on the Hudson River. The court held that, there being at that time no rule of law prescribing the speed a boat might use or the swell it might make or how near it might pass to another, any reasonable speed was justifiable, and 17 knots, the steamer's regular speed, was proper.[50] But in a case a few years later, court interpretation changed this rule considerably, and for more than one hundred years now the opposite view has prevailed: namely, that a large power-driven vessel that proceeds at such speed as to create a swell causing injury to another vessel properly in the waters she is navigating and properly handled, is liable for such injury, even if that speed is only 5 or 6 miles an hour.[51] In the case referred to, the offending steamer, with her engines on dead slow ahead, passed within a few feet of a scow loading at an icehouse and was held for failure to stop her engine entirely. In another case two or three years later, an ocean liner passed a river tug with a scow in tow on each side in upper New York Bay, and her swells caused the tug to seriously damage one of the scows. She passed the two within half a mile at a speed of about 11 knots. But notwithstanding the steamship's argument that the tug contributed to the damage by failure to present her stern to the swell of the overtaking vessel, the latter was found fully liable. (Incidentally, while a meeting vessel is under obligation to head into a swell with her tow if it will lessen damage, the courts have steadfastly refused to require an overtaken vessel, which is privileged, thus to alter her course.) The opinion of the court included the following excellent statement of the rule, which still holds:

> Such waters are not to be appropriated to the exclusive use of any class of vessels. We do not mean to hold that ocean steamers are to accommodate their movements to craft unfit to navigate the bay, either from inherent weakness, or overloading, or improper handling, or which are carelessly navigated. But of none of these is there any proof here, and in the absence of such proof we do hold that craft such as the libelant's have the right to navigate there without anticipation of any abnormal dangerous condition, produced solely by the wish of the owners of exceptionally large craft to run them at such a rate of speed as

[50]The *Daniel Drew* (CC NY 1876) Fed. Cas. No. 3565.
[51]The *New York* (NY 1888) 34 F 757.
[52]The *Majestic* (CCA NY 1891) 48 F 730.

will insure the quickest passage. To hold otherwise would be virtually to exclude smaller vessels, engaged in a legitimate commerce, from navigating the same waters.[52]

Even in lower New York Bay substantially the same rule has been applied. As said by the circuit court of appeals when a Cunard liner, coming in at night, caused a swell that damaged two scows in a tow by making one scow override the other:

The rule that large vessels navigating New York Bay must so regulate their speed as not to injure by their swells small craft, which are seaworthy and properly loaded and navigated, is also applicable to the lower bay, though not with the same strictness. Owing to its less crowded condition and nearer proximity to the sea, incoming steamers may there proceed at greater speed, provided the channel is free, but not when it is full of boats, at night, or in a fog.[53]

In this case the tug was found contributorily negligent in having tow lines only 6 feet in length between the scows, and only half damages were allowed.

In a more recent case, a barge loading on the St. Lawrence River just west of Quebec was injured by pounding on the bottom when she was struck by a heavy swell from the Cunard steamer *Andania*, which passed her at a speed over the ground of about 17 knots, though there is a legal speed limit in that part of the river of 9 knots. Ascertaining the extent of the damage after the barge had completed a voyage to New York, the owner sued the Cunard Line *in personam*, and was met with the interesting defense that the steamship was in charge of a compulsory pilot, and that the company through its agent, the master, was therefore not liable. However, the court held that the master, who was also on the bridge, was negligent in not exercising his superior authority and ordering the pilot to slow down in conformity with the government speed regulation.[54]

Excessive speed can also cause damage and death in more open waters where no government speed limits apply. Outside Dover harbor in the English Channel, a pilot was boarding a vessel when his launch rolled violently in a choppy sea and he was crushed. Finding that the wake of a passing ship contributed to the unexpected severity of the roll, the court said

. . . those in charge of the *Maid of Kent* should have realized from the time of clearing the breakwater in Dover Harbour that, if they passed *Dunedin Star* at a

[53]The *Campania* (NY 1913) 203 F 855.
[54]The *Emma Grimes* (NY 1933) 2 F Supp 319; *see also* the *Hendrick Hudson* (NY 1933) 3 F Supp 317.

distance of about half a mile and at nearly 20 knots, the wash might create a danger for the pilot launch if it proceeded alongside the *Dunedin Star*.[55]

As a matter of practical seamanship it is well to remember that to reduce the swell of a speeding power-driven vessel, it is necessary to slow down a considerable distance before reaching the vessel it is intended to protect. In a New York case where the passing steamer in a narrow channel did not slow sufficiently or in time, it is reported that the tow was actually broken up by the swell that piled up *ahead* before the steamer had come abeam.[56]

Commanding officers of naval vessels who find themselves under the necessity of making high-speed trial or post-repair runs in more or less confined waters will be interested in cases where the following defenses have *not* been accepted by the courts: (1) that a vessel's waves did not render navigation more perilous than would a high wind;[57] (2) that a vessel was navigating at a speed customary for ships of her class,[57] (3) that other vessels passed on that or other similar occasions were not injured;[58] (4) that the vessel injured did not sound a warning signal to the other vessel to slow down.[59] (5) that the vessel injured might have saved herself by taking unusual precautions.[60]

(2) *Speed excessive under particular conditions.* It has long been held by the Supreme Court that a power-driven vessel in a crowded harbor or river should not be operated at a higher speed than will keep her under perfect control. This is, of course, merely a rule of common sense. In a collision between two early steamships off the Battery, with numerous other vessels in the immediate vicinity either underway or at anchor, a speed of 6 knots was held excessive;[61] and in another collision the next year between a steamship and a schooner, near the same spot under like conditions, 7 knots was held excessive. In the last case, the schooner was standing over to the Jersey shore to anchor and await a fair tide up the East River; and the old steamship *City of Paris*, bound out at 7 or 8 knots, headed to pass through a 300-foot opening between a brig and a sailing ship, and did not see the schooner until the latter passed at right angles under the stern of the brig. The steamship immediately reversed, but although the schooner luffed slightly in a futile effort to escape, she was almost cut in two by the impact and sank so quickly as to imperil the lives of all her crew. In

[55]The *Maid of Kent* [1974] 1 Lloyd's Rep. 435.
[56]The *Luke* (CCA NY 1927) 19 F (2d) 925.
[57]The *New York* (NY 1888) 34 F 757.
[58]The *Asbury Park* (NY 1905) 144 F 553; the *Hendrick Hudson* (NY 1908) 159 F 581.
[59]The *Chester W. Chapin* (NY 1907) 155 F 854.
[60]The *Emma Grimes* (NY 1933) 2 F Supp 319.
[61]The *Corsica* (NY 1870) 19 L Ed. 804.

upholding both lower courts in their condemnation of the steamship, the Supreme Court said:

> She ought not to have entered upon the narrow track between the ship and the brig without being very careful first to see that her passage would involve no danger to any approaching vessel in its transit. The results proved that the speed of the steamer was higher than was consistent with the safety of other vessels in so crowded a thoroughfare and hence higher than she was warranted to assume.[62]

The high court dismissed the argument that the schooner contributed to the disaster by luffing, in the following brief, but pointed, comment:

> The acts complained of were done in the excitement of the moment and *in extremis*. Whether they were wise it is not material to inquire. If unwise they were errors and not faults. In such cases the law in its wisdom gives absolution.

Similarly, in numerous other decisions vessels in collision have been held liable for excessive speed where it was found that they approached other vessels in restricted waters at speeds that the results showed were imprudent. In a collision between two steamships on the Patapsco River near Baltimore on a clear day, one of them, the *Acilia*, attempted to blow two blasts for a starboard-to-starboard meeting, and the whistle cord stuck, causing a single whistle of 5 or 6 minutes' duration. The other vessel, the *Crathorne*, which was making about 6 knots, tried to pass port to port; the *Acilia* executed left rudder, both vessels reversed, and they came together with some $50,000 damage to the *Crathorne*. The *Acilia* was found solely liable for the damage for going at her ordinary cruising speed of 10 knots, the circuit court of appeals remarking that:

> Full speed in these dredged channels when about to pass other vessels is undeniably a fault which increases every risk of navigation.[63]

It is a well-known fact that vessels in shallow water have a tendency to sheer and become unmanageable, and that if they attempt to pass too closely they are likely to be brought into collision by suction. Hence, vessels colliding from either of these causes are often convicted of excessive speed. A typical case occurred at Horeshoe Bend on the Delaware River between the steamship *Saratoga*, going down light and the steamship *Taunton* coming up from sea. It was a clear day; the vessels saw each other 2 miles apart, signaled a port-to-port meeting when a mile apart, and were about to clear each other in the usual manner, when the *Saratoga* touched a mud bank at the side of the channel, the existence of

[62]The Liverpool, New York, and Philadelphia SS. Co. v. Henry P. Simmons (NY 1870) 19 L Ed. 751; *see also* the *George H. Jones* (CCA NY 1928) 27 F (2d) 665.

[63]The *Acilia* (CCA Md 1903) 120 F 455.

which was well known and marked, and sheered into the other vessel before she could be stopped. On a showing that the *Taunton* was properly on her own side of the channel, but that the *Saratoga*'s speed of not less than 8 knots caused her to "smell the bottom" as she rounded the buoy, the latter was found solely liable for the collision.[64]

In another case a steamship 314 feet long anchored for the night in Brewerton channel below Baltimore, at a point where the channel is 600 feet wide, so that when she swung around her stern was about 100 feet from one side and her anchor chain extended toward the other. Another steamship, heavily loaded with iron, coming up from sea, attempted to pass under her stern at 8 knots, and as she reached the shallow edge of the channel, took an uncontrollable sheer toward the anchored vessel. While the anchored steamship was held liable for unnecessarily obstructing a navigable channel, in violation of the Act of March 3, 1899, the colliding vessel was held equally at fault for a speed that prevented her overcoming the effects of a sheer that might reasonably have been expected.[65]

A more recent case of excessive speed in a river was in the Schelde, Belgium, where interaction between two ships passing resulted in the overtaken ship taking the ground.

> It was not seamanlike for *Ore Chief* to overtake *Olympic Torch* . . . because the two ships were large vessels entering a narrow and shallow stretch of river after having made a turn . . . she should have reduced speed for good seamanship required that overtaking should have been postponed. . . .[66]

The question of good seamanship is also involved when vessels make their way at too great a speed along a city waterfront and fail to keep a safe distance off the pierheads. Many harbors have local regulations prohibiting such movements within a specific distance of the piers, and in New York a statute requires vessels navigating the East River to go up and down in mid-channel. But regardless of such local rules, the courts have again and again held vessels at fault that collided with vessels properly emerging from their slips. Vessels maneuvering around piers are under special circumstances, and the greatest caution must be observed when there is a possibility of encountering them. This doctrine was enunciated by the United States Supreme Court as long ago as 1873, when a case was carried up involving a collision between a side-wheel excursion steamer hugging the Brooklyn piers at a speed of 8 knots to avoid a tug with tow, and a ferry that emerged from her slip, saw the steamer bearing down from port, and reversed in a frantic, though perhaps mistaken, attempt to

[64]The *Saratoga* (Pa 1910) 180 F 620; *see also* Appleby v. the *Kate Irving* (Md 1880) 2 F 919.
[65]The *Caldy* (CCA Md 1907) 153 F 837.
[66]The *Ore Chief* [1974] 2 Lloyd's Rep. 432, 433.

escape disaster. The decision of the Supreme Court, condemning the excursion vessel and exonerating the ferry, included much that would apply in greater or lesser degree to any busy harbor:

> In the East River, vessels cannot with safety run across the mouths of ferry slips in going to or from their wharves, but they should occupy as near as possible the middle of the river. . . . If the middle of the river be previously occupied and the ship is obliged to go nearer to shore in order to avoid other vessels pursuing the same track she must run at such a slow rate of speed as to be easily stopped, so as not to endanger boats pursuing their regular and accustomed occupation.[67]

And in clearing the ferry on the charge of failure to hold her course and speed, the court reiterated the opinion expressed three years earlier:[68]

> In a moment of sudden danger, caused by the misconduct of the colliding vessel, the law will not hold the pilot of the injured vessel, acting in good faith, guilty of a fault, if it should turn out after the event that he chose the wrong means to avoid the collision, unless his seamanship was clearly unskillful.[69]

In concluding this point it may perhaps be unnecessary to point out that it is no defense to a collision a few feet off the piers to argue that the speed was less than the statutory limit. As said by the court in finding a side-wheel steamer solely liable for a collision while navigating within a ship length of the piers at 9 knots:

> A statute imposing a penalty for running along the piers of the East River at a speed exceeding 10 knots does not necessarily render a less rate of speed prudent. The speed must be regulated by the dangers attending the navigation under the particular circumstances of the case.[70]

Among other decisions inculpating vessels for excessive speed under the rule of good seamanship may be mentioned the case of a steamer approaching a sailing vessel beating through a 300-foot channel in the Penobscot River, failing to anticipate the sailing vessel's tack and holding to a speed of 8 knots;[71] the case of a steam yacht that approached the blind bend at Horn's Hook near New York at 15 knots and was in collision with a tow coming down the river;[72] the case of a steamship that approached a confusion of lights, part of them improper, which turned out to be on a single tow of two tugs and seven barges off Governor's Island, and in trying to go through them at more than 4 knots, sank two of them;[73] the

[67]The *Favorita* v. Union Ferry Co. (NY 1873) 21 L Ed. 856; *see also* the *Shady Side* (NY 1899) 93 F 507.
[68]The *Sif* (NY 1920) 266 F 166.
[69]*See* note 67 above.
[70]Greenman v. Narragansett (NY 1880) 4 F 244.
[71]The *Northern Warrior* (Me 1870) Fed. Cas. No. 10,325.
[72]The *Hoquendaqua* (CCA NY 1918) 251 F 562.
[73]The *Corsica* (NY 1870) 19 L Ed. 804.

case of a government lighthouse tender coming down the East River at night that approached at 10 knots a group of three vessels crossing the river both ways ahead of her;[74] and the case of a Puget Sound steamer that entered Port Townsend Harbor at full speed, and in approaching her wharf, failed to distinguish the flickering lantern of an anchored bark with its background of bright city lights, until collision was inevitable.[75]

Even in less restricted waters, with good weather and no laid down speed limit, ships can be castigated for excessive speed:

> . . . it was not . . . seamanlike for her to try to work up to full speed while still in the roadstead.[76]

Vessels Must Undock at Proper Time

Despite the cases mentioned earlier in this chapter, where vessels proceeding along rivers were faulted for unseamanlike speed after colliding with vessels *properly* leaving their berths, it does not follow that vessels undocking can blindly cast off. All ships must dock or undock at a proper time, having any regard to any ship navigating in the stream. Obviously it requires a good lookout,[77] for as has been said:

> . . . it is the duty of the vessel that is docking or undocking to give way to the vessel which is lawfully proceeding up or down river. . . . she has to carry out her operation without embarrassing passing traffic, and that involves that she must choose the proper moment for embarking on the adventure of docking or undocking.[78]

> What is wrong for the vessel entering the main channel from the side is to do so at such a time and in such a manner as to require that upcoming ship to take drastic action.[79]

Equally the vessel in the stream, as we saw before, should be handled with consideration for the difficulties of the undocking vessel. When the *Jan Laurenz*, proceeding down river, saw the *St. William* undocking close ahead of her, she maintained her speed in what was described as a "race for the viaduct." However, the court also found that:

> *St. William* was to blame for leaving her berth at an improper time . . . and persisting in going down the canal and not holding back.[80]

[74]U.S. v. the *Mineola* (NY 1879) Fed. Cas. No 15,799a.
[75]Fristad v. the *Premier* (Wash 1892) 51 F 766.
[76]The *Savina* [1974] 2 Lloyd's Rep. 325.
[77]The *Aracelio Iglesias* [1968] 1 Lloyd's Rep. 131.
[78]The *Hopper R.G.* [1952] 2 Lloyd's Rep. 352.
[79]The *Adellen* [1954] 1 Lloyd's Rep. 144.
[80]The *Jan Laurenz* [1972] 1 Lloyd's Rep. 404.

Good Seamanship May Require Holding Back at a Bend

It is axiomatic that a good seaman will know the state of the tide and of the current, particularly in rivers and narrow straits. If stemming the current, and the blind-bend signal is heard from another vessel coming downstream, it will often be prudent to wait until the other has passed.[81] However, unless there are local regulations, the duty to wait only applies if a current is actually running at the time.[82] Even then, the duty is not absolute, for it is applicable only when good seamanship decrees it proper to stop and wait, a question that clearly depends on the circumstances of the case.[83] Thus, in a narrow channel with a swift current in one direction, it has been held in several cases that when vessels meet, the one moving with the current is the favored vessel.[84] In a collision between a tug and tow on a long hawser coming down the Hudson River and rounding the sharp bend at West Point, where the current sweeps rapidly toward the opposite bank, and a similar tug and tow coming up the river, this rule was applied in favor of the former and against the latter.[85] It was applied when two steamships met in the Delaware River, in the narrow channel above Horseshoe Buoy, when one of them, running light and stemming the tide, tried to cross the bow of the other, deeply laden and coming with the tide. The former was held solely liable. The Supreme Court, applying it to a collision between a steamer and three barges in a tow that the steamer sank in Hell Gate with the tide running 7 knots, thus definitely stated the rule:

> Where two steamers about to meet are running one with and the other against the tide, if it be necessary that one or the other should stop in order to avoid a collision, the one proceeding against the tide should stop.[86]

Formal recognition of this doctrine exists in local rules throughout the rivers of the world. For U.S. inland waters, it is coverd by Inland Rule 9(a)(ii), where downbound vesels in narrow channels or fairways of the Great Lakes, Western Rivers, and other specified waters, have right-of-way over upbound vessels and the duty of proposing the manner and place of passing. Although, therefore, geographical limits have been placed on the applicability of the rule, the practice of good seamanship may extend it to other narrow channels, as the cases above demonstrate.

Vessels Must be Properly Manned and Operated

Apart from the owner's responsibility properly to provide for and man

[81]The *Talabot* (1890) 15 P.D. 194; the *Ezardian* (1911) P. 92.
[82]The *Sagaporack* & *Durham Castle* v. Hontestroom (1927) 25 Ll. L. Rep. 377.
[83]The *Backworth* (1927) 43 T.L.R. 644.
[84]The *Marshall* (NY 1882) 12 F 921.
[85]The *Scots Greys* v. the *Santiago de Cuba* (CC Pa 1883) 19 F 213.
[86]The *Galatea* (NY 1876) 23 L Ed. 727.

a ship, good seamanship requires the master to organize his watch-standers properly and insist that the correct standards are maintained.

> It appeared . . . clear from . . . evidence that the state of discipline on board the ship was inclined to be lax, and that the attitude of those on the bridge to their watchkeeping duties was lacking in a proper sense of responsibility.[87]

In commenting on the general slackness in the running of the ships that were involved in a serious collision in the North Sea, the court said owners should ensure that the masters understood their duties and appreciated that they were to run an efficient ship. The court produced a long list of what factors constituted a well-run ship.[88] In a separate case, it was said that in certain circumstances, such as thick fog and dense traffic:

> . . . if the master . . . needed a rest from the bridge, his place should have been taken by another officer, such as the chief officer.[89]

At all times, there must be a sufficiency of trained crew on board and, when necessary, on watch. A tug was faulted because she could not sound her whistle while the mate on watch was handling the helm and there was no other person available.[90] When a small schooner, moored at New Orleans in a gale, sought to change her position in the absence of the captain, with only a man and a boy to handle her, she was held responsible for her own injuries when impaled on the bow of a steamer and sunk.[91] In a collision between a schooner and a steamer off Sea Girt, New Jersey, at night in clear weather, the erratic actions of the schooner were adjudged due to the incompetence of her helmsman, who managed to display on the witness stand a profound ignorance of the duties of a seaman, and the schooner was held solely at fault for the collision.[92] And similarly, when a vessel in a tow on the Saint Mary's River turned toward an approaching steamer because the helmsman made a mistake in executing the master's orders—putting the wheel hard right instead of hard left—she was found liable for the resulting damage.[93]

It is often necessary to supplement an emergency maneuver with the dropping of one or both anchors. In order to do this, a man must be standing by. As said by a district court:

> In waters well frequented by small tows, a ship must have a competent person standing by in the forecastle ready at a moment's notice to let go the anchors.[94]

[87]The *Sea Star* [1976] 1 Lloyd's Rep. 122.
[88]The "*Eleni V* and the *Roseline*," *Seaways*, August 1981.
[89]The *Zaglebie Dabrowskie* [1978] 1 Lloyd's Rep. 570.
[90]The *Murdoch* [1953] 1 Lloyd's Rep. 440.
[91]The *Sarah* v. *Bellais* (CCA La 1892) 52 F 233.
[92]The *Tallahassee* (CAA NY 1903) 125 F 1005.
[93]The *Sitka* (NY 1904) 132 F 861.
[94]River Terminals Corp. v. U.S. (DC La 1954) 121 F Supp 98.

However, in one unusual case in a busy river, one ship was:

> . . . not guilty of negligence in failing to let go her anchors because the orders to let them go could not be heard because of the noise of [the other ship's] anchors running out.[95]

Vessel Should Not Sail with Defective Equipment

Vessels have been found liable for a collision where they have chosen to get underway even though there was a defective compass[96] or lights that were not operating correctly. In some cases vessels have been convicted because a failure occurred a second time, such as a steering failure, and the cause was not determined and corrected after the first failure. An unexpected failure of equipment will not be excused if the equipment has not been submitted to periodic inspection and preventive maintenance.

Every self-propelled vessel of 1,600 or more gross tons must have a marine radar system, an illuminated magnetic steering compass with an up-to-date deviation card, a gyrocompass that the helmsman can read, an illuminated rudder angle indicator in the wheelhouse, an echo-sounder, and equipment on the bridge for plotting relative motion.[97] Similar equipment requirements are laid down internationally in chapter V of the 1974 SOLAS. If during the voyage some equipment fails, the voyage may be completed subject to the nearest CG authority being informed as soon as possible. Similarly, a failure of the bridge-to-bridge radiotelephone should be "given consideration in the navigation of the vessel".[98]

Navigator Must Know His Vessel and the Conditions to Be Encountered

Particularly in close quarters, a navigator must be familiar with the advance and transfer that is characteristic of his vessel for the speed at which he is traveling. He must also be knowledgeable of the effectiveness of a backing bell depending on the speed. Allowance must be made, in both cases, for the effect of wind and current. As expressed by the court:

> A navigator is chargeable with knowledge of the maneuvering capacity of his vessel. He is bound to know the character of his vessel and how she would turn in ordinary conditions.[99]

A vessel will not be excused for a collision caused by weather conditions if those conditions could have been avoided. As stated by a district court in a recent case:

[95]The *Oldekerk* [1974] 1 Lloyd's Rep. 95.
[96]Greater New Orleans Expressway v. Tug *Claribel,* 222 F Supp 521 (E.D. La) 1963.
[97]33 CFR 164.35 (a)–(j).
[98]33 USC 1205 (Supp V 1975).
[99]City of New York v. Morania No. 12 Inc. (1973) 357 F Supp 234.

> Tug's obligation includes responsibility to utilize available weather reports so that it can operate in manner consistent with foreseeable risk and captain of tug is chargeable with knowledge of weather predictions whether he knows them or not.[100]

Similarly, a vessel cannot be excused for striking a navigational hazard if the cause of the navigator's ignorance is failure to carry up-to-date charts or if he goes against the considered advice of the pilot and harbor master.[101] Equally, ignorance is no excuse for failure to comply with a traffic separation scheme, even though in the past it was not binding on the particular flag-state's ships:

> Her master should have known of the scheme and obeyed it as a matter of good seamanship.[102]

IMCO-adopted traffic separation schemes are now, of course, binding to all ships on the high seas. Even when some distance from them, a ship can be affected by failure to take into account the implications to traffic flow:

> Good seamanship required *Garden City* to steer a more westerly course so as to line herself up on the entrance to the southbound traffic route and avoid unnecessary encounters with ships proceeding northwards in the other traffic route.[103]

Leniency Toward Salvors

Damages arising from salvage efforts are regarded leniently. In the *St. Blane* case, it was not accepted that the collision that occurred while going alongside the disabled yacht *Ariadne* was due to negligent seamanship, though the master of the *St. Blane* was guilty of negligence in failing to take all reasonable care to avoid further damage to the yacht after the crew had been taken off.

> It is well established that the Court takes a lenient view of the conduct of salvors or would be salvors, and is slow to find that those who try their best, in good faith, to save life or property in peril at sea, and make mistakes, or errors of judgement in doing so, have been guilty of negligence . . . salvors should not in general be criticized if, faced with an actual or potential conflict between saving life on the one hand, and preserving property on the other, they err on the side of the former at the expense of the latter.[104]

Thus, despite the loss of the *Ariadne* and the unseamanlike precautions to minimize damage when the two vessels were alongside each other, the *St. Blane* was not found liable.

[100]M. P. Howlett Inc. v. Tug *Dalzellido* (DC NY 1971) 324 F Supp 912.
[101]Texada Mines v. *AFOVOS* [1974] 2 Lloyd's Rep. 175.
[102]The *Genimar* [1977] 2 Lloyd's Rep. 17.
[103]The *Zaglebie Dabrowskie* [1978] 1 Lloyd's Rep. 571.
[104]The *St. Blane* [1974] 1 Lloyd's Rep. 563.

First Rule of Good Seamanship

In the final analysis, good seamanship becomes a factor in collision prevention in so far as the navigator exercises it, by observing the Rules of the Road, and by conducting his vessel where it is safe, when it is safe, and in the manner of a prudent seaman. Such is the test placed upon him by the courts, as typically expressed by the circuit court of appeals:

> No man is infallible, and there are certain errors for which the law does not hold a navigator liable; but he is liable for an error of judgment *which a careful and prudent navigator would not have made.*[105]

SUMMARY

Good seamanship as defined in the rules of the road means the practice of every precaution that may be required by the ordinary practice of seamen. The first test of good seamanship is obedience to the rules, but it must be remembered that:

> The Collision Regulations do not contain the whole wisdom of the sea . . .[106]

Timely preparation and insistence on proper standards, allied to forehandedness, are the mark of a good seaman and so often can avert the situation that would otherwise lead to disaster.

Under the rule of good seamanship there is always a presumption of fault against a vessel anchored improperly; against a moving vessel that collides with a vessel moored or at anchor; and sometimes against a vessel going up a swift stream that collides with a vessel going downstream.

A vessel may be held at fault for damage done by her swell, or for maintaining such a speed that, because of the proximity of piers or of crowded traffic, collision results. Nonetheless, it is the responsibility of the vessel departing to ensure she undocks at a proper time. It is an obligation of good seamanship, indeed of seaworthiness, that a vessel be properly equipped, manned, and operated, including to be prepared to let go the anchors at short notice under certain conditions. A vessel must not leave the dock with essential equipment that is defective, and a navigator is charged with knowledge of the maneuvering characteristics of his vessel, the weather, and information from up-to-date publications and charts. To repeat, good seamanship is a factor in collision prevention as it influences the mariner in observing the Rules of the Road, and in conducting his vessel where it is safe, when it is safe, and in the manner of a prudent seaman.

[105]The *Old Reliable* (CCA Pa 1921) 269 F 725.
[106]The *Hardwick Grange* (1940) 67 Ll. L.Rep. 359.

20

A Proper Lookout

Requirement of Proper Lookout Positive

In the naval service there is probably no rule of the road more conscientiously observed than the admonition of Rule 5 international and inland rules to keep a proper lookout. In the merchant service, where vessels of corresponding tonnage carry much smaller crews, the relative scarcity of men results in many more cases of collision attributable at least in part, according to the case books, to improper lookout. It is interesting to note that of the many collision cases in the records very few involved naval vessels. As might be expected, about three times as many cases occur in inland waters as on high seas, a fact no doubt due to the relative congestion of shipping and not to the maintenance of a less efficient lookout in crowded waters. However, as evidenced by specific findings of faulty lookout on the high seas, there is apparently some tendency by navigators to let down on the requirements once a vessel is clear of the land.

Nothing could be more positive than the obligation to keep a proper lookout, as construed by the civil courts and, it might be added, by naval courts and boards. As early as 1833 an American sailing vessel was held liable for a collision with another sailing vessel having the right-of-way, because she was navigating at sea, in daylight in clear weather, with no watch on deck but the man at the wheel.[1] As a requisite of common seamanship, our courts have been enforcing careful vigilance by those entrusted with the navigation of vessels for more than one hundred and fifty years. It is an obligation that applies to all vessels of a size capable of

[1]The *Rebecca* (NY 1833) Fed. Cas. No. 11,168.

committing injuries.[2] As held in a collision between two small vessels on a clear night many years ago:

> The failure to keep a lookout is a violation of the general rule to prevent collisions between vessels, and nothing can exonerate a vessel from such failure, unless it should appear that the collision would have occurred notwithstanding such failure. This rule is undoubtedly as applicable to the boats of the motor class as to ocean vessels.[3]

The obligation should be regarded as applying at all times when underway, day or night, and even, under some circumstances, when at anchor. For while the rule does not specify a watch on a vessel at anchor, and a vessel securely anchored in a safe harbor, with proper lights, in ordinary weather need not have one, yet it was held in an old case that a schooner at anchor inside the Delaware breakwater during a storm, when numerous vessels were seeking shelter, was in fault for not having a watch on deck, and when she was sunk by another vessel that was properly navigated, she was unable to recover damages. The decision was affirmed by the United States Supreme Court.[4] Again, in a Massachusetts case it was held that when a vessel is at anchor in a place where other vessels are frequently passing, and where navigation is difficult and dangerous because of shoals and a channel only a mile and a half wide, special care and vigilance are required, and she must have a good lookout and an anchor light of the regular pattern lighted and burning. To have a watch on deck is not sufficient if there is no one on lookout at the time of the collision to warn off an approaching vessel.[5] And, of course, the well-known liability of a vessel for any damage that it may do to another through dragging anchor makes a proper lookout imperative whenever conditions create a risk of such an occurence, as when a vessel was anchored in New York Harbor during a winter gale that reached a velocity of 88 miles an hour.[6]

Lookout Defined

A lookout has been defined by the federal court as a person who is specially charged with the duty of observing the lights, sounds, echoes, or any obstruction to navigation with the thoroughness that the circumstances permit.[7] The words *specially charged* imply that such person shall have no other duties that detract in any way from the keeping of a proper lookout. Thus, it has been held in numerous cases that because the lookout must devote his attention to this duty, the officer of the deck or

[2]The *Harry Lynn* (Wash 1893) 56 F 271.
[3]Brindle v. the *Eagle* (Alaska 1922) 6 Alaska 503.
[4]The *Clara* (NY 1880) 26 L Ed. 145.
[5]The *Henry Warner* (Mass 1886) 29 F 601.
[6]The *Forde* (CCA 1919) 262 F 127.
[7]The *Tillicum* (Wash 1914) 217 F 976.

the helmsman cannot properly serve as lookout.[8] Even on a slow-moving tug with a tow, the officer in charge of navigation is not legally complying with the duty by keeping a lookout from the pilot house.[9] Where the captain of a steamer is acting at the same time as pilot and lookout, the vessel has not a proper lookout, and the owners may be liable for the damage caused by such omission.[10] A seaman who had been dividing his attention between looking out and reefing sail was held not to be a vigilant lookout,[11] and where the only two men on the deck of a schooner navigating at night were engaged in taking down sail, it was held that neither one nor both seamen constituted a proper lookout, and accordingly, the schooner was at fault for colliding with another schooner having the right-of-way.[12] A vessel backing out from a pier in a fog was at fault for requiring the lookout in her bow to take in the bow line, since this interfered with the degree of vigilance required under such conditions.[13] In still another case, where the lookout on a car float alongside a tug was directed to concentrate his attention on the East River piers, so that he did not see a large block of floating ice that forced tug and tow to collide with a steamer alongside a pier, it was held that the tug should have detailed an additional man for the general lookout duty.[14] However, in a very early decision the Supreme Court found that the man blowing the foghorn on a sailing vessel was a proper lookout,[15] and in 1911 the Circuit Court of Appeals of New York rendered a similar decision in refusing to hold a schooner guilty of contributory fault for colliding with a steamer when it was shown that the mate, in dense fog, was acting as lookout and sounding the foghorn at the same time, but that all his duties were properly done.[16] Absence of key personnel from the bridge will invariably lead to a charge of improper lookout. Thus, the *Statue of Liberty* was at fault for her bad lookout one night with clear visibility, when for 12 minutes before the collision, only the second officer was on her bridge.[17] And the other ship, the *Andulo*, was faulted for not taking more accurate observations of compass bearings, which constituted inadequate lookout. Similarly:

[8]The *Kaga Maru* (Wash 1927) 18 F (2d) 295; the *Donau* (Wash 1931) 49 F (2d) 799.
[9]The *City of Philadelphia* (Pa 1894) 62 F 617; the *Sea Breeze* Fed. Cas. No. 12,572a.
[10]Bill v. Smith (1872) 39 Conn 206; Dahlmer v. Bay State Dredging & Contracting Co. (CCA Mass 1928) 26 F (2d) 603.
[11]The *Twenty-one Friends* (Pa 1887) 33 190.
[12]The *Fannie Hayden* (Me 1905) 137 F 280.
[13]The *Albatross* (Mass 1921) 273 F 285.
[14]New York and Oriental SS. Co. v. N.Y., N.H., and H. Ry. (NY 1906) 143 F 991.
[15]The *Nacoochee* v. Mosely (NY 1890) 34 L Ed. 687.
[16]The *Pallanza* (NY 1811) 189 F 43.
[17]The *Statue of Liberty* [1971] 2 Lloyd's Rep. 277.

> The fault of the *Horta Barbosa* was due to a most serious defect of lookout on board her, amounting virtually to an abdication of responsibility by her second officer, who was absent from the bridge 6–7 minutes before the collision.[18]

Even when a bridge team is closed-up, many collisions in clear weather have been found to have an underlying theme of bad lookout that led to a poor appreciation of the situation. The *Homer* was at fault for bad aural and visual lookout, because she did not hear two separate sound signals or observe the other ship's alteration of course to starboard.[19] In another collision, at night in clear weather where both ships were again cited for bad lookout, one, the *Ziemia*:

> . . . did not appreciate what lights *Djerada* was carrying, or what course she was on, until one minute before the collision . . . that could have only been due to his extremely bad lookout.[20]

In a river a good lookout must be kept for down-going and up-coming vessels before starting a turn[21] as well as during the turn.[22]

In addition to maintaining "a proper lookout by sight and hearing," the rules also require a proper lookout "by all available means appropriate in the prevailing circumstances and conditions so as to make a full appraisal of the situation and of the risk of collision." Therefore, under certain circumstances on the high seas and in inland waters, the use of radar information is required in order to have a proper lookout.

Rule 5 places greater emphasis than hitherto on the need for a proper lookout; the requirement is positive rather than implied. Guidance on the maintenance of a proper lookout is given in the IMCO *Recommendations on Navigational Watchkeeping*, Section I, paragraph 11.[23] This guidance distills much of the wisdom the courts have pronounced over the years on this subject.

The expression "all available means appropriate" means that effective use must be made of suitable instruments and equipment and is not confined to the use of radar only to supplement a visual and aural watch. Binoculars should be used, not only by the lookout, but by the bridge personnel, and if necessary, used on the bridge wing or through an open window.

[18]The *Sea Star* [1976] 2 Lloyd's Rep. 478; the *Devotion II* [1979] 1 Lloyd's Rep. 513; the *Johanna O.D. 18*, LM & CLQ, Feb 1980, p. 97.

[19]The *Homer* [1973] 1 Lloyd's Rep. 50; also the *Troll River* [1974] 2 Lloyd's Rep. 181; the *Glenfalloch* [1979] 1 Lloyd's Rep. 255; the *Stella Antares* [1978] 1 Lloyd's Rep. 52; the *City of Leeds* [1978] 2 Lloyd's Rep. 356.

[20]The *Djerada* [1976] 1 Lloyd's Rep. 56; also the *Sestriere* [1976] 1 Lloyd's Rep. 1; the *Auriga* [1977] 1 Lloyd's Rep. 396; the *Savina* [1975] 2 Lloyd's Rep. 141.

[21]The *Francesco Nullo* [1973] 1 Lloyd's Rep. 72.

[22]The *Sabine* [1974] 1 Lloyd's Rep. 465.

[23]*See* Appendix J.

It is difficult, in my view, in any event, to understand why he did not use binoculars on seeing the approaching *Gorm*. Apparently he remained behind closed windows in the wheelhouse.[24]

In addition to keeping a radar lookout, vessels should not neglect to listen to the radio:

I find that the *Antonio Carlos* was at fault for bad look-out in the broadest sense; namely, faulty appreciation of VHF information and total absence of radar look-out.[25]

Where shore-based radar stations are operating, it has been held that use should be made of them:

. . . these facilities of radar advice are made and supplied and established for the greater safety of shipping in general and for greater accuracy in navigation. . . . A vessel which deliberately disregards such an aid when available is exposing not only herself, but other shipping to undue risks, that is, risks which with seamanlike prudence could, and should, be eliminated. . . . there is a duty upon shipping to use such aids when readily available—and if they elect to disregard such aids they do so at their own risk.[26]

Clear visibility does not dispense with the need to keep a radar lookout. In the United States courts, ships have been found at fault for not using radar as a general lookout at night, after colliding with offshore oil-drilling rigs, even though the visibility was clear.

Other courts have ruled similarly:

. . . it was clear that those on board *Kylix* had made wholly erroneous estimates of both the relative course and the distance of *Rustringen* and that those errors arose primarily from their relying solely on visual observations and estimates instead of making use of radar which was available to them. The pilot and master both said that, since it was a clear night, there was no need to use the radar. The advice . . . which I accept, is that radar should have been used.[27]

However, in a more recent case, the admiralty court found that

Esso Chittagong was also negligent in failing to observe the course change of *Thomaseverett* but her failure to make radar observations was not negligent in that *Thomaseverett* had been seen and visibility was good.[28]

Clearly, whether radar should be used to supplement a proper visual lookout will depend on the circumstances and the adequacy of the lookout. Sometimes, a vessel is not obliged to use radar, even in restricted

[24]The *Gorm* [1961] 1 Lloyd's Rep. 196.

[25]The *Bovenkerk* [1973] 1 Lloyd's Rep. 70.; *see also* the *Anco Princess* [1978] 1 Lloyd's Rep. 296.

[26]The *Vechtstroom* [1964] 1 Lloyd's Rep. 118

[27]The *Kylix* [1979] 1 Lloyd's Rep. 139.

[28]The *Thomaseverett* [1981] 1 Lloyd's Rep. 1.

visibility, if it is not working properly, though active efforts should be made to have repairs completed as soon as possible.

> There might well be times when the continued use of radar by a navigator who was uncertain of the results he was observing and unwilling to place reliance thereon might well be foolhardy and hazardous.[29]

However, Judge Medina in the U.S. Appeals Court, when reviewing this case in 1959, gave the following caution:

> This does not mean that, in the face of the fact that a properly functioning radar will give useful and necessary information, the master had a discretion to decide that it will not give such information and turn off his radar. A master has no more discretion to disregard this aid to navigation than he has to disregard the use of charts, current tables and soundings where the circumstances require the use thereof.

The use of a radar as a general lookout does not dispense with the need to maintain a visual lookout:

> The question . . . on this occasion was: "Was it seamanlike for the *Arietta* to rely on relative motion radar observation only and to have no visual look-out?" and the answer was: "No."[30]

Lookout comprises not only the proper use of sight and hearing, augmented by equipment such as radar, but also a proper appreciation of a situation by the person in charge of the watch. The officer of the deck must be alert to what is happening in his own vessel, checking the steering, the correct functioning of equipment and, not least, that the correct lights continue to be shown at night. Several collisions in recent years have occurred due to dilatory discovery of equipment defects.

> Where, in my judgement, she was at fault, was in having a very bad look-out, and a bad look-out in every possible sense of the term. It seems to me that it comes within the term "bad look-out" when I say that she was at fault for failing to take proper precautions to meet the situation in the event of the compass breaking down again, as it in fact did. It was, in my judgement, bad look-out on the part of this young third officer in failing to appreciate, long before he did appreciate it, what was happening, namely that his vessel was falling off to starboard, and in failing to appreciate what the probable cause of the falling off was. It was bad look-out on the part of the quartermaster, when he knew perfectly well that the compass had stuck again, not to report the matter at once to the officer in charge. It was bad look-out on the part of the officer to take no steps himself, whether by going to the standard compass or otherwise, to check up on what was happening and what was the course of his vessel.[31]

[29]The Pocahontas Steamship Co. v. Esso Aruba (1950).

[30]The *Arietta* [1970] 1 Lloyd's Rep. 70: also the *Anneliese* [1970] 1 Lloyd's Rep. 36; the *Miguel de Larrinaga* [1956] 2 Lloyd's Rep. 538.

[31]The *Staffordshire* (1948) 81 Ll. L. Rep. 141: also the *Anna Salem* [1954] 1 Lloyd's Rep. 488.

Many similar judgments have followed in subsequent cases.[32] The advent of modern aids in ships seems sometimes to lead to an unfounded belief in their reliability and a too-casual attitude toward keeping a proper lookout. In the Mediterranean, in 1964, the cargo ship *Trentbank* suffered a failure of her automatic steering as she was overtaking the tanker *Fogo* and swung across the bow of the latter.

> I ought not to leave this part of the case without observing how lamentable was the attitude of the master of the *Trentback* and her chief officer towards the system of automatic steering. The master had given no orders to ensure that somebody was on the look-out all the time. The chief officer, according to his own story, saw nothing wrong in undertaking a clerical task and giving only an occasional glance forward when he knew that there was other shipping about and that he was the only man on board his ship who was keeping any semblance of a look-out at all. Automatic steering is a most valuable invention if properly used. It can lead to disaster when it is left to look after itself while vigilance is relaxed. It is on men that safety at sea depends and they cannot make a greater mistake than to suppose that machines can do all their work for them.[33]

Immaterial Absence of Proper Lookout Not a Fault

It is not meant to imply that whenever two vessels collide, the mere proof of improper lookout on either vessel, in the technical sense, will *ipso facto* condemn that vessel for the collision. On the contrary, it has been held by the Supreme Court that the absence of a lookout is unimportant where the approaching vessel was seen long before the collision occurred;[34] that it is immaterial where it does not appear that the collision could in any way be attributed to his absence;[35] and that the absence of a lookout stationed where he should be will not render a vessel in fault for a collision where she was navigated exactly as she should have been had there been a lookout reporting the situation.[36] As said by the circuit court of appeals in a later case, where an overtaking vessel rammed the vessel ahead:

> Absence of a lookout is not entitled to weight in cases where the proof is satisfactory that the vessel in fault saw the other in time to have taken every precaution it was its duty to take, and which, if taken, would have avoided the collision.[37]

But the difficulty in practice is, of course, to overcome the presumption

[32]The *Chusan* [1955] 2 Lloyd's Rep. 685; the *Esso Plymouth* [1955] 1 Lloyd's Rep. 429; the *Indus* [1957] 1 Lloyd's Rep. 335; the *Greathope* [1957] 2 Lloyd's Rep. 197; the *British Tenacity* [1963] 2 Lloyd's Rep. 1; the *Salaverry* [1968] 1 Lloyd's Rep. 53.

[33]The *Trentback* [1967] 2 Lloyd's Rep. 208.

[34]The *Dexter* (Md 1875) 23 L Ed. 84; the *George W. Elder* (CCA 1918) 249 F 9656.

[35]The *Tacoma* (Wash 1888) 36 L Ed. 469.

[36]Elcoate v. the *Plymothian* (Va 1897) 42 L Ed. 519.

[37]The *M. J. Rudolph* (NY 1923) 292 F 740; the *Lehigh* (NY 1935) 12 F Supp 75.

of fault that the absence of a proper lookout entails and to furnish such satisfactory proof. It is only when it can be made clear that the lack of a lookout could not have contributed to the collision that it will be excused.[38] Two illustrative cases may be mentioned here. In one of them, a dredge at work in a channel during a dense fog, while sounding a fog bell at intervals of less than a minute that could be heard many times the distance of visibility, was run down by a stern-wheel steamer making 15 knots, and the presence or absence of an efficient lookout on the dredge was held immaterial.[39] In another case, in a crossing situation on the East River, a burdened ferry stopped well off the course of a privileged tug to let her pass, but was rammed when the latter suddenly changed her course; it was held with some degree of reason that the want of a proper lookout on the burdened vessel was not a contributory fault.[40]

Notwithstanding such occasional exceptions, the navigator should always adhere to the general admiralty rule that:

> The strict performance of a vessel's duty to maintain proper lookout is required and failure to do so, especially when other craft are known to be in the vicinity, is culpable negligence.[41]

Many court decisions on the subject indicate that the strict performance referred to means, at least in most circumstances, not only that lookouts shall be free from other duties but that they shall be (1) qualified by a certain amount of experience as seamen, (2) vigilant and alert, (3) properly stationed, and (4) in such numbers as circumstances require in order that the vessel may avoid risk of collision.[42]

Lookout Should Be Experienced Seaman

No definite minimum experience requirement has been laid down by the courts as qualifying a man for duty as lookout, but several decisions have shown the necessity of some attention to this point. In an early case the District Court of New York held that the steward, who was standing by the companion way, and was no mariner, and had not been stationed as a lookout, was not a proper lookout;[43] and fourteen years later the district court of Pennsylvania held that it is doubtful whether a steward is a competent lookout, and he certainly is not when his attention is divided between such duty and the duties belonging to his employment as steward, such as serving coffee to the crew.[44] The circuit court of appeals has

[38]The *Titan* (CC 1885) 23 F 413; *see also* Dion v. United States (D. Me 1961) 199 F Supp 707.
[39]The *Bailey Gatzert* (Ore 1910) 179 F 44.
[40]The *N and W No. 2* (NY 1903) 122 F 171.
[41]The *Kaga Maru* (Wash 1927) 18 F (2d) 295.
[42]For a discussion of court interpretation of "risk of collision," see chapter 16.
[43]The *Gratitude* (NY 1868) Fed. Cas. No. 5,704.
[44]The *Bessie Morris* (Pa 1882) 13 F 397.

said that besides watching for lights ahead and on crossing courses a lookout should also be watchful for things adrift, such as a disabled launch, so near as to be likely to drift against his vessel, and a tug with a long tow must extend this watchfulness the full length of the tow;[45] in another decision the circuit court found that a steamer should have a "trustworthy" lookout.[46] It was held a fault rendering a steamship liable for a collision with a schooner to have as the only lookout in a dense fog, on a frequented part of the Atlantic coast, a boy of sixteen years who had been on the water but a few weeks.[47] When the lookout on a moving vessel confused the lights of an anchored vessel with others on the shore five miles distant, the vessel was not relieved from liability for collision with the anchored vessel.[48] In two other decisions implying that a lookout must have some knowledge of his responsibilities that comes with experience, the circuit court of appeals in New York held:

> The failure of the lookout of a steamer to report a vessel when discovered is negligence, though the master and pilot were on the bridge.[49]
> A lookout's duty is to report as soon as he sees any vessel with which there is danger of collision or which in any way may affect the navigation of his own; and he cannot speculate on the probabilities of collision, such responsibility being for the master.[50]

This does not suggest that a lookout should report everything he sees; in crowded waters he could not be expected to cope. In such a case consideration should be given to placing additional lookouts. Certainly the courts will take into account the number of seamen available on board when considering whether a proper lookout was maintained.[51] Even if this is done, a lookout must use his discretion to some extent in reporting what he sees. As was said in one case:

> You cannot report every light you see in the River Thames. You have to watch until you see a light, which, perhaps, you have seen before, becoming material, because if you are going to report every light in Gravesend Reach when coming up the Thames the confusion would be something appalling to those in charge of the navigation; but you have to have a look-out to report every material light as soon as it becomes material.[52]

[45]Cook v. Moran Towing Co. (CCA NY 1911) 193 F 48.
[46]The *Pilot Boy* (SC 1902) 115 F 873.
[47]The *Pottsville* (Pa 1882) 12 F 631.
[48]The *John G. McCullough* (CCA Va 1916) 239 F 111.
[49]The *Hansa* (CCA NY 1870) Fed Cas. No. 6,036.
[50]The *Madison* (CCA NY 1918) 250 F 850.
[51]The *Spirality* [1954] 2 Lloyd's Rep. 59; also the *Saxon Queen* [1954] 2 Lloyd's Rep. 286; the *Mode* [1954] 2 Lloyd's Rep. 26.
[52]The *Shakkeborg* (1911) Sh. Gaz. Apr. 11.

This heightens the need for a lookout to be experienced, well trained, and thoroughly briefed on every occasion before taking up his duty.

In the naval service it is suggested that except on the smallest vessels the lookout might well be a petty officer. On a capital ship, when we consider the value of the property at risk, the intelligence that should be demanded in reporting various kinds of lights, with correct bearings, and the legal importance attached to a proper performance, the duty should by no means be regarded as beneath the dignity of a chief petty officer. It goes without saying that the choice of a competent lookout is only half the requirement, the other half being an insistence by the officer of the deck that reports to him be made promptly and correctly, at times of good visibility as well as bad, so that in darkness or in thick weather they will be rendered as a matter of habit. Perhaps it is significant that more than 70 percent of the cases cited in this chapter were clear weather collisions.

Lookout Should Be Vigilant

That a high degree of vigilance is constantly required of the lookout is evident from the findings in numerous cases—a high degree, though not an unreasonable degree. Thus, the District Court of Maryland declined to condemn a steamer for failure to discover a sailing vessel without lights on a dark night, merely because the sailing vessel might have been discovered in time to avoid collision if an officer had been constantly observing the horizon with a good pair of glasses.[53] But a vessel's failure to see the lights on another vessel, properly set and burning, due to want of vigilance, renders it liable for the resulting collision;[54] and a vessel's failure to discover the lights of a passing vessel in time to avoid collision is tantamount to having no lookout.[55] An overtaking vessel without lights in a convoy during the war was liable for a collision with the ship ahead, because the lookout failed to keep under close observation the other vessel, and failed to give warning of the close approach in time to avoid collision.[56]

That it is incumbent on a vessel navigating New York Harbor and vicinity, even in the daytime, to maintain a vigilant lookout, is a requirement that the Supreme Court declined to review.[57] In another New York case, this time in the harbor of Buffalo, it was said by the district court:

> It is the imperative duty of a steamship when making a landing at a dock in a river where other vessels are constantly passing, to maintain an efficient look-

[53]The *Leversons* (Md 1882) 10 F 753.
[54]The *Buenos Aires* (CCA NY 1924) 5 F (2d) 425.
[55]Pendleton Bros. v. Morgan (Md 1926) 11 F (2d) 67.
[56]The *War Pointer* (CCA Va 1921) 277 F 718.
[57]The *Transfer No. 15* (CCA NY 1917) 243 F 174.

out, and the absence of such lookout cannot be excused on the ground that all the crew were otherwise engaged.[58]

And in a case in Alaska where a tug negligently allowed a loaded barge to be cast upon the rocks during a snowstorm, the Supreme Court affirmed the decision of the circuit court of appeals awarding a decree to the barge owners, which held that:

> The strict rules with respect to the necessity of having a lookout properly stationed and devoting his whole attention to the situation ahead is not limited to vessels navigating harbors, but applies as well to a vessel navigating along the coast, where danger from striking the land is as great as the danger of collision in harbor.[59]

In a collision between a steamship and a sailing vessel, the circuit court of appeals found that the duty of the steamship to maintain a constant lookout is especially imperative in favor of a vessel that under the rules has the right-of-way.[60] Other decisions emphasize the fact that the degree of vigilance exercised by the lookout is quite likely to be judged by the single standard of its effectiveness in preventing collision:

> The failure of a steamer to see a sailing vessel which she ought to have discovered, in time to give her sufficient room, is a fault rendering the steamer liable if it results from insufficient lookout.[61]
>
> The excuse that a vessel is unable to determine (by reason of the darkness and the direction of the wind) from the lights of the other vessel, on what course she is sailing, and which is the privileged vessel, cannot be invoked where, by reason of the inefficiency of her lookout, she failed to discover the approaching vessel until the two were in close proximity, and she had no time to study the situation.[62]
>
> A burdened vessel which fails, through the inexcusable absence of her lookout, to maintain it steadily, and thus causes a collision, is liable.[63]
>
> Where the evidence leaves no doubt that two blasts of a whistle were given by one steamer, which were heard on the other as only a single blast, the distance being such that both ought to have been heard, the court must conclude, in the absence of other explanation, that the officers and lookout were inattentive.[64]
>
> It is a fault for the lookout of a vessel to leave his post after reporting the light of another vessel.[65]

[58]The *Northland* (NY 1903) 125 F 58.
[59]The British Columbia Mills Tug and Barge Co. v. Mylroie (1922) 66 L Ed. 807.
[60]The *Dorchester* (1908) 167 F 124.
[61]The *Belgenland* (1885) 29 L Ed. 152.
[62]The *Queen Elizabeth* (NY 1903) 100 F 874 (reversed on other grounds), 122 F 406.
[63]The *Robert Graham Dun* (CCA 1895) 70 F 270.
[64]The *Ottoman* (CCA Mass 1896) 74 F 316.
[65]Wilders SS. Co. v. Low (Hawaii 1911) 112 F 161; the *Havre* (NY 1868) Fed. Cas. 6,232.

Lookout Specially Necessary in Fog

The degree of vigilance required is, of course, greatly increased underway in foggy weather, and absence or insufficiency of lookout under such conditions underway can never be justified by the plea that visibility was so low as to render a lookout useless. If he cannot see, at least he can hear. Under the general admiralty rules it is the duty of every vessel, when navigating in a fog, to maintain a lookout in a proper position, who shall be charged with no other duty.[66] A local custom cannot excuse a vessel from observing this rule.[67] As said by the circuit court of appeals:

> The denser the fog and the worse the weather are greater cause for vigilance, and a vessel cannot excuse failure to maintain lookout on the ground that the weather was so thick that another vessel could not be seen until actually in collision.[68]

However, as brought out by the circuit court in the collision of the *Bailey Gatzert* with a Colombia River dredge, already cited, this injunction apparently does not apply with the same force to a vessel at anchor in fog, provided she is making proper fog signals.[69]

The degree of vigilance required is in no way lessened because a vessel in a collision situation may happen to have the right-of-way. In a number of very early decisions it was held that if a vessel having the right-of-way held her course, it was all an approaching vessel had a right to require, and in the case of a small sailing schooner run down by a large steamship in Chesapeake Bay, the Supreme Court held that whether she had a proper lookout or not was immaterial.[70] But as early as 1874 the district court of Maine established the rule that:

> A vessel having the right of way must keep a proper lookout and use proper seamanship to avoid collision.[71]

Accordingly, as said by the federal court in the *Kaga Maru*, a vessel is not relieved of her obligation to maintain a proper lookout because she is the preferred vessel, if prudent navigation with the aid of a good lookout would have avoided the collision,[72] and in a New York Harbor collision decided in 1928, a steamer without a lookout was held equally at fault with an army dredge on the wrong side of the channel with which it collided.[73] Similarly, the failure of a tug to keep a lookout rendered it liable for the

[66]The *Wilbert L. Smith* (Wash 1914) 217 F 981.
[67]The *Tillicum* (Wash 1914) 217 F 976.
[68]The *Sagamore* (Mass 1917) 247 F 743.
[69]The *Bailey Gatzert* (CCA Ore 1910) 179 F 44.
[70]The *Fannie* (Md 1871) 20 L Ed. 114.
[71]The *Mary C* (Me 1874) Fed. Cas. No. 9,201.
[72]The *Kaga Maru* (Wash 1927) 18 F (2d) 295.
[73]A. H. Bull SS. Co. v. U.S. (NY 1928) 29 F (2d) 765.

death of a man in a rowboat on the Delaware River, on the showing that notwithstanding the latter's contributory negligence in attempting to cross the tug's course, a vigilant lookout would have discovered his peril in time for the tug to avoid him. In this case the tug was flanked on either side by a loaded car float, neither float having a lookout.[74]

The old idea that to be efficient a seaman must be uncomfortable is happily disappearing in the naval service. In this connection it might be well to remember that a lookout who is freezing from exposure in a cold wind can scarcely be expected to be vigilant. After a few moments there would be little thought given to approaching lights, or of anything else except getting in out of the weather. A canvas dodger would have worked wonders, as indeed it will with the lookout on the nose of a present-day cruiser, heading into a breeze at 20 knots. The officer of the deck who expects his lookouts to be alert will do well to assure himself that the men performing this duty are adequately clothed and as well protected from the weather as conditions will permit.

Not only must visual and aural lookouts be closed-up in restricted visibility, but so must a proper radar lookout be maintained:

> *John C. Pappas* was at fault because she took no special precaution by way of extra look-outs on encountering fog, while the *Almizar* was guilty of poor radar look-out and should have realized that radar echo was of a large vessel on a steady bearing. *John C. Pappas* was also at fault for not hearing fog signals forward of her beam.[75]

Time and time again, ships have been faulted for a bad radar lookout and faulty appreciation of the information available.

> . . . if a proper radar lookout had been kept during the approach period, it should have been appreciated that a close-quarters situation with risk of collision was developing. Both ships were guilty of defective appreciation of radar.[76]

Lookout Should Be Properly Stationed

Although the rule is silent as to the specific location of the lookout, a long line of court decisions has well established his proper position to be as low down and as far forward in the ship as conditions allow. As said by the circuit court of appeals:

> He is required by good navigation to be placed at the point best suited for the purpose alike of hearing and observing the approach of objects likely to be

[74] Klutt v. Philadelphia and Reading Ry. Co. (CCA Pa 1906) 142 F 394.
[75] The *Almizar* [1971] 2 Lloyd's Rep. 290.
[76] The *Ercole* [1979] 1 Lloyd's Rep. 539; also the *Hagen* [1973] 1 Lloyd's Rep. 257; the *Elazig* [1972] 1 Lloyd's Rep. 355; the *Bovenkerk* [1973] 1 Lloyd's Rep. 62; the *Zaglebie Dabrowskie* [1978] 1 Lloyd's Rep. 571.

brought into collision with the vessel, having regard to the circumstances of the case and the conditions of the weather.[77]

Accordingly, the *Vedamore,* a large ocean steamship navigating Chesapeake Bay at night in fog, was at fault because her only lookout was in the crow's nest 60 feet above the deck and 100 feet from the stem. It was held to be the duty of another ocean steamer passing out of the Delaware at night to maintain her lookout as far forward and as near the water as possible.[78] In a collision in Boston Harbor in which a ferryboat was sunk by a mud scow, the pilothouse was declared not a proper place for the former's lookout;[79] and in another case, stationing the lookout on top of the pilothouse 140 feet from the bow of a steamship navigating in a dense fog, with visibility of 100 feet, was a gross fault.[80] A lookout on the bridge without any in the bow is insufficient;[81] and in another collision in fog previously cited, the steamship *Sagamore* was held at fault for not maintaining a lookout from the forecastle head despite lookouts both in the crow's nest and on the bridge.[82] In a collision on a clear night at sea, when the usual order was reversed by having a sailing vessel sink a steamship, the circuit court of appeals held that every steamer must have at least one lookout in the eyes of the ship.[83] In a collision in a dense fog on Puget Sound, the court found a tug at fault for having the lookout by the pilothouse, but only 12 feet from the stem.[84] In the *Winnisimmet,* mentioned above, the judicial attitude was thus summarized:

> the courts have been rigid in holding vessels to maintaining lookout as far forward and as near the water as possible. Especially where the water is dark, with otherwise a fairly clear night, it is important that the lookout should be as near it as possible, in order that his eye may follow the surface, and thus be in position to detect anything low down which may be approaching.[85]

A United States submarine was held solely at fault for sinking a small schooner near the western end of the Cape Cod Canal shortly after World War I, on a clear night with smooth sea and all lights burning brightly, a charge of improper lookout being sustained. In its decision, the court recognized the extremely limited space available for a lookout on the bow of an *R*-boat, but declared that every moving vessel must maintain a competent, careful, and efficient lookout stationed on the forward part of

[77]The *Vedamore* (1905) 137 F 844.
[78]The *Prinz Oskar* (1915) 219 F 483.
[79]Eastern Dredging Co. v. *Winnisimmet* 162 F 860.
[80]The *Campania* (La 1927) 21 F (2d) 233.
[81]Neally v. the *Michigan* 63 F 280.
[82]The *Sagamore* (CCA Mass 1917) 247 F 743.
[83]The *Stifinder* (CAA NY 1921) 275 F 271.
[84]The *Kaga Maru* (Wash 1927) 18 F (2d) 295.
[85]See note 79 above.

the vessel, and that the rule applies with equal force to submarines or other naval vessels, in the absence of statutory exception.[86]

When physical conditions prevent, of course, a lookout need not be kept forward of the bridge. An ocean tug whose lookout was in the pilothouse because green seas were sweeping over the bow, and a steamship navigating the Atlantic on a clear night with her lookout on the bridge because of the coldness of the weather and the freezing spray forward [87] were both absolved from blame. If the weather is clear and the lookout is sufficiently vigilant, it really matters little where he is stationed; but the practical catch that goes with failure to place him away forward is that the burden of proof devolves upon the offender to show that the lookout functioned as well as he would have done if properly placed. As an example in point, a steamer at sea on a clear night with a lookout stationed on the bridge instead of forward was held not to be at fault in a collision with a fishing vessel, since the evidence showed that each vessel reasonably discovered the other and kept her under continuous observation.[88] But in another case, involving a tug and a sailing vessel on the Hudson River, it was held that:

> The position of the captain of a schooner abaft the wheel is not a proper position for a lookout, when sailing full and free with a strong wind; and in case of a conflict of testimony, observation reported from such a position must be deemed partial, interrupted, and incomplete, and entitled to far less weight than that of a lookout properly stationed.[89]

The decisions are clear in requiring that the lookout have unobstructed visibility ahead and on both bows. Hence, it has been held that an old rule of the former Board of Supervising Inspectors, requiring passenger steamers and ferryboats to keep one of the crew on watch in or near the pilothouse could not supersede the general rule requiring a lookout forward.[90] In the case of the *Scandinavia* it was pointed out that the duty to maintain a lookout on the lower deck of a ferryboat is not statutory but is imposed by the general maritime law.[91] The same principle was established in several decisions requiring that a tug must keep a lookout at the bow of a tow alongside where it projects beyond the tug;[92] and where other traffic may be expected, a vessel towing astern of a tug should maintain as careful a lookout as the tug herself, and be prepared, if

[86]U.S. v. Black (CCA 1936) 82 F (2d) 394.
[87]The *Caro* (NY 1884) 23 F 734; the *Kaiserin Marie Theresa* (CCA NY 1906) 147 F 97.
[88]The *Lake Monroe* (CCA 1921) 271 F 474.
[89]The *Excelsior* (NY 1882) 12 F 195.
[90]The *Tillicum* (Wash 1914) 217 F 976.
[91]The *Scandinavia* (NY 1918) 11 F (2d) 542.
[92]The *Pennsylvania* (NY 1878) Fed. Cas. No. 10, 949; the *A. P. Skidmore* (NY 1901) 108 F 972.

necessary to avoid collision, to sheer out or cut her hawser.[93] Because the tow cannot always see ahead, the vessel towing should keep a specially vigilant lookout. In addition to having an unobstructed view, a further reason for placing the lookout forward is to keep his hearing as unimpaired as much as is possible by the noise from the engines, especially in diesel and gas turbine-powered ships.[94]

However, should a collision occur, it is well to bear in mind that clear visibility, rather than technical location, is the essential thing, and that after all, as expressed by the District Court of Virginia in a collision between a government steam launch and a fishing steamer near Norfolk:

> All that the law requires with respect to a lookout is that there shall be someone properly stationed to best observe, see, and hear the approach of other vessels; and a small launch, only 61 feet long, having her pilothouse, in which her navigator stood, well forward, with open windows all around, and other members of the crew on the deck, cannot be held in fault for a collision in the daytime, in fair weather, because she did not have a lookout specially stationed where she was the privileged vessel, entitled to keep her course and speed, and the absence of such lookout did not contribute to the collision.[95]

While in the absence of special conditions the place for the lookout is at the bow,[96] there are circumstances that require a lookout also at the stern. Thus, it has been held that an ocean steamer starting her propeller in order to leave her slip, in which there are other vessels, should have a lookout at the stern to give warning of danger to such vessels from the motion of the propeller;[97] that a vessel backing out of her slip must keep a lookout astern;[98] as must a tug floating downstream bow towards the shore,[99] a vessel drifting backwards with the tide,[100] or in fact a vessel making sternway for any purpose, as when backing and filling to turn around,[101] or when backing into her slip.[102] And a steam tug that towed a barge past a vessel at anchor so close as to cause a collision between her tow and the vessel without keeping a proper lookout at the stern, was solely liable.[103]

In general, an overtaken vessel is under no duty to keep a lookout aft to

[93]The *Virginia Ehrman* and the *Agnese* 24 L Ed. 890; the *American* 102 F 767.

[94]*Cabo Santo Tome* (1933) 46 Ll. L. Rep. 165.

[95]The *Pocomoke* (Va 1906) 150 F 193.

[96]Yamashita Kisen Kabushiki Kaisha v. McCormick Intercoastal SS. Co. (CCA Cal 1927) 20 F (2d) 25.

[97]The *Nevada* v. Quick (NY 1882) 27 L Ed. 149.

[98]The *Luzerne* (CCA NY 1912) 197 F 162; the *Herbert L. Pontin* (CCA NY 1931) 50 F (2d) 177.

[99]The *Mary J. Kennedy* (CCA 1924) 11 F (2d) 169.

[100]The *Senator D. C. Chase* (CCA NY 1901) 108 F 110.

[101]The *Deutschland* (CCA NY 1904) 137 F 1018.

[102]Greenwood v. the *William Fletcher* and the *Grapeshot* (NY 1889) 38 F 156.

[103]The *Cement Rock* and the *Venture* (NY 1876) Fed. Cas. No. 2,544.

prevent being run down by the overtaking vessel, but has a right to act on the presumption that the latter will keep clear.[104] The circuit court of appeals in New York has gone so far as to rule that in inland waters, where the overtaking vessel has no right to pass without a signal being given and answered, the overtaken vessel is not required to discover her before she changes course, however abruptly; if the overtaking vessel comes so close without signaling that a sudden change of course by the vessel ahead brings about a collision, the fault is that of the overtaking vessel.[105] However, good seamanship would seem to go a step further here, and require the precaution of always at least looking aft before changing course or reducing speed, because of the possibility of embarrassing an overtaking vessel. As said by the Supreme Court in the case of the *Illinois*, when a sailing vessel—failing to notice a following steamship—tacked to avoid floating ice in Delaware Bay and was sunk by the steamship:

> While a man stationed at the stern as a lookout is not at all times necessary, no vessel should change her course materially without having first made such an observation in all directions as will enable her to know how what she is about to do will affect others in her immediate vicinity.[106]

In a collision between the USS *Bell* and a steam lighter, both vessels were leaving Boston Harbor in a thick fog, with the destroyer leading. The *Bell* stopped without warning when it appeared that the gate in a steel submarine net stretched across the north channel during the war was closed; and the lighter, attempting to sheer out, collided with the depth-charge sponson, injuring both vessels. The circuit court of appeals found both at fault: the lighter for excessive speed in striking a vessel that was dead in the water, and the destroyer for not keeping a proper lookout astern when she knew another vessel was following, although she had two depth-charge men aft who had not been instructed to report approaching vessels.[107]

Lookouts Must Be Adequate for Circumstances

There can be no doubt from the decisions that under some circumstances more than one lookout is required, although ordinarily one with that exclusive duty will be sufficient. In a collision at sea in a dense fog at night where, besides a man forward, stationed as a lookout, there were two persons on watch in the pilothouse of a large ocean steamer, the lookout was held sufficient.[108] On the other hand, in a decision of the

[104]The *Greystoke Castle* (Cal 1912) 199 F 521.
[105]The *Merrill C. Hart* (CCA 1911) 188 F 49; the *M. J. Rudolph* 292 F 740.
[106]The *Illinois* (1881) 26 L Ed. 563.
[107]Boston Sand and Gravel Co. v. U.S. (CCA 1925) 7 F (2d) 278.
[108]Watts v. U.S. (1903) 123 F 105.

Supreme Court relating to a collision on the Columbia River between a passenger steamship and a dredge at anchor, it was held that:

> Where the circumstances require more than ordinary care, as in the case of a steamer running at a speed of 15 miles an hour, on a dark night, in a narrow channel, where there is a great probability of meeting other vessels, a deck watch composed of the river pilot in command, stationed upon the bridge just above the pilot house, a man at the wheel, and a lookout upon the forecastle head, is insufficient, and prudent navigation requires a lookout to be stationed on either bow.[109]

In a collision during the Spanish-American War a few miles off Fire Island at night in a thick fog, the armored cruiser *Columbia* sank a British freighter. The *Columbia* was making 6 knots, and in accordance with the orders of the squadron commander, showing no lights and making no fog signals. The freighter was making 3½ knots and sounding regular fog signals. In an action several years later the cruiser was found solely at fault, the court commenting on the fact that under such circumstances unusual vigilance was required, and that her lookouts were quite insufficient, being only those usually maintained in clear weather—one at each end of the bridge, 94 feet from the stem and 38 feet above the water, and two farther aft.[110] While it is true that in this case the real fault lay in not having the forward lookouts out on the bow, yet it is significant that here was an instance of improper lookout with four men detailed to exclusive lookout duty. On a large vessel in thick weather there is a positive obligation to have lookouts stationed so as to give the earliest warning of approaching vessels from whatever direction they may come.

It may seem to the reader that the courts have given undue emphasis to the necessity of lookouts when their function is a duty already laid on the officer of the deck. It may be argued that an officer of the deck who is worth his salt will be the first to discover the approach of anything that might endanger his vessel. The explanation can only be the obvious one that the commissioned or licensed watch stander has other concurrent duties that he cannot neglect, and the law contemplates that every vessel underway shall exercise vigilance *which is continuous and unbroken*, both for her own protection and that of other vessels. However, it is doubtful if the civil courts are more exacting in this regard than are the naval courts and boards, which must be encountered by an officer in the service whenever a naval vessel is involved in collision. With them the question of proper lookout in collision cases is inevitable. In a recent hearing that followed a collision between a naval vessel and a merchant ship, the board of investigation found the commanding officer at fault for improper lookout on

[109] The *Oregon* (Ore 1895) 39 L Ed. 943.
[110] Watts v. U.S. (1903) 123 F 105.

the technical point that the quartermaster of the watch was serving as lookout while the vessel was leaving port, although the collision occurred in broad daylight with excellent visibility, and the testimony showed that the approaching vessel, as she emerged from a channel entrance, was sighted by the commanding officer, the officer of the deck, the man at the wheel, and the lookout himself. This illustrates the importance that may be attached by one's brother officers to maintaining a technically proper lookout even when visibility is unquestionably good.

Importance of Lookouts

From the standpoint of the navigator who is interested in avoiding liability for collision, which usually means avoiding the collision itself, perhaps there could be no more appropriate conclusion to these remarks on a proper lookout than the following words of the United States Supreme Court, delivered in a case that arose out of a collision between a steamship and a brig outside New York Harbor on a foggy night in 1865:

> The duty of the lookout is of the highest importance. Upon nothing else does the safety of those concerned so much depend. A moment's negligence on his part may involve the loss of his vessel with all the property and the lives of all on board. The same consequence may ensue to the vessel with which his shall collide. In the performance of this duty the law requires indefatigable care and sleepless vigilance. . . . It is the duty of all courts charged with the administration of this branch of our jurisprudence to give it the fullest effect whenever the circumstances are such as to call for its application. Every doubt as to the performance of the duty and the effect of nonperformance, should be resolved against the vessel sought to be inculpated until she vindicates herself by testimony to the contrary.[111]

SUMMARY

A lookout has been defined by the federal court as a person who is specially charged with the duty of observing the lights, sounds, echoes, or any obstruction to navigation with that thoroughness that the circumstances permit. Any neglect to keep a proper lookout has caused the courts to hold a vessel in collision without a proper lookout at fault unless it can be proved that the other vessel was discovered as soon as a proper lookout would have discovered her. As technology progressed, the concept of lookout expanded to include not only the person keeping a visual and aural lookout, but the entire bridge team and particularly that competent person who maintained a continuous surveillance with radar as well as monitoring VHF radio. All the data received from various instruments must be intelligently interpreted, and this interpretation constitutes part of keeping a good lookout.

[111]The *Ariadne* (1872) 13 Wall 475.

Numerous court decisions have built up a considerable doctrine with reference to what constitutes a proper lookout. Such a lookout must have no other duties, such as conning or steering the vessel; he must be constantly alert and vigilant, he must have had a reasonable amount of experience as a seaman; he must report what he sees or hears to the officer of the watch; and he must ordinarily be stationed as low down and as far forward on the vessel as circumstances permit. In conditions of crowded traffic and in thick weather, enough lookouts must be posted to detect the approach of another vessel from any direction. What is required is the maintenance of sufficient lookout, both visually, aurally, and by radar when circumstances warrant.

21

Inevitable Accident

Collision Liability under American Admiralty Law

In a previous chapter it was pointed out that whenever two vessels are in a collision that results in litigation, there are precisely four possibilities, under American maritime law, as to the judicial determination of liability: (1) vessel *A* may be held solely at fault and liable for all the damage to both vessels; (2) the same may be true of vessel *B*; (3) both may be at fault, with the invariable result of a division of damages; (4) or neither may be at fault, in which case each must bear her own damage, be it great or small.[1] While the question of legal liability does not ordinarily arise in a collision between two naval vessels that have a common owner, it does arise, and is settled in the same manner as when a naval vessel collides with a merchant vessel; and even in a collision between two vessels of the Navy, the rules of the road and governing court decisions that apply in a civil action are to a large extent followed by naval courts and boards in fixing the culpability of officers involved in the collision. A proper understanding of collision law as administered by the courts therefore becomes of as much importance to officers in the naval service as it is to ship operators and to officers of the Merchant Marine.

A casual examination of the case books will show that of the over 8,000 cases that have come before our courts, only about 70 have fallen in the fourth-mentioned category, or less than 1 percent of the total. There are

[1]The ability of the wronged vessel to collect may, of course, be modified by the Limited Liability Acts, which limit the liability of a vessel to her value after the collision plus pending freight—i.e., earnings collected or collectible for the voyage on which the collision occurs—and which limit aggregate recovery for personal injury or loss of life to $15 per gross ton, where value of the vessel plus pending freight is not more.

numerous collisions in which only one vessel is at fault, but the surprising fact to the ininitiated is that in the majority of cases *both* vessels are at fault.

Cases of Inevitable Accident Are Rare

It is well known that collisions are not brought about intentionally, and that comparatively few of them, moreover, occur because of thick fog. Despite the wide-spread practice of maintaining excessive speed with low visibility, vessels navigating in fog successfully dodge each other most of the time. Probably at least three-fourths of the collisions take place in clear weather, with each vessel or her lights plainly visible to the other for a considerable time before the casualty, and are the result of some misunderstanding. When we consider that the rules of the road were carefully drawn up for the express purpose of preventing not only actual collision but even serious risk of collision, perhaps the prevalence of double culpability is not surprising. It would seem to indicate that the rules as interpreted are almost air-tight; that if they do not always prevent collision, at least they make it impossible in most cases for a collision to occur unless both vessels fail to obey them. This is, indeed, the opinion of the admiralty courts, which rightly or wrongly are the final arbiters of collision liability. As pointed out by the circuit court of appeals in the well-known case of the *West Hartland*, when a stand-on ocean freighter making not over 6 knots sank a 15-knot passenger liner that attempted to cross her bow on a clear night in Puget Sound, with the loss of several lives, largely because the liner's pilot mistook the freighter's port light for a more distant pier light, and the master of the freighter overestimated his distance from the liner:

> There can seldom be a collision in the open sea in clear weather, where there is no obstruction and the vessels are plainly visible to each other for a long distance, without fault on the part of both vessels.[2]

From this introduction the reader will realize that the present chapter considers an extremely rare type of marine collision. Nevertheless, it is well for the professional mariner to know the characteristics of what the courts call inevitable, or unavoidable, accident for two reasons: (1) an appreciation of its rarity should incline him to a stricter observance of the rules; and (2) if he is ever involved in such a case he can take what comfort there may be in the judicial assurance that:

> The civil law, the common law, the maritime law, and the law of Great Britain and the United States agree that where a collision takes place by unavoidable accident, without blame being imputable to either party, the consequences of the misfortune must be borne by the party upon whom it happens to fall.[3]

[2]The *West Hartland* (CCA 1924) 2 F (2d) 834.
[3]The *Olympia* (CCA 1894) 61 F 120, aff. 52 F 985.

Under the common law, an accident was said to be inevitable when it was not occasioned in any degree, either remotely or directly, by the want of such care and skill as the law holds every man bound to exercise. Similarly, under the maritime law:

> An inevitable accident is something that human skill and foresight could not, in the exercise of ordinary prudence, have provided against.[4]

> The term "inevitable accident" as applied to a collision means a collision which occurs when both parties have endeavored by every means in their power, with due care and caution and a proper display of nautical skill, to prevent the occurrence of the accident, and where the proofs show that it occurred in spite of everything that nautical skill, care, and precaution could do to keep the vessels from coming together.[5]

The matter of establishing the vessel's innocence by satisfactory proofs is even more emphatically stated in a later decision holding that:

> To sustain the defense of inevitable accident in a suit for collision, the defendant has the burden of proof and must show either what was the cause of the accident, and that cause was inevitable, or he must show all the possible causes and in regard to every one of such possible causes that the result could not have been avoided.[6]

It is not enough for a vessel to show that all that could be done was done as soon as the need to take action to avoid collision was determined. The point is whether actions should not have been taken earlier. When two ships are shown to have been in a position in which collision became inevitable, the question is, by whose fault, if there was fault, did the vessels get into such a position?[7]

In the light of these interpretations of inevitable accident, it will readily be understood that most of the cases determined by the courts have been due either to (1) *vis major*, or superior force of the elements, or (2) to a mechanical failure, in a collision situation, of steering gear or other machinery that due diligence could not prevent. The few cases that remain may for want of a better term be classed as miscellaneous, due either to a special combination of circumstances, or to causes that could not be ascertained from the evidence and were therefore not chargeable to either vessel.

[4]The *Drum Craig* (1904) 133 F 804; the *Pennsylvania* (1861) 16 L Ed. 699.

[5]New York and Oriental SS. Co. V. N.Y., N.H., and H. Ry. Co. (NY 1906) 143 F 991; the *Mabey* and *Cooper* (1872) 20 L Ed. 881.

[6]The *Edmund Moran* (CCA NY 1910) 180 F 700, adopted from the *Merchant Prince* (1892) P 179, 189; *see also* Southport Corporation v. Esso Petroleum Company Ltd. [1955] 2 Lloyd's Rep. 655.

[7]Marsdens *Collisions at Sea*, 11th Ed. p. 8, p. 9.

Inevitable Accident Due to "Vis Major"

Vis major has been defined as an irresistible, natural cause that cannot be guarded against by the ordinary exertions of human skill and prudence.[8] An injury caused by a *vis major* is equivalent to an act of God.[9] It has been held that, as respects the liability of carriers for loss or damage of goods, the term *vis major* is used in the civil law in the same way that the words "act of God" are used in the common law, meaning inevitable accident or casualty.[10] Thus, storms of great and unexpected violence, unusual tidal currents, abnormal river floods, unpredictable ice conditions, or very dense fog may bring about marine collisions classed as inevitable. In this connection, it is fortunate that the word inevitable is considered as a relative term and construed not absolutely, but reasonably,with regard to the circumstances of the particular case.[11] In the case just cited, the steamer *Anna C. Minch* was broken from her moorings in the Buffalo River by a spring freshet, during which a huge mass of ice subjected her to such pressure that all her lines were parted at once and she was carried down stream at 8 to 10 knots against other moored vessels. It being shown that her lines were sufficient under any conditions ordinarily to be anticipated, the accident was held inevitable.

Similarly, in a New York case, a tier of six canal boats was knocked from its moorings by another group of boats that had gone adrift as a combined result of a fresh wind and a strong tide, and one of the boats, the *Nora Costello*, struck and damaged a vessel at another pier and was libeled. In dismissing the libel, the court held that:

> Vessels in making fast to piers are bound to provide only against ordinary contingencies such as they can anticipate; that they are not bound to make fast by lines so strong or numerous as to resist the impact of such a fleet of vessels as got adrift in this case; and that as there was no negligence in the *Costello* as to her mode of fastening, the libel against her should be dismissed.[12]

Again, when a sudden and extraordinary flood in the Monongahela River tore a fleet of water craft from its mooring at night, and crashed it at 10 knots or more into a fleet of coal boats, the courts dismissed a libel charging negligent moorings, and cited the following opinion of the Supreme Court:

> Inevitable accident is where a vessel is pursuing a lawful avocation in a lawful manner using the proper precautions against danger, and an accident occurs. The highest degree of caution that can be used is not required. It is enough that

[8]Evans v. Wabash Ry. Co. 12 SW (2d) 767.
[9]Southern Pacific v. Schuyler (1905) 135 F 1015.
[10]Lehman, Stern, and Co. v. Morgan's La. and Tex. R.R. and SS. Co. 38 South 873.
[11]The *Anna C. Minch* (CCA 1921) 271 F 192.
[12]The *Nora Costello* (1890) 46 F 869.

> it is reasonable under the circumstances, such as is usual in similar cases, and has been found by long experience to be sufficient to answer the end in view—the safety of life and property. Neel v. Blythe (1890) 42 F 457, citing the *Grace Girdler* (1869) 19 L Ed. 113.

The Virginia Dollar broke from her moorings in Genoa harbor and drifted on to the *Storborg* and damaged her. It was shown that a steel bollard on the dock side had broken. The owners of the *Virginia Dollar* succeeded in their defense of inevitable accident.[13]

Other Cases of Vis Major at Anchor

In still another case, when a car float moored in a slip in the East River at the beginning of an extraordinary blizzard demonstrated the security of her mooring by remaining through two full tides, but was finally wrenched free by a floe of ice of a size not reasonably to be anticipated coming up the river, the court held that damage done by the car float to another vessel was due to inevitable accident.[14] As a final illustration, when the bow lines of a steamship were carried away by a storm in New York Harbor that gave no more warning than the usual thunderstorn, but the wind actually reached a velocity for half a minute of 125 miles an hour, a canal boat was damaged by the steamship, but the latter was held free from fault.[15]

But an early case in Galveston was decided quite differently. It seems that three vessels—a brig, a schooner, and a bark—lay moored at a wharf when a heavy storm arose, breaking the brig adrift from the wharf. By quick work she was brought up by her anchors 100 yards or less from the schooner. Not long afterward, the bark was driven against the schooner, crushing in her stern and putting her in a sinking condition. The master of the schooner, to save her from sinking with her cargo in the deep water at the wharf, cut her adrift; whereupon, despite all efforts to the contrary, she struck and injured the brig. In the action brought against her by the brig, the court denied the schooner's plea of inevitable accident inasmuch as the act of her master in cutting her adrift had been a voluntary one.[16]

When vessels at anchor fail to hold their position and are brought into collision with other vessels through dragging, it is very unusual for the courts to excuse them on the plea of inevitable accident. It is apparent from the decisions that in such cases a heavy burden is put upon the offending vessel to show that she was properly anchored. In a December gale in New York the steamship *Bragdo*, at anchor off Staten Island with 45 fathoms out, dragged across the chain of the steamship *British Isles* and

[13]The *Virginia Dollar* (1926) 25 Ll. L. Rep. 227.
[14]The *Transfer No. 2* and *Car Float No. 12* (1893) 56 F 313.
[15]The *Campanello* (NY 1917) 244 F 312.
[16]Sherman v. Mott (1871) Fed. Cas. No. 12,767 (affirmed by CCA, 1873).

set her adrift. The plea of the *Bragdo* when libeled for the damage was that she did not come to anchor with more chain out for fear of hitting piers half a length astern. The gale reached hurricane strength and in finding her liable for the collision, the circuit court of appeals cited the following passage from Knight's *Seamanship*:

> It is a common rule to give, under ordinary circumstances, a length of cable equal to seven times the depth of water. This is perhaps enough for a ship riding steadily and without any great tension of her cable, but it should be promptly increased if, for any reason, she begins to sheer about or jump, for it is always easier to prevent an anchor from dragging than to make it hold after it has once begun to drag.[17]

In the *Djerissa*, an ocean steamship of that name was at anchor in the James River when she was damaged by the *Neva*, a second steamship, that anchored subsequently and dragged into her. The storm which caused the mishap was evidently approaching for several hours, and aside from the fact that the first of two vessels to anchor is naturally the favored vessel, the courts found the *Neva* at fault for not dropping a second anchor.[18] An ocean tug left two seagoing barges at anchor in a storm off the Rhode Island coast for thirty-six hours while she took shelter in Newport Harbor. One barge dragged into the other and was held at fault when the evidence showed that although she was forewarned as to the storm, she had only one anchor out until after she began to drag. Incidentally, on a showing that the tug could have stood by without danger to herself, the tug was held to share the liability because of contributory negligence.[19]

An excellent case to illustrate the distinction between negligence and necessary prudence in a similar situation was the *Herm*, decided by the circuit court of appeals after being carried up from the district court of Virginia. Three barges, anchored off the Newport News coal piers, after dragging for more than an hour under the pressure of drifting ice, struck and damaged a vessel moored at one of the piers with her stern projecting 40 feet into the fairway. Two of the barges were held at fault for failure to drop a second anchor, while the third barge, which had lost her second anchor the night before and had been unable to replace it, was exonerated. The moored vessel recovered only half damages, however, for although it is not ordinarily a fault for a vessel's stern to project into the fairway beyond a pier end unless she obstructs navigation, it was contribu-

[17]The *British Isles* (CCA 1920) 264 F 318.
[18]The *Djerissa* (CCA 1920) 267 F 115.
[19]The *Sea King* (NY 1926) 14 F (2d) 684.

tory negligence in this case where anchored barges subject to the action of drifting ice were in plain sight.[20]

However, in a unique collision between two anchored steamships in New York Harbor in a thick fog, no fault was found with either vessel. The White Star liner *Adriatic* left her pier on the East River late on a November afternoon intending to proceed to sea, there being at the time sufficient visibility to see across the river. Near the Statue of Liberty the liner encountered fog of unprecedented density and was compelled to come to anchor. She proceeded very cautiously to a position off the Battery, dropped her hook, and swung around on it so as to contact the stem of the *Saint Michael*, an anchored vessel, about 125 feet forward of her own stern, with damage to both vessels. Each vessel libeled the other, the *Adriatic* being charged with negligence for leaving her pier in fog and for failing to try to return to it, and the *Saint Michael* being charged for lack of promptness in paying out her chain when the *Adriatic* appeared. Nevertheless, both the district and the appellate courts ruled that under all the attendant circumstances the accident was inevitable.[21]

Cases of Vis Major Underway

Not all cases of *vis major*, of course, occur while vessels are moored or at anchor, though they are less likely to happen to vessels with way on and fully capable of maneuvering. For the same reason not a few of these cases are responsible for collisions of tugs and of tows with which they are encumbered. Thus, two long tows of vessels with a combined length of 3,200 feet were in the act of passing starboard to starboard, as customary, about 200 feet apart below the Poughkeepsie Bridge on the Hudson River, when a terriffic windstorm came up and drove some of the boats in the tows together.[22] A similar case of inevitable accident under the *vis major* rule occurred when a flotilla of vessels crossing Newark Bay towed by two capable tugs was set down on two steamers lying at the National Drydock pier by the combined action of wind and tide of unpredictable violence.[23] A tug, maneuvering a barge in the ice off Delaware Breakwater, was carried by tide and ice into an anchored schooner and, having convinced the court that she was managed with ordinary care and skill under the circumstances, was exonerated on the same grounds.[24] A tug was libeled by a vessel in her tow that was forced ashore in the Hudson when a large mass of ice, caught on a point near Piermont, was turned by

[20]The *Herm* (CCA 1920) 267 F 373.
[21]The *Adriatic* (CCA 1922) 287 F 259.
[22]The *Cornell* (1905) 134 F 694.
[23]The *Mary Tracy* (CCA 1925) 8 F (2d) 591.
[24]The *Harold* (NY 1922) 287 F 757.

the tide and pocketed the slow-moving tug before it could escape, but the tug was also exonerated.[25]

A steamer overtook and passed another in a river clogged with ice. She then maintained her lead for some time until the amount of ice increased so as to impede her progress; it finally forced her to stop, and she immediately sounded the danger signal. The other steamer, although she was keeping a sharp lookout and although she reversed full speed and executed left full rudder, was unable because of the ice to get out of the way. The damage, which was slight, was held due to inevitable accident.[26]

When vessels navigating with due caution and without fault in thick fog collide, the accident may be said to result from a form of *vis major*. A ferryboat is a class of vessel that because of the nature of her service is permitted to leave her wharf in the thickest weather. A New York ferry, while feeling her way into her slip, with a lookout properly stationed, drifted into a car float that had been moored outside her tug at the face of a pier, along with several other tugs and tows, all of which had been forced to take refuge there because of a unusually heavy fog. In the collision the car float was knocked from her moorings, whence she drifted down the river and collided with several barges moored ten piers below. Both collisions under the circumstances were held inevitable.[27]

In another New York Harbor case, the tug *Edwin Hawley*, learning that a ferry was disabled out in the river, put out in the fog to rescue her. Finding that the ferry had lost her rudder, the tug took her in tow on a bridle secured to the port and starboard bitts, respectively. In a collision a few days before, a steamship had been sunk in the harbor, and her steel masts formed an obstruction in the channel. The tug laid a course that would have cleared these masts, but when almost up with them the ferry took a sheer in the strong tide and knocked off one of the masts. The owners of the steamship libeled the tug. In finding the accident inevitable the court held that:

> There was no negligence on the part of the tug in going to the aid of the disabled vessel, notwithstanding the fog; there was no negligence in her method of towing, which was the only method possible; and as the sheer of the tug which was the immediate and sole cause of the accident was occasioned by the disabled vessel and the latter was unable to prevent it, or the tug to anticipate or withstand it, the tug was not liable for the damage.[28]

Vis Major Not Excuse if Incurred Through Negligence

However, the mere presence of *vis major* does not excuse a vessel if she

[25]The *General William McCandless* (NY 1879) Fed. Cas. No. 5,322.
[26]The *Erandio* (1908) 163 F 435.
[27]Wright and Cobb Lighterage Co. v. New England Nav. Co. (CCA 1913) 204 F 762.
[28]The *Edwin Hawley* (NY 1890) 41 F 606.

has been negligent in bringing herself into a critical situation. A number of very old decisions shows that such has long been the policy of the federal courts. As long ago as 1868, a tug tied up at a pier in New York with her stern projecting several feet into a ferry slip was struck, due to a strong tide, by a ferry trying to get into her slip. Both vessels were held at fault, the tug for obstructing the slip, and the ferry for attempting a landing under the circumstances.[29] About a year later the British steamship *Russia*, passing through the heavy tide rips that occur because the ebb begins to run out of the North River an hour and a half before it runs out of the East River, took a sheer and rammed and sank an Austrian ship at anchor off the Battery. As a pilot is presumed to know the action of the tides, this collision was held solely the fault of the *Russia*.[30] A schooner was aground and partially athwart a 150-foot channel in Hatteras Inlet and a steamer, whose pilot had taken soundings the day before and thought he could get around her, scraped bottom and sheered into the schooner. Notwithstanding that she reversed immediately she began to sheer, the steamer was found at fault for going at such speed as to put a hole in the schooner.[31] An ice boat owned by the city of Baltimore, in coming alongside a steamer with an engineer and stores that she had taken off a disabled tug, miscalculated the strength of the ice and the effect of backing her starboard screw and struck the steamship just forward of the poop. Although the ice boat was performing a purely gratuitous service, the court held the city of Baltimore liable for the damage.[32] In the case of the *Columbia*, a large steam elevator having a high tower that presented a large surface to the wind, she attempted to get alongside some barges on the exposed side of the river and sank one of them. The court held that while inevitable accident may arise from sudden gusts of wind, the evidence in this case showed lack of sufficient caution by the pilot, since the wind was rising and a 10-knot wind was known to make handling this elevator dangerous.[33]

Numerous attempts have been made to invoke inevitable accident under the *vis major* rule as an excuse for collision, but the courts have found one or both vessels guilty of contributory fault. An Army transport proceeding down the Elizabeth River from Norfolk Navy Yard at 8 knots, after dodging a tug with barges in tow, sagged into the Southern Railway docks at Town Point and sank a lighter loaded with valuable cargo. The immediate cause of the accident was failure of the steamship to answer her rudder promptly because of the combined effect of a wind on her

[29]The *Baltic* (NY 1868) Fed. Cas. No. 823.
[30]The *Russia* (NY 1869) Fed. Cas. No. 12,168.
[31]The *Ellen S. Terry* (NY 1874) Fed. Cas. No. 4,378.
[32]The *F. C. Latrobe* (Md 1886) 28 F 377.
[33]The *Columbia* (NY 1891) 48 F 325.

port bow and an ebb tide on her starboard quarter, but the court of claims called attention to local regulations limiting speed to 4 knots and pointed out that:

> Inevitable accident cannot be maintained as a defense unless it be shown that the master acted reasonably, that he did everything which an experienced mariner could do, and that the collision ensued in spite of ordinary caution and his exertions.[34]

A tug with a car float was proceeding through San Francisco Bay in a dense fog at night, with lookout properly stationed and alert, at a speed of about 7 knots, when she collided with a properly anchored barkentine whose fog bell she mistook for the bell on the mole. The district court accepted the tug's plea of inevitable accident or inscrutable fault but was reversed by the circuit court of appeals, which found contributory fault because of excessive speed.[35] A steamship that was very light was being docked in Mobile Harbor by two tugs whose movements were directed from her bridge. She struck and damaged another vessel moored at a pier, and the court found that while a wind squall was undoubtedly the proximate cause of the collision, the weather conditions were well known, storm warnings had been hoisted, and the landing should not have been attempted.[36] As was said in a similar case in New Orleans where the damage, however, was done by a steamer attempting a landing under her own power:

> In a collision case defense of inevitable accident will not avail unless the vessel was free from fault, and such defense cannot be maintained if a vessel voluntarily put herself in a situation where she received the effect of natural forces, the result of which should have been foreseen and might reasonably have been anticipated.[37]

The same attitude was shown in a case in New York, where failure of a tug to make preparations for an impending storm, or even to observe the approach of the storm, precluded a plea of inevitable accident, when, after the storm broke, her tow was swept into collision with another tow.[38]

Court Treatment of Mechanical Failure

Inevitable accident due to mechanical failure Inevitable accident may result at a critical moment from unforeseen casualty to main engines or steering gear, the latter being more likely, of course, to cause collision with another vessel. One case of inevitable accident occurred in Craighill Channel,

[34]The Southern Ry. Co. v. U.S. (1910) 45 Ct. Cl. 322.
[35]The *Fullerton* (Cal 1914) 211 F 833.
[36]Coello v. U.S. (La 1925) 9 F (2d) 931.
[37]The *Mendocino* (La 1929) 34 F (2d) 783.
[38]The *Patrick A. Dee* (CCA 1931) 50 F (2d) 393.

Baltimore Harbor, between the passenger steamer *City of Baltimore* and the tanker *Beacon*. The outbound *City of Baltimore* and the inbound *Beacon* approached each other near Buoy 9 at a combined speed of 23 knots—each on the proper side of the 600-foot channel—and exchanged one-blast signals for the usual port-to-port passing. When the vessels were not more than four to five lengths apart the *City of Baltimore*, from a cause later discovered to be the slipping of the yoke on a steering engine valve stem, suddenly began sheering to port. She immediately reversed full speed and sounded the danger signal. The *Beacon* did not reply to her signals or alter her course but promptly reversed her own engines, and approximately a minute and a half after her first sheer began the *City of Baltimore*, almost dead in the water, was struck by the *Beacon* at about 6 knots just abaft her chain locker, and both vessels were damaged. Although both omitted the three-blast reversing signals, it was established to the satisfaction of the court that the omission did not affect the action of either vessel and that the mechanical failure, which had never occurred before, was of a nature so unpredictable as to free the vessel from fault.[39]

But in a similar collision some years earlier on the St. Clair River, not only did the sheering vessel fail to justify her involuntary change of course but the steamer and towed barge with which she collided were held for contributory fault for not taking steps to avoid the disaster when, after exchanging meeting signals, the disabled vessel, then nearly half a mile distant, began to sheer and twice sounded the danger signal. While the court accepted testimony that failure of her steering gear was due either to fouling of steering cables by cargo in the hold or because some foreign substance had been carried into a steam valve, preventing it from closing, it held that this could not be classed as inevitable accident since the steering gear had stuck within half an hour of the collision and the vessel had proceeded without any attempt to ascertain the cause.[40]

Somewhat similar was the line of reasoning followed in the case of a collision between the USS *O-7* and the passenger vessel *Lexington*, which was en route from New York to Providence one evening shortly after World War I. The submarine was rounding Hallet's Point, when the *Lexington*, after sounding the bend signal of one long blast, appeared well clear on the port hand. The *O-7*'s electric steering gear jammed, causing her to sheer sharply to port. She blew the danger signal, stopped both engines, threw in the hand gear, and backed on her motors, and although the *Lexington*, which had stopped her engines before sighting the submarine when she heard the bend signal, also backed full speed, the submarine struck her near the stern, doing considerable damage. The

[39]The *Beacon* (Md 1934) 6 F Supp 779.
[40]Australia Transit Co. v. Lehigh Valley Transp. Co. (1916) 235 F 53.

government contended that the trouble was caused by the dynamic contact breaker in the panel, fusing and sticking, but on the admission of the commanding officer that similar trouble had been experienced before (though not since the contactor panel had recently been overhauled) and the nature of the trouble had been reported to the Bureau of Construction and Repair for the preceding two quarters, it was held that it was a fault to attempt the crowded and turbulent waters of Hell Gate without shifting to hand gear, and the accident was not therefore inevitable.[41]

In a second collision near the end of the first World War involving a submarine, this time the *R-19*, and a barge anchored in San Francisco Bay, the plea of the submarine that she was rendered helpless by the blowing out of a fuse, which made her electric steering gear inoperable and enabled the tide to throw her against the barge, was held not to establish the defense of inevitable accident. The night was clear, and the court declared that the vessel need not have been maneuvered in such a manner as to be suddenly put up against a strong running flood tide, throwing on the steering apparatus a load too great for the fuse to carry. She was accordingly found solely liable for the collision.[42]

In the famous English case of the *Marchant Prince*, that vessel came down the River Mersey in a gale of wind and approached the ship *Catalonia*, which was at anchor and because of the direction of the wind was lying somewhat athwart the channel. Due to a kink in some new chain in her tiller lines, the rudder of the *Merchant Prince* jammed so as to make her sheer into the *Catalonia*. She pleaded inevitable accident, but on appeal Lord Esher, speaking for the higher court, called attention to the fact that the tendency of new chain to stretch with use is well known and to be guarded against by taking in the slack often enough that kinking will not occur. To quote the decision:

> If that is so, is not that stretching of the chain a thing which they could have foreseen, which they ought to have foreseen, and which if they had foreseen—not that it would do it but that it might do it—ought not they to have taken means on that morning to have had the other steerages ready to act in a moment, even if they ought not to have used those other steerages, and those other steerages alone? It seems to me in this case, from what one can see of the facts proved of the conduct of the ships here, to show a probable cause, and if that was the cause it could have been avoided.[43]

Even if the cause of mechanical breakdown is found to have not been caused by negligence, a plea of inevitable accident does not necessarily follow. In the *Norwalk Victory* the steering gear of the vessel failed without

[41]Colonial Nav. Co. v. U.S. (NY 1926) 14 F (2d) 480.
[42]U.S. v. King Coal Co. (CCA 1925) 5 F (2d) 780.
[43]The *Merchant Prince* (1892) Eng. Law Rep. (PD) 179.

any negligence on her part, and she came into collision with another vessel she met on the bend of the River Scheldt. She was, nevertheless, held at fault for failure to monitor her helm indicator and for not detecting earlier the fact she had a breakdown, and for not letting go an anchor. It was held that the collision itself was not inevitable.[44] Similarly, a vessel that was disabled by the failure of her steering gear was held at fault for not giving warning of the fact in due time, and the other vessel was blamed for bad lookout, which resulted in her failing to take action to avoid the disabled vessel.[45]

When Breaking of Tiller Rope Is Inevitable Accident

A number of collisions have been caused by the sudden breaking of a vessel's tiller ropes, and in these cases, too, the question is not merely one of proving the fact but of showing that the breaking was unavoidable. The usual attitude of the courts in such circumstances was well expressed by the circuit court in the old case of the schooner *John Sherman* and the steamer *Olympia*, which collided on the Detroit River in 1891:

> When a collision results from the breaking of a steamer's tiller rope, the burden is upon her to rebut the presumption of negligence, either by showing the cause which broke the rope, and that the result of that cause was inevitable, or by showing all the possible causes which might have produced the break and then showing that the result of each one of them could not have been avoided.[46]

The court enumerated four possible causes of the breaking of a tiller rope as follows: (1) patent defects, due either to original unfitness or to use; (2) mismanagement of steering engine, too sudden spinning of the the wheel, etc.; (3) extraordinary strain, as unexplained caprice of steam, or obstruction encountered by rudder; (4) latent defects due to negligence of manufacturer or to use, not discoverable by such examination as can ordinarily be made. On the showing that in this case the tiller rope was of charcoal iron wire of suitable size, of the usual kind, and externally sound; that it had been bought from a reputable outfitter and used less than two seasons, while the usual life of such a wire was three years; that it had been inspected a few hours before the accident, and that when the broken ends were seen afterwards by witnesses there were no indications of defects; and finally, that although the steering engines were capable of putting a severe strain on the tiller ropes, in this instance the wheel was not suddenly handled, it was held that the collision was due to inevitable accident and not to the steamer's fault.

[44] The *Norwalk Victory* (1949) 82 Ll. L. Rep. 539.
[45] The *Nevitina* (1946) 79 Ll. L. Rep. 531.
[46] The *Olympia* (1894) 61 F 120.

In the case of the *Jumna*, three tugs were engaged in turning a steamship in the East River a short distance above the Brooklyn Bridge. Signals had been exchanged with an up-bound tow to pass starboard to starboard, when without warning the hawser with which one of the tugs was pulling on the stem of the steamship parted, allowing the tug to crash into the tows, which in turn struck and injured the piers on the Manhattan side. The court held that the collisions resulted solely from the breaking of the hawser, which was attributable in a legal sense to inevitable accident.[47]

Mechanical Failure Not Excuse if Due to Negligence

The breaking of a hawser by a tug, however, was not inevitable accident when it was due to unskillfullness on the part of the tug in getting her tow into strong tide rips, knowing they were present, so that the tows would sheer violently. The tug was liable for damage done immediately afterward by her tow to an anchored dredge that it failed to clear. This decision the Supreme Court, on appeal, declined to review.[48] In another case, a tug with a tow of twenty-seven boats suddenly rounded to, putting such a heavy strain on the steel wire towing bridle that the port hawser broke. The tug was held liable for collision damage to vessels in her tow. As held by the circuit court of appeals:

> To sustain defense of inevitable accident, defendant must prove, if the cause of the accident is shown, that he did not, by want of care and skill, contribute to it, or that he could not have prevented it by the exercise of such care and skill; and if the cause is not shown, he must show all possible causes, and that he was not responsible for any of them.[49]

Hence the courts have declined to attribute collision to inevitable accident in cases where a tug's wheel jammed as the result of spinning it too suddenly to avoid a schooner that a proper lookout would have discovered in time to make such action unnecessary;[50] where a steering cable parted when a give-way vessel went hard right within 60 to 70 feet of a stand-on canal boat;[51] where a steering gear shaft broke under similar circumstances;[52] or where a steering gear failed because of a set screw coming loose, when it had not been inspected for more than two years.[53]

A similar treatment occurs in cases where a collision follows failure of the main engines. A steamboat with a side-wheel engine was approaching

[47]The *Jumna* (NY 1906) 149 F 171.
[48]Petition of Red Star Towing and Transp. Co. (CCA 1929) 30 F (2d) 454.
[49]The *Osceola* (CCA 1927) 18 F (2d) 418.
[50]Brigham v. Luckenbach (Me 1905) 140 F 322.
[51]The *Edmund Moran* (CCA 1910) 180 F 700.
[52]The *Stimson* (Va 1919) 257 F 762.
[53]Van Eyken v. Erie R.R. Co. (NY 1902) 117 F 712.

her wharf at a speed of about 2 knots, and when the signal to reverse was given, the exhaust valve failed to operate and the steamer struck a ferry that was just leaving her slip. No trouble with the engine was experienced either before or after the accident, and it was thought probable that it was caused by sediment in the steam. Recognizing that while failure of an exhaust valve in that type of engine may happen but it cannot be foreseen or prevented, the court exonerated the steamer.[54] But where a tug was maneuvering to turn around in Port Arthur, Texas, and when a bolt broke in the reversing machinery she struck a steamer moored at a wharf, on a showing that the bolt had not been inspected for six years, the circuit court of appeals held the tug liable.[55]

Miscellaneous Cases of Inevitable Accident

Inevitable accident due to combination of circumstances A considerable variety of cases occurs under this category, a few of which will be given by way of illustration. Perhaps they may be appropriately characterized by those stirring words in the old melodramas, "It was the hand of Fate!" It was doubtless on this theory that in a very early case a tug was exonerated from liability for the destruction of her tow when the latter was brought into contact with an uncharted rock while navigating a regularly used river channel.[56] In another case a tug was coming through Hell Gate on a fair tide with a tow of five boats and a length of 800 feet when she suddenly came upon a stranded steamer with a derrick alongside lightering cargo. She was prevented from swinging into the Harlem River by an approaching schooner, and was forced to try to go between the stern of the wrecked vessel and Hallet's Point, a passage perhaps 500 feet in width. Despite all her efforts to avoid disaster, the tug was unable to prevent the tide from carrying her tow into the derrick, which was sunk along with the lightered cargo. This was quite properly held to be inevitable accident.[57] In an old case occurring in Boston Harbor in fair weather and in broad daylight, the Cunard steamship *Java* was approaching her wharf and was about to pass a large school ship at anchor, when a small schooner that had been dropped by her tug and was completely concealed from the liner by the school ship while she hoisted her sails, suddenly drifted into the path of the Cunarder. The latter had lookouts properly posted, was making only 2 knots, and was immediately reversed, but nevertheless she struck and seriously damaged the schooner. The Supreme Court upheld the district court and reversed the circuit court of appeals, holding that the

[54]The *Rose Standish* (Mass 1928) 26 F (2d) 480.

[55]The *J.N. Gilbert* (CCA 1915) 222 F 37.

[56]The *Angelina Corning* (NY 1867) Fed. Cas. No. 384; also the *America* (NY 1872).

[57]Merritt and Chapman Derrick and Wrecking Co. v. Cornell Steamboat Co. (CCA 1911) 185 F 261.

circumstances justified the contention of inevitable accident.[58] As a final illustration, consider the unusual case of the schooner *Southern Home*, en route from Santo Domingo to New York with most of her crew stricken with yellow fever. The captain and the mate were confined to their bunks, the steward had died, and the second mate had turned in exhausted after standing a continuous watch for thirty hours, leaving only a convalescent seaman at the wheel and in charge of the deck. There was no one left to stand lookout, and when the man at the wheel sighted the lights of a privileged sailing vessel close aboard he was too weak to put over the wheel, though his shouts of alarm were successful in arousing the second mate. The ensuing collision was, of course, proximately due to failure of the burdened schooner to keep out of the way, and it was argued by the libelants that the stricken schooner should have come to anchor off the Jersey coast instead of attempting to proceed. However, it was shown that the weakened crew, while they could have dropped anchor, would have been unable afterward to heave it up, and the court, considering all the circumstances, held that reasonable precautions had been taken and the accident was inevitable.[59]

Collision can be inevitable as far as a ship sued is concerned, where the fault lies elsewhere—such as in a case where a ship is thrust or rolled against another by the swell of a passing vessel or by a third vessel fouling her[60] or causing her suddenly to alter course.[61]

SUMMARY

Inevitable accident has been defined as something that human skill and foresight could not, in the exercise of ordinary prudence, have provided against. The term as applied to a marine collision means one that occurs when both parties have tried by every means in their power, with due care and caution and a proper display of nautical skill, to prevent the occurrence of the accident, and it occurred in spite of everything that nautical skill, care, and precaution could do to keep the vessels from coming together. Thus, a collision between vessels falls in this category only when neither vessel has violated a rule and both vessels are free from fault. Such collisions are rare, numbering fewer than 1 percent of the cases coming into court; when they occur, each vessel bears her own damage and has no recourse against the other.

Most collisions of this kind are due to *vis major*, i.e., superior force of the elements, or to mechanical failure of engines or steering gear that due

[58]The *Java* (1872) 20 L Ed. 834.
[59]The *Southern Home* (CC NY 1879) Fed. Cas. No. 13,187.
[60]The *Sisters* (1876) 1 P.D. 117; the *Hibernia* (1858) 4 Jur. (NS) 1244.
[61]The *Schwan* and the *Albano* (1892) p. 419; the *Thames* (1874) 2 Asp. MC 512.

diligence could not anticipate or prevent. *Vis major* is illustrated when a storm of great violence, a spring freshet, an extraordinary fog, or an unexpected heavy movement of ice brings one vessel into collision with another. Mechanical failure as a cause of inevitable accident is illustrated where a steering gear, despite originally proper construction and frequent periodical inspection, carries away in the crucial moment of an approaching situation and precipitates a collision. In none of these cases can the doctrine of inevitable accident be invoked where legal fault, which could have contributed to the collision, can be imputed to either vessel. The defense of inevitable accident is never an easy argument to mount. The whole sequence of events that led up to the collision will be examined, and the defense will fail if negligence is shown at any point.

It is significant that many of the cases of inevitable accident shown in this chapter were decided over fifty years ago, and that there are very few recent cases where the plea of inevitable accident has been successful.

APPENDIX A

COLREGS Demarcation Lines[1]

PART 82—Colregs Demarcation Lines[2]

General

Sec.
82.01 General basis and purpose of demarcation lines.

Atlantic Coast

82.105 Calais, ME to Cape Small, ME.
82.110 Casco Bay ME.
82.115 Portland Head, ME to Cape Ann, MA.
82.120 Cape Ann, MA to Marblehead Neck, MA.
82.125 Marblehead Neck, MA to Winthrop Head, MA.
82.130 Boston Harbor Entrance.
82.135 Point Allerton, MA to Race Point, MA.
82.140 Race Point, MA to Marthas Vineyard, MA.
82.145 Marthas Vineyard, MA to Watch Hill, RI.
82.150 Block Island, RI.
82.305 Watch Hill, RI to Montauk Point, NY.
82.310 Montauk Point, NY to Atlantic Beach, NY.
82.315 New York Harbor.
82.320 Sandy Hook, NJ to Cape May, NJ.
82.325 Delaware Bay.
82.505 Cape Henlopen, DL to Cape Charles, VA.
82.510 Chesapeake Bay Entrance, VA.
82.515 Cape Henry, VA to Cape Hatteras, NC.
82.520 Cape Hatteras, NC to Cape Lookout, NC.
82.525 Cape Lookout, NC to Cape Fear, NC.
82.530 Cape Fear, NC to New River Inlet, NC.

[1]The former "Lines of Demarcation" in 33 CFR 82 have been republished in 46 CFR 7 as "Boundary Lines." These boundary lines are applicable only to certain specified statutes such as the Vessel Bridge-to-Bridge Radiotelephone Communications Act.
[2]Authority: Rule 1, 72 COLREGS; 33 CFR 82.

82.703 Little River Inlet, SC to Cape Romain, SC.
82.707 Cape Romain, SC to Sullivans Island, SC.
82.710 Charleston Harbor, SC.
82.712 Morris Island, SC to Hilton Head Island, SC.
82.715 Savannah River.
82.717 Tybee Island, GA to St. Simons Island, GA.
82.720 St. Simons Island, Ga to Amelia Island, FL.
82.723 Amelia Island, FL to Cape Canaveral, FL.
82.727 Cape Canaveral, FL to Miami Beach, FL.
82.730 Miami Harbor, FL.
82.735 Miami, FL to Long Key, FL.

Puerto Rico and Virgin Islands

82.738 Puerto Rico and Virgin Islands.

Gulf Coast

82.740 Long Key, FL to Cape Sable, FL.
82.745 Cape Sable, FL to Cape Romano, FL.
82.748 Cape Romano, FL to Sanibel Island, FL.
82.750 Sanibel Island, FL to St. Petersburg, FL.
82.753 St. Petersburg, FL to the Anclote, FL.
82.755 Anclote, FL to the Suncoast Keys, FL.
82.757 Suncoast Keys, FL to Horseshoe Point, FL.
82.760 Horseshoe Point, FL to Rock Island, FL.
82.805 Rock Island, FL to Cape San Blas, FL.
82.810 Cape San Blas, FL to Perdido Bay, FL.
82.815 Mobile Bay, AL to the Chandeleur Islands, LA.
82.820 Mississippi River.
82.825 Mississippi Passes, LA.
82.830 Mississippi Passes, LA to Point Au Fer, LA.
82.835 Point Au Fer, LA to Calcasieu Pass, LA.
82.840 Sabine Pass, TX to Galveston, TX.
82.845 Galveston, TX to Freeport, TX.
82.850 Brazos River, TX to the Rio Grande, TX.

Pacific Coast

82.1105 Santa Catalina Island, CA.
82.1110 San Diego Harbor, CA.
82.1115 Mission Bay, CA.
82.1120 Oceanside Harbor, CA.
82.1125 Dana Point Harbor, CA.
82.1130 Newport Bay, CA.
82.1135 San pedro Bay–Anaheim Bay, CA.
82.1140 Redondo Harbor, CA.
82.1145 Marina Del Rey, CA.
82.1150 Port Hueneme, CA.
82.1155 Channel Islands Harbor, CA.
82.1160 Ventura Marina, CA.
82.1165 Santa Barbara Harbor, CA.
82.1205 San Luis Obispo Bay, CA.
82.1210 Estero–Morro Bay, CA.

82.1215 Monterey Harbor, CA.
82.1220 Moss Landing Harbor, CA.
82.1225 Santa Cruz Harbor,CA.
82.1230 Pillar Point Harbor, CA.
82.1250 San Francisco Harbor, CA.
82.1255 Bodega and Tomales Bay, CA.
82.1260 Albion River, CA.
82.1265 Noyo River, CA.
82.1270 Arcato–Humboldt Bay, CA.
82.1275 Crescent City Harbor, CA.
82.1305 Chetco River, OR.
82.1310 Rogue River, OR.
82.1315 Coquille River, OR.
82.1320 Coos Bay, OR.
82.1325 Umpqua River, OR.
82.1330 Sinslaw River, OR.
82.1335 Alsea Bay, OR.
82.1340 Yaquina Bay, OR.
82.1345 Depoe Bay, OR.
82.1350 Netarts Bay, OR.
82.1360 Nehalem River, OR.
82.1365 Columbia River Entrance, OR/WA.
82.1370 Willapa Bay, WA.
82.1375 Grays Harbor, WA.
82.1380 Quillayute River, WA.
82.1385 Strait of Juan de Fuca.
82.1390 Haro Strait and Strait of Georgia.

Pacific Islands

82.1410 Hawaiian Island Exemption from General Rule.
82.1420 Mamala Bay, Oahu, HI.
82.1430 Kaneohe Bay, Oahu, HI.
82.1440 Port Allen, Kauai, HI.
82.1450 Nawiliwili Harbor, Kauai, HI.
82.1460 Kahului Harbor, Maui, HI.
82.1470 Kawaihae Harbor, Hawaii, HI.
82.1480 Hilo Harbor, Hawaii, HI.
82.1490 Apra Harbor, U.S. Territory of Guam.
82.1495 U.S. Pacific Island Possessions.

Alaska

82.1705 Canadian (BC) and United States (AK) borders to Cape Muzon, AK.
82.1710 Cape Muzon, AK to Cape Bartolome, AK.
82.1715 Cape Bartolome, AK to Cape Ulitka, AK.
82.1720 Cape Ulitka, AK to Cape Ommaney, AK.
82.1725 Cape Ommaney, AK to Cape Edgecumbe, AK.
82.1730 Cape Edgecumbe, AK to Cape Spencer, AK.
82.1735 Cape Spencer, AK to Point Whitshed, AK.
82.1740 Prince William Sound, AK.
82.1750 Alaska West and North of Prince William Sound.

General

General Basis and Purpose of Demarcation Lines

SEC. 82.01 (a) The regulations in this part establish the lines of demarcation delineating those waters upon which mariners must comply with the International Regulations for Preventing Collisions at Sea, 1972 (72 COLREGS) and those waters upon which mariners must comply with the Navigation Rules for Harbors, Rivers, and Inland Waters (Inland Rules).

(b) The waters inside the line are INLAND RULES WATERS. The waters outside the lines are COLREGS WATERS.

(c) The regulations in this part do not apply to the Great Lakes or their connecting and tributary waters as described in Part 90 of this Chapter, or the Western Rivers as described in Part 95 of this Chapter.

Atlantic Coast

Calais, ME to Cape Small, ME

SEC. 82.105 The 72 COLREGS shall apply on the harbors, bays, and inlets on the east coast of Maine from International Bridge at Calais, ME to the southwesternmost extremity of Bald Head at Cape Small.

Casco Bay, ME

SEC. 82.110 (a) A line drawn from the southwesternmost extremity of Bald Head at Cape Small to the southernmost extremity of Ragged Island; thence to the southern tangent of Jaquish Island thence to Little Mark Island Monument Light; thence to the northernmost extremity of Jewell Island.

(b) A line drawn from the tower on Jewell Island charted in approximate position latitude 43°40.6′ N. longitude 70°05.9′ W. to the northeasternmost extremity of Outer Green Island.

(c) A line drawn from the southwesternmost extremity of Outer Green Island to Ram Island Ledge Light to Portland Head Light.

Portland Head, ME to Cape Ann, MA

SEC. 82.115 (a) Except inside lines specifically described in this section, the 72 COLREGS shall apply on the harbors, bays, and inlets on the east coast of Maine, New Hampshire, and Massachusetts from Portland Head to Halibut Point at Cape Ann.

(b) A line drawn from the southernmost tower on Gerrich Island charted in approximate position latitude 43°04.0′ N. longitude 70°41.2′ W. to Whaleback Light; thence to the northeasternmost extremity of Frost Point.

(c) A line drawn from the northernmost extremity of Farm Point to Annisquam Harbor Light.

Cape Ann, MA to Marblehead Neck, MA

SEC. 82.120 (a) Except inside lines specifically described in this section, the 72 COLREGS shall apply on the harbors, bays and inlets on the east coast of Massachusetts from Halibut Point at Cape Ann to Marblehead Neck.

(b) A line drawn from Gloucester Harbor Breakwater Light to the twin towers charted in approximate position latitude 42°35.1′ N. longitude 70°41.6′ W.

(c) A line drawn from the westernmost extremity of Gales Point to the easternmost extremity of House Island; thence to Bakers Island Light; thence to Marblehead Light.

Marblehead Neck, MA to Winthrop Head, MA

SEC. 82.125 The 72 COLREGS shall apply on the bays, harbors and inlets on the east coast of Massachusetts from Marblehead Neck to Winthrop Head.

Boston Harbor Entrance

SEC 82.130 A line drawn from the standpipe on Winthrop Head charted in approximate position latitude 42°22.1′ N. longitude 70°58.1′ W. to Great Faun Bar Daybeacon; thence to Boston Light; thence to the tower on Point Allerton charted in approximate position latitude 42°18.4′ N. Longitude 70°53.1′ W.

Point Allerton, MA to Race Point, MA

SEC 82.135 (a) Except inside lines specifically described in this section, the 72 COLREGS shall apply on the harbors, bays and inlets on the east coast of Massachusetts from Point Allerton to Race Point on Cape Cod.

(b) A line drawn from Cape Cod Canal Breakwater Light south to the shoreline.

Race Point, MA to Marthas Vineyard, MA

SEC. 82.140 (a) The 72 COLREGS apply to the harbors, bays and inlets along the coast of Cape Cod from Race Point to the southernmost extremity of Nauset Beach.

(b) A line drawn from the southernmost extremity of Nauset Beach to the northernmost extremity of Monomoy Island.

(c) A line drawn from the abandoned lighthouse tower on the southern end of Monomoy Island to Nantucket (Great Point Light).

(d) A line drawn from the westernmost extremity of Nantucket Island to the southernmost tangent of Wasque Point on Marthas Vineyard.

Marthas Vineyard, MA to Watch Hill, RI

SEC. 82.145 (a) Except lines specifically described in this section, the 72 COLREGS shall apply on the harbors, bays and inlets on the south coast of Massachusetts and Rhode Island from Marthas Vineyard to Watch Hill.

Watch Hill, RI to Montauk Point, NY

SEC. 82.305 (a) A line drawn from Watch Hill Light to East Point on Fishers Island.

(b) A line drawn from Race Point to Race Rock Light; thence to Little Gull Island Light thence to East Point on Plum Island.

(c) A line drawn from Plum Island Harbor East Dolphin Light and Plum Island Harbor West Dolphin Light.

(d) A line drawn from Plum Island Light to Orient Point Light; thence to Orient Point.

(e) A line drawn from the lighthouse ruins at the southwestern end of Long Beach Point to Cornelius Point.

(f) A line drawn from Coecles Harbor Entrance Light to Sungie Point.

(g) A line drawn from Nichols Point to Cedar Island Light.

(h) A line drawn from Three Mile Harbor West Breakwater Light to Three Mile Harbor East Breakwater Light.
(i) A line drawn from Montauk West Jetty Light to Montauk East Jetty Light.

Montauk Point, NY to Atlantic Beach, NY

SEC. 82.310 (a) A line drawn from Shinnecock Inlet East Breakwater Light to Shinnecock West Breakwater Light.
(b) A line drawn from Moriches Inlet East Breakwater Light to Moriches Inlet West Breakwater Light.
(c) A line drawn from Fire Island Inlet Breakwater Light 348° true to the southernmost extremity of the spit of land at the western end of Oak Beach.
(d) A line drawn from Jones Inlet Light 142° true across the southwest tangent of the island on the north side of Jones Inlet to the shoreline.

New York Harbor

SEC. 82.315 A line drawn from East Rockaway Inlet Breakwater Light to Sandy Hook Light.

Sandy Hook, NJ to Cape May, NJ

SEC. 82.320 (a) A line drawn from Shark River Inlet North Breakwater Light to Shark River Inlet South Breakwater Light.
(b) A line drawn from Manasquan Inlet North Breakwater Light to Manasquan Inlet South Breakwater Light.
(c) A line drawn from Barnegat Inlet North Breakwater Light to Barnegat Inlet South Breakwater Light. Lines formed by the submerged Barnegat Breakwaters.
(d) A line drawn from the seaward tangent on Long Beach Island to the seaward tangent to Pullen Island across Beach Haven and Little Egg Inlets.
(e) A line drawn from the seaward tangent of Pullen Island and Brigantine Island across Brigantine Inlet.
(f) A line drawn from the seaward extremity of Absecon Inlet North Jetty and Atlantic City Light.
(g) A line drawn from the southernmost point of Longport at latitude 39°18.2′ N. longitude 74°32.2′ W. to the northeasternmost point of Ocean City at latitude 39°17.6′ N. longitude 74°33.1′ W. across Great Egg Harbor Inlet.
(h) A line formed by the centerline of the Townsend Inlet Highway Bridge.
(i) A line formed by the shoreline of Seven Mile Beach and Hereford Inlet Light.
(j) A line drawn from Cape May Inlet East Jetty Light to Cape May Inlet West Jetty Light.

Delaware Bay

SEC. 82.325 A line drawn from Cape May Light to Harbor of Refuge Light; thence to the northernmost extremity of Cape Henlopen.

Cape Henlopen, DL to Cape Charles, VA

SEC. 82.505 (a) A line drawn from Indian River Inlet North Jetty Light to Indian River Inlet South Jerry Light.
(b) A line drawn from Ocean City Inlet Light 6 234° true across Ocean City Inlet to the submerged south breakwater.

(c) A line drawn from Assateague Beach Tower Light to the tower charted at latitude 37°52.6′ N. longitude 75°26.7′ W.

(d) A line formed by the range of Wachapreague Inlet Light 3 and Parramore Beach Lookout Tower drawn across Wachapreague Inlet.

(e) A line drawn from the lookout tower charted on the northern end of Hog Island to the seaward tangent of Parramore Beach.

(f) A line drawn 207° true from the lookout tower charted on the southern end of Hog Island across Great Machipongo Inlet.

(g) A line formed by the range of the two cupolas charted on the southern end of Cobb Island drawn across Sand Shoal Inlet.

(h) Except as provided elsewhere in this section from Cape Henlopen to Cape Charles, lines drawn parallel with the general trend of the highwater shoreline across the entrances to small bays and inlets.

Chesapeake Bay Entrance, VA

SEC. 82.510 A line drawn from Cape Charles Light to Cape Henry Light.

Cape Henry, VA to Cape Hatteras, NC

SEC. 82.515 (a) A line drawn from Rudee Inlet Jetty Light 2 to Rudee Inlet Jetty Light 1.

(b) A line formed by the centerline of the highway bridge across Oregon Inlet.

Cape Hatteras, NC to Cape Lookout, NC

SEC. 82.520 (a) A line drawn from Hatteras Inlet Light 255° to the eastern end of Ocracoke Island.

(b) A line drawn from the westernmost extremity of Ocracoke Island at latitude 35°04.0′ N. longitude 76°00.8′ W. to the northeastern extremity of Portsmouth Island at latitude 35°03.7′ N. longitude 76°02.3′ W.

(c) A line drawn across Drum Inlet parallel with the general trend of the highwater shoreline.

Cape Lookout, NC to Cape Fear, NC

SEC. 82.525 (a) A line drawn from Cape Lookout Light to the seaward tangent of the southeastern end of Shackleford Banks.

(b) A line drawn from Morehead City Channel Range Front Light to the seaward extremity of the Beaufort Inlet west jetty.

(c) A line drawn from the southernmost extremity of Bogue Banks at latitude 34°38.7′ N. longitude 77°06.0′ W. across Bogue Inlet to the northernmost extremity of Bear Beach at latitude 34°38.5′ N. longitude 77°07.1′ W.

(d) A line drawn from the tower charted in approximate position latitude 34°31.5′ N. longitude 77°20.8′ W. to the seaward tangent of the shoreline on the northeast side of New River Inlet.

(e) A line drawn across New Topsail Inlet between the closest extremities of the shore on either side of the inlet from latitude 34°20.8′ N. longitude 77°39.2′ W. to latitude 34°20.6′ N. longitude 77°39.6′ W.

(f) A line drawn from the seaward extremity of the jetty on the northeast side of Masonboro Inlet west to the shoreline approximately 0.6 mile southwest of the inlet.

(g) Except as provided elsewhere in this section from Cape Lookout to Cape

Fear, lines drawn parallel with the general trend of the highwater shoreline across the entrance of small bays and inlets.

Cape Fear, NC to New River Inlet, NC

SEC. 82.530 (a) A line drawn from the abandoned lighthouse charted in approximate position latitude 33°52.4′ N. longitude 78°00.1′ W. across the Cape Fear River Entrance to Oak Island Light.

(b) Except as provided elsewhere in this section from Cape Fear to New River Inlet, lines drawn parallel with the general trend of the highwater shoreline across the entrance to small inlets.

Little River Inlet, SC to Cape Romain, SC

SEC. 82.703 (a) A line drawn from the westernmost extremity of the sand spit on Bird Island to the easternmost extremity of Waties Island across Little River Inlet.

(b) Lines drawn parallel with the general trend of the highwater shoreline across Hog Inlet, Muriels Inlet, Midway Inlet, Pawleys Inlet, and North Inlet.

(c) A line drawn from the charted position of Winyah Bay North Jetty End Buoy 2N south to the Winyah Bay South Jetty.

(d) A line drawn from Santee Point to the seaward tangent of Cedar Island.

(e) A line drawn from Cedar Island Point west to Murphy Island.

(f) A north-south line (longitude 79°20.3′ W.) line drawn from the southern extremity of Murphy Island to the northernmost extremity of Cape Island Point.

Cape Romain, SC to Sullivans Island, SC

SEC. 82.707 (a) A line drawn from the western extremity of Cape Romain 292° true to Raccoon Key on the west side of Raccoon Creek.

(b) A line drawn from the northwesternmost extremity of Sandy Point across Bull Bay to the northernmost extremity of Northeast Point.

(c) A line drawn from the southernmost extremity of Bull Island to the easternmost extremity of Capers Island.

(d) A line formed by the overhead power cable from Capers Island to Dewees Island.

(e) A line formed by the overhead power cable from Dewees Island to Isle of Palms.

(f) A line formed by the centerline of the highway bridge between Isle of Palms and Sullivans Island over Beach Inlet.

Charleston Harbor, SC

SEC. 82.710 (a) A line drawn from across the seaward extremity of the Charleston Harbor Jetties.

(b) A line drawn from the west end of the South Jetty across the South Entrance to Charleston Harbor to shore on a line formed by the submerged south jetty.

Morris Island, SC to Hilton Head Island, SC

SEC. 82.712 (a) A line drawn from the Folly Island Loran Tower charted in approximate position latitude 32°41.0′ N. longitude 79°53.2′ W. to the abandoned lighthouse tower on the northside of Lighthouse Inlet; thence west to the shoreline of Morris Island.

(b) A straight line drawn from the seaward tangent of Folly Island through Folly River Daybeacon 10 across Stono River to the shoreline of Sandy Point.

(c) A line drawn from the southernmost extremity of Seabrook Island 257° true across the North Edisto River Entrance to the shore of Botany Bay Island.

(d) A line drawn from the microwave antenna tower on Edisto Beach charted in approximate position latitude 32°29.3′ N. longitude 80°19.2′ W. across St. Helena Sound to the abandoned lighthouse tower on Hunting Island.

(e) A line formed by the centerline of the highway bridge between Hunting Island and Fripp Island.

(f) A line drawn from the westernmost extremity of Bull Point on Capers Island to Port Royal Sound Channel Rear Range Light; thence 245° true to the easternmost extremity of Hilton Head at latitude 32°13.2′ N. longitude 80°40.1′ W.

Savannah River

SEC. 82.715 A line drawn from the southernmost tank on Hilton Head Island charted in approximate position latitude 32°06.7′ N. longitude 80°49.3′ W. to Bloody Point Range Rear Light; thence to Tybee (Range Rear) Light.

Tybee Island, GA to St. Simons Island, GA

SEC. 82.717 (a) A line drawn from the southernmost extremity of Savannah Beach on Tybee Island 255° true across Tybee Inlet to the shore of Little Tybee Island south of the entrance to Buck Hammock Creek.

(b) A straight line drawn from the northeasternmost extremity of Wassaw Island 031° true through Tybee River Daybeacon 1 to the shore of Little Tybee Island.

(c) A line drawn approximately parallel with the general trend of the highwater shorelines from the seaward tangent of Wassaw Island to the seaward tangent of Bradley Point on Ossabaw Island.

(d) A north-south line (longitude 81°8.4′ W.) drawn from the southernmost extremity of Ossabaw Island to St. Catherines Island.

(e) A north-south line (longitude 81°10.6′ W.) drawn from the southernmost extremity of St. Catherines Island to Northeast Point on Blackbeard Island.

(f) A north-south line (longitude 81°16.9′ W.) drawn from the southwesternmost point on Sapelo Island to Wolf Island.

(g) A north-south line (longitude 81°17.1′ W.) drawn from the southeasternmost point of Wolf Island to the northeasternmost point on Little St. Simons Island.

(h) A line drawn from the northeastern extremity of Sea Island 045° true to Little St. Simons Island.

St. Simons Island, GA to Amelia Island, FL

SEC. 82.720 (a) A line drawn from St. Simons Light to the northernmost tank on Jekyll Island charted in approximate position latitude 31°05.9′ N. longitude 81°24.5′ W.

(b) A line drawn from the southernmost tank on Jekyll Island charted in approximate position latitude 31°01.6′ N. longitude 81°25.2′ W. to coordinate latitude 30°59.4′ N. longitude 81°23.7′ W. (0.5 nautical mile east of the charted position of St. Andrew Sound Lighted Buoy 32); thence to the abandoned lighthouse tower on the north end of Little Cumberland Island charted in approximate position latitude 30°58.5′ N. longitude 81°24.8′ W.

(c) A line drawn across the seaward extremity of the St. Marys River Entrance Jetties.

Amelia Island, FL to Cape Canaveral, FL

SEC. 82.723 (a) A line drawn from the southernmost extremity of Amelia Island to the northeasternmost extremity of Little Talbot Island.

(b) A line drawn across the seaward extremity of the St. Johns River Entrance Jetties.

(c) A line drawn across the seaward extremity of the St. Augustine Inlet Jetties.

(d) A line formed by the centerline of the highway bridge over Matanzas Inlet.

(e) A line drawn across the seaward extremity of the Ponce de Leon Inlet Jetties.

Cape Canaveral, FL to Miami Beach, FL

SEC.82.727 (a) A line drawn across the seaward extremity of the Port Canaveral Entrance Channel Jetties.

(b) A line drawn across the seaward extremity of the Sebastian Inlet Jetties.

(c) A line drawn across the seaward extremity of the Fort Pierce Inlet Jetties.

(d) A north-south line (longitude 80°09.8′ W.) drawn across St. Lucie Inlet through St. Lucie Inlet Entrance Range Front Daybeacon.

(e) A line drawn from the seaward extremity of Jupiter Inlet North Jetty to the northeast extremity of the concrete apron on the south side of Jupiter Inlet.

(f) A line drawn across the seaward extremity of the Lake Worth Inlet Jetties.

(g) A line drawn across the seaward extremity of the South Lake Worth Inlet Jetties.

(h) A line drawn from Boca Raton Inlet North Jetty Light 2 to Boca Raton Inlet South Jetty Light 1.

(i) A line drawn from Hillsboro Inlet Entrance Light 2 to Hillsboro Inlet Entrance Light 1; thence west to the shoreline.

(j) A line drawn across the seaward extremity of the Port Everglades Entrance Jetties.

(k) A line formed by the centerline of the highway bridge over Bakers Haulover Inlet.

Miami Harbor, FL

SEC. 82.730 A line drawn across the seaward extremity of the Miami Harbor Government Cut Jetties.

Miami, FL to Long Key, FL

SEC. 82.735 (a) A line drawn from the southernmost extremity of Fisher Island 211° true to the point latitude 25°45.1′ N. longitude 80°08.6′ W. on Virginia Key.

(b) A line formed by the centerline of the highway bridge between Virginia Key and Key Biscayne.

(c) A line drawn from the abandoned lighthouse tower on Cape Florida to Biscayne Channel Light 8; thence to the northernmost extremity on Soldier Key.

(d) A line drawn from the southernmost extremity on Soldier Key to the northernmost extremity of the Ragged Keys.

(e) A line drawn from the Ragged Keys to the southernmost extremity of Angelfish Key following the general trend of the seaward shoreline.

(f) A line drawn on the centerline of the Overseas Highway (U.S. 1) and bridges from latitude 25°19.3′ N. longitude 80°16.0′ W. at Little Angelfish Creek to the radar dome charted on Long Key at approximate position latitude 24°49.3′ N. longitude 80°49.2′ W.

Puerto Rico and Virgin Islands

SEC. 82.738 (a) Except inside lines specifically described in this section, the 72 COLREGS shall apply on all other bays, harbors, and lagoons of Puerto Rico and the U.S. Virgin Islands.

(b) A line drawn from Puerto San Juan Light to Cabras Light across the entrance of San Juan Harbor.

Gulf Coast

Long Key, FL to Cape Sable, FL

SEC. 82.740 A line drawn from the radar dome charted on Long Key at approximate position latitude 24°49.3′ N. longitude 80°49.2′ W. to Long Key Light 2; thence to Arsenic Bank Light 1; thence to Arsenic Bank Light 2; thence to Sprigger Bank Light 5; thence to Schooner Bank Light 6; thence to Oxfoot Bank Light 10; thence to East Cape Light 2; thence through East Cape Daybeacon 1A to the shoreline at East Cape.

Cape Sable, FL to Cape Romano, FL

SEC. 82.745 (a) A line drawn following the general trend of the mainland, highwater shoreline from Cape Sable at East Cape to Little Shark River Light 1; thence to westernmost extremity of Shark Point; thence following the general trend of the mainland, highwater shoreline crossing the entrances of Harney River, Broad Creek, Broad River, Rodgers River First Bay, Chatham River, Huston River, to the shoreline at coordinate latitude 25°41.8′ N. longitude 81°17.9′ W.

(b) The 72 COLREGS shall apply to the waters surrounding the Ten Thousand Islands and the bays, creeks, inlets, and rivers between Chatham Bend and Marco Island except inside lines specifically described in this part.

(c) A north-south line drawn at longitude 81°20.2′ W. across the entrance to Lopez River.

(d) A line drawn across the entrance to Turner River parallel to the general trend of the shoreline.

(e) A line formed by the centerline of Highway 92 Bridge at Goodland.

Cape Romano, FL to Sanibel Island, FL

SEC. 82.748 (a) Lines drawn across Big Marco Pass parallel to the general trend of the seaward, highwater shoreline.

(b) A line drawn through Capri Pass Daybeacons 2A and 3 across Capri Pass.

(c) Lines drawn across Hurricane and Little Marco Passes parallel to the general trend of the seaward, highwater shoreline.

(d) A straight line drawn from Gordon Pass Light 4 through Daybeacon 5 to the shore.

(e) A line drawn across the seaward extremity of Doctors Pass Jetties.

(f) Lines drawn across Wiggins, Big Hickory, New, and Big Carlos Passes parallel to the general trend of the seaward, highwater shoreline.

(g) A straight line drawn from Sanibel Island Light through Matanzas Pass Channel Light 2 to the shore of Estero Island.

Sanibel Island, FL to St. Petersburg, FL

SEC. 82.750 (a) Lines drawn across Redfish and Captiva Passes parallel to the general trend of the seaward, highwater shorelines.

(b) A line drawn from La Costa Test Pile North Light to Port Boca Grande Light.

(c) Lines drawn across Gasparilla and Stump Passes parallel to the general trend of the seaward, highwater shorelines.

(d) A line across the seaward extremity of Venice Inlet Jetties.

(e) A line drawn across Midnight Pass parallel to the general trend of the seaward, highwater shoreline.

(f) A line drawn from Big Sarasota Pass Light 14 to the southernmost extremity of Lido Key.

(g) A line drawn through Sarasota Bay Channel Light 7A across New Pass parallel to the seaward, highwater shoreline of Longboat Key.

(h) A line drawn across Longboat Pass parallel to the seaward, highwater shoreline.

(i) A line drawn from the northwesternmost extremity of Bean Point to the southeasternmost extremity of Egmont Key.

(j) A straight line drawn from Egmont Key Light through Egmont Channel Range Rear Light to the shoreline on Mullet Key.

(k) A line drawn from the northernmost extremity of Mullet Key across Bunces Pass and South Channel to Pass-a-Grille Daybeacon 9; thence to the southwesternmost extremity of Long Key.

St. Petersburg, FL to the Anclote, FL

SEC. 82.753 (a) A line drawn across Blind Pass parallel with the general trend of the seaward, highwater shoreline.

(b) Lines formed by the centerline of the highway bridges over Johns and Clearwater Passes.

(c) A line drawn across Dunedin and Hurricane Passes parallel with the general trend of the seaward, highwater shoreline.

(d) A line drawn from the northernmost extremity of Honeymoon Island to Anclote Anchorage South Entrance Light 7; thence to Anclote Keys Light; thence a straight line through Anclote River Cut B Range Rear Light to the shoreline.

Anclote, FL to the Suncoast Keys, FL

SEC. 82.755 (a) Except inside lines specifically described in this section, the 72 COLREGS shall apply on the bays, bayous, creeks, marinas, and rivers from Anclote to the Suncoast Keys.

(b) A north-south line drawn at longitude 82°38.3′ W. across the Chassahowitgka River Entrance.

Suncoast Keys, FL to Horsehoe Point, FL

SEC. 82.757 (a) Except inside lines specifically described in this section, the 72 COLREGS shall apply on the bays, bayous, creeks, and marinas from the Suncoast to Horseshoe Point.

(b) A line formed by the centerline of Highway 44 Bridge over the Salt River.

(c) A north-south line drawn through Crystal River Entrance Daybeacon 25 across the river entrance.

(d) A north-south line drawn through the Cross Florida Barge Canal Daybeacon 38 across the canal.

(e) A north-south line drawn through Withlacoochee River Daybeacon 40 across the river.

(f) A line drawn from the westernmost extremity of South Point north to the shoreline across the Waccasassa River Entrance.

(g) A line drawn from position latitude 29°16.6′ N. longitude 83°06.7′ W. 300° true to the shoreline of Hog Island.

(h) A north-south line drawn through Suwanee River West Pass Daybeacons 27 and 28 across the Suwannee River.

Horseshoe Point, FL to Rock Islands, FL

SEC. 82.760 (a) Except inside lines specifically described provided in this section, the 72 COLREGS shall apply on the bays, bayous, creeks, marinas, and rivers from Horseshoe Point to the Rock Islands.

(b) A north-south line drawn through Steinhatchee River Light 21.

(c) A line drawn from Fenholloway River Approach Light FR east across the entrance to Fenholloway River.

Rock Island, FL to Cape San Blas, FL

SEC. 82.805 (a) A south-north line drawn from the Econfina River Light to the opposite shore.

(b) A line drawn from Gamble Point Light to the southernmost extremity of Cabell Point.

(c) A line drawn from St. Marks (Range Rear) Light to St. Marks Channel Light 11; thence to Live Oak Point; thence to Ochlockonee Point; thence to Bald Point.

(d) A line drawn from the south shore of Southwest Cape at longitude 84°22.7′ W. to Dog Island Reef East Light 1; thence to Turkey Point Light 2; thence to the easternmost extremity of Dog Island.

(e) A line drawn from the westernmost extremity of Dog Island to the easternmost extremity of St. George Island.

(f) A line drawn across the seaward extremity of the St. George Island Channel Jetties.

(g) A line drawn from the northwesternmost extremity of Sand Island to West Pass Light 7.

(h) A line drawn from the westernmost extremity of St. Vincent Island to the southeast, highwater shoreline of Indian Peninsula at longitude 85°13.5′ W.

Cape San Blas, FL to Perdido Bay, FL

SEC. 82.810 (a) A line drawn from St. Joseph Range A Rear Light through St. Joseph Range B Front Light to St. Joseph Point.

(b) A line drawn across the mouth of Salt Creek as an extension of the general trend of the shoreline.

(c) A line drawn from the northernmost extremity of Crooked Island 000° T. to the mainland.

(d) A line drawn from the easternmost extremity of Shell Island 120° true to the shoreline across the east entrance to St. Andrews Bay.

(e) A line drawn between the seaward end of the St. Andrews Bay Entrance Jetties.

(f) A line drawn between the seaward end of the Choctawhatchee Bay Entrance Jetties.

(g) A west-east line drawn from Fort McGee Leading Light across the Pensacola Bay Entrance.

(h) A line drawn between the seaward end of the Perdido Pass Jetties.

Mobile Bay, AL to the Chandeleur Islands, LA

SEC. 82.815 (a) A line drawn across the inlets to Little Lagoon as an extension of the general trend of the shoreline.

(b) A line drawn from Mobile Point Light to Dauphin Island Spit Light to the eastern corner of Fort Gaines at Pelican Point.

(c) A line drawn from the westernmost extremity of Dauphin Island to the easternmost extremity of Petit Bois Island.

(d) A line drawn from Horn Island Pass Entrance Range Front Light on Petit Bois Island to the easternmost extremity of Horn Island.

(e) An east-west line (latitude 30°14.7′ N.) drawn between the westernmost extremity of Horn Island to the easternmost extremity of Ship Island.

(f) A curved line drawn following the general trend of the seaward, highwater shoreline of Ship Island.

(g) A line drawn from Ship Island Light; thence to Chandeleur Light; thence in a curved line following the general trend of the seaward, highwater shorelines of the Chandeleur Islands to the island at coordinate latitude 29°31.1′ N. longitude 89°05.7′ W.; thence to Breton Island Light located at latitude 29°29.1′ N. longitude 89°09.7′ W.

Mississippi River

SEC. 82.820 The Pilot Rules for Western Rivers are to be followed in the Mississippi River and its tributaries above the Huey P. Long Bridge.

Mississippi Passes, LA

SEC. 82.825 (a) A line drawn from Breton Island Light to coordinate latitude 29°21.5′ N. thence to coordinate latitude 29°21.5′ N. longitude 89°11.7′ W.

(b) A line drawn from coordinate latitude 29°21.5′ N. longitude 89°117′ W. following the general trend of the seaward, highwater shoreline in a southeasterly direction to coordinate latitude 29°12.4′ N. longitude 89°06.0′ W.; thence following the general trend of the seaward, highwater shoreline in a northeasterly direction to coordinate latitude 29°13.0′ N. longitude 89°01.3′ W. located on the northwest bank of North Pass.

(c) A line drawn from coordinate latitude 29°13.0′ N. longitude 89°01.3′ W.; thence coordinate latitude 29°12.7′ N. longitude 89°0.9′ W.; thence coordinate latitude 29°10.6′ N. longitude 88°59.8′ W.; thence coordinate latitude 29°03.5′ N. longitude 89°59.8′ W.; thence coordinate latitude 29°03.5′ N. longitude 89°03.7′ W., thence Mississippi River South Pass East Jetty Light 4.

(d) A line drawn from Mississippi River South Pass East Jetty Light 4; thence following the general trend of the seaward, highwater shoreline in a northwesterly direction to coordinate latitude 29°03.4′ N. longitude 89°13.0′ W.; thence following the general trend of the seaward, highwater shoreline in a southwesterly direction to Mississippi River Southwest Pass Entrance Light.

(e) A line drawn from Mississippi River Southwest Pass Entrance Light; thence to the seaward extremity of the Southwest Pass West Jetty located at coordinate latitude 28°54.5′ N. longitude 89°26.1′ W.

Mississippi Passes, LA to Point Au Fer, LA

SEC. 82.830 (a) A line drawn from the seaward extremity of the Southwest Pass West Jetty located at coordinate latitude 28°54.5′ N. longitude 89°26.1′ W.; thence following the general trend of the seaward, highwater jetty and shoreline in a north-northeasterly direction to Old Tower latitude 28°58.8′ N. longitude 89°23.3′ W.; thence to West Bay Light; thence to coordinate latitude 29°05.2′ N. longitude 89°24.3′ W.; thence a curved line following the general trend of the highwater shoreline to Point Au Fer Island except as otherwise described in this section.

(b) A line drawn across the seaward extremity of the Empire Waterway (Bayou Fontanelle) entrance jetties.

(c) A line drawn from Barataria Bay Light to the Grand Isle Fishing Jetty Light.

(d) A line drawn between the seaward extremity of the Belle Pass Jetties.

(e) A line drawn from the westernmost extremity of the Timbolier Island to the easternmost extremity of Isles Dernieres.

(f) A south-north line drawn from Caillou Bay Light 13 across Caillou Boca.

(g) A line drawn 107° true from Caillou Bay Boat Landing Light across the entrances to Grand Bayou du Large and Bayou Grand Caillou.

(h) A line drawn on an axis of 103° true through Taylors Bayou Light across the entrances to Jack Stout Bayou, Taylors Bayou, Pelican Pass, and Bayou de West.

Point Au Fer, LA to Calcasieu Pass, LA

SEC. 82.835 (a) A line drawn from Point Au Fer to Atchafalaya Channel Light 32; thence Point Au Fer Reef Light; Atchafalaya Bay Pipeline Light D latitude 29°25.0′ N. longitude 91°31.7′ W.; thence Atchafalaya Bay Light 1 latitude 29°25.3′ N. longitude 91°35.8′ W.; thence South Point.

(b) Lines following the general trend of the highwater shoreline drawn across the bayou canal inlet from the Gulf of Mexico between South Point and Calcasieu Pass except as otherwise described in this section.

(c) A line drawn on a axis of 130° T. through Vermillion Bay Light 2 across Southwest Pass.

(d) A line drawn across the seaward extremity of the Freshwater Bayou Canal Entrance Jetties.

(e) A line drawn from Mermentau River Channel Light 4 to Mermentau River Channel Light 5.

(f) A line drawn from the radio tower in approximate position latitude 29°45.7′ N. longitude 93°06.3′ W. 160° true across Mermentau.

(g) A line drawn across the seaward extremity of the Calcasieu Pass Jetties.

Sabine Pass, TX to Galveston, TX

SEC. 82.840 (a) A line drawn from the Sabine Pass East Jetty Light to the seaward end of the Sabine Pass West Jetty.

(b) Lines drawn across the small boat passes through the Sabine Pass East and West Jetties.

(c) A line formed by the centerline of the highway bridge over Rollover Pass at Gilchrist.

Galveston, TX to Freeport, TX

SEC. 82.845 (a) A line drawn from Galveston North Jetty Light to Galveston South Jetty Light.

(b) A line formed by the centerline of the highway bridge over San Luis Pass.

(c) Lines formed by the centerlines of the highway bridges over the inlets to Christmas Bay (Cedar Cut) and Drum Bay.

(d) A line drawn from the seaward extremity of the Freeport North Jetty to Freeport Entrance Light 6; thence Freeport Entrance Light 7; thence the seaward extremity of Freeport South Jetty.

Brazos River, TX to the Rio Grande, TX

SEC. 82.850 (a) Except as otherwise described in this section lines drawn continuing the general trend of the seaward, highwater shorelines across the inlets to Brazos River Diversion Channel, San Bernard River, Cedar Lakes, Brown

Cedar Cut, Colorado River, Matagorda Bay Cedar Bayou, Corpus Christi Bay, and Laguna Madre.

(b) A line drawn across the seaward extremity of Matagorda Ship Channel North Jetties.

(c) A line drawn from the seaward tangent of Matagorda Peninsula at Decros Point to Matagorda Daybeacon 2; thence to Matagorda Light.

(d) A line drawn across the seaward extremity of the Aransas Pass Jetties.

(e) A line drawn across the seaward extremity of the Port Mansfield Entrance Jetties.

(f) A line drawn across the seaward extremity of the Brazos Santiago Pass Jetties.

Pacific Coast

Santa Catalina Island, CA

SEC. 82.1105 The 72 COLREGS shall apply to the harbors on Santa Catalina Island.

San Diego Harbor, CA

SEC. 82.1110 A line drawn from Zunica Jetty Light "V" to Zunica Jetty Light "Z"; thence to Point Loma Light.

Mission Bay, CA

SEC. 82.1115 A line drawn from Mission Bay South Jetty Light 2 to Mission Bay North Jetty Light 1.

Oceanside Harbor, CA

SEC. 82.1120 A line drawn from Oceanside South Jetty Light 4 to Oceanside Breakwater Light 3.

Dana Point Harbor, CA

SEC. 82.1125 A line drawn from Dana Point Jetty Light 6 to Dana Point Breakwater Light 5.

Newport Bay, CA

SEC. 82.1130 A line drawn from Newport Bay East Jetty Light 4 to Newport Bay West Jetty Light 3.

San Pedro Bay—Anaheim Bay, CA

SEC. 82.1135 (a) A line drawn from Anaheim Bay East Jetty Light 6 to Anaheim Bay West Jetty Light 5; thence to Long Beach Breakwater East End Light.

(b) A line drawn from Long Beach Channel Entrance Light 2 to Long Beach Light.

(c) A line drawn from Los Angeles Main Entrance Channel Light 2 to Los Angeles Light.

Redondo Harbor, CA

SEC. 82.1140 A line drawn from Redondo Beach East Jetty Light 2 to Redondo Beach West Jetty Light 3.

Marina Del Rey, CA

SEC. 82.1145 (a) A line drawn from Marina Del Rey Breakwater South Light 1 to Marina Del Rey Light 4.

(b) A line drawn from Marina Del Rey Breakwater North Light 2 to Marina Del Rey Light 3.

(c) A line drawn from Marina Del Rey Light 4 to the seaward extremity of the Ballona Creek South Jetty.

Port Hueneme, CA

SEC. 82.1150 A line drawn from Port Hueneme East Jetty Light 4 to Port Hueneme West Jetty Light 3.

Channel Islands Harbor, CA

SEC. 82.1155 (a) A line drawn from Channel Islands Harbor South Jetty Light 2 to Channel Island Harbor Breakwater South Light 1.

(b) A line drawn from Channel Islands Harbor Breakwater North Light to Channel Islands Harbor North Jetty Light 5.

Ventura Marina, CA

SEC. 82.1160 A line drawn from Ventura Marina South Jetty Light 2 to Ventura Marina Breakwater South Light 1; thence to Ventura Marina North Jetty Light 3.

Santa Barbara Harbor, CA

SEC. 82.1165 A line drawn from Santa Barbara Harbor Light 4 to Santa Barbara Harbor Breakwater Light.

San Luis Obispo Bay, CA

SEC. 82.1205 A line drawn from the southernmost extremity of Fossil Point to the seaward extremity of Whaler Island Breakwater.

Estero—Morro Bay, CA

SEC. 82.1210 A line drawn from the seaward extremity of the Morro Bay East Breakwater to the Morro Bay West Breakwater Light.

Monterey Harbor, CA

SEC. 82.1215 A line drawn from Monterey Harbor Breakwater Light to the northern extremity of Monterey Municipal Wharf 2.

Moss Landing Harbor, CA

SEC. 82.1220 A line drawn from the seaward extremity of the pier located 0.3 mile south of Moss Landing Harbor Entrance to the seaward extremity of the Moss Landing Harbor North Breakwater.

Santa Cruz Harbor, CA

SEC. 82.1225 A line drawn from the seaward extremity of the Santa Cruz Harbor East Jetty to the seaward extremity of the Santa Cruz Harbor West Jetty; thence to Santa Cruz Light.

Pillar Point Harbor, CA

SEC. 82.1230 A line drawn from Pillar Point Harbor Light 6 to Pillar Point Harbor Light 5.

San Francisco Harbor, CA

SEC. 82.1250 A straight line drawn from Point Bonita Light through Mile Rocks Light to the shore.

Bodega and Tomales Bay, CA

SEC. 82.1255 (a) An east-west line drawn through Tomales Bay Daybeacon 3 from Sand Point to Avalis Beach.

(b) A line drawn from the seaward extremity of Bodega Harbor North Breakwater to Bodega Harbor Entrance Light 1.

Albion River, CA

SEC. 82.1260 A line drawn on an axis of 030° true through Albion River Light 1 across Albion Cove.

Noyo River, CA

SEC. 82.1265 A line drawn from Noyo River Entrance Daybeacon 4 to Noyo River Entrance Light 5.

Arcata—Humboldt Bay, CA

SEC. 82.1720 A line drawn from Humboldt Bay Entrance Light 4 to Humboldt Bay Entrance Light 3.

Crescent City Harbor, CA

SEC. 82.1275 A line drawn from Crescent City Outer Breakwater Light to the southeasternmost extremity of Whaler Island.

Chetco River, OR

SEC. 82.1305 A line drawn from the seaward extremity of the Chetco River South Jetty to Chetco River Entrance Light 5.

Rogue River, OR

SEC. 82.1310 A line drawn from the seaward extremity of the Rogue River Entrance South Jetty to Rogue River North Jetty Light 3.

Coquille River, OR

SEC. 82.1315 A line drawn across the seaward extremity of the Coquille River Entrance Jetties.

Coos Bay, OR

SEC. 82.1320 A line drawn across the seaward extremity of the Coos Bay Entrance Jetties.

Umpqua River, OR

SEC. 82.1325 A line drawn across the seaward extremity of the Umpqua River Entrance Jetties.

Siuslaw River, OR

SEC. 82.1330 A line drawn from the seaward extremity of the Siuslaw River Entrance South Jetty to Siuslaw River Light 9.

Alsea Bay, OR

SEC. 82.1335 A line drawn from the seaward shoreline on the north of the Alsea Bay Entrance 165° true across the channel entrance.

Yaquina Bay, OR

SEC. 82.1340 A line drawn from the seaward extremity of Yaquina Bay Entrance South Jetty to Yaquina Bay North Jetty Light 5.

Depoe Bay, OR

SEC. 82.1345 A line drawn across the Depoe Bay Channel entrance parallel with the general trend of the highwater shoreline.

Netarts Bay, OR

SEC. 82.1350 A line drawn from the northernmost extremity of the shore on the south side of Netarts Bay north to the opposite shoreline.

Tillamook Bay, OR

SEC. 82.1355 A north-south line drawn from the lookout tower charted on the north side of the entrance to Tillamook Bay south to the Tillamook Bay South Jetty.

Nehalem River, OR

SEC. 82.1360 A line drawn approximately parallel with the general trend of the highwater shoreline across the Nehalem River Entrance.

Columbia River Entrance, OR/WA

SEC. 82.1365 A line drawn from the seaward extremity of the Columbia River North Jetty (above water) 155° true to the seaward extremity of the Columbia River South Jetty (above water).

Willapa Bay, WA

SEC. 82.1370 A line drawn from Willapa Bay Light 171° true to the westernmost tripod charted 1.6 miles south of Leadbetter Point.

Grays Harbor, WA

SEC. 82.1375 A line drawn from across the seaward extremity, (above water) of the Grays Harbor Entrance Jetties.

Quillayute River, WA

SEC. 82.1380 A line drawn from the seaward extremity of the Quillayute River Entrance East Jetty to the overhead power cable tower charted on James Island; thence a straight line through Quillayute River Entrance Light 3 to the shoreline.

Strait of Juan de Fuca

SEC. 82.1385 (a) The 72 COLREGS shall apply on Neah Bay and the waters inside Ediz Hook (Port Angeles Harbor).

(b) A line drawn from New Dungeness Light through Puget Sound Traffic Lane Entrance Lighted Buoy S to Rosario Strait Traffic Lane Entrance Lighted Horn Buoy R; through Hein Bank Lighted Bell Buoy to Cattle Point Light.

Haro Strait and Strait of Georgia

SEC. 82.1390 (a) The 72 COLREGS shall apply on the bays of the southwest coast of San Juan Island from Cattle Point Light to Lime Kiln Light.

(b) A line drawn from Lime Kiln Light to Kellett Bluff Light; thence to Turn Point Light; thence to Skipjack Island Light; thence to Sucia Island Daybeacon 1.

(c) A line drawn from the shoreline of Sucia Island at latitude 48°46.1′ N. longitude 122°53.5′ W. through Clements Reef Buoy 2 to Alden Bank Lighted Gong Buoy A; thence to the westernmost tip of Birch Point at latitude 48°56.6′ N. longitude 122°49.2′ W.

(d) The 72 COLREGS shall apply in Semiamoo Bay and Drayton Harbor.

Pacific Islands

Hawaiian Island Exemption from General Rule

SEC. 82.1410 Except as provided elsewhere in this part of Mamala Bay and Kaneohe Bay on Oahu; Port Allen and Nawiliwili Bay on Kauai; Kahului Harbor on Maui; and Kawailae and Hilo Harbors on Hawaii, the 72 COLREGS shall apply on all other bays, harbors, and lagoons of the Hawaiian Islands (including Midway).

Mamala Bay, Oahu, HI

SEC. 82.1420 A line drawn from Barbers Point Light to Diamond Head Light.

Kaneohe Bay, Oahu, HI

SEC. 82.1430 A straight line drawn from Pyramid Rock Light across Kaneohe Bay through the center of Mokolii Island to the shoreline.

Port Allen, Kauai, HI

SEC. 82.1440 A line drawn from Hanapepe Light to Hanapepe Bay Breakwater Light.

Nawiliwili Harbor, Kauai, HI

SEC. 82.1450 A line drawn from Nawiliwili Harbor Breakwater Light to Kukii Point Light.

Kahului Harbor, Maui, HI

SEC. 82.1460 A line drawn from Kahului Harbor Entrance East Breakwater Light to Kahului Harbor Entrance West Breakwater Light.

Kawaihae Harbor, Hawaii, HI

SEC. 82.1470 A line drawn from Kawaihae Light to the seaward extremity of the Kawaihae South Breakwater.

Hilo Harbor, Hawaii, HI

SEC. 82.1480 A line drawn from the seaward extremity of the Hilo Breakwater 265° true (as an extension of the seaward side of the breakwater) to the shoreline 0.2 nautical mile north of Alealea Point.

Apra Harbor, U.S. Territory of Guam

SEC. 82.1490 A line drawn from the westernmost extremity of Orote Island to the westernmost extremity of Glass Breakwater.

U.S. Pacific Island Possessions

SEC. 82.1495 The 72 COLREGS shall apply on the bays, harbors, lagoons, and waters surrounding the U.S. Pacific Island Possessions of American Samoa, Baker, Canton, Howland, Jarvis, Johnson, Palmyra, Swains and Wake Island. (The Trust Territory of the Pacific Islands is not a U.S. possession, and therefore PART 82 does not apply thereto.)

Alaska

Canadian (BC) and United States (AK) borders to Cape Muzon, AK

SEC. 82.1705 (a) A line drawn from the northeasternmost extremity of Point Mansfield, Sitklan Island 040° true to the mainland.

(b) A line drawn from the southernmost extremity of Sitklan Island to the southernmost extremity of Garnet Point, Kanagunut Island.

(c) A line drawn from the westernmost extremity of Tingbeg Island to the southwesternmost extremity of Tongass Island.

(d) A line drawn from the northern shoreline of Tongass Island at longitude 130°44.6′ W. to Tongass Reef Daybeacon; thence to Boat Rock Light; thence to the shoreline.

(e) A line drawn from Tree Point Light to Barren Island Light; thence to Cape Chacon Light; thence to Cape Muzon Light.

Cape Muzon, AK to Cape Bartolome, AK

SEC. 82.1710 (a) The 72 COLREGS shall apply on the harbors and bays of the west coast of Doll Island from Cape Muzon to Cape Lookout.

(b) A line drawn from the westernmost extremity of Cape Lookout to Diver Islands Light; thence to the southernmost extremity of Cape Felix; thence to Cape Bartolome Light.

Cape Bartolome, AK to Cape Ulitka, AK

SEC. 82.1715 A line drawn from the westernmost extremity of Outer Point on Baker Island to the southernmost extremity of St. Nicholas Point on Noyes Island.

Cape Ulitka, AK to Cape Ommaney, AK

SEC. 82.1720 (a) A line drawn from Cape Ulitka Light to the southwesternmost extremity of St. Joseph Island.

(b) A line drawn from south-north line (longitude 133°42.8′ W.) from the northernmost extremity of St. Joseph Island to the southernmost extremity of the Wood Islands.

(c) A line drawn from the northwesternmost extremity of Wood Island to Cape Lynch Light; thence to the southwesternmost extremity of Boot Point on Warren Island.

(d) A line drawn from the northwesternmost extremity of Point Borlase on Warren Island to the northeastern extremity of the Spanish Islands.

(e) A line drawn from Spanish Islands Light to Cape Decision Light; thence through Cape Ommaney Light to the shoreline.

(f) The 72 COLREGS shall apply on the bays and harbors of Coronation Island.

Cape Ommaney, AK to Cape Edgecumbe, AK

SEC. 82.1725 (a) The 72 COLREGS shall apply on the bays, inlets, and harbors of the west coast of Baranof Island from Cape Ommaney to Cape Burunof.

(b) A line drawn from the westernmost extremity of Cape Burunof to Kulichkof Rock; thence to Vitskari Island Light; thence to the southeasternmost extremity of Shoals Point on Kruzof Island.

Cape Edgecumbe, AK to Cape Spencer, AK

SEC. 82.1730 (a) The 72 COLREGS shall apply on the bays and harbors of the south and west coasts of Kruzof Island from Shoals Point to Cape Georgiana.

(b) A line drawn from the northwesternmost extremity of Cape Georgiana on Kruzof Island to Klokachef Island Light.

(c) A line drawn from the northernmost extremity of Fortuna Point on Klokachef Island 055° true to the shoreline of Khaz Peninsula.

(d) The 72 COLREGS shall apply on the bays, inlets and harbors of the west coast of Chichogof Island from Fortuna Strait to Easter Island.

(e) A line drawn from Lisianski Strait Entrance Light to the southernmost extremity of Point Theodore on Yakobi Island.

(f) The 72 COLREGS shall apply on the bays and harbors of the west coast of Yakobi Island from Point Theodore to Soapstone Point.

(g) A line drawn from Lisianski Inlet Light to Cape Spencer Light; thence to the southernmost extremity of Cape Spencer.

Cape Spencer, AK to Point Whitshed, AK

SEC. 82.1735 The 72 COLREGS shall apply on the bays and harbors from Cape Spencer to Point Whitshed on the coast of Alaska Mainland.

Prince William Sound, AK

SEC. 82.1740 (a) Hawkins Island Cutoff: A line drawn from Point Whitshed on the Alaska Mainland at position 60°26.7′ N. 145°52.7′ W. westsouthwesterly to Point Bentinck aerobeacon on Hinchinbrook Island.

(b) Hinchinbrook Entrance: A line drawn from Cape Hinchinbrook Light northerly to Schooner Rock Light.

(c) Montague Strait: A line drawn from a point on the western end of Montague Island at position 59°50.2′ N. 147°54.4′ W. northwesterly to Point Elrington Light on Elrington Island thence due west to the Alaska Mainland at Cape Puget.

Alaska west and north of Prince William Sound

SEC. 82.1750 The 72 COLREGS shall apply on the sounds, bays, inlets, and harbors of Alaska west of Cape Puget, Kodiak Island, Aleutian Islands, and the west and north coasts of Alaska.

APPENDIX B

Convention on the International Regulations for Preventing Collisions at Sea, 1972[1]

The Convention, as signed on October 20, 1972, and as rectified on December 1, 1973, together with the International Regulations attached thereto are as follows:

CONVENTION ON THE INTERNATIONAL REGULATIONS FOR PREVENTING COLLISIONS AT SEA, 1972

The Parties to the present Convention,

Desiring to maintain a high level of safety at sea,

Mindful of the need to revise and bring up to date the International Regulations for Preventing Collisions at Sea annexed to the Final Act of the International Conference on Safety of Life at Sea, 1960,

Having considered those Regulations in the light of developments since they were approved,

Have agreed as follows:

Article I

General Obligations

The Parties to the present Convention undertake to give effect to the Rules and other Annexes constituting the International Regulations for Preventing Collisions at Sea, 1972, (hereinafter referred to as "the Regulations") attached hereto.

Article II

Signature, Ratification, Acceptance, Approval and Accession

1. The present Convention shall remain open for signature until 1 June 1973 and shall thereafter remain open for accession.
2. States Members of the United Nations, or of any of the Specialized Agencies,

[1]The President deposited U.S. acceptance with IMCO effective 23 November 1976. The International Regulations for Preventing Collisions at Sea, 1972, became effective for U.S. mariners on 15 July 1977. The International Navigational Rules Act of 1977, enacted by Congress on 27 July 1977 (Public Law 95-75, 33 USC 1601), implemented the convention.

or the International Atomic Energy Agency, or Parties to the Statute of the International Court of Justice may become Parties to this Convention by:

(a) signature without reservation as to ratification, acceptance or approval;

(b) signature subject to ratification, acceptance or approval followed by ratification, acceptance or approval; or

(c) accession.

3. Ratification, acceptance, approval or accession shall be effected by the deposit of an instrument to that effect with the Inter-Governmental Maritime Consultative Organization (hereinafter referred to as "the Organization") which shall inform the Governments of States that have signed or acceded to the present Convention of the deposit of each instrument and of the date of its deposit.

Article III

Territorial Application

1. The United Nations in cases where they are the administering authority for a territory or any Contracting Party responsible for the international relations of a territory may at any time by notification in writing to the Secretary-General of the Organization (hereinafter referred to as "the Secretary-General"), extend the application of this Convention to such a territory.

2. The present Convention shall, upon the date of receipt of the notification or from such other date as may be specified in the notification, extend to the territory named therein.

3. Any notification made in accordance with paragraph 1 of this Article may be withdrawn in respect of any territory mentioned in that notification and the extension of this Convention to that territory shall cease to apply after one year or such longer period as may be specified at the time of the withdrawal.

4. The Secretary-General shall inform all Contracting Parties of the notification of any extension or withdrawal of any extension communicated under this Article.

Article IV

Entry into Force

1. (a) The present Convention shall enter into force twelve months after the date on which at least 15 States, the aggregate of whose merchant fleets constitutes not less than 65 percent by number or by tonnage of the world fleet of vessels of 100 gross tons and over have become Parties to it, whichever is achieved first.

(b) Notwithstanding the provisions in subparagraph (a) of this paragraph, the present Convention shall not enter into force before 1 January 1976.

2. Entry into force for States which ratify, accept, approve or accede to this Convention in accordance with Article II after the conditions prescribed in subparagraph 1 (a) have been met and before the Convention enters into force, shall be on the date of entry into force of the Convention.

3. Entry into force for States which ratify, accept, approve or accede after the date on which this Convention enters into force, shall be on the date of deposit of an instrument in accordance with Article II.

4. After the date of entry into force of an amendment to this Convention in accordance with paragraph 4 of Article VI, any ratification, acceptance, approval or accession shall apply to the Convention as amended.

5. On the date of entry into force of this Convention, the Regulations replace and abrogate the International Regulations for Preventing Collisions at Sea, 1960.

6. The Secretary-General shall inform the Governments of States that have signed or acceded to this Convention of the date of its entry into force.

Article V

Revision Conference

1. A Conference for the purpose of revising this Convention or the Regulations or both may be convened by the Organization.
2. The Organization shall convene a Conference of Contracting Parties for the purpose of revising this Convention or the Regulations or both at the request of not less than one-third of the Contracting Parties.

Article VI

Amendments to the Regulations

1. Any amendment to the Regulations proposed by a Contracting Party shall be considered in the Organization at the request of that Party.
2. If adopted by a two-thirds majority of those present and voting in the Maritime Safety Committee of the Organization, such amendment shall be communicated to all Contracting Parties and Members of the Organization at least six months prior to its consideration by the Assembly of the Organization. Any Contracting Party which is not a Member of the Organization shall be entitled to participate when the amendment is considered by the Assembly.
3. If adopted by a two-thirds majority of those present and voting in the Assembly, the amendment shall be communicated by the Secretary-General to all Contracting parties for their acceptance.
4. Such an amendment shall enter into force on a date to be determined by the Assembly at the time of its adoption unless, by a prior date determined by the Assembly at the same time, more than one-third of the Contracting Parties notify the Organization of their objection to the amendment. Determination by the Assembly of the dates referred to in this paragraph shall be by a two-thirds majority of those present and voting.
5. On entry into force any amendment shall, for all Contracting Parties which have not objected to the amendment, replace and supersede any previous provision to which the amendment refers.
6. The Secretary-General shall inform all Contracting Parties and Members of the Organization of any request and communication under this Article and the date on which any amendment enters into force.

Article VII

Denunciation

1. The present Convention may be denounced by a Contracting Party at any time after the expiry of five years from the date on which the Convention entered into force for that Party.
2. Denunciation shall be effected by the deposit of an instrument with the Organization. The Secretary-General shall inform all other Contracting Parties of the receipt of the instrument of denunciation and of the date of its deposit.
3. A denunciation shall take effect one year, or such longer period as may be specified in the instrument, after its deposit.

Article VIII

Deposit and Registration

1. The present Convention and the Regulations shall be deposited with the

Organization, and the Secretary-General shall transmit certified true copies thereof to all Governments of States that have signed this Convention or acceded to it.

2. When the present Convention enters into force, the text shall be transmitted by the Secretary-General to the Secretariat of the United Nations for registration and publication in accordance with Article 102 of the Charter of the United Nations.

Article IX

Languages

The present Convention is established, together with the Regulations, in a single copy in the English and French languages, both texts being equally authentic. Official translations in the Russian and Spanish languages shall be prepared and deposited with the signed original.

IN WITNESS WHEREOF the undersigned being duly authorized by their respective Governments for that purpose have signed the present Convention.

DONE AT LONDON this twentieth day of October one thousand nine hundred and seventy-two.

Procès-Verbal of Rectification

Whereas a Convention on the International Relations for Preventing Collisions at Sea was done at London on 20 October 1972 and is deposited with the Inter-Governmental Maritime Consultative Organization; and

Whereas certain errors in English and in French have been discovered in the original signed copy of the said Convention and brought to the notice of the interested Governments; and

Whereas no objection to the correction of these errors having been raised by any of the Governments which were represented at the International conference on Revision of the International Regulations for Preventing Collisions at Sea, 1972, which adopted the Convention, the said errors should be corrected as indicated. . . .:

Now, therefore, I the undersigned, Colin Goad, Secretary-General of the Inter-Governmental Maritime Consultative Organization, acting for the depositary of the Convention on the International Regulations for Preventing Collisions at Sea, 1972, have caused the original text of the Convention to be modified by the corrections indicated above, and initialled in the margin thereof.

In witness thereof, I have signed the present Procès-Verbal at the Headquarters of the Organization this first day of December 1973, in the English and French languages, in a single copy which shall be kept in the archives of the Organization with the original signed copy of the Convention on the International Regulations for Preventing Collisions at Sea, 1972.

A certified copy of this Procès-Verbal shall be communicated to each Government which has signed or acceded to the aforementioned Convention.

Colin Goad.

INTERNATIONAL REGULATIONS FOR PREVENTING COLLISIONS AT SEA, 1972 (As rectified by Procès-Verbal of December 1, 1973)

PART A—GENERAL

Application

Rule 1 (a) These Rules shall apply to all vessels upon the high seas and in all waters connected therewith navigable by seagoing vessels.

(b) Nothing in these Rules shall interfere with the operation of special rules made by an appropriate authority for roadsteads, harbours, rivers, lakes or inland waterways connected with the high seas and navigable by seagoing vessels. Such special rules shall conform as closely as possible to these Rules.

(c) Nothing in these Rules shall interfere with the operation of any special rules made by the Government of any State with respect to additional station or signal lights or whistle signals for ships of war and vessels proceeding under convoy, or with respect to additional station or signal lights for fishing vessels engaged in fishing as fleet. These additional station or signal lights or whistle signals shall, so far as possible, be such that they cannot be mistaken for any light or signal authorized elsewhere under these Rules.

(d) Traffic separation schemes may be adopted by the Organization for the purpose of these Rules.

(e) Whenever the Government concerned shall have determined that a vessel of special construction or purpose cannot comply fully with the provisions of any of these Rules with respect to the number, position, range or arc of visibility of lights or shapes, as well as to the disposition and characteristics of sound-signalling appliances, without interfering with the special function of the vessel, such vessel shall comply with such other provisions in regard to the number, position, range or arc of visibility of lights or shapes, as well as to the disposition and characteristics of sound-signalling appliances, as her Government shall have determined to be the closest possible compliance with these Rules in respect to that vessel.

Responsibility

Rule 2 (a) Nothing in these Rules shall exonerate any vessel, or the owner, master or crew thereof, from the consequences of any neglect to comply with these Rules or of the neglect of any precaution which may be required by the ordinary practice of seamen, or by the special circumstances of the case.

(b) In construing and complying with these Rules due regard shall be had to all dangers of navigation and collision and to any special circumstances, including the limitations of the vessels involved which may make a departure from these Rules necessary to avoid immediate danger.

General Definitions

RULE 3 For the purpose of these Rules, except where the context otherwise requires:

(a) The word "vessel" includes every description of water craft, including non-displacement craft and seaplanes, used or capable of being used as a means of transportation on water.

(b) The term "power-driven vessel" means any vessel propelled by machinery.

(c) The term "sailing vessel" means any vessel under sail provided that propelling machinery, if fitted, is not being used.

(d) The term "vessel engaged in fishing" means any vessel fishing with nets, lines, trawls or other fishing apparatus which restrict manoeuvrability, but does not include a vessel fishing with trolling lines or other fishing apparatus which do not restrict manoeuvrability.

(e) The word "seaplane" includes any aircraft designed to manoeuvre on the water.

(f) The term "vessel not under command" means a vessel which through some exceptional circumstance is unable to manoeuvre as required by these Rules and is therefore unable to keep out of the way of another vessel.

(g) The term "vessel restricted in her ability to manoeuvre" means a vessel which from the nature of her work is restricted in her ability to manoeuvre as required by these Rules and is therefore unable to keep out of the way of another vessel.

The following vessels shall be regarded as vessels restricted in their ability to manoeuvre:

(i) a vessel engaged in laying, servicing or picking up a navigation mark, submarine cable or pipeline;

(ii) a vessel engaged in dredging, surveying or underwater operations;

(iii) a vessel engaged in replenishment or transferring persons, provisions or cargo while underway;

(iv) a vessel engaged in the launching or recovery of aircraft;

(v) a vessel engaged in minesweeping operations;

(vi) a vessel engaged in a towing operation such as severely restricts the towing vessel and her tow in their ability to deviate from their course.

(h) The term "vessel constrained by her draught" means a power-driven vessel which because of her draught in relation to the available depth of water is severely restricted in her ability to deviate from the course she is following.

(i) The word "underway" means that a vessel is not at anchor, or made fast to the shore, or aground.

(j) The words "length" and "breadth" of a vessel mean her length overall and greatest breadth.

(k) Vessels shall be deemed to be in sight of one another only when one can be observed visually from the other.

(l) The term "restricted visibility" means any condition in which visibility is restricted by fog, mist, falling snow, heavy rainstorms, sandstorms or any other similar causes.

PART B—STEERING AND SAILING RULES

Section I—Conduct of Vessels in Any Condition of Visibility

Application

RULE 4 Rules in this Section apply in any condition of visibility.

Look-out

RULE 5 Every vessel shall at all times maintain a proper look-out by sight and hearing as well as by all available means appropriate in the prevailing circumstances and conditions so as to make a full appraisal of the situation and of the risk of collision.

Safe Speed

RULE 6 Every vessel shall at all times proceed at a safe speed so that she can take proper and effective action to avoid collision and be stopped within a distance appropriate to the prevailing circumstances and conditions.

In determining a safe speed the following factors shall be among those taken into account:

(a) By all vessels:

(i) the state of visibility;

(ii) the traffic density including concentrations of fishing vessels or any other vessels;

(iii) the manoeuvrability of the vessel with special reference to stopping distance and turning ability in the prevailing conditions;

(iv) at night the presence of background light such as from shore lights or from back scatter of her own lights;

(v) the state of wind, sea and current, and the proximity of navigational hazards;

(vi) the draught in relation to the available depth of water.

(b) Additionally, by vessels with operational radar:

(i) the characteristics, efficiency and limitations of the radar equipment;

(ii) any constraints imposed by the radar range scale in use;

(iii) the effect on radar detection of the sea state, weather and other sources of interference;

(iv) the possibility that small vessels, ice and other floating objects may not be detected by radar at an adequate range;

(v) the number, location and movement of vessels detected by radar;

(vi) the more exact assessment of the visibility that may be possible when radar is used to determine the range of vessels or other objects in the vicinity.

Risk of Collision

RULE 7 (a) Every vessel shall use all available means appropriate to the prevailing circumstances and conditions to determine if risk of collision exists. If there is any doubt such risk shall be deemed to exist.

(b) Proper use shall be made of radar equipment if fitted and operational, including long-range scanning to obtain early warning of risk of collision and radar plotting or equivalent systematic observation of detected objects.

(c) Assumptions shall not be made on the basis of scanty information, especially scanty radar information.

(d) In determining if risk of collision exists the following considerations shall be among those taken into account:

(i) such risk shall be deemed to exist if the compass bearing of an approaching vessel does not appreciably change;

(ii) such risk may sometimes exist even when an appreciable bearing change is evident, particularly when approaching a very large vessel or a tow or when approaching a vessel at close range.

Action to Avoid Collision

Rule 8 (a) Any action taken to avoid collision shall, if the circumstances of the case admit, be positive, made in ample time and with due regard to the observance of good seamanship.

(b) Any alteration of course and/or speed to avoid collision shall, if the circumstances of the case admit, be large enough to be readily apparent to another vessel observing visually or by radar; a succession of small alterations of course and/or speed should be avoided.

(c) If there is sufficient sea room, alteration of course alone may be the most effective action to avoid a close-quarters situation provided that it is made in good time, is substantial and does not result in another close-quarters situation.

(d) Action taken to avoid collision with another vessel shall be such as to result in passing at a safe distance. The effectiveness of the action shall be carefully checked until the other vessel is finally past and clear.

(e) If necessary to avoid collision or allow more time to assess the situation, a vessel shall slacken her speed or take all way off by stopping or reversing her means of propulsion.

Narrow Channels

Rule 9 (a) A vessel proceeding along the course of a narrow channel or fairway shall keep as near to the outer limit of the channel or fairway which lies on her starboard side as is safe and practicable.

(b) A vessel of less than 20 metres in length or a sailing vessel shall not impede the passage of a vessel which can safely navigate only within a narrow channel or fairway.

(c) A vessel engaged in fishing shall not impede the passage of any other vessel navigating within a narrow channel or fairway.

(d) A vessel shall not cross a narrow channel or fairway if such crossing impedes the passage of a vessel which can safely navigate only within such channel or fairway. The latter vessel may use the sound signal prescribed in Rule 34(d) if in doubt as to the intention of the crossing vessel.

(e)(i) In a narrow channel or fairway when overtaking can take place only if the vessel to be overtaken has to take action to permit safe passing, the vessel intending to overtake shall indicate her intention by sounding the appropriate signal prescribed in Rule 34(c)(i). The vessel to be overtaken shall, if in agreement, sound the appropriate signal prescribed in Rule 34(c) (ii) and take steps to permit safe passing. If in doubt she may sound the signals prescribed in Rule 34(d).

(ii) This Rule does not relieve the overtaking vessel of her obligation under Rule 13.

(f) A vessel nearing a bend or an area of narrow channel or fairway where other vessels may be obscured by an intervening obstruction shall navigate with particular alertness and caution and shall sound the appropriate signal prescribed in Rule 34(e).

(g) Any vessel shall, if the circumstances of the case admit, avoid anchoring in a narrow channel.

Traffic Separation Schemes

Rule 10 (a) This Rule applies to traffic separation schemes adopted by the Organization.

(b) A vessel using a traffic separation scheme shall:

(i) proceed in the appropriate traffic lane in the general direction of traffic flow for that lane;

(ii) so far as practicable keep clear of a traffic separation line or separation zone;

(iii) normally join or leave a traffic lane at the termination of the lane, but when joining or leaving from the side shall do so at as small an angle to the general direction of traffic flow as practicable.

(c) A vessel shall so far as practicable avoid crossing traffic lanes, but if obliged to do so shall cross as nearly as practicable at right angles to the general direction of traffic flow.

(d) Inshore traffic zones shall not normally be used by through traffic which can safely use the appropriate traffic lane within the adjacent traffic separation scheme.

(e) A vessel, other than a crossing vessel, shall not normally enter a separation zone or cross a separation line except:

(i) in case of emergency to avoid immediate danger;

(ii) to engage in fishing within a separation zone.

(f) A vessel navigating in areas near the terminations of traffic separation schemes shall do so with particular caution.

(g)A vessel shall so far as practicable avoid anchoring in a traffic separation scheme or in areas near its terminations.

(h) A vessel not using a traffic separation scheme shall avoid it by as wide a margin as is practicable.

(i) A vessel engaged in fishing shall not impede the passage of any vessel following a traffic lane.

(j) A vessel of less than 20 metres in length or a sailing vessel shall not impede the safe passage of a power-driven vessel following a traffic lane.

Section II—Conduct of Vessels in Sight of One Another

Application

Rule 11 Rules in this Section apply to vessels in sight of one another.

Sailing Vessels

RULE 12 (a) When two sailing vessels are approaching one another, so as to involve risk of collision, one of them shall keep out of the way of the other as follows:

(i) when each has the wind on a different side, the vessel which has the wind on the port side shall keep out of the way of the other;

(ii) when both have the wind on the same side, the vessel which is to windward shall keep out of the way of the vessel which is to leeward;

(iii) if a vessel with the wind on the port side sees a vessel to windward and cannot determine with certainty whether the other vessel has the wind on the port or on the starboard side, she shall keep out of the way of the other.

(b) For the purposes of this Rule the windward side shall be deemed to be the side opposite to that on which the mainsail is carried or, in the case of a square-rigged vessel, the side opposite to that on which the largest fore-and-aft sail is carried.

Overtaking

RULE 13 (a) Notwithstanding anything contained in the Rules of this Section any vessel overtaking any other shall keep out of the way of the vessel being overtaken.

(b) A vessel shall be deemed to be overtaking when coming up with another vessel from a direction more than 22.5 degrees abaft her beam, that is, in such a position with reference to the vessel she is overtaking, that at night she would be able to see only the sternlight of that vessel but neither of her sidelights.

(c) When a vessel is in any doubt as to whether she is overtaking another, she shall assume that this is the case and act accordingly.

(d) Any subsequent alteration of the bearing between the two vessels shall not make the overtaking vessel a crossing vessel within the meaning of these Rules or relieve her of the duty of keeping clear of the overtaken vessel until she is finally past and clear.

Head-on Situation

RULE 14 (a) When two power-driven vessels are meeting on reciprocal or nearly reciprocal courses so as to involve risk of collision each shall alter her course to starboard so that each shall pass on the port side of the other.

(b) Such a situation shall be deemed to exist when a vessel sees the other ahead or nearly ahead and by night she could see the masthead lights of the other in a line or nearly in a line and/or both sidelights and by day she observes the corresponding aspect of the other vessel.

(c) When a vessel is in any doubt as to whether such a situation exists she shall assume that it does exist and act accordingly.

Crossing Situation

RULE 15 When two power-driven vessels are crossing so as to involve risk of collision, the vessel which has the other on her own starboard side shall keep out of the way and shall, if the circumstances of the case admit, avoid crossing ahead of the other vessel.

Action by Give-Way Vessel

RULE 16 Every vessel which is directed to keep out of the way of another vessel shall, so far as possible, take early and substantial action to keep well clear.

Action by Stand-On Vessel

RULE 17 (a)(i) Where one of two vessels is to keep out of the way the other shall keep her course and speed.

(ii) The latter vessel may however take action to avoid collision by her manoeuvre alone, as soon as it becomes apparent to her that the vessel required to keep out of the way is not taking appropriate action in compliance with these Rules.

(b) When, from any cause, the vessel required to keep her course and speed finds herself so close that collision cannot be avoided by the action of the give-way vessel alone, she shall take such action as will best aid to avoid collision.

(c) A power-driven vessel which takes action in a crossing situation in accordance with sub-paragraph (a)(ii) of this Rule to avoid collision with another power-driven vessel shall, if the circumstances of the case admit, not alter course to port for a vessel on her own port side.

(d) This Rule does not relieve the give-way vessel of her obligation to keep out of the way.

Responsibilities Between Vessels

RULE 18 Except where Rules 9, 10 and 13 otherwise require:

(a) A power-driven vessel underway shall keep out of the way of:

(i) a vessel not under command;
(ii) a vessel restricted in her ability to manoeuvre;
(iii) a vessel engaged in fishing;
(iv) a sailing vessel.

(b) A sailing vessel underway shall keep out of the way of:

(i) a vessel not under command;
(ii) a vessel restricted in her ability to manoeuvre;
(iii) a vessel engaged in fishing.

(c) A vessel engaged in fishing when underway shall, so far as possible, keep out of the way of:

(i) a vessel not under command;
(ii) a vessel restricted in her ability to manoeuvre.

(d)(i) Any vessel other than a vessel not under command or a vessel restricted in her ability to manoeuvre shall, if the circumstances of the case admit, avoid impeding the safe passage of a vessel constrained by her draught, exhibiting the signals in Rule 28.

(ii) A vessel constrained by her draught shall navigate with particular caution having full regard to her special condition.

(e) A seaplane on the water shall, in general, keep well clear of all vessels and avoid impeding their navigation. In circumstances, however, where risk of collision exists, she shall comply with the Rules of this Part.

Section III—Conduct of Vessels in Restricted Visibility

Conduct of Vessels in Restricted Visibility

RULE 19 (a) This Rule applies to vessels not in sight of one another when navigating in or near an area of restricted visibility.

(b) Every vessel shall proceed at a safe speed adapted to the prevailing circumstances and conditions of restricted visibility. A power-driven vessel shall have her engines ready for immediate manoeuvre.

(c) Every vessel shall have due regard to the prevailing circumstances and conditions of restricted visibility when complying with the Rules of Section I of this Part.

(d) A vessel which detects by radar alone the presence of another vessel shall determine if a close-quarters situation is developing and/or risk of collision exists. If so, she shall take avoiding action in ample time, provided that when such action consists of an alteration of course, so far as possible the following shall be avoided:

(i) an alteration of course to port for a vessel forward of the beam, other than for a vessel being overtaken;

(ii) an alteration of course towards a vessel abeam or abaft the beam.

(e) Except where it has been determined that a risk of collision does not exist, every vessel which hears apparently forward of her beam the fog signal of another vessel, or which cannot avoid a close-quarters situation with another vessel forward of her beam, shall reduce her speed to the minimum at which she can be kept on her course. She shall if necessary take all her way off and in any event navigate with extreme caution until danger of collision is over.

PART C—LIGHTS AND SHAPES

Application

RULE 20 (a) Rules in this Part shall be complied with in all weathers.

(b) The Rules concerning lights shall be complied with from sunset to sunrise, and during such times no other lights shall be exhibited, except such lights as cannot be mistaken for the lights specified in these Rules or do not impair their visibility or distinctive character, or interfere with the keeping of a proper lookout.

(c) The lights prescribed by these Rules shall, if carried, also be exihibited from sunrise to sunset in restricted visibility and may be exhibited in all other circumstances when it is deemed necessary.

(d) The Rules concerning shapes shall be complied with by day.

(e) The lights and shapes specified in these Rules shall comply with the provisions of Annex I to these Regulations.

Definitions

RULE 21 (a) "Masthead light" means a white light placed over the fore and aft centreline of the vessel showing an unbroken light over an arc of the horizon of 225 degrees and so fixed as to show the light from right ahead to 22.5 degrees abaft the beam on either side of the vessel.

(b) "Sidelights" means a green light on the starboard side and a red light on the port side each showing an unbroken light over an arc of the horizon of 112.5 degrees and so fixed as to show the light from right ahead to 22.5 degrees abaft the beam on its respective side. In a vessel of less than 20 metres in length the sidelights may be combined in one lantern carried on the fore and aft centreline of the vessel.

(c) "Sternlight" means a white light placed as nearly as practicable at the stern showing an unbroken light over an arc of the horizon 135 degrees and so fixed as to show the light 67.5 degrees from right aft on each side of the vessel.

(d) "Towing light" means a yellow light having the same characteristics as the "sternlight" defined in paragraph (c) of this Rule.

(e) "All-round light" means a light showing an unbroken light over an arc of the horizon of 360 degrees.

(f) "Flashing light" means a light flashing at regular intervals at a frequency of 120 flashes or more per minute.

Visibility of Lights

RULE 22 The lights prescribed in these Rules shall have an intensity as specified in Section 8 of Annex I to these Regulations so as to be visible at the following minimum ranges:

(a) In vessels of 50 metres or more in length:
a masthead light, 6 miles;
a sidelight, 3 miles;
a sternlight, 3 miles;
a towing light, 3 miles;
a white, red, green or yellow all-round light, 3 miles.

(b) In vessels of 12 metres or more in length but less than 50 metres in length:
a masthead light, 5 miles; except that where the length of the vessel is less than 20 metres, 3 miles;
a sidelight, 2 miles;
a sternlight, 2 miles;
a towing light, 2 miles;
a white, red, green or yellow all-round light, 2 miles.

(c) In vessels of less than 12 metres in length:
a masthead light, 2 miles;
a sidelight, 1 mile;
a sternlight, 2 miles;
a towing light, 2 miles;
a white, red, green or yellow all-round light, 2 miles.

Power-Driven Vessels Underway

RULE 23 (a) A power-driven vessel underway shall exhibit:

(i) a masthead light forward;

(ii) a second masthead light abaft of and higher than the forward one; except that a vessel of less than 50 metres in length shall not be obliged to exhibit such light but may do so;

(iii) sidelights;

(iv) a sternlight.

(b) An air-cushion vessel when operating in the nondisplacement mode shall, in addition to the lights prescribed in paragraph (a) of this Rule, exhibit an all-round flashing yellow light.

(c) A power-driven vessel of less than 7 metres in length and whose maximum speed does not exceed 7 knots may, in lieu of the lights prescribed in paragraph (a) of this Rule, exhibit an all-round white light. Such vessel shall, if practicable, also exhibit sidelights.

Towing and Pushing

RULE 24 (a) A power-driven vessel when towing shall exhibit:

(i) instead of the light prescribed in Rule 23(a)(i), two masthead lights forward in a vertical line. When the length of the tow, measuring from the stern of the towing vessel to the after end of the tow exceeds 200 metres, three such lights in a vertical line;

(ii) sidelights;

(iii) a sternlight;

(iv) a towing light in a vertical line above the sternlight;

(v) when the length of the tow exceeds 200 metres, a diamond shape where it can best be seen.

(b) When a pushing vessel and a vessel being pushed ahead are rigidly connected in a composite unit they shall be regarded as a power-driven vessel and exhibit the lights prescribed in Rule 23.

(c) A power-driven vessel when pushing ahead or towing alongside, except in the case of a composite unit, shall exhibit:

(i) instead of the light prescribed in Rule 23(a)(i), two masthead lights forward in a vertical line;

(ii) sidelights;

(iii) a sternlight.

(d) A power-driven vessel to which paragraphs (a) and (c) of this Rule apply shall also comply with Rule 23(a)(ii).

(e) A vessel or object being towed shall exhibit:

(i) sidelights;

(ii) a sternlight;

(iii) when the length of the tow exceeds 200 meters, a diamond shape where it can best be seen.

(f) Provided that any number of vessels being towed alongside or pushed in a group shall be lighted as one vessel,

(i) a vessel being pushed ahead, not being part of a composite unit, shall exhibit at the forward end, sidelights;

(ii) a vessel being towed alongside shall exhibit a sternlight and at the forward end, sidelights.

(g) Where from any sufficient cause it is impracticable for a vessel or object being towed to exhibit the lights prescribed in paragraph (e) of this Rule, all possible measures shall be taken to light the vessel or object towed or at least to indicate the presence of the unlighted vessel or object.

Sailing Vessels Underway and Vessels Under Oars

RULE 25 (a) A sailing vessel underway shall exhibit:

(i) sidelights;

(ii) a sternlight.

(b) In sailing vessel of less than 12 metres in length the lights prescribed in paragraph (a) of this Rule may be combined in one lantern carried at or near the top of the mast where it can best be seen.

(c) A sailing vessel underway may, in addition to the lights prescribed in paragraph (a) of this Rule, exhibit at or near the top of the mast, where they can best be seen, two all-round lights in a vertical line, the upper being red and the lower green, but these lights shall not be exhibited in conjunction with the combined lantern permitted by paragraph (b) of this Rule.

(d)(i) A sailing vessel of less than 7 metres in length shall, if practicable, exhibit the lights prescribed in paragraph (a) or (b) of this Rule, but if she does not, she shall have ready at hand an electric torch or lighted lantern showing a white light which shall be exhibited in sufficient time to prevent collision.

(ii) A vessel under oars may exhibit the lights prescribed in this Rule for sailing vessels, but if she does not, she shall have ready at hand an electric torch or lighted lantern showing a white light which shall be exhibited in sufficient time to prevent collision.

(e) A vessel proceeding under sail when also being propelled by machinery shall exhibit forward where it can best be seen a conical shape, apex downwards.

Fishing Vessels

RULE 26 (a) A vessel engaged in fishing, whether underway or at anchor, shall exhibit only the lights and shapes prescribed in this Rule.

(b) A vessel when engaged in trawling, by which is meant the dragging through the water of a dredge net or other apparatus used as a fishing appliance, shall exhibit:

(i) two all-round lights in a vertical line, the upper being green and the lower white, or a shape consisting of two cones with their apexes together in a vertical line one above the other; a vessel of less than 20 metres in length may instead of this shape exhibit a basket;

(ii) a masthead light abaft of and higher than the all-round green light; a vessel of less than 50 metres in length shall not be obliged to exhibit such a light but may do so;

(iii) when making way through the water, in addition to the lights prescribed in this paragraph, sidelights and a sternlight.

(c) A vessel engaged in fishing, other than trawling, shall exhibit:

(i) two all-round lights in a vertical line, the upper being red and the lower white, or a shape consisting of two cones with apexes together in a vertical line one above the other; a vessel of less than 20 metres in length may instead of this shape exhibit a basket;

(ii) when there is outlying gear extending more than 150 metres horizontally from the vessel, an all-round white light or a cone apex upwards in the direction of the gear;

(iii) when making way through the water, in addition to the lights prescribed in this paragraph, sidelights and a sternlight.

(d) A vessel engaged in fishing in close proximity to other vessels engaged in fishing may exhibit the additional signals described in Annex II to these Regulations.

(e) A vessel when not engaged in fishing shall not exhibit the lights or shapes prescribed in this Rule, but only those prescribed for a vessel of her length.

Vessels Not Under Command or Restricted in Their Ability to Manoeuvre

RULE 27 (a) A vessel not under command shall exhibit:

(i) two all-round red lights in a vertical line where they can best be seen;

(ii) two balls or similar shapes in vertical line where they can best be seen;

(iii) when making way through the water, in addition to the lights prescribed in this paragraph, sidelights and a sternlight.

(b) A vessel restricted in her ability to manoeuvre, except a vessel engaged in minesweeping operations, shall exhibit:

(i) three all-round lights in a vertical line where they can best be seen. The highest and lowest of these lights shall be red and the middle light shall be white;

(ii) three shapes in a vertical line where they can best be seen. The highest and lowest of these shapes shall be balls and the middle one a diamond;

(iii) when making way through the water, masthead lights, sidelights and a sternlight, in addition to the lights prescribed in subparagraph (i);

(iv) when at anchor, in addition to the lights or shapes prescribed in subparagraphs (i) and (ii), the light, lights or shape prescribed in Rule 30.

(c) A vessel engaged in a towing operation such as renders her unable to deviate from her course shall, in addition to the lights or shapes prescribed in subparagraph (b)(i) and (ii) of this Rule, exhibit the lights or shape prescribed in Rule 24(a).

(d) A vessel engaged in dredging or underwater operations, when restricted in her ability to manoeuvre, shall exhibit the lights and shapes prescribed in paragraph (b) of this Rule and shall in addition, when an obstruction exists, exhibit:

(i) two all-round red lights or two balls in a vertical line to indicate the side on which the obstruction exists;

(ii) two all-round green lights or two diamonds in a vertical line to indicate the side on which another vessel may pass;

(iii) when making way through the water, in addition to the lights prescribed in this paragraph, masthead lights, sidelights and a sternlight;

(iv) a vessel to which this paragraph applies when at anchor shall exhibit the lights or shapes prescribed in subparagraphs (i) and (ii) instead of the lights or shape prescribed in Rule 30.

(e) Whenever the size of a vessel engaged in diving operations makes it impracticable to exhibit the shapes prescribed in paragraph (d) of this Rule, a rigid replica of the International Code flag "A" not less than 1 metre in height shall be exhibited. Measures shall be taken to ensure all-round visibility.

(f) A vessel engaged in minesweeping operations shall, in addition to the lights prescribed for a power-driven vessel in Rule 23, exhibit three all-round green lights or three balls. One of these lights or shapes shall be exhibited at or near the foremast head and one at each end of the fore yard. These lights or shapes indicate that it is dangerous for another vessel to approach closer than 1,000 metres astern or 500 metres on either side of the minesweeper.

(g) Vessels of less than 7 metres in length shall not be required to exhibit the lights prescribed in this Rule.

(h) The signals prescribed in this Rule are not signals of vessels in distress and requiring assistance. Such signals are contained in Annex IV to these Regulations.

Vessels Constrained by Their Draught

RULE 28 A vessel constrained by her draught may, in addition to the lights prescribed for power-driven vessels in Rule 23, exhibit where they can best be seen three all-round red lights in a vertical line, or a cylinder.

Pilot Vessels

RULE 29 (a) A vessel engaged on pilotage duty shall exhibit:

(i) at or near the masthead, two all-round lights in a vertical line, the upper being white and the lower red;

(ii) when underway, in addition, sidelights and a sternlight;

(iii) when at anchor, in addition to the lights prescribed in subparagraph (i), the anchor light, lights or shape.

(b) A pilot vessel when not engaged on pilotage duty shall exhibit the lights or shapes prescribed for a similar vessel of her length.

Anchored Vessel and Vessels Aground

RULE 30 (a) A vessel at anchor shall exhibit where it can best be seen:

(i) in the fore part, an all-round white light or one ball;

(ii) at or near the stern and at a lower level than the light prescribed in sub-paragraph (i), an all-round white light.

(b) A vessel of less than 50 metres in length may exhibit an all-round white light where it can best be seen instead of the lights prescribed in paragraph (a) of this Rule.

(c) A vessel at anchor may, and a vessel of 100 metres and more in length shall, also use the available working or equivalent lights to illuminate her decks.

(d) A vessel aground shall exhibit the lights prescribed in paragraph (a) or (b) of this Rule and in addition, where they can best be seen:

(i) two all-round red lights in a vertical line;

(ii) three balls in a vertical line.

(e) A vessel of less than 7 metres in length, when at anchor or aground, not in or near a narrow channel, fairway or anchorage, or where other vessels normally navigate, shall not be required to exhibit the lights or shapes prescribed in paragraph (a), (b) or (d) of this rule.

Seaplanes

RULE 31 Where it is impracticable for a seaplane to exhibit lights and shapes of the characteristics or in the positions prescribed in the Rules of this Part she shall exhibit lights and shapes as closely similar in characteristics and position as is possible.

PART D—SOUND AND LIGHT SIGNALS

Definitions

RULE 32 (a) The world "whistle" means any sound signalling appliance capable of producing the prescribed blasts and which complies with the specifications in Annex III to these Regulations.

(b) The term "short blast" means a blast of about one second's duration.

(c) The term "prolonged blast" means a blast of from four to six seconds' duration.

Equipment for Sound Signals

RULE 33 (a) A vessel of 12 metres or more in length shall be provided with a whistle and a bell and a vessel of 100 metres or more in length shall, in addition, be provided with a gong, the tone and sound of which cannot be confused with that of the bell. The whistle, bell and gong shall comply with the specifications in Annex III to these Regulations. The bell or gong or both may be replaced by other equipment having the same respective sound characteristics, provided that manual sounding of the required signals shall always be possible.

(b) A vessel of less than 12 metres in length shall not be obliged to carry the sound signalling appliances prescribed in paragraph (a) of this Rule but if she does not, she shall be provided with some other means of making an efficient sound signal.

Manoeuvring and Warning Signals

RULE 34 (a) When vessels are in sight of one another, a power-driven vessel underway, when manoeuvring as authorized or required by these Rules, shall indicate that manoeuvre by the following signals on her whistle:

One short blast to mean "I am altering my course to starboard";

–two short blasts to mean "I am altering my course to port";

–three short blasts to mean "I am operating astern propulsion."

(b) Any vessel may supplement the whistle signals prescribed in paragraph (a) of this Rule by light signals, repeated as appropriate, whilst the manoeuvre is being carried out:

(i) these light signals shall have the following significance:

–one flash to mean "I am altering my course to starboard";

–two flashes to mean "I am altering my course to port";

–three flashes to mean "I am operating astern propulsion";

(ii) the duration of each flash shall be about one second, the interval between flashes shall be about one second, and the interval between successive signals shall be not less than ten seconds;

(iii) the light used for this signal shall, if fitted, be an all-round white light, visible at a minimum range of 5 miles, and shall comply with the provisions of Annex I.

(c) When in sight of one another in a narrow channel or fairway;

(i) a vessel intending to overtake another shall in compliance with Rule 9(e)(i) indicate her intention by the following signals on her whistle:

–two prolonged blasts followed by one short blast to mean "I intend to overtake you on your starboard side";

–two prolonged blasts followed by two short blasts to mean "I intend to overtake you on your port side".

(ii) the vessel about to be overtaken when acting in accordance with Rule 9(e)(i) shall indicate her agreement by the following signal on her whistle:

–one prolonged, one short, one prolonged and one short blast, in that order.

(d) When vessels in sight of one another are approaching each other and from any cause either vessel fails to understand the intentions or actions of the other, or is in doubt whether sufficient action is being taken by the other to avoid collision, the vessel in doubt shall immediately indicate such doubt by giving at least five short and rapid blasts on the whistle. Such signal may be supplemented by a light signal of at least five short and rapid flashes.

(e) A vessel nearing a bend or an area of a channel or fairway where other vessels may be obscured by an intervening obstruction shall sound one prolonged blast. Such signal shall be answered with a prolonged blast by any approaching vessel that may be within hearing around the bend or behind the intervening obstruction.

(f) If whistles are fitted on a vessel at a distance apart of more than 100 metres, one whistle only shall be used for giving manoeuvring and warning signals.

Sound Signals in Restricted Visibility

RULE 35 In or near an area of restricted visibility, whether by day or night, the signals prescribed in this Rule shall be used as follows:

(a) A power-driven vessel making way through the water shall sound at intervals of not more than 2 minutes one prolonged blast.

(b) A power-driven vessel underway but stopped and making no way through the water shall sound at intervals of not more than 2 minutes two prolonged blasts in succession with an interval of about 2 seconds between them.

(c) A vessel not under command, a vessel restricted in her ability to manoeuvre, a vessel constrained by her draught, a sailing vessel, a vessel engaged in fishing and a vessel engaged in towing or pushing another vessel shall, instead of the signals prescribed in paragraphs (a) or (b) of this Rule, sound at intervals of not more than 2 minutes three blasts in succession, namely one prolonged followed by two short blasts.

(d) A vessel towed or if more than one vessel is towed the last vessel of the tow, if manned, shall at intervals of not more than 2 minutes sound four blasts in succession, namely one prolonged followed by three short blasts. When practicable, this signal shall be made immediately after the signal made by the towing vessel.

(e) When a pushing vessel and a vessel being pushed ahead are rigidly connected in a composite unit they shall be regarded as a power-driven vessel and shall give the signals prescribed in paragraphs (a) or (b) of this Rule.

(f) A vessel at anchor shall at intervals of not more than one minute ring the bell rapidly for about 5 seconds. In a vessel of 100 metres or more in length the bell shall be sounded in the forepart of the vessel and immediately after the ringing of the bell the gong shall be sounded rapidly for about 5 seconds in the after part of the vessel. A vessel at anchor may in addition sound three blasts in succession, namely, one short, one prolonged and one short blast, to give warning of her position and of the possibility of collision to an approaching vessel.

(g) A vessel aground shall give the bell signal and if required the gong signal prescribed in paragraph (f) of this Rule and shall, in addition, give three separate and distinct strokes on the bell immediately before and after the rapid ringing of the bell. A vessel aground may in addition sound an appropriate whistle signal.

(h) A vessel of less than 12 metres in length shall not be obliged to give the above-mentioned signals but, if she does not, shall make some other efficient sound signal at intervals of not more than 2 minutes.

(i) A pilot vessel when engaged on pilotage duty may in addition to the signals prescribed in paragraph (a), (b) or (f) of this Rule sound an identity signal consisting of four short blasts.

Signals to Attract Attention

RULE 36 If necessary to attract the attention of another vessel any vessel may make light or sound signals that cannot be mistaken for any signal authorized elsewhere in these Rules, or may direct the beam of her searchlight in the direction of the danger, in such a way as not to embarrass any vessel.

Distress Signals

RULE 37 When a vessel is in distress and requires assistance she shall use or exhibit the signals prescribed in Annex IV to these Regulations.

PART E—EXEMPTIONS

Exemptions

RULE 38 Any vessel (or class of vessels) provided that she complies with the requirements of the International Regulations for Preventing Collisions at Sea, 1960, the keel of which is laid or which is at a corresponding stage of construction before the entry into force of these Regulations may be exempted from compliance therewith as follows:

(a) The installation of lights with ranges prescribed in Rule 22, until four years after the date of entry into force of these Regulations.

(b) The installation of lights with colour specifications as prescribed in Section 7 of Annex I to these Regulations, until four years after the date of entry into force of these Regulations.

(c) The repositioning of lights as a result of conversion from Imperial to metric units and rounding off measurement figures, permanent exemption.

(d)(i) The repositioning of masthead lights on vessels of less than 150 metres in length, resulting from the prescriptions of Section 3(a) of Annex I, permanent exemption.

(ii) The repositioning of masthead lights on vessels of 150 metres or more in length, resulting from the prescriptions of Section 3(a) of Annex I to these Regulations, until nine years after the date of entry into force of these Regulations.

(e) The repositioning of masthead lights resulting from the prescriptions of Section 2(b) of Annex I, until nine years after the date of entry into force of these Regulations.

(f) The repositioning of sidelights resulting from the prescriptions of Sections 2(g) and 3(b) of Annex I, until nine years after the date of entry into force of these Regulations.

(g) The requirements for sound signal appliances prescribed in Annex III, until nine years after the date of entry into force of these Regulations.

ANNEX I

Positioning and Technical Details of Lights and Shapes

1. Definition

The term "height above the hull" means height above the uppermost continuous deck.

2. Vertical Positioning and Spacing of Lights

(a) On a power-driven vessel of 20 metres or more in length the masthead lights shall be placed as follows:

(i) the forward masthead light, or if only one masthead light is carried,then that light, at a height above the hull of not less than 6 metres, and, if the breadth of the vessel exceeds 6 metres, then at a height above the hull not less than such breadth, so however that the light need not be placed at a greater height above the hull than 12 metres;

(ii) when two masthead lights are carried the after one shall be at least 4.5 metres vertically higher than the forward one.

(b) The vertical separation of masthead lights of power-driven vessels shall be such that in all normal conditions of trim the after light will be seen over and separate from the forward light at a distance of 1000 metres from the stem when viewed from sea level.

(c) The masthead light of a power-driven vessel of 12 metres but less than 20 metres in length shall be placed at a height above the gunwale of not less than 2.5 metres.

(d) A power-driven vessel of less than 12 metres in length may carry the uppermost light at a height of less than 2.5 metres above the gunwale. When however a masthead light is carried in addition to sidelights and a sternlight, then such masthead light shall be carried at least 1 metre higher than the sidelights.

(e) One of the two or three masthead lights prescribed for a power-driven vessel when engaged in towing or pushing another vessel shall be placed in the same position as the forward masthead light of a power-driven vessel.

(f) In all circumstances the masthead light or lights shall be so placed as to be above and clear of all other lights and obstructions.

(g) The sidelights of a power-driven vessel shall be placed at a height above the hull not greater than three quarters of that of the forward masthead light. They shall not be so low as to be interfered with by deck lights.

(h) The sidelights, if in a combined lantern and carried on a power-driven vessel of less than 20 metres in length, shall be placed not less than 1 metre below the masthead light.

(i) When the Rules prescribe two or three lights to be carried in a vertical line, they shall be spaced as follows:

(i) on a vessel of 20 metres in length or more such lights shall be spaced not less than 2 metres apart, and the lowest of these lights shall, except where a towing light is required, not be less than 4 metres above the hull;

(ii) on a vessel of less than 20 metres in length such lights shall be spaced not less than 1 metre apart and the lowest of these lights shall, except where a towing light is required, not be less than 2 metres above the gunwale;

(iii) when three lights are carried they shall be equally spaced.

(j) The lower of the two all-round lights prescribed for a fishing vessel when engaged in fishing shall be at a height above the sidelights not less than twice the distance between the two vertical lights.

(k) The forward anchor light,when two are carried, shall not be less than 4.5 metres above the after one. On a vessel of 50 metres or more in length this forward anchor light shall not be less than 6 metres above the hull.

3. Horizontal Positioning and Spacing of Lights

(a) When two masthead lights are prescribed for a power-driven vessel, the horizontal distance between them shall not be less than one half of the length of the vessel but need not be more than 100 metres. The forward light shall be placed not more than one quarter of the length of the vessel from the stem.

(b) On a vessel of 20 metres or more in length the sidelights shall not be placed in front of the forward masthead lights. They shall be placed at or near the side of the vessel.

4. Details of Location of Direction-Indicating Lights for Fishing Vessels, Dredgers and Vessels Engaged in Underwater Operations

(a) The light indicating the direction of the outlying gear from a vessel engaged in fishing as prescribed in Rule 26(c)(ii) shall be placed at a horizontal distance of not less than 2 metres and not more than 6 metres away from the two all-round red and white lights. This light shall be placed not higher than the all-round white light prescribed in Rule 26(c)(i) and not lower than the sidelights.

(b) The lights and shapes on a vessel engaged in dredging or underwater operations to indicate the obstructed side and/or the side on which it is safe to pass, as prescribed in Rule 27(d)(i) and (ii), shall be placed at the maximum practical horizontal distance, but in no case less than 2 metres, from the lights or shapes prescribed in Rule 27(b)(i) and (ii). In no case shall the upper of these lights or shapes be at a greater height than the lower of the three lights or shapes prescribed in Rule 27(b)(i) and (ii).

5. Screens for Sidelights

The sidelights shall be fitted with inboard screens painted matt black, and meeting the requirements of Section 9 of this Annex. With a combined lantern, using a single vertical filament and a very narrow division between the green and red sections, external screens need not be fitted.

6. Shapes

(a) Shapes shall be black and of the following sizes:
(i) a ball shall have a diameter of not less than 0.6 metres;
(ii) a cone shall have a base diameter of not less than 0.6 metre and a height equal to its diameter;
(iii) a cylinder shall have a diameter of at least 0.6 metre and a height of twice its diameter;
(iv) a diamond shape shall consist of two cones as defined in (ii) above having a common base.
(b) The vertical distance between shapes shall be at least 1.5 metre.
(c) In a vessel of less than 20 metres in length shapes of lesser dimensions but commensurate with the size of the vessel may be used and the distance apart may be correspondingly reduced.

7. Colour Specification of Lights

The chromaticity of all navigation lights shall conform to the following standards, which lie within the boundaries of the area of the diagram specified for each colour by the International Commission on Illumination (CIE).

The boundaries of the area for each colour are given indicating the corner co-ordinates, which are follows:

(i) *White*							
	x	0.525	0.525	0.452	0.310	0.310	0.443
	y	0.382	0.440	0.440	0.348	0.283	0.382
(ii) *Green*							
	x	0.028	0.009	0.300	0.203		
	y	0.385	0.735	0.511	0.356		
(iii) Red							
	x	0.680	0.660	0.735	0.721		
	y	0.320	0.320	0.264	0.259		
(iv) *Yellow*							
	x	0.612	0.618	0.575	0.575		
	y	0.382	0.382	0.425	0.406		

8. Intensity of Lights

(a) the minimum luminous intensity of lights shall be calculated by using the formula:

$$I = 3.43 \times 10^6 \times T \times D^2 \times K^{-D}$$

where I is luminous intensity in candelas under service conditions.

T is threshold factor 2×10^{-7} lux,

D is range of visibility (luminous range) of light in nautical miles

K is atmospheric transmissivity. For prescribed lights the value of K shall be 0.8, corresponding to a meteorological visibility of approximately 13 nautical miles.

(b) A section of figures derived from the formula is given in the following table:[1]

Range of visibility (luminous range) of light in nautical miles *D*	*Luminous intensity of light in candelas for K = 0.8* *I*
1	0.9
2	4.3
3	12.0
4	27.0
5	52.0
6	94.0

9. Horizontal Sectors

(a)(i) In the forward direction, sidelights as fitted on the vessel must show the minimum required intensities. The intensities must decrease to reach practical cut-off between 1 degree and 3 degrees outside the prescribed sectors.

(ii) For sternlights and masthead lights and at 22.5 degrees abaft the beam for sidelights, the minimum required intensities shall be maintained over the arc of the horizon up to 5 degrees within the limits of the sectors prescribed in Rule 21. From 5 degrees within the prescribed sectors the intensity may decrease by 50 percent up to the prescribed limits; it shall decrease steadily to reach practical cut-off at not more than 5 degrees outside the prescribed limits.

(b) All-round lights shall be so located as not to be obscured by masts, topmasts or structures within angular sectors of more than 6 degrees, except anchor lights, which need not be placed at an impracticable height above the hull.

10. Vertical Sectors

(a) The vertical sectors of electric lights, with the exception of lights on sailing vessels shall ensure that:

(i) at least the rquired minimum intensity is maintained at all angles from 5 degrees above to 5 degrees below the horizontal;

(ii) at least 60 percent of the required minimum intensity is maintained from 7.5 degrees above to 7.5 degrees below the horizontal.

(b) In the case of sailing vessels the vertical sectors of electric lights shall ensure that:

(i) at least the required minimum intensity is maintained at all angles from 5 degrees above to 5 degrees below the horizontal;

(ii) at least 50 percent of the required minimum intensity is maintained from 25 degrees above to 25 degrees below the horizontal.

(c) In the case of lights other than electric these specifications shall be met as closely as possible.

[1]The maximum luminous intensity of navigation lights should be limited to avoid undue glare.

11. Intensity of Non-Electric Lights

Non-electric lights shall so far as practicable comply with the minimum intensities, as specified in the Table given in Section 8 of this Annex.

12. Manoeuvring Light

Notwithstanding the provisions of paragraph 2(f) of this Annex the manoeuvring light described in Rule 34(b) shall be placed in the same fore and aft vertical plane as the masthead light or lights and, where practicable, at a minimum height of 2 metres vertically above the forward masthead light, provided that it shall be carried not less than 2 metres vertically above or below the after masthead light. On a vessel where only one masthead light is carried the manoeuvring light, if fitted, shall be carried where it can best be seen, not less than 2 metres vertically apart from the masthead light.

13. Approval

The construction of lanterns and shapes and the installation of lanterns on board the vessel shall be to the satisfaction of the appropriate authority of the State where the vessel is registered.

ANNEX II

Additional Signals for Fishing Vessels Fishing in Close Proximity

1. General

The lights mentions herein shall, if exhibited in pursuance of Rule 26(d), be placed where they can best be seen. They shall be at least 0.9 metre apart but at a lower level than lights prescribed in Rule 26(b)(i) and (c)(i). The lights shall be visible all round the horizon at a distance of at least 1 mile but at a lesser distance than the lights prescribed by these Rules for fishing vessels.

2. Signals for Trawlers

(a) Vessels when engaged in trawling, whether using demersal or pelagic gear, may exhibit:

(i) when shooting their nets: two white lights in a vertical line;

(ii) when hauling their nets: one white light over one red light in a vertical line;

(iii) when the net has come fast upon an obstruction: two red lights in a vertical line.

(b) Each vessel engaged in pair trawling may exhibit:

(i) by night, a searchlight directed forward and in the direction of the other vessel of the pair;

(ii) when shooting or hauling their nets or when their nets have come fast upon an obstruction, the lights prescribed in 2(a) above.

3. Signals for Purse Seiners

Vessels engaged in fishing with purse seine gear may exhibit two yellow lights in a vertical line. These lights shall flash alternately every second and with equal light and occultation duration. These lights may be exhibited only when the vessel is hampered by its fishing gear.

ANNEX III

Technical Details of Sound Signal Appliances

1. Whistles

(a) *Frequencies and range of audibility*—The fundamental frequency of the signal shall lie within the range 70–700Hz.

The range of audibility of the signal from a whistle shall be determined by those frequencies, which may include the fundamental and/or one or more higher frequencies, which lie within the range 180–700 Hz (± 1 percent) and which provide the sound pressure levels specified in paragraph 1(c) below.

(b) *Limits of fundamental frequencies*—To ensure a wide variety of whistle characteristics, the fundamental frequency of a whistle shall be between the following limits:

(i) 70–200 Hz, for a vessel 200 metres or more in length;
130–350 Hz, for a vessel 75 metres but less than 200 metres in length;
(iii) 250–700 Hz, for a vessel less than 75 metres in length.

(c) *Sound signal intensity and range of audibility*—A whistle fitted in a vessel shall provide, in the direction of maximum intensity of the whistle and at a distance of 1 metre from it, a sound pressure level in at least ⅓-octave band within the range of frequencies 180–700 Hz (± 1 percent) of not less than the appropriate figure given in the table below.

Length of vessel in meters	⅓d-octave band level at 1 meter in dB referred to 2×10^{-5} N/m^2	Audibility range in nautical miles
200 or more	153	2.0
75 but less than 200	138	1.5
20 but less than 75	130	1.0
Less than 20	120	.5

The range of audibility in the table above is for information and is approximately the range at which a whistle may be heard on its forward axis with 90 percent probability in conditions of still air on board a vessel having average background noise level at the listening posts (taken to be 68 dB in the octave band centred on 250 Hz and 63 dB in the octave band centred on 500 Hz).

In practice the range at which a whistle may be heard is extremely variable and depends critically on weather conditions; the values given can be regarded as typical but under conditions of strong wind or high ambient noise level at the listening post the range may be much reduced.

(d) *Directional properties*—The sound pressure level of a directional whistle shall not be more than 4 dB below the sound pressure level on the axis at any direction in the horizontal plane within ± 45 degrees of the axis. The sound pressure level at any other direction in the horizontal plane shall be not more than 10 dB below the ground pressure level on the axis, so that the range in any direction will be at least half the range on the forward axis. The sound pressure level shall be measured in that ⅓rd-octave band which determines the audibility range.

(e) *Positioning of whistles*—When a directional whistle is to be used as the only whistle on a vessel, it shall be installed with its maximum intensity directed straight ahead.

A whistle shall be placed as high as practicable on a vessel, in order to reduce interception of the emitted sound by obstructions and also to minimize hearing damage risk to personnel. The sound pressure level of the vessel's own signal at listening posts shall not exceed 110 dB (A) and so far as practicable should not exceed 100 dB (A).

(f) *Fitting of more than one whistle*—If whistles are fitted at a distance apart of more than 100 metres, it shall be so arranged that they are not sounded simultaneously.

(g) *Combined whistle systems*—If due to the presence of obstructions the sound field of a single whistle or of one of the whistles referred to in paragraph 1(f) above is likely to have a zone of greatly reduced signal level, it is recommended that a combined whistle system be fitted so as to overcome this reduction. For the purposes of the Rules a combined whistle system is to be regarded as a single whistle. The whistles of a combined system shall be located at a distance apart of not more than 100 metres and arranged to be sounded simultaneously. The frequency of any one whistle shall differ from those of the others by at least 10Hz.

2. Bell or Gong

(a) Intensity of signal—A bell or gong, or other device having similar sound characteristics shall produce a sound pressure level of not less than 110 dB at 1 metre.

(b) *Construction*—Bells and gongs shall be made of corrosion-resistant material and designed to give a clear tone. The diameter of the mouth of the bell shall be not less than 300 mm for vessels of more than 20 metres in length, and shall be not less than 200 mm for vessels of 12 to 20 metres in length. Where practicable, a power-driven bell striker is recommended to ensure constant force but manual operation shall be possible. The mass of the striker shall be not less than 3 percent of the mass of the bell.

3. Approval

The construction of sound signal appliances, their performance and their installation on board the vessel shall be to the satisfaction of the appropriate authority of the State where the vessel is registered.

ANNEX IV

Distress Signals

1. The following signals, used or exhibited either together or separately, indicate distress and need of assistance:

(a) a gun or other explosive signal fired at intervals of about a minute;

(b) a continuous sounding with any fog-signalling apparatus;

(c) rockets or shells, throwing red stars fired one at a time at short intervals;

(d) a signal made by radiotelegraphy or by any other signalling method consisting of the group . . . - - - . . . (SOS) in the Morse Code;

(e) a signal sent by radiotelephony consisting of the spoken word "Mayday";

(f) the International Code Signal of distress indicated by N.C.;

(g) a signal consisting of a square flag having above or below it a ball or anything resembling a ball;

(h) flames on the vessel (as from a burning tar barrel, oil barrel, etc.)

(i) a rocket parachute flare or a hand flare showing a red light;

(j) a smoke signal giving off orange-coloured smoke;

(k) slowly and repeatedly raising and lowering arms outstretched to each side;

(l) the radiotelegraph alarm signal;
(m) the radiotelephone alarm signal;
(n) signals transmitted by emergency position—indicating radio beacons.

2. The use or exhibition of any of the foregoing signals except for the purpose of indicating distress and need of assistance and the use of other signals which may be confused with any of the above signals is prohibited.

3. Attention is drawn to the relevant sections of the International Code of Signals, the Merchant Ship Search and Rescue Manual and the following signals:

(a) a piece of orange-coloured canvas with either a black square and circle or other appropriate symbol (for identification from the air);
(b) a dye marker.

APPENDIX C

72 COLREGS: Interpretive Rules[1]

PART 88

§ 88.1 Purpose.

This part contains the interpretative rules concerning the 72 COLREGS that are adopted by the Coast Guard for the guidance of the public.

§ 88.3 Pushing vessel and vessel being pushed: Composite unit.

Rule 24(b) of the 72 COLREGS states that when a pushing vessel and a vessel being pushed ahead are rigidly connected in a composite unit, they are regarded as a power-driven vessel and must exhibit the lights under Rule 23. A "composite unit" is interpreted to be a pushing vessel that is rigidly connected by mechanical means to a vessel being pushed so they react to sea and swell as one vessel. "Mechanical means" does not include the following:

(a) Lines.
(b) Hawsers.
(c) Wires.
(d) Chains.

(Convention on the International Regulations for Preventing Collisions at Sea, 1972 (as rectified); E.O. 11964 (42 FR 4327); 40 CFR 1.46(b).)

[1]Code of Federal Regulations: Title 33—Navigation and Navigable Waters; Part 88—72 COLREGS: Interpretive Rules.

APPENDIX D

Amendments to the International Regulations for Preventing Collisions at Sea, 1972

The IMCO Maritime Safety Committee adopted the following amendments to the International Regulations for Preventing Collisions at Sea, 1972, in accordance with Article VI (2). The proposed amendments were presented to the full assembly in November 1981 for consideration and adoption. The proposed amendments were adopted and are expected to enter into force in midsummer 1983.

1. Rule 1(c)

Amend to read:

"(c) Nothing in these Rules shall interfere with the operation of any special rules made by the Government of any State with respect to additional station or signal lights, shapes or whistle signals for ships of war and vessels proceeding under convoy, or with respect to additional station or signal lights or shapes for fishing vessels engaged in fishing as a fleet. These additional station or signal lights, shapes or whistle signals shall, so far as possible, be such that they cannot be mistaken for any light, shape or signal authorized elsewhere under these Rules."

2. Rule 3(g)

Replace the sentence immediately before sub-paragraphs (i) to (vi) by the following:

"The term 'vessels restricted in their ability to manoeuvre' shall include but not be limited to:".

3. Rule 3(g)(v)

Replace the word "minesweeping" by the word "mineclearance".

4. Rule 10(b)(iii)

Replace the words "when joining or leaving from the side" by the words "when joining or leaving from either side".

5. Rule 10(d)

Add the following sentence to the present text:

"However, vessels of less than 20 metres in length and sailing vessels may under all circumstances use inshore traffic zones."

6. Rule 10(e)

Amend to read:

"(e) A vessel other than a crossing vessel or a vessel joining or leaving a lane shall not normally enter . . .".

7. Rule 10(k)

Add the following new paragraph:

"(k) A vessel restricted in her ability to manoeuvre when engaged in an operation for the maintenance of safety of navigation in a traffic separation scheme is exempted from complying with this Rule to the extent necessary to carry out the operation."

8. Rule 10(1)

Add the following new paragraph:

"(1) A vessel restricted in her ability to manoeuvre when engaged in an operation for the laying, servicing or picking up of a submarine cable, within a traffic separation scheme, is exempted from complying with this Rule to the extent necessary to carry out the operation."

9. Rule 13(a)

Amend to read:

"(a) Notwithstanding anything contained in the Rules of Part B, Sections I and II . . ."

10. Rule 22(d)

Add a new paragraph:

"(d) In inconspicuous, partly submerged vessels or objects being towed:—a white all-round light, 3 miles."

Rule 23(c)

Amend to read:

"(c)(i) A power-driven vessel of less than 12 metres in length may in lieu of the lights prescribed in paragraph (a) of this Rule exhibit an all-round white light and sidelights;

(ii) a power-driven vessel of less than 7 metres in length whose maximum speed does not exceed 7 knots may in lieu of the lights prescribed in paragraph (a) of this Rule exhibit an all-round white light and shall, if practicable, also exhibit sidelights;

(iii) the masthead light or all-round white light on a power-driven vessel of less than 12 metres in length may be displaced from the fore and aft centreline of the vessel if centerline fitting is not practicable, provided that the sidelights are combined in one lantern which shall be carried on the fore and aft centreline of the vessel or located as nearly as practicable in the same fore and aft line as the masthead light or the all-round white light."

12. Rule 24(a)(i) and (c)(i)

Insert "or (a)(ii)" after "in Rule 23 (a)(i)" and delete "forward".

13. Rule 24(d)

Replace the words "paragraphs (a) and (c)" in the first line by the words "paragraph (a) or (c)".

14. Rule 24(e)

Amend the lead-in sentence to read:

"A vessel or object being towed, other than those mentioned in paragraph (g) of this Rule, shall exhibit:"

15. Rule 24(g)

Insert the following new paragraph (g):

"(g) An inconspicuous, partly submerged vessel or object, or combination of such vessels or objects being towed, shall exhibit:

(i) if it is less than 25 metres in breadth, one all-round white light at or near the forward end and one at or near the after end except that dracones need not exhibit a light at or near the forward end;

(ii) if it is 25 metres or more in breadth, two additional all-round white lights at or near the extremities of its breadth;

(iii) if it exceeds 100 metres in length, additional all-round white lights between the lights prescribed in sub-paragraphs (i) and (ii) so that the distance between the lights shall not exceed 100 metres;

(iv) a diamond shape at or near the aftermost extremity of the last vessel or object being towed and if the length of the tow exceeds 200 metres an additional diamond shape where it can best be seen and located as far forward as is practicable."

16. Rule 24(h)

Reletter existing paragraph (g), which becomes paragraph (h), and amend it to read:

"(h) Where from any sufficient cause it is impracticable for a vessel or object being towed to exhibit the lights or shapes prescribed in paragraph (e) or (g) of this Rule, all possible measures shall be taken to light the vessel or object towed or at least to indicate the presence of such vessel or object."

17. Rule 24(i)

Add the following new paragraph:

"(i) Where from any sufficient cause it is impracticable for a vessel not normally engaged in towing operations to display the lights prescribed in paragraph (a) or (c) of this Rule, such vessel shall not be required to exhibit those lights when engaged in towing another vessel in distress or otherwise in need of assitance. All possible measures shall be taken to indicate the nature of the relationship between the towing vessel and the vessel being towed as authorized by Rule 36, in particular by illuminating the towline."

18. Rule 25(b)

Amend "12 metres" to read "20 metres".

19. Rule 27(b) (preamble)

Replace the word "minesweeping" by the word "mineclearance" in the first sentence.

20. Rule 27(b)(iii)

Replace the words "masthead lights" by the words "a masthead light or lights".

21. Rule 27(c)

Amend to read:

"A power-driven vessel engaged in a towing operation such as severely restricts the towing vessel and her tow in their ability to deviate from their course shall, in addition to the lights or shapes prescribed in Rule 24(a), exhibit the lights or shapes prescribed in sub-paragraphs (b)(i) and (ii) of this Rule."

22. Rule 27(d)

—Replace the words "paragraph (b)" by the words "sub-paragraphs (b)(i), (ii) and (iii)";

—delete existing sub-paragraph (iii);

—renumber existing sub-paragraph (iv) which becomes (iii) and amend it to read:

"(iii) when at anchor, the lights or shapes prescribed in this paragraph instead of the lights or shape prescribed in Rule 30."

23. Rule 27(e)

Amend to read:

"(e) Whenever the size of a vessel engaged in diving operations makes it impracticable to exhibit all lights and shapes prescribed in paragraph (d) of this Rule, the following shall be exhibited:

(i) three all-round lights in a vertical line where they can best be seen. The highest and lowest of these lights shall be red and the middle light shall be white;

(ii) a rigid replica of the International Code flag "A" not less than 1 metre in height. Measures shall be taken to ensure its all-round visibility."

24. Rule 27(f)

Amend to read:

"A vessel engaged in mineclearance operations shall in addition to the lights prescribed for a power-driven vessel in Rule 23 or to the lights or shape prescribed for a vessel at anchor in Rule 30 as appropriate, exhibit three all-round green lights or three balls. One of these lights or shapes shall be exhibited near the foremast head and one at each end of the fore yard. These lights or shapes indicate that it is dangerous for another vessel to approach within 1000 metres of the mineclearance vessel."

25. Rule 27(g)

Amend to read:

"(g) Vessels of less than 12 metres in length, except those engaged in diving operations, shall not be required to exhibit the lights and shapes prescribed in this Rule."

26. Rule 29(a)(iii)

Amend to read:

"(a)(iii) when at anchor, in addition to the lights prescribed in sub-paragraph (i), the light, lights or shape prescribed in Rule 30 for vessels at anchor."

27. Rule 30(e)

Delete "or aground" and amend "shapes prescribed in paragraphs (a), (b) or (d) or this Rule" to read:

"shape prescribed in paragraphs (a) and (b) of this Rule."

28. Rule 30(f)

Add the following new paragraph:
"(f) A vessel of less than 12 metres in length, when aground, shall not be required to exhibit the lights or shapes prescribed in sub-paragraphs (d)(i) and (ii) of this Rule."

29. Rule 33(a)

In the last line replace "required" by "prescribed".

30. Rule 34(b)(iii)

Add "to these Regulations" after the words "Annex I".

31. Rule 35(d)

Insert a new paragraph (d) and re-letter existing paragraphs (d) to (i) which become (e) to (j), as appropriate:
"(d) A vessel engaged in fishing, when at anchor, and a vessel restricted in her ability to manoeuvre when carrying out her work at anchor, shall instead of the signals prescribed in paragraph (g) of this Rule sound the signal prescribed in paragraph (c) of this Rule."

32. Rule 36

Add the following at the end of the present text:
"Any light to attract the attention of another vessel shall be such that it cannot be mistaken for any aid to navigation. For the purpose of this Rule the use of high intensity intermittent or revolving lights, such as strobe lights, shall be avoided."

33. Rule 37

Replace the word "prescribed" by "described".

34. Rule 38

Insert "to these Regulations" after the words "Annex I" in paragraphs (d)(i), (e), (f) and after the words "Annex III" in paragraph (g).

35. Rule 38(h)

Add the following new paragraph:
"(h) The repositioning of all-round lights resulting from the prescription of Section 9(b) of Annex I to these Regulations, permanent exemption."

36. Annex I, Section 1

Add the following sentence to the present text of the definition:
"This height shall be measured from the position vertically beneath the location of the light."

37. Annex I, Section 2(e)

Amend to read as follows:
"One of the two or three masthead lights prescribed for a power-driven vessel when engaged in towing or pushing another vessel shall be placed in the same position as either the forward masthead light or the after masthead light; provided that, if carried on the aftermast, the lowest after masthead light shall be at least 4.5 metres vertically higher than the forward masthead light."

38. Annex I, Section 2(f)

Amend to read:

"(f)(i) The masthead light or lights prescribed in Rule 23(a) shall be so placed as to be above and clear of all other lights and obstructions except as described in sub-paragraph (ii).

(ii) When it is impracticable to carry the all-round lights prescribed by Rule 27(b)(i) or Rule 28 below the masthead lights, they may be carried above the after masthead light(s) or vertically in between the forward masthead light(s) and after masthead lights(s), provided that in the latter case the requirement of Section 3(c) of this Annex shall be complied with."

39. Annex I, Section 2(i)(i)

Replace all words of this sub-paragraph after the word "required" in the penultimate line by the following:

"be placed at a height of not less than 4 metres above the hull".

40. Annex I, Section 2(i)(ii)

Replace all words of this sub-paragraph after the word "required" in the penultimate line by the following:

"be placed at a height of not less than 2 metres above the hull".

41. Annex I, Section 2(j)

Delete "fishing" before "vessel".

42. Annex I, Section 2(k)

Insert "prescribed in Rule 30(a)(i)" between "light" and ", when two are carried".

Replace all words after "shall" in the second sentence by "be placed at a height of not less than 6 metres above the hull."

43. Annex I, Section 3(b)

In the first line replace "On a vessel" by "On a power-driven vessel".

44. Annex I, Section 3(c)

Add the following new paragraph:

"(c) When the lights prescribed in Rule 27(b)(i) or Rule 28 are placed vertically between the forward masthead light(s) and the after masthead light(s) these all-round lights shall be placed at a horizontal distance of not less than 2 metres from the fore and aft centreline of the vessel in the athwartship direction."

45. Annex I, Section 5

Insert in the first line after "The sidelights" the words "of vessels of 20 metres or more in length" and add the following sentence after the first sentence:

"On vessels of less than 20 metres in length the sidelights, if necessary to meet the requirements of Section 9 of this Annex, shall be fitted with inboard matt black screens."

46. Annex I, Section 8

Add the following sentence to the Note at the end of this section:

"This shall not be achieved by a variable control of the luminous intensity".

47. Annex I, Section 9(a)(i)

Replace "must" by "shall".

48. Annex I, Section 9(a)(ii), last line

Replace "limits" by "sectors".

49. Annex I, Section 9(b)

Insert "prescribed in Rule 30" between "lights" and ", which need not be. . . .".

50. Annex I, Section 10(a) and (b)

Insert "as fitted" after "electric lights" in the introductory sentences of Section 10(a) and (b).

51. Annex I, Section 13

Amend to read as follows:

"The construction of lights and shapes and the installation of lights on board the vessel shall be to the satisfaction of the appropriate authority of the State whose flag the vessel is entitled to fly."

52. Annex III, Section 1(d)

Replace "4dB below the sound pressure" by "4dB below the prescribed sound pressure" and replace "10dB below the sound pressure" by "10dB below the prescribed sound pressure".

53. Annex III, Section 2(a)

Replace the words "1 metre" by the words "a distance of 1 metre from it".

54. Annex III, Section 2(b)

Amend the second sentence to read:

"the diameter of the mouth of the bell shall be not less than 300 mm for vessels of 20 metres or more in length, and shall be not less than 200 mm for vessels of 12 metres or more but of less than 20 metres in length."

55. Annex III, Section 3

Replace "the State where the vessel is registered" by "the State whose flag the vessel is entitled to fly".

56. Rule 35(b) (French text)

Insert "à propulsion mécanique" between "navire" and "faisant route".

APPENDIX E

Inland Navigational Rules Act of 1980

PART A—GENERAL

Rule 1

Application

(a) These Rules apply to all vessels upon the inland waters of the United States, and to vessels of the United States on the Canadian waters of the Great Lakes to the extent that there is no conflict with Canadian law.[1]

(b)(i) These Rules constitute special rules made by an appropriate authority within the meaning of Rule 1(b) of the International Regulations.

(ii) All vessels complying with the construction and equipment requirements of the International Regulations are considering to be in compliance with these Rules.

(c) Nothing in these Rules shall interfere with the operation of any special rules made by the Secretary of the Navy with respect to additional station or signal lights and shapes or whistle signals for ships of war and vessels proceeding under convoy, or by the Secretary with respect to additional station or signal lights and shapes for fishing vessels engaged in fishing as a fleet. These additional station or signal lights and shapes or whistle signals shall, so far as possible, be such that they cannot be mistaken for any light, shape, or signal authorized elsewhere under these Rules. Notice of such special rules shall be published in the Federal Register and, after the effective date specified in such notice, they shall have effect as if they were a part of these Rules.

(d) Vessel traffic service regulations may be in effect in certain areas.

(e) Whenever the Secretary determines that a vessel or class of vessels of special construction or purpose cannot comply fully with the provisions of any of these Rules with respect to the number, position, range, or arc of visibility of lights or shapes, as well as to the disposition and characteristics of sound-signaling appliances, without interfering with the special function of the vessel, the vessel shall comply with such other provisions in regard to the number, position, range, or arc of visibility of lights or shapes, as well as to the disposition and characteristics of

[1]The rules went into effect on all United States inland waters except on the Great Lakes on 24 December 1981. As of this writing, the date for the Great Lakes is expected to be 1 March 1983.

sound-signaling appliances, as the Secretary shall have determined to be the closest possible compliance with these Rules. The Secretary may issue a certificate of alternative compliance for a vessel or class of vessels specifying the closest possible compliance with these Rules. The Secretary of the Navy shall make these determinations and issue certificates of alternative compliance for vessels of the Navy.

(f) The Secretary may accept a certificate of alternative compliance issued by a contracting party to the International Regulations if he determines that the alternative compliance standards of the contracting party are substantially the same as those of the United States.

Rule 2

Responsibility

(a) Nothing in these Rules shall exonerate any vessel, or the owner, master, or crew thereof, from the consequences of any neglect to comply with these Rules or of the neglect of any precaution which may be required by the ordinary practice of seamen, or by the special circumstances of the case.

(b) In construing and complying with these Rules due regard shall be had to all dangers of navigation and collision and to any special circumstances, including the limitations of the vessels involved, which may make a departure from these Rules necessary to avoid immediate danger.

Rule 3

General Definitions

For the purpose of these Rules and this Act, except where the context otherwise requires:

(a) The word "vessel" includes every description of water craft, including nondisplacement craft and seaplanes, used or capable of being used as a means of transportation on water;

(b) The term "power-driven vessel" means any vessel propelled by machinery;

(c) The term "sailing vessel" means any vessel under sail provided that propelling machinery, if fitted, is not being used;

(d) The term "vessel engaged in fishing" means any vessel fishing with nets, lines, trawls, or other fishing apparatus which restricts maneuverability, but does not include a vessel fishing with trolling lines or other fishing apparatus which do not restrict maneuverability;

(e) The word "seaplane" includes any aircraft designed to maneuver on the water;

(f) The term "vessel not under command" means a vessel which through some exceptional circumstance is unable to maneuver as required by these Rules and is therefore unable to keep out of the way of another vessel;

(g) The term "vessel restricted in her ability to maneuver" means a vessel which from the nature of her work is restricted in her ability to maneuver as required by these Rules and is therefore unable to keep out of the way of another vessel; vessels restricted in their ability to maneuver include, but are not limited to:

(i) a vessel engaged in laying, servicing, or picking up a navigation mark, submarine cable, or pipeline;

(ii) a vessel engaged in dredging, surveying, or underwater operations;

(iii) a vessel engaged in replenishment or transferring persons, provisions, or cargo while underway;

(iv) a vessel engaged in the launching or recovery of aircraft;

(v) a vessel engaged in minesweeping operations; and

(vi) a vessel engaged in a towing operation such as severely restricts the towing vessel and her tow in their ability to deviate from their course.

(h) The word "underway" means that a vessel is not at anchor, or made fast to the shore, or aground;

(i) The words "length" and "breadth" of a vessel mean her length overall and greatest breadth;

(j) Vessels shall be deemed to be in sight of one another only when one can be observed visually from the other;

(k) The term "restricted visibility" means any condition in which visibility is restricted by fog, mist, falling snow, heavy rainstorms, sandstorms, or any other similar causes;

(l) "Western Rivers" means the Mississippi River, its tributaries, South Pass, and Southwest Pass, to the navigational demarcation lines dividing the high seas from harbors, rivers, and other inland waters of the United States, and the Port Allen-Morgan City Alternate Route, and that part of the Atchafalaya River above its junction with the Port Allen-Morgan City Alternate Route including the Old River and the Red River;

(m) "Great Lakes" means the Great Lakes and their connecting and tributary waters including the Calumet river as far as the Thomas J. O'Brien Lock and Controlling Works (between mile 326 and 327), the Chicago River as far as the east side of the Ashland Avenue Bridge (between mile 321 and 322), and the Saint Lawrence River as far east as the lower exit of Saint Lambert Lock;

(n) "Secretary" means the Secretary of the department in which the Coast Guard is operating;

(o) "Inland Waters" means the navigable waters of the United States shoreward of the navigational demarcation lines dividing the high seas from harbors, rivers, and other inland waters of the United States and the waters of the Great Lakes on the United States side of the International Boundary.

(p) "Inland Rules" or "Rules" mean the Inland Navigational Rules and the annexes thereto, which govern the conduct of vessels and specify the lights, shapes, and sound signals that apply on inland waters; and

(q) "International Regulations" means the International Regulations for Preventing Collisions at Sea, 1972, including annexes currently in force for the United States.

PART B—STEERING AND SAILING RULES
SUBPART I—CONDUCT OF VESSELS IN ANY CONDITION OF VISIBILITY

Rule 4

Application

Rules in this subpart apply in any condition of visibility.

Rule 5

Look-out

Every vessel shall at all times maintain a proper look-out by sight and hearing as well as by all available means appropriate in the prevailing circumstances and conditions so as to make a full appraisal of the situation and of the risk of collision.

Rule 6

Safe Speed

Every vessel shall at all times proceed at a safe speed so that she can take proper and effective action to avoid collision and be stopped within a distance appropriate to the prevailing circumstances and conditions.

In determining a safe speed the following factors shall be among those taken into account:

(a) By all vessels:

(i) the state of visibility;

(ii) the traffic density including concentration of fishing vessels or any other vessels;

(iii) the maneuverability of the vessel with special reference to stopping distance and turning ability in the prevailing conditions;

(iv) at night the presence of background light such as from shore lights or from back scatter of her own lights;

(v) the state of wind, sea, and current, and the proximity of navigational hazards;

(vi) the draft in relation to the available depth of water.

(b) Additionally, by vessels with operational radar:

(i) the characteristics, efficiency and limitations of the radar equipment;

(ii) any constraints imposed by the radar range scale in use;

(iii) the effect on radar detection of the sea state, weather, and other sources of interference;

(iv) the possibility that small vessels, ice and other floating objects may not be detected by radar at an adequate range;

(v) the number, location, and movement of vessels detected by radar; and

(vi) the more exact assessment of the visibility that may be possible when radar is used to determine the range of vessels or other objects in the vicinity.

Rule 7

Risk of Collision

(a) Every vessel shall use all available means appropriate to the prevailing circumstances and conditions to determine if risk of collision exists. If there is any doubt such risk shall be deemed to exist.

(b) Proper use shall be made of radar equipment if fitted and operational, including long-range scanning to obtain early warning of risk of collision and radar plotting or equivalent systematic observation of detected objects.

(c) Assumptions shall not be made on the basis of scanty information, especially scanty radar information.

(d) In determining if risk of collision exists the following considerations shall be among those taken into account:

(i) such risk shall be deemed to exist if the compass bearing of an approaching vessel does not appreciably change; and

(ii) such risk may sometimes exist even when an appreciable bearing change is evident, particularly when approaching a very large vessel or a tow or when approaching a vessel at close range.

Rule 8

Action To Avoid Collision

(a) Any action taken to avoid collision shall, if the circumstances of the case admit, be positive, made in ample time and with due regard to the observance of good seamanship.

(b) Any alteration of course or speed to avoid collision shall, if the circumstances of the case admit, be large enough to be readily apparent to another vessel observing visually or by radar; a succession of small alterations of course or speed should be avoided.

(c) If there is sufficient sea room, alteration of course alone may be the the most effective action to avoid a close-quarters situation provided that it is made in good time, is substantial and does not result in another close-quarters situation.

(d) Action taken to avoid collision with another vessel shall be such as to result in passing at a safe distance. The effectiveness of the action shall be carefully checked until the other vessel is finally past and clear.

(e) If necessary to avoid collision or allow more time to assess the situation, a vessel shall slacken her speed or take all way off by stopping or reversing her means of propulsion.

Rule 9

Narrow Channels

(a)(i) A vessel proceeding along the course of a narrow channel or fairway shall keep as near to the outer limit of the channel or fairway which lies on her starboard side as is safe and practicable.

(ii) Nothwithstanding paragraph (a)(i) and Rule 14(a), a power-driven vessel operating in narrow channels or fairways on the Great Lakes, Western Rivers, or waters specified by the Secretary, and proceeding downbound with a following current shall have the right-of-way over an upbound vessel, shall propose the manner and place of passage, and shall initiate the maneuvering signals prescribed by Rule 34(a)(i), as appropriate. The vessel proceeding upbound against the current shall hold as necessary to permit safe passing.

(b) A vessel of less than 20 meters in length or a sailing vessel shall not impede the passage of a vessel that can safely navigate only within a narrow channel or fairway.

(c) A vessel engaged in fishing shall not impede the passage of any other vessel navigating within a narrow channel or fairway.

(d) A vessel shall not cross a narrow channel or fairway if such crossing impedes the passage of a vessel which can safely navigate only within that channel or fairway. The latter vessel shall use the danger signal prescribed in Rule 34(d) if in doubt as to the intention of the crossing vessel.

(e)(i) In a narrow channel or fairway when overtaking, the vessel intending to overtake shall indicate her intention by sounding the appropriate signal prescribed in Rule 34(c) and take steps to permit safe passing. The overtaken vessel, if in agreement, shall sound the same signal. If in doubt she shall sound the danger signal prescribed in Rule 34(d).

(ii) This Rule does not relieve the overtaking vessel of her obligation under Rule 13.

(f) A vessel nearing a bend or an area of a narrow channel or fairway where other vessels may be obscured by an intervening obstruction shall navigate with particular alertness and caution and shall sound the appropriate signal prescribed in Rule 34(e).

(g) Every vessel shall, if the circumstances of the case admit, avoid anchoring in a narrow channel.

Rule 10

Vessel Traffic Services

Each vessel required by regulation to participate in a vessel traffic service shall comply with the applicable regulations.

SUBPART II—CONDUCT OF VESSELS IN SIGHT OF ONE ANOTHER

Rule 11

Application

Rules in this subpart apply to vessels in sight of one another.

Rule 12

Sailing Vessels

(a) When two sailing vessels are approaching one another, so as to involve risk of collision, one of them shall keep out of the way of the other as follows:

(i) when each has the wind on a different side, the vessel which has the wind on the port side shall keep out of the way of the other;

(ii) when both have the wind on the same side, the vessel which is to windward shall keep out of the way of the vessel which is to leeward; and

(iii) if a vessel with the wind on the port side sees a vessel to windward and cannot determine with certainty whether the other vessel has the wind on the port or on the starboard side, she shall keep out of the way of the other.

(b) For the purpose of this Rule the windward side shall be deemed to be the side opposite to that on which the mainsail is carried or, in the case of a square-rigged vessel, the side opposite to that on which the largest fore-and-aft sail is carried.

Rule 13

Overtaking

(a) Notwithstanding anything contained in Rules 4 through 18, any vessel overtaking any other shall keep out of the way of the vessel being overtaken.

(b) A vessel shall be deemed to be overtaking when coming up with another vessel from a direction more than 22.5 degrees abaft her beam; that is, in such a position with reference to the vessel she is overtaking, that at night she would be able to see only the sternlight of that vessel but neither of her sidelights.

(c) When a vessel is in any doubt as to whether she is overtaking another, she shall assume that this is the case and act accordingly.

(d) Any subsequent alteration of the bearing between the two vessels shall not make the overtaking vessel a crossing vessel within the meaning of these Rules or relieve her of the duty of keeping clear of the overtaken vessel until she is finally past and clear.

Rule 14

Head-on Situation

(a) When two power-driven vessels are meeting on reciprocal or nearly reciprocal courses so as to involve risk of collision each shall alter her course to starboard so that each shall pass on the port side of the other.

(b) Such a situation shall be deemed to exist when a vessel sees the other ahead or nearly ahead and by night she could see the masthead lights of the other in a line or nearly in a line or both sidelights and by day she observes the corresponding aspect of the other vessel.

(c) When a vessel is in any doubt as to whether such a situation exists she shall assume that it does exist and act accordingly.

Rule 15

Crossing Situation

(a) When two power-driven vessels are crossing so as to involve risk of collision, the vessel which has the other on her starboard side shall keep out of the way and shall, if the circumstances of the case admit, avoid crossing ahead of the other vessel.

(b) Notwithstanding paragraph (a), on the Great Lakes, Western Rivers, or water specified by the Secretary, a vessel crossing a river shall keep out of the way of a power-driven vessel ascending or descending the river.

Rule 16

Action by Give-Way Vessel

Every vessel which is directed to keep out of the way of another vessel shall, so far as possible, take early and substantial action to keep well clear.

Rule 17

Action by Stand-on Vessel

(a)(i) Where one of two vessels is to keep out of the way, the other shall keep her course and speed.

(ii) The latter vessel may, however, take action to avoid collision by her maneuver alone, as soon as it becomes apparent to her that the vessel required to keep out of the way is not taking appropriate action in compliance with these Rules.

(b) When, from any cause, the vessel required to keep her course and speed finds herself so close that collision cannot be avoided by the action of the give-way vessel alone, she shall take such action as will best aid to avoid collision.

(c) A power-driven vessel which takes action in a crossing situation in accordance with subparagraph (a)(ii) of this Rule to avoid collision with another power-driven vessel shall, if the circumstances of the case admit, not alter course to port for a vessel on her own port side.

(d) This Rule does not relieve the give-way vessel of her obligation to keep out of the way.

Rule 18

Responsibilities Between Vessels

Except where Rules 9, 10, and 13 otherwise require:

(a) A power-driven vessel underway shall keep out of the way of:

(i) a vessel not under command;

(ii) a vessel restricted in her ability to maneuver;

(iii) a vessel engaged in fishing; and

(iv) a sailing vessel.

(b) A sailing vessel underway shall keep out of the way of:

(i) a vessel not under command;

(ii) a vessel restricted in her ability to maneuver; and

(iii) a vessel engaged in fishing.

(c) A vessel engaged in fishing when underway shall, so far as possible, keep out of the way of:

(i) a vessel not under command; and

(ii) a vessel restricted in her ability to maneuver.

(d) A seaplane on the water shall, in general, keep well clear of all vessels and avoid impeding their navigation. In circumstances, however, where risk of collision exists, she shall comply with the Rules of this Part.

SUBPART III—CONDUCT OF VESSELS IN RESTRICTED VISIBILITY

Rule 19

Conduct of Vessels in Restricted Visibility

(a) This Rule applies to vessels not in sight of one another when navigating in or near an area of restricted visibility.

(b) Every vessel shall proceed at a safe speed adapted to the prevailing circumstances and conditions of restricted visibility. A power-driven vessel shall have her engines ready for immediate maneuver.

(c) Every vessel shall have due regard to the prevailing circumstances and conditions of restricted visibility when complying with Rules 4 through 10.

(d) A vessel which detects by radar alone the presence of another vessel shall determine if a close-quarters situation is developing or risk of collision exists. If so, she shall take avoiding action in ample time, provided that when such action consists of an alteration of course, so far as possible the following shall be avoided:

(i) an alteration of course to port for a vessel forward of the beam, other than for a vessel being overtaken; and

(ii) an alteration of course toward a vessel abeam or abaft the beam.

(e) Except where it has been determined that a risk of collision does not exist, every vessel which hears apparently forward of her beam the fog signal of another vessel, or which cannot avoid a close-quarters situation with another vessel forward of her beam, shall reduce her speed to the minimum at which she can be kept on course. She shall if necessary take all her way off and, in any event, navigate with extreme caution until danger of collision is over.

PART C—LIGHTS AND SHAPES

Rule 20

Application

(a) Rules in the part shall be complied with in all weathers.

(b) The Rules concerning lights shall be complied with from sunset to sunrise, and during such times no other lights shall be exhibited, except such lights as cannot be mistaken for the lights specified in these Rules or do not impair their visibility or distinctive character, or interfere with the keeping of a proper lookout.

(c) The lights prescribed by these Rules shall, if carried, also be exhibited from sunrise to sunset in restricted visibility and may be exhibited in all other circumstances when it is deemed necessary.

(d) The Rules concerning shapes shall be complied with by day.

(e) the lights and shapes specified in these Rules shall comply with the provisions of Annex I of these Rules.

Rule 21

Definitions

(a) "Masthead light" means a white light placed over the fore and aft centerline of the vessel showing an unbroken light over an arc of the horizon of 225 degrees and so fixed as to show the light from right ahead to 22.5 degrees abaft the beam on either side of the vessel, except that on a vessel of less than 12 meters in length the masthead light shall be placed as nearly as practicable to the fore and aft centerline of the vessel.

(b) "Sidelights" mean a green light on the starboard side and a red light on the port side each showing an unbroken light over an arc of the horizon of 112.5 degrees and so fixed as to show the light from right ahead to 22.5 degrees abaft the beam on its respective side. On a vessel of less than 20 meters in length the side lights may be combined in one lantern carried on the fore and aft centerline of the vessel, except that on a vessel of less than 12 meters in length the sidelights when combined in one lantern shall be placed as nearly as practicable to the fore and aft centerline of the vessel.

(c) "Sternlight" means a white light placed as nearly as practicable at the stern showing an unbroken light over an arc of the horizon of 135 degrees and so fixed as to show the light 67.5 degrees from right aft on each side of the vessel.

(d) "Towing light" means a yellow light having the same characteristics as the "sternlight" defined in paragraph (c) of this Rule.

(e) "All-round light" means a light showing an unbroken light over an arc of the horizon of 360 degrees.

(f) "Flashing light" means a light flashing at regular intervals at a frequency of 120 flashes or more per minute.

(g) "Special flashing light" means a yellow light flashing at regular intervals at a frequency of 50 to 70 flashes per minute, placed as far forward and as nearly as practicable on the fore and aft centerline of the tow and showing an unbroken light over an arc of the horizon of not less than 180 degrees nor more than 225 degrees and so fixed as to show the light from right ahead to abeam and no more than 22.5 degrees abaft the beam on either side of the vessel.

Rule 22

Visibility of Lights

The lights prescribed in these Rules shall have an intensity as specified in Annex I to these Rules, so as to be visible at the following minimum ranges:

(a) In a vessel of 50 meters or more in length:
a masthead light, 6 miles;
a sidelight, 3 miles;
a sternlight, 3 miles;
a towing light, 3 miles;
a white, red, green or yellow all-round light, 3 miles; and
a special flashing light, 2 miles.

(b) In a vessel of 12 meters or more in length but less than 50 meters in length:
a masthead light, 5 miles; except that where the length of the vessel is less than 20 meters, 3 miles;
a sidelight, 2 miles;
a sternlight, 2 miles;
a towing light, 2 miles;
a white, red, green or yellow all-round light, 2 miles; and
a special flashing light, 2 miles.

(c) In a vessel of less than 12 meters in length:
a masthead light, 2 miles;
a sidelight, 1 mile;
a sternlight, 2 miles;
a towing light, 2 miles;

a white, red, green or yellow all-round light, 2 miles; and
a special flashing light, 2 miles.
(d) In an inconspicuous, partly submerged vessel or object being towed:
a white all-round light, 3 miles.

Rule 23

Power-Driven Vessels Underway

(a) A power-driven vessel underway shall exhibit:
(i) a masthead light forward; except that a vessel of less than 20 meters in length need not exhibit this light forward of amidships but shall exhibit it as far forward as is practicable.
(ii) a second masthead light abaft of and higher than the forward one; except that a vessel of less than 50 meters in length shall not be obliged to exhibit such light but may do so:
(iii) sidelights; and
(iv) a sternlight.
(b) An air-cushion vessel when operating in the nondisplacement mode shall, in addition to the lights prescribed in paragraph (a) of this Rule, exhibit an all-round flashing yellow light where it can best be seen.
(c) A power-driven vessel of less than 12 meters in length may, in lieu of the lights prescribed in paragraph (a) of this Rule, exhibit an all-round white light and sidelights.
(d) A power-driven vessel when operating on the Great Lakes may carry an all-round white light in lieu of the second masthead light and sternlight prescribed in paragraph (a) of this Rule. The light shall be carried in the position of the second masthead light and be visible at the same minimum range.

Rule 24

Towing and Pushing

(a) A power-driven vessel when towing astern shall exhibit:
(i) instead of the light prescribed either in Rule 23 (a)(i) or 23(a)(ii), two masthead lights in a vertical line. When the length of the tow, measuring from the stern of the towing vessel to the after end of the tow exceeds 200 meters, three such lights in a vertical line;
(ii) sidelights;
(iii) a sternlight;
(iv) a towing light in a vertical line above the sternlight; and
(v) when the length of the tow exceeds 200 meters, a diamond shape where it can best be seen.
(b) When a pushing vessel and a vessel being pushed ahead are rigidly connected in a composite unit they shall be regarded as a power-driven vessel and exhibit the lights prescribed in Rule 23.
(c) A power-driven vessel when pushing ahead or towing alongside, except as required by paragraphs (b) and (i) of this Rule, shall exhibit:
(i) instead of the light prescribed either in Rule 23(a)(i) or 23(a)(ii), two masthead lights in a vertical line;
(ii) sidelights, and
(iii) two towing lights in a vertical line.
(d) A power-driven vessel to which paragraphs (a) or (c) of this Rule apply shall also comply with Rule 23(a)(i) and 23(a)(ii).

(e) A vessel or object other than those referred to in paragraph (g) of this Rule being towed shall exhibit:

(i) sidelights;

(ii) a sternlight; and

(iii) when the length of the tow exceeds 200 meters, a diamond shape where it can best be seen.

(f) Provided that any number of vessels being towed alongside or pushed in a group shall be lighted as one vessel:

(i) a vessel being pushed ahead, not being part of a composite unit, shall exhibit at the forward end sidelights, and a special flashing light; and

(ii) a vessel being towed alongside shall exhibit a sternlight and at the forward end sidelights.

(g) An inconspicuous, partly submerged vessel or object being towed shall exhibit:

(i) if it is less than 25 meters in breadth, one all-round white light at or near each end;

(ii) if it is 25 meters or more in breadth, four all-round white lights to mark its length and breadth;

(iii) if it exceeds 100 meters in length, additional all-round white lights between the lights prescribed in subparagraphs (i) and (ii) so that the distance between the lights shall not exceed 100 meters: *Provided*, That all vessels or objects being towed alongside each other shall be lighted as one vessel or object;

(iv) a diamond shape at or near the aftermost extremity of the last vessel or object being towed; and

(v) the towing vessel may direct a searchlight in the direction of the tow to indicate its presence to an approaching vessel.

(h) Where from any sufficient cause it is impracticable for a vessel or object being towed to exhibit the lights prescribed in paragraph (e) or (g) of this Rule, all possible measures shall be taken to light the vessel or object towed or at least to indicate the presence of the unlighted vessel or object.

(i) Nothwithstanding paragraph (c), on the Western Rivers and on waters specified by the Secretary, a power-driven vessel when pushing ahead or towing alongside, except as paragraph (b) applies, shall exhibit:

(i) sidelights; and

(ii) two towing lights in a vertical line.

(j) Where from any sufficient cause it is impracticable for a vessel not normally engaged in towing operations to display the lights prescribed by paragraph (a), (c) or (i) of this Rule, such vessel shall not be required to exhibit those lights when engaged in towing another vessel in distress or otherwise in need of assistance. All possible measures shall be taken to indicate the nature of the relationship between the towing vessel and the vessel being assisted. The searchlight authorized by Rule 36 may be used to illuminate the tow.

Rule 25

Sailing Vessels Underway and Vessels Under Oars

(a) A sailing vessel underway shall exhibit:

(i) sidelights; and

(ii) a sternlight.

(b) In a sailing vessel of less than 20 meters in length the lights prescribed in paragraph (a) of this Rule may be combined in one lantern carried at or near the top of the mast where it can best be seen.

(c) A sailing vessel underway may, in addition to the lights prescribed in paragraph (a) of this Rule, exhibit at or near the top of the mast, where they can best be seen, two all-round lights in a vertical line, the upper being red and the lower green, but these lights shall not be exhibited in conjunction with the combined lantern permitted by paragraph (b) of this Rule.

(d)(i) A sailing vessel of less than 7 meters in length shall, if practicable, exhibit the lights prescribed in paragraph (a) or (b) of this Rule, but if she does not, she shall have ready at hand an electric torch or lighted lantern showing a white light which shall be exhibited in sufficient time to prevent collision.

(ii) A vessel under oars may exhibit the lights prescribed in this Rule for sailing vessels, but if she does not, she shall have ready at hand an electric torch or lighted lantern showing a white light which shall be exhibited in sufficient time to prevent collision.

(e) a vessel proceeding under sail when also being propelled by machinery shall exhibit forward where it can best be seen a conical shape, apex downward. A vessel of less than 12 meters in length is not required to exhibit this shape, but may do so.

Rule 26

Fishing Vessels

(a) A vessel engaged in fishing, whether underway or at anchor, shall exhibit only the lights and shapes prescribed in this Rule.

(b) A vessel when engaged in trawling, by which is meant the dragging through the water of a dredge net or other apparatus used as a fishing appliance, shall exhibit:

(i) two all-round lights in a vertical line, the upper being green and the lower white, or a shape consisting of two cones with their apexes together in a vertical line one above the other; a vessel of less than 20 meters in length may instead of this shape exhibit a basket;

(ii) a masthead light abaft of and higher than the all-round green light; a vessel of less than 50 meters in length shall not be obliged to exhibit such a light but may do so; and

(iii) when making way through the water, in addition to the lights prescribed in this paragraph, sidelights and a sternlight

(c) A vessel engaged in fishing, other than trawling, shall exhibit:

(i) two all-round lights in a vertical line, the upper being red and the lower white, or a shape consisting of two cones with apexes together in a vertical line one above the other; a vessel of less than 20 meters in length may instead of this shape exhibit a basket;

(ii) when there is outlying gear extending more than 150 meters horizontally from the vessel, an all-round white light or a cone apex upward in the direction of the gear; and

(iii) when making way through the water, in addition to the lights prescribed in this paragraph, sidelights and a sternlight.

(d) A vessel engaged in fishing in close proximity to other vessels engaged in fishing may exhibit the additional signals described in Annex II to these Rules.

(e) A vessel when not engaged in fishing shall not exhibit the lights or shapes prescribed in this Rule, but only those prescribed for a vessel of her length.

Rule 27

Vessels Not Under Command or Restricted in Their Ability to Maneuver

(a) A vessel not under command shall exhibit:

(i) two all-round red lights in a vertical line where they can best be seen;

(ii) two balls or similar shapes in a vertical line where they can best be seen; and

(iii) when making way through the water, in addition to the lights prescribed in this paragraph, sidelights and a sternlight.

(b) A vessel restricted in her ability to maneuver, except a vessel engaged in minesweeping operations, shall exhibit:

(i) three all-round lights in a vertical line where they can best be seen. The highest and lowest of these lights shall be red and the middle light shall be white;

(ii) three shapes in a vertical line where they can best be seen. The highest and lowest of these shapes shall be balls and the middle one a diamond;

(iii) when making way through the water, masthead lights, sidelights and a sternlight, in addition to the lights prescribed in subparagraph (b)(i); and

(iv) when at anchor, in addition to the lights or shapes prescribed in subparagraphs (b) (i) and (ii), the light, lights or shapes prescribed in Rule 30.

(c) A vessel engaged in a towing operation which severely restricts the towing vessel and her tow in their ability to deviate from their course shall, in addition to the lights or shapes prescribed in subparagraphs (b) (i) and (ii) of this Rule, exhibit the lights or shape prescribed in Rule 24.

(d) A vessel engaged in dredging or underwater operations, when restricted in her ability to maneuver, shall exhibit the lights and shapes prescribed in subparagraphs (b) (i), (ii), and (iii) of this Rule and shall in addition, when an obstruction exists, exhibit:

(i) two all-round red lights or two balls in a vertical line to indicate the side on which the obstruction exists;

(ii) two all-round green lights or two diamonds in a vertical line to indicate the side on which another vessel may pass; and

(iii) when at anchor, the lights or shape prescribed by this paragraph, instead of the lights or shapes prescribed in Rule 30 for anchored vessels.

(e) Whenever the size of a vessel engaged in diving operations makes it impracticable to exhibit all lights and shapes prescribed in paragraph (d) of this Rule, the following shall instead be exhibited:

(i) Three all-round lights in a vertical line where they can best be seen. The highest and lowest of these lights shall be red and the middle light shall be white.

(ii) A rigid replica of the international Code flag "A" not less than 1 meter in height. Measures shall be taken to insure its all-round visibility.

(f) A vessel engaged in minesweeping operations shall, in addition to the lights prescribed for a power-driven vessel in Rule 23, exhibit three all-round green lights or three balls. One of these lights or shapes shall be exhibited near the foremast head and one at each end of the fore yard. These lights or shapes indicate that it is dangerous for another vessel to approach closer than 1,000 meters astern or 500 meters on either side of the minesweeper.

(g) A vessel of less than 12 meters in length, except when engaged in diving operations, is not required to exhibit the lights or shapes prescribed in this Rule.

(h) The signals prescribed in this Rule are not signals of vessels in distress and requiring assistance. Such signals are contained in Annex IV to these Rules.

Rule 28
[Reserved]

Rule 29

Pilot Vessels

(a) A vessel engaged on pilotage duty shall exhibit:

(i) at or near the masthead, two all-round lights in a vertical line, the upper being white and the lower red;

(ii) when underway, in addition, sidelights and a sternlight; and

(iii) when at anchor, in addition to the lights prescribed in subparagraph (i), the anchor light, lights, or shape prescribed in Rule 30 for anchored vessels.

(b) A pilot vessel when not engaged on pilotage duty shall exhibit the lights or shapes prescribed for a vessel of her length.

Rule 30

Anchored Vessels and Vessels Aground

(a) A vessel at anchor shall exhibit where it can best be seen:

(i) in the fore part, an all-round white light or one ball; and

(ii) at or near the stern and at a lower level than the light prescribed in subparagraph (i), an all-round white light.

(b) A vessel of less than 50 meters in length may exhibit an all-round white light where it can best be seen instead of the lights prescribed in paragraph (a) of this Rule.

(c) A vessel at anchor may, and a vessel of 100 meters or more in length shall, also use the available working or equivalent lights to illuminate her decks.

(d) A vessel aground shall exhibit the lights prescribed in paragraph (a) or (b) of this Rule and in addition, if practicable, where they can best be seen:

(i) two all-round red lights in a vertical line; and

(ii) three balls in a vertical line.

(e) A vessel of less than 7 meters in length, when at anchor, not in or near a narrow channel, fairway, anchorage, or where other vessels normally navigate, shall not be required to exhibit the lights or shape prescribed in paragraphs (a) and (b) of this Rule.

(f) A vessel of less than 12 meters in length when aground shall not be required to exhibit the lights or shapes prescribed in subparagraphs (d)(i) and (ii) of this Rule.

(g) A vessel of less than 20 meters in length, when at anchor in a special anchorage area designated by the Secretary, shall not be required to exhibit the anchor lights and shapes required by this Rule.

Rule 31

Seaplanes

Where it is impracticable for a seaplane to exhibit lights and shapes of the characteristics or in the positions prescribed in the Rules of this Part she shall exhibit lights and shapes as closely similar in characteristics and position as is possible.

PART D—SOUND AND LIGHT SIGNALS

Rule 32

Definitions

(a) The word "whistle" means any sound signaling appliance capable of producing the prescribed blasts and which complies with specifications in Annex III to these Rules.

(b) The term "short blast" means a blast of about 1 second's duration.

(c) The term "prolonged blast" means a blast of from 4 to 6 seconds' duration.

Rule 33

Equipment for Sound Signals

(a) A vessel of 12 meters or more in length shall be provided with a whistle and a bell and a vessel of 100 meters or more in length shall, in addition, be provided with a gong, the tone and sound of which cannot be confused with that of the bell. The whistle, bell and gong shall comply with the specifications in Annex III to these Rules. The bell or gong or both may be replaced by other equipment having the same respective sound characteristics, provided that manual sounding of the prescribed signals shall always be possible.

(b) A vessel of less than 12 meters in length shall not be obliged to carry the sound signaling appliances prescribed in paragraph (a) of this Rule but if she does not, she shall be provided with some other means of making an efficient sound signal.

Rule 34

Maneuvering and Warning Signals

(a) When power-driven vessels are in sight of one another and meeting or crossing at a distance within half a mile of each other, each vessel underway, when maneuvering as authorized or required by these Rules:

(i) shall indicate that maneuver by the following signals on her whistle: one short blast to mean "I intend to leave you on my port side"; two short blasts to mean "I intend to leave you on my starboard side"; and three short blasts to mean "I am operating astern propulsion".

(ii) upon hearing the one or two blast signal of the other shall, if in agreement, sound the same whistle signal and take the steps necessary to effect a safe passing. If, however, from any cause, the vessel doubts the safety of the proposed maneuver, she shall sound the danger signal specified in paragraph (d) of this Rule and each vessel shall take appropriate precautionary action until a safe passing agreement is made.

(b) A vessel may supplement the whistle signals prescribed in paragraph (a) of this Rule by light signals:

(i) These signals shall have the following significance: one flash to mean "I intend to leave you on my port side"; two flashes to mean "I intend to leave you on my starboard side"; three flashes to mean "I am operating astern propulsion";

(ii) The duration of each flash shall be about 1 second; and

(iii) The light used for this signal shall, if fitted, be one all-round white or yellow light, visible at a minimum range of 2 miles, synchronized with the whistle, and shall comply with the provisions of Annex I to these Rules.

(c) When in sight of one another:

(i) a power-driven vessel intending to overtake another power-driven vessel shall indicate her intention by the following signals on her whistle: one short blast to mean "I intend to overtake you on your starboard side"; two short blasts to mean "I intend to overtake you on your port side"; and

(ii) the power-driven vessel about to be overtaken shall, if in agreement, sound a similar sound signal. If in doubt she shall sound the danger signal prescribed in paragraph (d).

(d) When vessels in sight of one another are approaching each other and from any cause either vessel fails to understand the intentions or actions of the other, or is in doubt whether sufficient action is being taken by the other to avoid collision, the vessel in doubt shall immediately indicate such doubt by giving at least five short and rapid blasts on the whistle. This signal may be supplemented by a light signal of at least five short and rapid flashes.

(e) A vessel nearing a bend or an area of a channel or fairway where other vessels may be obscured by an intervening obstruction shall sound one prolonged blast. This signal shall be answered with a prolonged blast by any approaching vessel that may be within hearing around the bend or behind the intervening obstruction.

(f) If whistles are fitted on a vessel at a distance apart of more than 100 meters, one whistle only shall be used for giving maneuvering and warning signals.

(g) When a power-driven vessel is leaving a dock or berth, she shall sound one prolonged blast.

(h) A vessel that reaches agreement with another vessel in a meeting, crossing, or overtaking situation by using the radiotelephone as prescribed by the Bridge-to-Bridge Radiotelephone Act (85 Stat. 165; 33 U.S.C. 1207), is not obliged to sound the whistle signals prescribed by this Rule, but may do so. If agreement is not reached, then whistle signals shall be exchanged in a timely manner and shall prevail.

Rule 35

Sound Signals in Restricted Visibility

In or near an area of restricted visibility, whether by day or night, the signals prescribed in this Rule shall be used as follows:

(a) A power-driven vessel making way through the water shall sound at intervals of not more than 2 minutes one prolonged blast.

(b) a power-driven vessel underway but stopped and making no way through the water shall sound at intervals of not more than 2 minutes two prolonged blasts in succession with an interval of about 2 seconds between them.

(c) A vessel not under command; a vessel restricted in her ability to maneuver, whether underway or at anchor; a sailing vessel; a vessel engaged in fishing, whether underway or at anchor; and a vessel engaged in towing or pushing another vessel shall, instead of the signals prescribed in paragraphs (a) or (b) of this Rule, sound at intervals of not more than 2 minutes, three blasts in succession; namely, one prolonged followed by two short blasts.

(d) A vessel towed or if more than one vessel is towed the last vessel of the tow, if manned, shall at intervals of not more than 2 minutes sound four blasts in succession; namely, one prolonged followed by three short blasts. When practicable, this signal shall be made immediately after the signal made by the towing vessel.

(e) When a pushing vessel and a vessel being pushed ahead are rigidly connected in a composite unit they shall be regarded as a power-driven vessel and shall give the signals prescribed in paragraphs (a) or (b) of this Rule.

(f) A vessel at anchor shall at intervals of not more than 1 minute ring the bell rapidly for about 5 seconds. In a vessel of 100 meters or more in length the bell shall be sounded in the forepart of the vessel and immediately after the ringing of the bell the gong shall be sounded rapidly for abut 5 seconds in the after part of the vessel. A vessel at anchor may in addition sound three blasts in succession; namely, one short, one prolonged and one short blast, to give warning of her position and of the possibility of collision to an approaching vessel.

(g) A vessel aground shall give the bell signal and if required the gong signal prescribed in paragraph (f) of this Rule and shall, in addition, give three separate and distinct strokes on the bell immediately before and after the rapid ringing of the bell. A vessel aground may in addition sound an appropriate whistle signal.

(h) A vessel of less than 12 meters in length shall not be obliged to give the above-mentioned signals but, if she does not, shall make some other efficient sound signal at intervals of not more than 2 minutes.

(i) A pilot vessel when engaged on pilotage duty may in addition to the signals prescribed in paragraphs (a), (b) or (f) of this Rule sound an identity signal consisting of four short blasts.

(j) The following vessels shall not be required to sound signals as prescribed in paragraph (f) of this Rule when anchored in a special anchorage area designated by the Secretary:

(i) a vessel of less than 20 meters in length; and

(ii) a barge, canal boat, scow, or other nondescript craft.

Rule 36

Signals To Attract Attention

If necessary to attract the attention of another vessel, any vessel may make light or sound signals that cannot be mistaken for any signal authorized elsewhere in these Rules, or may direct the beam of her searchlight in the direction of the danger, in such a way as not to embarrass any vessel.

Rule 37

Distress Signals

When a vessel is in distress and requires assistance she shall use or exhibit the signals described in Annex IV to these Rules.

PART E—EXEMPTIONS

Rule 38

Exemptions

Any vessel or class of vessels, the keel of which is laid or which is at a corresponding stage of construction before the date of enactment of this Act, provided that she complies with the requirements of—

(a) The Act of June 7, 1897 (30 Stat. 96), as amended (33 U.S.C. 154–232) for vessels navigating the waters subject to that statute;

(b) Section 4233 of the Revised Statutes (33 U.S.C. 301–356) for vessels navigating the waters subject to that statute;

(c) The Act of February 8, 1895 (28 Stat. 645), as amended (33 U.S.C. 241–295) for vessels navigating the waters subject to that statute; or

(d) Sections 3, 4, and 5 of the Act of April 25, 1940 (54 Stat. 163), as amended (46 U.S.C. 526 b, c, and d) for motorboats navigating the waters subject to that statute; shall be exempted from compliance with the technical Annexes to these Rules as follows:

(i) the installation of lights with ranges prescribed in Rule 22, until 4 years after the effective date of these Rules, except that vessels of less than 20 meters in length are permanently exempt;

(ii) the installation of lights with color specifications as prescribed in Annex I to these Rules, until 4 years after the effective date of these Rules, except that vessels of less than 20 meters in length are permanently exempt;

(iii) the repositioning of lights as a result of conversion to metric units and rounding off measurement figures, are permanently exempt; and

(iv) the horizontal repositioning of masthead lights prescribed by Annex I to these Rules:

(1) on vessels of less than 150 meters in length, permanent exemption.

(2) on vessels of 150 meters or more in length, until 9 years after the effective date of these Rules.

(v) the restructuring or repositioning of all lights to meet the prescriptions of Annex I to these Rules, until 9 years after the effective date of these Rules;

(vi) power-driven vessels of 12 meters or more but less than 20 meters in length are permanently exempt from the provisions of Rule 23(a)(i) and 23(a)(iv) provided that, in place of these lights, the vessel exhibits a white light aft visible all round the horizon; and

(vii) the requirements for sound signal appliances prescribed in Annex III to these Rules, until 9 years after the effective date of these Rules.

SEC. 3. The Secretary may issue regulations necessary to implement and interpret this Act. The Secretary shall establish the following technical annexes to these Rules: Annex I, Positioning and Technical Details of Lights and Shapes; Annex II, Additional Signals for Fishing Vessels Fishing in Close Proximity; Annex III, Technical Details of Sound Appliances; and Annex IV, Distress Signals. These annexes shall be as consistent as possible with the respective annexes to the International Regulations. The Secretary may establish other technical annexes, including local pilot rules.

SEC. 4. (a) Whoever operates a vessel in violation of this Act, or of any regulation issued thereunder, or in violation of a certificate of alternative compliance issued under Rule 1 is liable to a civil penalty of not more than $5,000 for each violation.

(b) Every vessel subject to this Act, other than a public vessel being used for noncommercial purposes, that is operated in violation of this Act, or of any regulation issued thereunder, or in violation of a certificate of alternative compliance issued under Rule 1 is liable to a civil penalty of not more than $5,000 for each violation, for which penalty the vessel may be seized and proceeded against in the district court of the United States of any district within which the vessel may be found.

(c) The Secretary may assess any civil penalty authorized by this section. No such penalty may be assessed until the person charged, or the owner of the vessel charged, as appropriate, shall have been given notice of the violation involved and an opportunity for a hearing. For good cause shown, the Secretary may remit, mitigate, or compromise any penalty assessed. Upon the failure of the person charged, or the owner of the vessel charged, to pay an assessed penalty, as it may have been mitigated or compromised, the Secretary may request the Attorney General to commence an action in the appropriate district court of the United States for collection of the penalty as assessed, without regard to the amount involved, together with such other relief as may be appropriate.

(d) The Secretary of the Treasury shall withhold or revoke, at the request of the Secretary, the clearance, required by section 4197 of the Revised Statutes of the United States (46 U.S.C. 91) of any vessel, the owner or operator of which is subject to any of the penalties in this section. Clearance may be granted in such cases upon the filing of a bond or other surety satisfactory to the Secretary.

SEC. 5. (a) The Secretary shall establish a Rules of the Road Advisory Council (hereinafter referred to as the Council) not exceeding 21 members. To assure balanced representation, members shall be chosen, insofar as practical, from the following groups: (1) recognized experts and leaders in organizations having an active interest in the Rules of the Road and vessel and port safety, (2) representatives of owners and operators of vessels, professional mariners, recreational boaters, and the recreational boating industry, (3) individuals with an interest in maritime law, and (4) Federal and State officials with responsibility for vessel and port safety. Additional persons may be appointed to panels of the Council to assist the Council in the performance of its functions.

(b) The Council shall advise, consult with, and make recommendations to the Secretary on matters relating to any major proposals for changes to the Inland rules. The Council may recommend changes to the Inland Rules and International Regulations to the Secretary. Any advice or recommendation made by the Council to the Secretary shall reflect the independent judgment of the Council on the matter concerned. The Council shall meet at the call of the Secretary, but in any event not less than once during each calendar year. All proceedings of the Council shall be public, and a record of the proceedings shall be made available for public inspection.

(c) The Secretary shall furnish to the Council an executive secretary and such secretarial, clerical, and other services as are deemed necessary for the conduct of its business. Members of the Council who are not officers or employees of the United States shall, while attending meetings of the Council or while otherwise engaged in the business of the Council, be entitled to receive compensation at a rate fixed by the Secretary, not exceeding the daily equivalent of the current rate of basic pay in effect for GS–18 of the General Schedule under section 5332 of title 5, United States Code, including travel-time; and while away from their home or regular place of business, they may be allowed travel expenses, including per diem in lieu of subsistence, as authorized by section 5703 of title 5, United States Code. Payments under this section shall not render members of the Council officers or employees of the United States for any purpose.

(d) Unless extended by subsequent Act of Congress, the Council shall terminate 5 years from the date of enactment of this Act.

SEC. 6. The International Navigational Rules Act of 1977 (91Stat. 308; 33 U.S.C. 1601), is amended as follows:

(1) in section 5 by amending subsection (a) to read as follows:

"The International Regulations do not apply to vessels while in the waters of the United States shoreward of the navigational demarcation lines dividing the high seas from harbors, rivers, and other inland waters of the United States";

(2) in section 6, by adding a new subsection (d) as follows:

"(d) A certification authorized by this section may be issued for a class of vessels.";

(3) in subsection (a) of section 9 by striking "$500" and inserting in lieu thereof "$5,000".

(4) in subsection (b) of section 9 by striking "$500" and inserting in lieu thereof "not more than $5,000".

SEC. 7. Sections 2, 4, 6(1), and 8(a) are effective 12 months after the date of enactment of this Act, except that on the Great Lakes, the effective date of sections 2 and 4 will be established by the Secretary. Section 5 is effective October 1, 1981.

SEC. 8. (a) The laws specified in the following schedules are repealed. Any prior rights or liabilities existing under these laws are not affected by their repeal.

Revised Statutes

SEC. 4233.
SEC. 4233A.
SEC. 4233B.
SEC. 4233C.

Date	Chapter	Statutes at Large, sections	Volume	Page
1890: Aug. 19	802		26	320
1893: Mar. 3	202		27	557
1895: Feb. 19	102	1, 3, 4	28	672
1897:				
Mar. 3	389	5, 12, 13	29	689
June 7	4		30	96
1900: Feb. 19	22		31	30
1905: Mar. 3	1457	10	33	1032
1914: May 25	98		38	381
1933: Mar. 1	157		47	1417
1935: Aug. 21	595	2, 3, 4, 5	49	669
1936: May 20	433		49	1367
1940:				
Apr. 22	128	1, 3	54	150
Apr. 25	155	3, 4, 5	54	164
1945: Dec. 3	511	1, 2	59	590
1948:				
Mar. 5	99		62	69
May 21	328		62	249
1953: Aug. 8	386		67	497
1956: June 4	353		70	228

Date	Public Law	Sections	Volume	Page
1858:				
Aug. 14	85–635		72	590
Aug. 14	85–656		72	612
1963:				
Aug. 5	88–84		77	116
Oct. 30	88–163		77	281
1966: Nov. 5	89–764	1, 2, 5, 6	80	1313

(b) The following laws are repealed when the Secretary establishes an effective date under section 7.

Date	Chapter	Statutes at Large, sections	Volume	Page
1895: Feb. 8	64		28	645
1928:				
May 17	600		45	592
May 17	601		45	593
1919: Feb. 28	370		45	1405
1932: May 9	175		47	152
1940: Apr. 22	128	2	54	150
1948: Mar. 18	138		62	82

Date	Public Law	Sections	Volume	Page
1958: Mar. 28	85–350		72	49
1966: Nov. 5	89–764	3, 4	80	1313

SEC. 9. Section 2(c) of the Act of February 19, 1895 (28 Stat. 672), as amended (33 U.S.C. 151), is amended by striking the words "the Canal Zone,".

Approved December 24, 1980.

ANNEX I: POSITIONING AND TECHNICAL DETAILS OF LIGHTS AND SHAPES[2]

Sec.
84.01 Definitions.
84.03 Vertical positioning and spacing of lights.
84.05 Horizontal positioning and spacing of lights.
84.07 Details of location of direction-indicating lights for fishing vessels, dredgers and vessels engaged in underwater operations.
84.09 Screens.
84.11 Shapes.
84.13 Color specification of lights.
84.15 Intensity of lights.
84.17 Horizontal sectors.
84.19 Vertical sectors.
84.21 Intensity of non-electric lights.
84.23 Maneuvering light.
84.25 Approval. [Reserved]

Authority: Sec. 3, Pub. L. 28–891, 53 U.S.C. 2071; 49 CFR 1.48(n)(14).

§84.01 Definitions.

(a) The term "height above the hull" means height above the uppermost

[2] 33 CFR Part 84.

continuous deck. This height shall be measured from the position vertically beneath the location of the light.

(b) The term "practical cut-off" means, for vessels 20 meters or more in length, 12.5 percent of the minimum luminous intensity (Table 84.15 (b)) corresponding to the greatest range of visibility for which the requirements of Annex I are met.

(c) The term "Rule" or "Rules" means the Inland Navigation Rules contained in sec. 2 of the Inland Navigational Rules Act of 1980 (Pub. L. 96–591, 94 Stat. 3415, 33 U.S.C. 2001, December 24, 1980) as amended.

§84.03 Vertical Positioning and Spacing of Lights.

(a) On a power-driven vessel of 20 meters or more in length the masthead lights shall be placed as follows:

(1) The forward masthead light, or if only one masthead light is carried, then that light, at a height above the hull of not less than 5 meters, and, if the breadth of the vessel exceeds 5 meters, then at a height above the hull not less than such breadth, so however that the light need not be placed at a greater height above the hull than 8 meters;

(2) When two masthead lights are carried the after one shall be at least 2 meters vertically higher than the forward one.

(b) The vertical separation of the masthead lights of power-driven vessels shall be such that in all normal conditions of trim the after light will be seen over and separate from the forward light at a distance of 1000 meters from the stem when viewed from water level.

(c) The masthead light of a power-driven vessel of 12 meters but less than 20 meters in length shall be placed at a height above the gunwale of not less than 2.5 meters.

(d) The masthead light, or the all-round light described in rule 23(c), of a power-driven vessel of less than 12 meters in length shall be carried at least one meter higher than the sidelights.

(e) One of the two or three masthead lights prescribed for a power-driven vessel when engaged in towing or pushing another vessel shall be placed in the same position as either the forward masthead light or the after masthead light, provided that the lowest after masthead light shall be at least 2 meters vertically higher than the highest forward masthead light.

(f)(1) The masthead light or lights prescribed in Rule 28(a) shall be so placed as to be above and clear of all other lights and obstructions except as described in paragraph (f)(2) of this section.

(2) When it is impracticable to carry the all-round lights prescribed in rule 27(b)(i) below the masthead lights, they may be carried above the after masthead light(s) or vertically in between the forward masthead light(s) and after masthead light(s), provided that in the latter case the requirement of § 84.05(d) shall be complied with.

(g) The sidelights of a power-driven vessel shall be placed at least one meter lower than the forward masthead light. They shall not be so low as to be interfered with by deck lights.

(h) [Reserved]

(i) When the Rules prescribe two or three lights to be carried in a vertical line, they shall be spaced as follows:

(1) On a vessel of 20 meters in length or more such lights shall be spaced not less than 1 meter apart, and the lowest of these lights shall, except where a towing light is required, be placed at a height of not less than 4 meters above the hull;

(2) On a vessel of less than 20 meters in length such lights shall be spaced not less than 1 meter apart and the lowest of these lights shall, except where a towing light is required, be placed at a height of not less than 2 meters above the hull;

(3) When three lights are carried they shall be equally spaced.

(j) The lower of the two all-round lights prescribed for a vessel when engaged in fishing shall be a height above the sidelights not less than twice the distance between the two vertical lights.

(k) The forward anchor light prescribed in rule 30(a)(i),when two are carried, shall not be less than 4.5 meters above the after one. On a vessel of 50 meters or more in length this forward anchor light shall be placed at a height or not less than 6 meters above the hull.

§ 84.05 Horizontal Positioning and Spacing of Lights.

(a) Except as specified in paragraph (b) of this section, when two masthead lights are prescribed for a power-driven vessel, the horizontal distance between them shall not be less than one quarter of the length of the vessel but need not be more than 50 meters. The forward light shall be placed not more than one half of the length of the vessel from the stem.

(b) On power-driven vessels 50 meters but less than 60 meters in length operated on the Western Rivers, the horizontal distance between masthead lights shall not be less than 10 meters.

(c) On a power-driven vessel of 20 meters or more in length the sidelights shall not be placed in front of the forward masthead lights. They shall be placed at or near the side of the vessel.

(d) When the lights prescribed in rule 27(b)(i) are placed vertically between the forward masthead light(s) and the after masthead light(s) these all-round lights shall be placed at a horizontal distance of not less than 2 meters from the fore and aft centerline of the vessel in the athwartship direction.

§ 84.07 Details of Location of Direction-Indicating Lights for Fishing Vessels, Dredgers and Vessels Engaged in Underwater Operations.

(a) The light indicating the direction of the outlying gear from a vessel engaged in fishing as prescribed in rule 26(c)(ii) shall be placed at a horizontal distance of not less than 2 meters and not more than 6 meters away from the two all-round red and white light prescribed in rule 26(c)(i) and not lower than the sidelights.

(b) The lights and shapes on a vessel engaged in dredging or underwater operations to indicate the obstructed side and/or the side on which it is safe to pass, as prescribed in rule 27(d)(i) and (ii), shall be placed at the maximum practical horizontal distance, but in no case less than 2 meters, from the lights or shapes prescribed in rule 27(b)(i) and (ii). In no case shall the upper of these lights or shapes be at a greater height than the lower of the three lights or shapes prescribed in rule 27(b)(i) and (ii).

§ 84.09 Screens.

(a) The sidelights of vessels of 20 meters or more in length shall be fitted with mat black inboard screens and meet the requirements of § 84.17. On vessels of less than 20 meters in length, the sidelights, if necessary to meet the requirements of §84.17, shall be fitted with mat black inboard screens. With a combined lantern, using a single vertical filament and a very narrow division between the green and red sections, external screens need not be fitted.

(b) On power-driven vessels less than 12 meters in length constructed after July 31, 1983, the masthead light, or the all-round light described in rule 23(c) shall be screened to prevent direct illumination of the vessel forward of the operator's position.

§ 84.11 Shapes.

(a) Shapes shall be black and of the following sizes:

(1) A ball shall have a diameter of not less than 0.6 meter;

(2) A cone shall have a base diameter of not less than 0.6 meter and a height equal to its diameter;

(3) A diamond shape shall consist of two cones (as defined in paragraph (a)(2) of this section) having a common base.

(b) The vertical distance between shapes shall be at least 1.5 meter.

(c) In a vessel of less than 20 meters in length shapes of lesser dimensions but commensurate with the size of the vessel may be used and the distance apart may be correspondingly reduced.

§ 84.13 Color Specification of Lights.

(a) The chromaticity of all navigation lights shall conform to the following standards, which lie within the boundaries of the area of the diagram specified for each color by the International Commission on Illumination (CIE), in the "Colors of Light Signals", which is incorporated by reference. It is Publication CIE No. 2.2. (TC–1.6), 1975, and is available from the Illumination Engineering Society, 345 East 47th Street, New York, NY 10017. It is also available for inspection at the Office of the Federal Register, Room 8401, 1100 L Street N.W., Washington, DC 20408. This incorporation by reference was approved by the Director of the Federal Register.

(b) The boundaries of the area for each color are given by indicating the corner co-ordinates, which are as follows:

(1) *Whites:*

x	0.525	0.525	0.452	0.310	0.310	0.443
y	0.382	0.440	0.440	0.348	0.283	0.382

(2) *Green:*

x	0.028	0.009	0.300	0.203
y	0.385	0.723	0.511	0.358

(3) *Red:*

x	0.680	0.660	0.735	0.721
y	0.320	0.320	0.265	0.259

(4) *Yellow:*

x	0.612	0.618	0.575	0.575
y	0.382	0.382	0.425	0.406

§ 84.15 Intensity of Lights.

(a) The minimum luminous intensity of lights shall be calculated by using the formula:

$$I = 3.43 \times 10^6 \times T \times D^2 \times K^{-D}$$

where I is luminous intensity in candelas under service conditions,

T is threshold factor 2×10^{-7} lux,

D is range of visibility (luminous range) of the light in nautical miles,

K is atmospheric transmissivity. For prescribed lights the value of K shall be 0.8, corresponding to a meteorological visibility of approximately 13 nautical miles.

(b) A selection of figures derived from the formula is given in Table 84.15(b)

Table 84.15(b)

Range of visibility (luminous range) of light in nautical miles D	Minimum luminous intensity of light in candelas for K = 0.8 1
1	0.9
2	4.3
3	12
4	27
5	52
6	94

§ 84.17 Horizontal Sectors.

(a)(i) In the forward direction, sidelights as fitted on the vessel shall show the minimum required intensities. The intensities shall decrease to reach practical cut-off between 1 and 3 degrees outside the prescribed sectors.

(2) For sternlights and masthead lights and at 22.5 degrees abaft the beam for sidelights, the minimum required intensities shall be maintained over the arc of the horizon up to 5 degrees within the limits of the sectors prescribed in rule 21. From 5 degrees within the prescribed sectors the intensity may decrease by 50 percent up to the prescribed limits; it shall decrease steadily to reach practical cut-off at not more than 5 degrees outside the prescribed sectors.

(b) All-round lights shall be so located as not to be obscured by masts, topmasts or structures within angular sectors of more than 6 degrees, except anchor lights prescribed in rule 30, which need not be placed at an impracticable height above the hull, and the all-round white light described in rule 23(d), which may not be obscured at all.

§ 84.19 Vertical Sectors.

(a) The vertical sectors of electric lights as fitted, with the exception of lights on sailing vessels and on unmanned barges, shall ensure that:

(1) At least the required minimum intensity is maintained at all angles from 5 degrees above to 5 degrees below the horizontal;

(2) At least 60 percent of the required minimum intensity is maintained from 7.5 degrees above to 7.5 degrees below the horizontal.

(b) In the case of sailing vessels the vertical sectors of electric lights as fitted shall ensure that:

(1) At least the required minimum intensity is maintained at all angles from 5 degrees above to 5 degrees below the horizontal;

(2) At least 50 percent of the required minimum intensity is maintained from 25 degrees above to 25 degrees below the horizontal.

(c) In the case of unmanned barges the minimum required intensity of electric lights as fitted shall be maintained on the horizontal.

(d) In the case of lights other than electric lights these specifications shall be met as closely as possible.

§ 84.21 Intensity of Non-Electric Lights.

Non-electric lights shall so far as practicable comply with the minimum intensities, as specified in the Table given in § 84.15.

§ 84.23 Maneuvering Light.

Notwithstanding the provisions of § 84.03(f), the maneuvering light described in rule 34(b) shall be placed approximately in the same fore and aft vertical plane as the masthead light or lights and, where practicable, at a minimum height of one-half meter vertically above the forward masthead light, provided that it shall be carried not less than one-half meter vertically above or below the after masthead light. On a vessel where only one masthead light is carried the maneuvering light, if fitted, shall be carried where it can best be seen, not less than one-half meter vertically apart from the masthead light.

§ 84.25 Approval [Reserved]

(Sec. 3. Pub. L. 96–591: 49 CFR 1.46(n)(14))

ANNEX II, ADDITIONAL SIGNALS FOR FISHING VESSELS FISHING IN CLOSE PROXIMITY[3]

85.1 General.
85.3 Signals for trawlers.
85.5 Signals for purse seiners.

Authority: Sec. 3. Pub. L. 96–591; 49 CFR 1.46(n)(14).

§ 85.1 General.

The lights mentioned herein shall, if exhibited in pursuance of Rule 26(d), be placed where they can best be seen. They shall be at least 0.9 meter apart but at a lower level than lights prescribed in Rule 26(b)(i) and (c)(i) contained in the Navigational Rules Act of 1980. The lights shall be visible all around the horizon at a distance of at least 1 mile but at a lesser distance from the lights prescribed by these Rules for fishing vessels.

§ 85.3 Signals for Trawlers

(a) Vessels when engaged in trawling, whether using demersal or pelagic gear, may exhibit:

(1) When shooting their nets: two white lights in a vertical line;

(2) When hauling their nets: one white light over one red light in a vertical line;

(3) When the net has come fast upon an obstruction: two red lights in a vertical line.

(b) Each vessel engaged in pair trawling may exhibit:

(1) By night, a search light directed forward and in the direction of the other vessel of the pair;

(2) When shooting or hauling their nets or when their nets have come fast upon an obstruction, the lights prescribed in paragraph (a) above.

§ 85.5 Signals for purse seiners.

Vessels engaged in fishing with purse seine gear may exhibit two yellow lights in a vertical line. These lights shall flash alternately every second and with equal light

[3]33 CFR Part 85.

and occultation duration. These lights may be exhibited only when the vessel is hampered by its fishing gear.

(Sec. 3. Pub. L. 96.591, 33 U.S.C. 2071; 49 CFR 1.46(n)(14))

ANNEX III: TECHNICAL DETAILS OF SOUND SIGNAL APPLIANCES[4]

SUBPART A—WHISTLES

Sec.
86.01 Frequencies and range of audibility.
86.03 Limits of fundamental frequencies.
86.05 Sound signal intensity and range of audibility.
86.07 Directional properties.
86.09 Positioning whistles.
86.11 Fitting of more than one whistle.
86.13 Combined whistle systems.
86.15 Towing vessel whistles.

SUBPART B—BELL OR GONG

86.21 Intensity of signal.
86.23 Construction.

SUBPART C—APPROVAL

86.31 Approval. [Reserved]
Authority: Sec. 3, Pub. L. 96–59L; 49 CFR 1.46(n)(14).

SUBPART A—WHISTLES

§ 86.01 Frequencies and Range of Audibility.

The fundamental frequency of the signal shall lie within the range 70–525 Hz. The range of audibility of the signal from a whistle shall be determined by those frequencies, which may include the fundamental and/or one or more higher frequencies, which lie within the frequency ranges and provide the sound pressure levels specified in § 86.05.

§ 86.03 Limits of fundamental frequencies.

To ensure a wide variety of whistle characteristics, the fundamental frequency of a whistle shall be between the following limits:

(a) 70–200 Hz, for a vessel 200 meters or more in length:
(b) 130–350 Hz, for a vessel 75 meters but less than 200 meters in length;
(c) 250–525 Hz, for a vessel less than 75 meters in length.

§ 86.05 Sound Signal Intensity and Range of Audibility

A whistle on a vessel shall provide, in the direction of the forward axis of the whistle and at a distance of 1 meter from it, a sound pressure level in at least one ⅓–octave band of not less than the appropriate figure given in Table 86.05 within the following frequency ranges (± 1 percent):

(a) 130–1200 Hz, for a vessel 75 meters or more in length;

[4]33 CFR Part 86.

(b) 250–1600 Hz, for a vessel 20 meters but less than 75 meters in length;
(c) 250–2100 Hz, for a vessel 12 meters but less than 20 meters in length.

Table 86.05

Length of vessel in meters	Fundamental frequency range (Hz)	For measured frequencies (Hz)	46 octave band level at 1 meter in Db referred to 2 × 10	Audibility range in nautical miles
200 or more	70–200	130–180	145	2
		180–250	143	
		250–1200	140	
75 but less than 200	130–350	130–180	140	
		180–250	136	1.5
		250–1200	134	
20 but less than 75	250–625	250–450	130	
		450–800	125	1.0
		800–1800	121	
12 but less than 20	250–525	250–450	120	
		450–800	115	0.5
		800–2100	111	

NOTE: The range of audibility in the table above is for information and is approximately the range at which a whistle may usually be heard on its forward axis in conditions of still air on board a vessel having average background noise level at the listening posts (taken to be 58 dB in the octave band centered on 250 Hz and 63 dB in the octave band centered on 500 Hz).

In practice the range at which a whistle may be heard is extremely variable and depends critically on weather conditions; the values given can be regarded as typical but under conditions of strong wind or high ambient noise level at the listening post the range may be much reduced.

§ 86.07 Directional Properties.

The sound pressure level of a directional whistle shall be not more than 4 dB below the sound pressure level specified in § 86.05 in any direction in the horizontal plane within ± 45 degrees of the forward axis. The sound pressure level of the whistle in any other direction in the horizontal plane shall not be more than 10 dB less than the sound pressure specified for the forward axis, so that the range of audibility in any direction will be at least half the range required on the forward axis. The sound pressure level shall be measured in that one-third octave band which determines the audibility range.

§ 86.09 Positioning of Whistles.

(a) When a directional whistle is to be used as the only whistle on the vessel and is permanently installed, it shall be installed with its forward axis directed forward.

(b) A whistle shall be placed as high as practicable on a vessel, in order to reduce interception of the emitted sound by obstructions and also to minimize hearing damage risk to personnel. The sound pressure level of the vessel's own signal at listening posts shall not exceed 110 dB(A) and so far as practicable should not exceed 100 dB(A).

§ 86.11 Fitting of More Than One Whistle.

If whistles are fitted at a distance apart of more than 100 meters, they shall not be sounded simultaneously.

§ 86.13 Combined Whistle Systems.

(a) A combined whistle system is a number of whistles (sound emitting sources) operated together. For the purposes of the Rules a combined whistle system is to be regarded as a single whistle.

(b) The whistles of a combined system shall—

(1) Be located at a distance apart of not more than 100 meters,

(2) Be sounded simultaneously,

(3) Each have a fundamental frequency different from those of the others by at least 10 Hz, and

(4) Have a tonal characteristic appropriate for the length of vessel which shall be evidenced by at least two-thirds of the whistles in the combined system having fundamental frequencies falling within the limits prescribed in § 86.03, or if there are only two whistles in the combined system, by the higher fundamental frequency falling within the limits prescribed in § 86.03.

Note.—If due to the presence of obstructions the sound field of a single whistle or of one of the whistles referred to in § 86.11 is likely to have a zone of greatly reduced signal level, a combined whistle system should be fitted so as to overcome this reduction.

§ 86.15 Towing Vessel Whistles.

A power-driven vessel normally engaged in pushing ahead or towing alongside may, at all times, use a whistle whose characteristic falls within the limits prescribed by § 86.03 for the longest customary composite length of the vessel and its tow.

SUBPART B—BELL OR GONG

§ 86.21 Intensity of Signal.

A bell or gong, or other device having similar sound characteristics shall produce a sound pressure level of not less than 110 dB at 1 meter.

§ 86.23 Construction.

Bells and gongs shall be made of corrosion-resistant material and designed to give a clear tone. The diameter of the mouth of the bell shall be not less than 300 mm for vessels of more than 20 meters in length, and shall be not less than 20 meters in length, and shall be not less than 200 mm for vessels of 12 to 20 meters in length. The mass of the striker shall be not less than 3 percent of the mass of the bell. The striker shall be capable of manual operation. Note: When practicable, a power-driven bell striker is recommended to ensure constant force.

SUBPART C—APPROVAL

§ 86.31 Approval. [Reserved]

(Sec. 3, Pub. L. 96–591; 49 CFR 1.46 (n)(14))

PART 87—ANNEX IV, DISTRESS SIGNALS[5]

§ 87.1 Need of Assistance.

The following signals, used or exhibited either together or separately, indicate distress and need of assistance:

(a) A gun or other explosive signal fired at intervals of about a minute.
(b) A continuous sounding with any fog-signaling apparatus;
(c) Rockets or shells, throwing red stars fired one at a time at short intervals;
(d) A signal made by radiotelegraphy or by any other signaling method consisting of the group . . . — — — . . . (SOS) in the Morse Code;
(e) A signal sent by radiotelephony consisting of the spoken word "Mayday";
(f) The International Code Signal of distress indicated by N.C.
(g) A signal consisting of a square flag having above or below it a ball or anything resembling a ball;
(h) Flames on the vessel (as from a burning tar barrel, oil barrel, etc.);
(i) A rocket parachute flare or a hand flare showing a red light;
(j) A smoke signal giving off orange-colored smoke;
(k) Slowly and repeatedly raising and lowering arms outstretched to each side;
(l) The radiotelegraph alarm signal;
(m) The radiotelephone alarm signal;
(n) Signals transmitted by emergency position-indicating radio beacons;
(o) A high intensity white light flashing at regular intervals from 50 to 70 times per minute.

§ 87.3 Exclusive Use.

The use or exhibition of any of the foregoing signals except for the purpose of indicating distress and need of assistance and the use of other signals which may be confused with any of the above signals is prohibited.

§ 87.5 Supplemental Signals.

Attention is drawn to the relevant sections of the International Code of Signals, the Merchant Ship Search and Rescue Manual and the following signals:

(a) A piece of orange-colored canvas with either a black square and circle or other appropriate symbol (for identification from the air);
(b) A dye marker.

PART 88—ANNEX V, PILOT RULES[6]

[5]Authority: Sec. 3, Pub. L. 96–591, 33 U.S.C. 2071; 49 CFR 1.46(n)(14).
[6]Authority: Sec. 3, Pub. L. 96–591, 33 U.S.C. 2071; 49 CFR 1.46(n)(14).

§ 88.01 Purpose and Applicability.

This part applies to all vessels operating on United States inland waters and to United States vessels operating on the Canadian waters of the Great Lakes to the extent there is no conflict with Canadian law.

§ 88.03 Definitions.

The terms used in this part have the same meaning as defined in the Inland Navigational Rules Act of 1980.

§ 88.05 Copy of Rules.

After January 1, 1983, the operator of each self-propelled vessel 12 meters or more in length shall carry on board and maintain for ready reference a copy of the Inland Navigational Rules.

§ 88.09 Temporary Exemption from Light and Shape Requirements When Operating Under Bridges.

A vessel's navigation lights and shapes may be lowered if necessary to pass under a bridge.

§ 88.11 Law Enforcement Vessels.

(a) Law enforcement vessels may display a flashing blue light when engaged in direct law enforcement activities. This light shall be located so that it does not interfere with the visibility of the vessel's navigation lights.

(b) The blue light described in this section may be displayed by law enforcement vessels of the United States and the States and their political subdivisions.

§ 88.13 Lights on Barges at Bank or Dock.

(a) The following barges shall display at night and if practicable in periods of restricted visibility the lights described in paragraph (b) of this section—

(1) Every barge projecting into a buoyed or restricted channel.

(2) Every barge so moored that it reduces the available navigable width of any channel to less than 80 meters.

(3) Barges moored in groups more than two barges wide or to a maximum width of over 25 meters.

(4) Every barge not moored parallel to the bank or dock.

(b) Barges described in paragraph (a) of this section shall carry two unobstructed white lights of an intensity to be visible for at least one mile on a clear dark night, and arranged as follows:

(1) On a single moored barge, lights shall be placed on the two corners farthest from the bank or dock.

(2) On barges moored in group formation, a light shall be placed on each of the upstream and downstream ends of the group, on the corners farthest from the bank or dock.

(3) Any barge in a group, projecting from the main body of the group toward the channel, shall be lighted as a single barge.

(c) Barges moored in any slip or slough which is used primarily for mooring purposes are exempt from the lighting requirements of this section.

(d) Barges moored in well-illuminated areas are exempt from the lighting requirements of this section. These areas are as follows:

Chicago Sanitary Ship Canal
(1) Mile 293.2 to 293.9
(3) Mile 295.2 to 296.1
(5) Mile 297.5 to 297.8
(7) Mile 298 to 298.2
(9) Mile 298.6 to 298.8
(11) Mile 299.3 to 299.4
(13) Mile 299.8 to 300.5
(15) Mile 303 to 303.2
(17) Mile 303.7 to 303.9
(19) Mile 305.7 to 305.8
(21) Mile 310.7 to 310.9
(23) Mile 311 to 311.2
(25) Mile 312.5 to 312.6
(27) Mile 313.8 to 314.2
(29) Mile 314.6
(31) Mile 314.8 to 315.3
(33) Mile 315.7 to 316
(35) Mile 316.8
(37) Mile 316.85 to 317.05
(39) Mile 317.5
(41) Mile 318.4 to 318.9
(43) Mile 318.7 to 318.8
(45) Mile 320 to 320.3
(47) Mile 320.6
(49) Mile 322.3 to 322.4
(51) Mile 322.8
(53) Mile 322.9 to 327.2

Calumet Sag Channel
(61) Mile 316.5

Little Calumet River
(71) Mile 321.2
(73) Mile 322.3

Calumet River
(81) Mile 328.5 to 328.7
(83) Mile 329.2 to 329.4
(85) Mile 330 west bank to 330.2
(87) Mile 331.4 to 331.6
(89) Mile 332.2 to 332.4
(91) Mile 332.6 to 332.8

Cumberland River
(101) Mile 126.8
(103) Mile 191

§ 88.15 Lights on Dredge Pipelines.

Dredge pipelines that are floating or supported on trestles shall display the following lights at night and in periods of restricted visibility.

(a) One row of yellow lights. The lights must be—

(1) Flashing 50 to 70 times per minute,
(2) Visible all around the horizon,
(3) Visible for at least 2 miles on a clear dark night,
(4) Not less than 1 and not more than 3.5 meters above the water,
(5) Approximately equally spaced, and
(6) Not more than 10 meters apart where the pipeline crosses a navigable channel. Where the pipeline does not cross a navigable channel the lights must be sufficient in number to clearly show the pipeline's length and course.

(b) Two red lights at each end of the pipeline, including the ends in a channel where the pipeline is separated to allow vessels to pass (whether open or closed). The lights must be—
(1) Visible all around the horizon, and
(2) Visible for at least 2 miles on a clear dark night, and
(3) One meter apart in a vertical line with the lower light at the same height above the water as the flashing yellow light.

APPENDIX F

Vessel Bridge-to-Bridge Radiotelephone Act[1]

TO REQUIRE A RADIOTELEPHONE ON CERTAIN VESSELS WHILE NAVIGATING UPON SPECIFIED WATERS OF THE UNITED STATES

Be it enacted by the Senate and House of Representatives of the United States of America in Congress assembled, That this Act may be cited as the "Vessel Bridge-to-Bridge Radiotelephone Act".

SEC. 2 It is the purpose of this Act to provide a positive means whereby the operators of approaching vessels can communicate their intentions to one another through voice radio, located convenient to the operator's navigation station. To effectively accomplish this, there is need for a specific frequency or frequencies dedicated to the exchange of navigational information, on navigable waters of the United States.

SEC. 3. (1) "Secretary" means the Secretary of the Department in which the Coast Guard is operating;

(2) "power-driven vessel" means any vessel propelled by machinery; and

(3) "towing vessel" means any commercial vessel engaged in towing another vessel astern, alongside, or by pushing ahead.

SEC. 4. (a) Except as provided in section 7 of this Act—

(1) every power-driven vessel of three hundred gross tons and upward while navigating;

(2) every vessel of one hundred gross tons and upward carrying one or more passengers for hire while navigating;

(3) every towing vessel of twenty-six feet or over in length while navigating; and

(4) every dredge and floating plant engaged in or near a channel or fairway in operations likely to restrict or affect navigation of other vessels—shall have a radiotelephone capable of operation from its navigational bridge or, in the case of a dredge, from its main control station and capable of transmitting and receiving on the frequency or frequencies within the 156–162 Mega-Hertz band using the classes of emissions designated by the Federal Communications Commission, after consultation with other cognizant agencies, for the exchange of navigational information.

[1]85 Stat. 164; 33 U.S.C. Sec. 1201–1208.

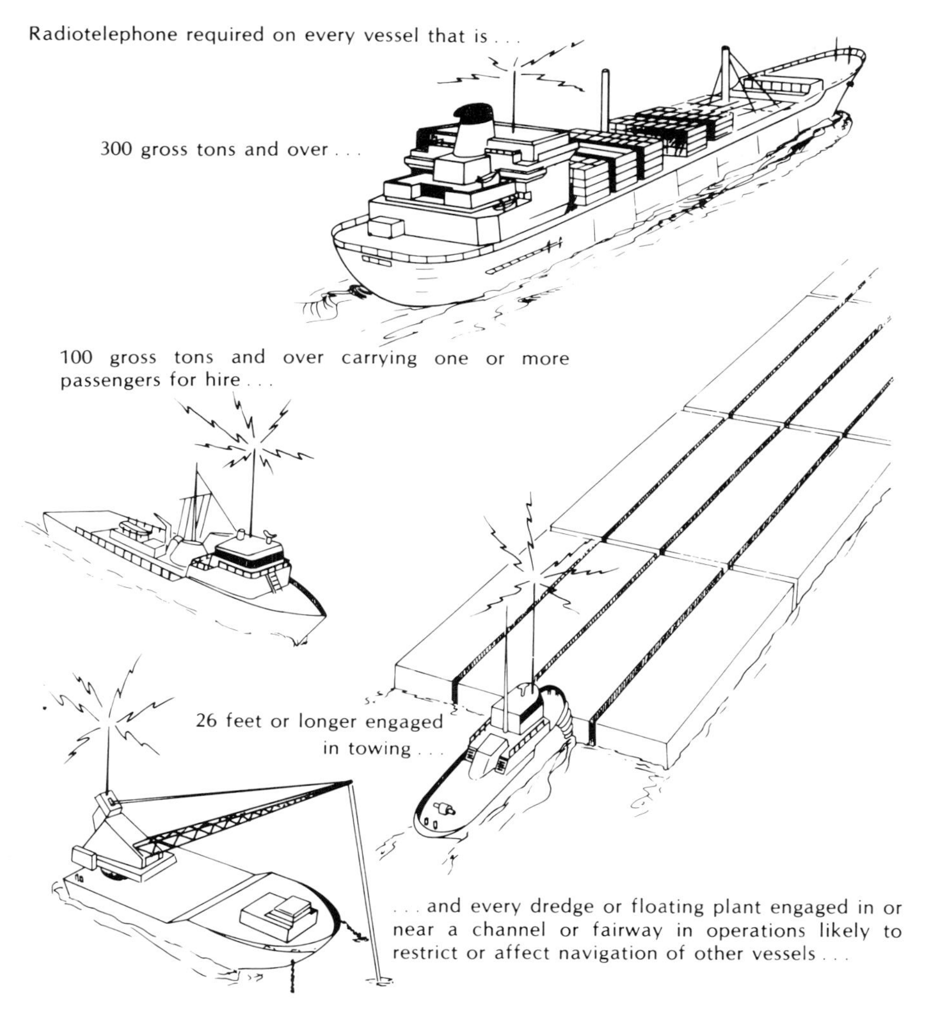

Fig. F 1. Some examples of vessels required to have radiotelephones.

(b) The radiotelephone required by subsection (a) shall be carried on board the described vessels, dredges, and floating plants upon the navigable waters of the United States inside the lines established pursuant to section 2 of the Act of February 19, 1895 (28 Stat. 672), as amended.

SEC. 5. The radiotelephone required by this Act is for the exclusive use of the master or person in charge of the vessel, or the person designated by the master or person in charge to pilot or direct the movement of the vessel, who shall maintain a listening watch on the designated frequency. Nothing contained herein shall be interpreted as precluding the use of portable radiotelephone equipment to satisfy the requirements of this Act.

SEC. 6. Whenever radiotelephone capability is required by this Act, a vessel's radiotelephone equipment shall be maintained in effective operating condition. If the radiotelephone equipment carried aboard a vessel ceases to operate, the master shall exercise due diligence to restore it or cause it to be restored to effective operating condition at the earliest practicable time. The failure of a vessel's radiotelephone equipment shall not, in itself, constitute a violation of this Act, nor shall it obligate the master of any vessel to moor or anchor his vessel; however, the loss of radiotelephone capability shall be given consideration in the navigation of the vessel.

SEC. 7. The Secretary may, if he considers that marine navigational safety will not be adversely affected or where a local communication system fully complies with the intent of this concept but does not conform in detail, issue exemptions from any provisions of this Act, on such terms and conditions as he considers appropriate.

SEC. 8. (a) The Federal Communications Commission shall, after consultation with other cognizant agencies, prescribe regulations necessary to specify operating and technical conditions and characteristics including frequencies, emission, and power of radiotelephone equipment required under this Act.

(b) The Secretary shall, subject to the concurrence of the Federal Communications Commission, prescribe regulations for the enforcement of this Act.

SEC. 9. (a) Whoever, being the master or person in charge of a vessel subject to this Act, fails to enforce or comply with this Act or the regulations hereunder; or

Whoever, being designated by the master or person in charge of a vessel subject to this Act to pilot or direct the movement of the vessel, fails to enforce or comply with this Act or the regulations hereunder—

Is liable to a civil penalty of not more than $500 to be assessed by the Secretary.

(b) Every vessel navigating in violation of this Act or the regulations hereunder is liable to a civil penalty of not more than $500 to be assessed by the Secretary for which the vessel may be proceeded against in any district court of the United States having jurisdiction.

(c) Any penalty assessed under this section may be remitted or mitigated by the Secretary upon such terms as he may deem proper.

SEC. 10. This Act shall become effective May 1, 1971, or six months after the promulgation of regulations which implement its provisions, whichever is later.

APPENDIX G

Vessel Bridge-to-Bridge Radiotelephone Regulations[1]

PART 26[2]

Sec.

Purpose

SEC. 26.01 (a) The purpose of this part is to implement the provisions of the Vessel Bridge-to-Bridge Radiotelephone Act. This part—

(1) Requires the use of the vessel bridge-to-bridge radiotelephone;

(2) Provides the Coast Guard's interpretation of the meaning of important terms in the Act;

(3) Prescribes the procedures for applying for an exemption from the Act and the regulations issued under the Act and a listing of exemptions.

(b) Nothing in this part relieves any person from the obligation of complying with the rules of the road and the applicable pilot rules.

Definitions

SEC 26.02 For the purpose of this part and interpreting the Act—

"Secretary" means the Secretary of the Department in which the Coast Guard is operating;

[1]Code of Federal Regulations: Title 33—Navigation and Navigable Waters; Part 26—Vessel bridge-to-bridge radiotelephone Regulations.

[2]Authority: The provisions of this Part 26 issued under 85 stat. 164: 33 U.S.C.A. secs. 1201–1208, 49 CFR 1.46(o)(2).

"Act" means the "Vessel Bridge-to-Bridge Radiotelephone Act", 33 U.S.C.A. sections 1201–1208;

"Length" is measured from end to end over the deck excluding sheer;

"Navigable waters of the United States inside the lines established pursuant to section 2 of the Act of February 19, 1895 (28 Stat. 672), as amended," means those waters governed by the Navigation Rules for Harbors, Rivers, and Inland Waters (33 U.S.C. sec. 151 et seq.), the Navigation Rules for Great Lakes and their Connecting and Tributary Waters (33 U.S.C. sec. 241 et seq.), and the Navigation Rules for Red River of the North and Rivers emptying into Gulf of Mexico and Tributaries (33 U.S.C. sec. 301 et seq.);

"Power-driven vessel" means any vessel propelled by machinery; and

"Towing vessel" means any commercial vessel engaged in towing another vessel astern, alongside, or by pushing ahead.

Radiotelephone Required

SEC. 26.03 (a) Unless an exemption is granted under Section 26.09 and except as provided in subparagraph (4) of this paragraph, section 4 of the Act provides that:

(1) Every power-driven vessel of 300 gross tons and upward while navigating:

(2) Every vessel of 100 gross tons and upward carrying one or more passengers for hire while navigating:

(3) Every towing vessel of 26 feet or over in length while navigating: and

(4) Every dredge and floating plant engaged in or near a channel or fairway in operations likely to restrict or affect navigation of other vessels: *Provided*, That an unmanned or intermittently manned floating plant under the control of a dredge need not be required to have separate radiotelephone capability:

Shall have a radiotelephone capable of operation from its navigational bridge, or in the case of a dredge, from its main control station, and capable of transmitting and receiving on the frequency or frequencies within the 156–162 Mega-Hertz band using the classes of emissions designated by the Federal Communications Commission, after consultation with other cognizant agencies, for the exchange of navigational information.

(b) The radiotelephone required by paragraph (a) of this section shall be carried on board the described vessels, dredges, and floating plants upon the navigable waters of the United States inside the lines established pursuant to section 2 of the Act of February 19, 1895 (28 Stat. 672), as amended.

Use of the Designated Frequency

SEC. 26.04 (a) No person may use the frequency designated by the Federal Communications Commission under section 8 of the Act, 33 U.S.C.A. section 1207(a), to transmit any information other than information necessary for the safe navigation of vessels or necessary tests.

(b) Each person who is required to maintain a listening watch under section 5 of the Act shall, when necessary, transmit and confirm, on the designated frequency, the intentions of his vessel and any other information necessary for the safe navigation of vessels.

(c) Nothing in these regulations may be construed as prohibiting the use of the designated frequency to communicate with shore stations to obtain or furnish information necessary for the safe navigation of vessels.[3]

[3]The Federal Communications Commission has designated the frequency 156.65 MHz for the use of bridge-to-bridge radiotelephone stations.

Use of Radiotelephone

SEC. 26.05 Section 5 of the Act states:

(a) The radiotelephone required by this Act is for the exclusive use of the master or person in charge of the vessel, or the person designated by the master or person in charge of the vessel, or the person designated by the master or person in charge to pilot or direct the movement of the vessel, who shall maintain a listening watch on the designated frequency. Nothing contained herein shall be interpreted as precluding the use of portable radiotelephone equipment to satisfy the requirements of this Act.

Maintenance of Radiotelephone; Failure of Radiotelephone

SEC. 26.06 Section 6 of the Act states:

(a) Whenever radiotelephone capability is required by this Act, a vessel's radiotelephone equipment shall be maintained in effective operating condition. If the radiotelephone equipment carried aboard a vessel ceases to operate, the master shall exercise due diligence to restore it or cause it to be restored to effective operating condition at the earliest practicable time. The failure of a vessel's radiotelephone equipment shall not, in itself, constitute a violation of this Act, nor shall it obligate the master of any vessel to moor or anchor his vessel; however, the loss of radiotelephone capability shall be given consideration in the navigation of the vessel.

English Language

SEC. 26.07 No person may use the services of, and no person may serve as a person required to maintain a listening watch under section 5 of the Act, 33 U.S.C.A. section 1204 unless he can speak the English language.

Exemption Procedures

SEC. 26.08 (a) Any person may petition for an exemption from any provision of the Act or this part;

(b) Each petition must be submitted in writing to U.S. Coast Guard (M), 400 Seventh Street SW., Washington, DC 20590, and must state:

(1) The provisions of the Act or this part from which an exemption is requested; and

(2) The reasons why marine navigation will not be adversely affected if the exemption is granted and if the exemption relates to a local communication system how that system would fully comply with the intent of the concept of the Act but would not conform in detail if the exemption is granted.

List of Exemptions

SEC. 26.09 (Reserved)

Penalties

SEC. 26.10 Section 9 of this Act states:

(a) Whoever, being the master or person in charge of a vessel subject to the Act, fails to enforce or comply with the Act or the regulations hereunder; or whoever, being designated by the master or person in charge of a vessel subject to the Act to pilot or direct the movement of a vessel fails to enforce or comply with the Act or the regulations hereunder—is liable to a civil penalty of not more than $500 to be assessed by the Secretary.

(b) Every vessel navigated in violation of the Act or the regulations hereunder is liable to a civil penalty of not more than $500 to be assessed by the Secretary, for which the vessel may be proceeded against in any District Court of the United States having jurisdiction.

(c) Any penalty assessed under this section may be remitted or mitigated by the Secretary, upon such terms as he may deem proper.

This amendment shall become effective January 1, 1973.

APPENDIX H

Port and Tanker Safety Act of 1978[1]

Be it enacted by the Senate and House of Representatives of the United States of America in Congress assembled, That this Act may be cited as the "Port and Tanker Safety Act of 1978"

SEC. 2. PORTS AND WATERWAYS SAFETY AND PROTECTION OF THE MARINE ENVIRONMENT.

The Ports and Waterways Safety Act of 1972 (Public Law 92–340, 86 Stat. 424) is amended to read as follows:

"SECTION 1. SHORT TITLE.

"This Act may be cited as the 'Ports and Waterways Safety Act'.

"SEC. 2. STATEMENT OF POLICY.

"The Congress finds and declares—

"(a) that navigation and vessel safety and protection of the marine environment are matters of major national importance;

"(b) that increased vessel traffic in the Nation's ports and waterways creates substantial hazard to life, property, and the marine environment;

"(c) that increased supervision of vessel and port operations is necessary in order to—

"(1) reduce the possibility of vessel or cargo loss, or damage to life, property, or the marine environment;

"(2) prevent damage to structures in, on, or immediately adjacent to the navigable waters of the United States or the resources within such waters;

"(3) insure that vessels operating in the navigable waters of the United States shall comply with all applicable standards and requirements for vessel construction, equipment, manning, and operational procedures; and

"(4) insure that the handling of dangerous articles and substances on the structures in, on, or immediately adjacent to the navigable waters of the United States is conducted in accordance with established standards and requirements; and

"(d) that advance planning is critical in determining proper and adequate protective measures for the Nation's ports and waterways and the marine

[1]Public Law 95–474, October 17, 1978 (Title 33 U.S. Code).

environment, with continuing consultation with other Federal agencies, State representatives, affected users, and the general public, in the development and implementation of such measures.

"SEC. 3. DEFINITIONS.—as used in this Act, unless the context otherwise requires—

"(1) 'Marine environment' means the navigable waters of the United States and the land and resources therein and thereunder; the waters and fishery resources of any area over which the United States asserts exclusive fishery management authority; the seabed and subsoil of the Outer Continental Shelf of the United States, the resources thereof and the waters superjacent thereto; and the recreational, economic, and scenic values of such waters and resources.

"(2) 'Secretary' means the Secretary of the department in which the Coast Guard is operating.

"(3) 'State' includes each of the several States of the United States, the District of Columbia, the Commonwealth of Puerto Rico, the Canal Zone, Guam, American Samoa, the United States Virgin Islands, the Trust Territories of the Pacific Islands, the Commonwealth of the Northern Marianas, and any other commonwealth, territory, or possession of the United States.

"(4) 'United States', when used in geographical context, means all the States thereof.

"SEC. 4. VESSEL OPERATING REQUIREMENTS.

"(a) IN GENERAL.—Subject to the requirements of section 5, the Secretary may—

"(1) in any port or place under the jurisdiction of the United States, in the navigable waters of the United States, or in any area covered by an international agreement negotiated pursuant to section 11, establish, operate, and maintain vessel traffic services, consisting of measures for controlling or supervising vessel traffic or for protecting navigation and the marine environment and may include, but need not be limited to one or more of the following: reporting and operating requirements, surveillance and communications systems, routing systems, and fairways;

"(2) require vessels which operate in an area of a vessel traffic service to utilize or comply with that service;

"(3) require vessels to install and use specified navigation equipment, communications equipment, electronic relative motion analyzer equipment, or any electronic or other device necessary to comply with a vessel traffic service or which is necessary in the interests of vessel safety: *Provided*, That the Secretary shall not require fishing vessels under 300 gross tons or recreational vessels 65 feet or less to possess or use the equipment or devices required by this subsection solely under the authority of this Act;

"(4) control vessel traffic in areas subject to the jurisdiction of the United States which the Secretary determines to be hazardous, or under conditions of reduced visibility, adverse weather, vessel congestion, or other hazardous circumstances by—

"(A) specifying times of entry, movement, or departure;

"(B) establishing vessel traffic routing schemes;

"(C) establishing vessel size, speed, draft limitations and vessel operating conditions; and

"(D) restricting operation, in any hazardous area or under hazardous conditions, to vessels which have particular operating characteristics or capabilities which he considers necessary for safe operation under the circumstances; and

"(5) require the receipt of prearrival messages from any vessel, destined for a port or place subject to the jurisdiction of the United States, in sufficient time to permit advance vessel traffic planning prior to port entry, which shall include any information which is not already a matter of record and which the Secretary determines necessary for the control of the vessel and the safety of the port or the marine environment.

"(b) SPECIAL POWERS.—The Secretary may order any vessel, in a port or place subject to the jurisdiction of the United States or in the navigable waters of the United States, to operate or anchor in a manner he directs if—

"(1) he has reasonable cause to believe such vessel does not comply with any regulation issued under this Act or any other applicable law or treaty;

"(2) he determines that such vessel does not satisfy the conditions for port entry set forth in section 9; or

"(3) by reason of weather, visibility, sea conditions, port congestion, other hazardous circumstances, or the condition of such vessel, he is satisfied that such directive is justified in the interest of safety.

"(c) PORT ACCESS ROUTES.—(1) In order to provide safe access routes for the movement of vessel traffic proceeding to or from ports or places subject to the jurisdiction of the United States, and subject to the requirements of paragraph (3) hereof, the Secretary shall designate necessary fairways and traffic separation schemes for vessels operating in the territorial sea of the United States and in high seas approaches, outside the territorial sea, to such ports or places. Such a designation shall recognize, within the designated area, the paramount right of navigation over all other uses.

"(2) No designation may be made by the Secretary pursuant to this subsection, if such a designation, as implemented, would deprive any person of the effective exercise of a right granted by a lease or permit executed or issued under other applicable provisions of law: *Provided*, That such right has become vested prior to the time of publication of the notice required by clause (A) of paragraph (3) hereof: *Provided further*, That the determination as to whether the designation would so deprive any such person shall be made by the Secretary, after consultation with the responsible official under whose authority the lease was executed or the permit issued.

"(3) Prior to making a designation pursuant to paragraph (1) hereof, and in accordance with the requirements of section 5, the Secretary shall—

"(A) within six months after date of enactment of this Act (and may, from time to time thereafter), undertake a study of the potential traffic density and the need for safe access routes for vessels in any area for which fairways or traffic separation schemes are proposed or which may otherwise be considered and shall publish notice of such undertaking in the Federal Register;

"(B) in consultation with the Secretary of State, the Secretary of the Interior, the Secretary of Commerce, the Secretary of the Army, and the Governors of affected States, as their responsibilities may require, take into account all other uses of the area under consideration (including, as appropriate, the exploration for, or exploitation of, oil, gas, or other mineral resources, the construction or operation of deepwater ports or other structures on or above the seabed or subsoil of the submerged lands or the Outer Continental Shelf of the United States, the establishment or operation of marine or estuarine sanctuaries, and activities involving recreational or commercial fishing); and

"(C) to the extent practicable, reconcile the need for safe access routes with the needs of all other reasonable uses of the area involved.

"(4) In carrying out his responsibilities under paragraph (3), the Secretary shall proceed expeditiously to complete any study undertaken. Thereafter, he shall promptly issue a notice of proposed rule-making for the designation contemplated or shall have published in the Federal Register a notice that no designation is contemplated as a result of the study and the reason for such determination.

"(5) In connection with a designation made pursuant to this subsection, the Secretary—

"(A) shall issue reasonable rules and regulations governing the use of such designated areas, including the applicability of rules 9 and 10 of the International Regulations for Preventing Collisions at Sea, 1972, relating to narrow channels and traffic separation schemes, respectively, in waters where such regulations apply;

"(B) to the extent that he finds reasonable and necessary to effectuate the purposes of the designation, make the use of designated fairways and traffic separation schemes mandatory for specific types and sizes of vessels, foreign and domestic, operating in the territorial sea of the United States and for specific types and sizes of vessels of the United States operating on the high seas beyond the territorial sea of the United States;

"(C) may, from time to time, as necessary, adjust the location or limits of designated fairways or traffic separation schemes, in order to accommodate the needs of other uses which cannot be reasonably accommodated otherwise: *Provided*, That such an adjustment will not, in the judgment of the Secretary, unacceptably adversely affect the purpose for which the existing designation was made and the need for which continues; and

"(D) shall, through appropriate channels, (i) notify cognizant international organizations of any designation, or adjustment thereof, and (ii) take action to seek the cooperation of foreign States in making it mandatory for vessels under their control to use any fairway or traffic separation scheme designated pursuant to this subsection in any area of the high seas, to the same extent as required by the Secretary for vessels of the United States.

"(d) EXCEPTION.—Except pursuant to international treaty, convention, or agreement, to which the United States is a party, this Act shall not apply to any foreign vessel that is not destined for, or departing from, a port or place subject to the jursidiction of the United States and that is in—

"(1) innocent passage through the territorial sea of the United States, or

"(2) transit through the navigable waters of the United States which form a part of an international strait.

"SEC. 5. CONSIDERATIONS BY SECRETARY.

"In carrying out his duties and responsibilities under section 4, the Secretary shall—

"(a) take into account all relevant factors concerning navigation and vessel safety and protection of the marine environment, incuding but not limited to—

"(1) the scope and degree of the risk or hazard involved;

"(2) vessel traffic characteristics and trends, including traffic volume, the sizes and types of vessels involved, potential interference with the flow of commercial traffic, the presence of any unusual cargoes, and other similar factors;

"(3) port and waterway configurations and variations in local conditions of geography, climate, and other similar factors:

"(4) the need for granting exemptions for the installation and use of equipment or devices for use with vessel traffic services for certain classes of small vessels, such as self-propelled fishing vessels and recreational vessels;

"(5) the proximity of fishing grounds, oil and gas drilling and production operations, or any other potential or actual conflicting activity;

"(6) environmental factors;

"(7) economic impact and effects;

"(8) existing vessel traffic services; and

"(9) local practices and customs, including voluntary arrangements and agreements within the maritime community; and

"(b) at the earliest possible time, consult with and receive and consider the views of representatives of the maritime community, ports and harbor authorities or associations, environmental groups, and other parties who may be affected by the proposed actions.

"SEC. 6. WATERFRONT SAFETY.

"(a) IN GENERAL.—The Secretary may take such action as is necessary to—

"(1) prevent damage to, or the destruction of, any bridge or other structure on or in the navigable waters of the United States, or any land structure or shore area immediately adjacent to such waters; and

"(2) protect the navigable waters and the resources therein from harm resulting from vessel or structure damage, destruction, or loss. Such action may include, but need not be limited to—

"(A) establishing procedures, measures, and standards for the handling, loading, unloading, storage, stowage, and movement on the structure (including the emergency removal, control, and disposition) of explosives or other dangerous articles and substances, including oil or hazardous material as those terms are defined in section 4417a of the Revised Statutes, as amended;

"(B) prescribing minimum safety equipment requirements for the structure to assure adequate protection from fire, explosion, natural disaster, and other serious accidents or casualties;

"(C) establishing water or waterfront safety zones, or other measures for limited, controlled, or conditional access and activity when necessary for the protection of any vessel, structure, waters, or shore area; and

"(D) establishing procedures for examination to assure compliance with the requirements prescribed under this section.

"(b) STATE LAW.—Nothing contained in this section, with respect to structures, prohibits a State or political subdivision thereof from prescribing higher safety equipment requirements or safety standards than those which may be prescribed by regulations hereunder.

"SEC. 7. PILOTAGE.

"The Secretary may require federally licensed pilots on any self-propelled vessel, foreign or domestic, engaged in the foreign trade, when operating in the navigable waters of the United States in areas and under circumstances where a pilot is not otherwise required by State law. Any such requirement shall be terminated when the State having jurisdiction over the area involved establishes a requirement for a State licensed pilot and has so notified the Secretary.

"SEC. 8. INVESTIGATORY POWERS.

"(a) SECRETARY.—The Secretary may investigate any incident, accident, or act involving the loss or destruction of, or damage to, any structure subject to this Act, or which affects or may affect the safety or environmental quality of the ports, harbors, or navigable waters of the United States.

"(b) POWERS.—In an investigation under this section, the Secretary may issue subpenas to require the attendance of witnesses and the production of documents or other evidence relating to such incident, accident, or act. If any person refuses to obey a subpena, the Secretary may request the Attorney General to invoke the aid of the appropriate district court of the United States to compel compliance with the subpena. Any district court of the United States may, in the case of refusal to obey a subpena, issue an order requiring compliance with the subpena, and failure to obey the order may be punished by the court as contempt. Witnesses may be paid fees for travel and attendance at rates not exceeding those allowed in a district court of the United States.

"SEC. 9. CONDITIONS FOR ENTRY TO PORTS OF THE UNITED STATES.

"(a) IN GENERAL.—No vessel, subject to the provisions of section 4417a of the Revised Statutes, as amended, shall operate in the navigable waters of the United States or transfer cargo or residue in any port or place under the jurisdication of the United States, if such vessel—

"(1) has a history of accidents, pollution incidents, or serious repair problems which, as determined by the Secretary, creates reason to believe that such vessel may be unsafe or may create a threat to the marine environment; or

"(2) fails to comply with any applicable regulation issued under this Act, under section 4417a of the Revised Statutes, as amended, or under any other applicable law or treaty; or

"(3) discharges oil or hazardous material in violation of any law of the United States or in a manner or quantities inconsistent with the provisions of any treaty to which the United States is a party; or

"(4) does not comply with any applicable vessel traffic service requirements; or

"(5) is manned by one or more officers who are licensed by a certificating state which the Secretary has determined, pursuant to section 4417a(11) of the Revised Statutes, as amended, does not have standards for licensing and certification of seafarers which are comparable to or more stringent than United States standards or international standards which are accepted by the United States; or

"(6) is not manned in compliance with manning levels as determined by the Secretary to be necessary to insure the safe navigation of the vessel; or

"(7) while underway, does not have at least one licensed deck officer on the navigation bridge who is capable of clearly understanding English.

"(b) EXCEPTIONS.—The Secretary may allow provisional entry of a vessel not in compliance with subsection (a), if the owner or operator of such vessel proves, to the satisfaction of the Secretary, that such vessel is not unsafe or a threat to the marine environment, and if such entry is necessary for the safety of the vessel or persons aboard. In addition, paragraphs (1), (2), (3), and (4) of subsection (a) shall not apply if the owner or operator of such vessel proves, to the satisfaction of the Secretary, that such vessel is no longer unsafe or a threat to the marine environment, and is no longer in violation of any applicable law, treaty, regulation or condition, as appropriate. Clauses (5) and (6) of subsection (a) shall become applicable eighteen months after the effective date of this section.

"SEC. 10. APPLICABILITY.

"This Act shall not apply to Panama Canal. The authority granted to the Secretary under sections 4, 5, 6, and 7 of this Act shall not be delegated with respect to the Saint Lawrence Seaway to any agency other than the Saint Lawrence Seaway Development Corporation. Any other authority granted the Secretary under this Act shall be delegated to the Saint Lawrence Seaway Development Corporation to the extent he determines such delegation is necessary for the proper operation of the Saint Lawrence Seaway.

"SEC. 11. INTERNATIONAL AGREEMENTS.

"(a) TRANSMITTAL OF REGULATIONS.—The Secretary shall transmit, via the Secretary of State, to appropriate international bodies or forums, any regulations issued under this Act, for consideration as international standards.

"(b) AGREEMENTS—The President is authorized and encouraged to—

"(1) enter into negotiations and conclude and execute agreements with neighboring nations, to establish compatible vessel standards and vessel traffic services, and to establish, operate, and maintain international vessel traffic services, in areas and under circumstances of mutual concern; and

"(2) enter into negotiations, through appropriate international bodies, and conclude and execute agreements to establish vessel traffic services in appropriate areas of the high seas.

"(c) OPERATIONS.—The Secretary, pursuant to any agreement negotiated under subsection (b) which is binding upon the United States in accordance with constitutional requirements, may—

"(1) require vessels in the vessel traffic service area to utilize or to comply with the vessel traffic service, including the carrying or installation of equipment and devices as necessary for the use of the service; and

"(2) waive, by order or regulation, the application of any United States law or regulation concerning the design, construction, operation, equipment, personnel qualifications, and manning standards for vessels operating in waters over which the United States exercises jurisdiction if such vessel is not en route to or from a United States port or place, and if vessels en route to or from a United States port or place are accorded equivalent waivers of laws and regulations of the neighboring nation, when operating in waters over which that nation exercises jurisdiction.

"SEC. 12. REGULATIONS.

"(a) IN GENERAL.—In accordance with the provisions of section 553 of title 5, United States Code, as amended, the Secretary shall issue, and may from time to time amend or repeal, regulations necessary to implement this Act.

"(b) PROCEDURES.—The Secretary, in the exercise of this regulatory authority, shall establish procedures for consulting with, and receiving and considering the views of all interested parties, including—

"(1) interested Federal departments and agencies,

"(2) officials of State and local governments,

"(3) representatives of the maritime community,

"(4) representatives of port and harbor authorities or associations,

"(5) representatives of environmental groups,

"(6) any other interested parties who are knowledgeable or experienced in dealing with problems involving vessel safety, port and waterways safety, and protection of the marine environment, and

"(7) advisory committees consisting of all interested segments of the public when the establishment of such committees is considered necessary because the issues involved are highly complex or controversial.

"SEC. 13. ENFORCEMENT.

"(a) CIVIL PENALTY.—(1) Any person who is found by the Secretary, after notice and an opportuity for a hearing, to have violated this Act or a regulation issued hereunder shall be liable to the United States for a civil penalty, not to exceed $25,000 for each violation. Each day of a continuing violation shall constitute a separate violation. The amount of such civil penalty shall be assessed by the Secretary, or his designee, by written notice. In determining the amount of such penalty, the Secretary shall take into account the nature, circumstances, extent and gravity of the prohibited acts committed and, with respect to the violator, the degree of culpability, any history of prior offenses, ability to pay, and such other matters as justice may require.

"(2) The Secretary may compromise, modify, or remit, with or without conditions, any civil penalty which is subject to imposition or which has been imposed under this section.

"(3) If any person fails to pay an assessment of a civil penalty after it has become final, the Secretary may refer the matter to the Attorney General of the United States, for collection in any appropriate district court of the United States.

(b) CRIMINAL PENALTY—(1) any person who willfully and knowingly violates this Act or any regulation issued hereunder shall be fined not more than $50,000 for each violation or imprisoned for not more than five years, or both.

"(2) Any person who, in the willfull and knowing violation of this Act or of any regulation issued hereunder, uses a dangerous weapon, or engages in conduct that causes bodily injury or fear of imminent bodily injury to any officer authorized to enforce the provisions of this Act or the regulations issued hereunder, shall, in lieu of the penalties prescribed in paragraph (1), be fined not more than $100,000, or imprisoned for not more than ten years, or both.

"(c) IN REM LIABILITY.—Any vessel subject to the provisions of this Act, which is used in violation of this Act, or any regulations issued hereunder, shall be liable in rem for any civil penalty assessed pursuant to subsection (a) and may be proceeded against in the United States district court for any district in which such vessel may be found.

"(d) INJUNCTION.—The United States district courts shall have jurisdiction to restrain violations of this Act or of regulations issued hereunder, for cause shown.

"(e) DENIAL OF ENTRY.—Except as provided in section 9, the Secretary may, subject to recognized principles of international law, deny entry into the navigable waters of the United States or to any port or place under the jurisdiction of the United States to any vessel not in compliance with the provisions of this Act or the regulations issued hereunder.

"(f) WITHHOLDING OF CLEARANCE.—The Secretary of the Treasury shall withhold or revoke, at the request of the Secretary, the clearance, required by section 4197 of the Revised Statutes of the United States, as amended (46 U.S.C. 91), of any vessel, the owner or operator of which is subject to any of the penalties in this section. Clearance may be granted in such cases upon the filing of a bond or other surety satisfactory to the Secretary.".

The remainder of this act, dealing with such matters as improved pilotage standards and vessels carrying certain kinds of bulk cargo, is not included herein.

APPENDIX I

IMCO Adopted Traffic Separation Schemes

ROUTEING SYSTEMS[1]

The Assembly

NOTING Article 16(i) of the Convention on the Inter-Governmental Maritime Consultative Organization concerning the functions of the Assembly,

RECALLING Regulation 8, Chapter V of the International Convention for the Safety of Life at Sea, 1960, and the amendment thereto adopted by Resolution A.205(VII),

RECALLING FURTHER Resolution A.228(VII) on observance of traffic separation schemes,

NOTING that the International Regulations for Preventing Collision at Sea, 1972, and, in particular Rules 1(d) and 10 thereof provide for adoption by the Organization of, and the behaviour of vessels in or near, traffic separation schemes,

RECOGNIZING that there is a need to bring the terms, definitions and general principles concerning traffic separation and routeing, as set out in Annex II to Resolution A.161(ES.IV), into harmony with the International Regulations for Preventing Collisions at Sea, 1972,

RECOGNIZING ALSO that the practice of complying with routeing measures adopted by IMCO for international use would contribute considerably to the avoidance of collisions between ships,

RECOGNIZING FURTHER that such practice would consequently reduce the risk of pollution of the marine environment and the risk of damage to marine life resulting from collisions or standings,

CONFIRMING that IMCO is recognized as the only international body for establishing and adopting routeing measures on an international level,

NOTING that the Ninth International Hydrographic Conference charged the International Hydrographic Bureau to deal with matters relating to presentation on the charts and in sailing directions, details of routeing provisions which have been considered, approved and adopted by IMCO for international use,

[1] IMCO Resolution A.284(VIII), adopted 20 Nov 73.

HAVING CONSIDERED the Recommendations by the Maritime Safety Committee at its twenty-fifth, twenty-seventh, and twenty-eight sessions,

RESOLVES:

(a) to adopt the general provisions pertaining to Ships' Routeing approved by the Maritime Safety Committee at its twenty-seventh and twenty-eighth sessions, the text of which appears at Annex I to this Resolution, as a substitute for the terms, definitions and general principles covering traffic separation and routeing set out in Annex II to Resolution A.161(ES.IV);

(b) to adopt the routeing measures approved by the Maritime Safety Committee at its twenty-fifth, twenty-seventh and twenty-eighth sessions, the text of which appears at Annex II to this Resolution,

REQUESTS the Maritime Safety Committee to revise and update as necessary the publication on "Ships' Routeing" to reflect the decisions taken in the foregoing part of this Resolution and to approve new routeing measures and revisions, cancellations and suspensions of routeing measures previously adopted by the Organization and to submit recommendations thereon to the Assembly for adoption,

INVITES the governments concerned to advise ships under their flag to comply with the adopted routeing measures,

URGES governments, when planning either to introduce new traffic separation schemes similar to those included in the IMCO publication on "Ships' Routeing" or to amend existing schemes in that publication, to consult the Organization in advance whenever practicable,

REQUESTS the Secretary-General to advise the International Hydrographic Bureau on details of the routeing provisions to facilitate the hydrographers' work on inclusion of this material in the appropriate nautical charts and related publications for the use of mariners,

REVOKES the following Resolution by which the Assembly adopted terms, definitions and general principles concerning traffic separation and various traffic separation schemes and areas to be avoided: Resolutions A.90(IV), A.161(ES.IV), A.186(VI), A.226(VII) and A.227(VII).

SHIPS' ROUTEING[2]

Part I General Provisions

Adoption and Recommendation

1. IMCO is recognized as the only international body responsible for establishing and recommending measures on an international level concerning ships' routeing.

2. In deciding whether or not to adopt a traffic separation scheme, IMCO will consider:

(a) whether the aids to navigation proposed will enable mariners to determine their position with sufficient accuracy to navigate in the scheme in accordance with the principles regarding the use of Routeing Schemes;

(b) whether or not the scheme complies with the established Methods of Routeing.

3. Having due regard to paragraph 5, a Government shall when establishing, reviewing or adjusting a routeing system, take due account of:

[2]'Ships' Routeing', Third Edition, 1973, Supplement 1975

(a) the rights and practices of Governments in respect of the exploitation of living and mineral resources of the high seas and of the sea-bed and subsoil underlying the high seas;

(b) the environment, traffic patterns or established routeing systems in the waters under such Government's jurisdiction;

(c) the aides to navigation already established in the area, and the effect the routeing system may have upon demands for hydrographic surveys and for improvements or adjustments in the navigation aids provided in the waters concerned.

4. IMCO shall not adopt or amend any scheme that is in the proximity of waters under a Government's jurisdiction without the agreement of that government, where that scheme may affect:

(a) the rights and practices of such Government in respect of the exploitation of living and mineral resources of the high seas and of the sea-bed and subsoil underlying the high seas;

(b) the environment, traffic patterns or established routeing systems in the waters under such Government's jurisdiction;

(c) demands for improvements or adjustments in the navigation aids provided in the waters concerned.

5. (a) A Government proposing a routeing system, any part of which lies within international waters, should consult with IMCO, so that such system may be adopted by IMCO for internation use.

(b) A Government may establish or adjust a routeing system lying partly within international waters, before consulting with IMCO, where local conditions require that early action be taken, with a view to later adoption by the Organization.

(c) A Government, when proposing or establishing a traffic separation scheme, should be guided by the following criteria, having due regard to the class of vessel for which the scheme is intended:

(i) the availability of visual aids to navigation, or

(ii) the possibility of position-fixing by the use of direction finder or radar.

6. When establishing areas to be avoided by certain ships, the necessity for creating such areas should be well established and the reasons stated. In general, these areas should be established only in places where inadequate survey or insufficient provision of aids to navigation may lead to danger of stranding, or where local knowledge is considered essential for safe passage or where there is the possibility of unacceptable damage to wildlife, which may result from a casualty. These areas shall not be regarded as prohibited areas unless specifically stated otherwise; the classes of ships which should avoid the areas should be considered in each particular case.

7. Routeing systems should be reviewed, resurveyed and adjusted as necessary, so as to maintain their effectiveness and compatibility with trade patterns, resource exploitation, changes in depth of water, and other developments.

8. Except where local conditions require that early action be taken, a routeing system adopted by IMCO should not come into force before a period of three months has elapsed since the date of adoption by the Assembly.

9. Nothing in the foregoing shall affect the rights, claims or views of any Government in regard to the limits of territorial waters.

Terminology and Symbols

1. The following terms and symbols are used . . . in connexion with matters related to ships' routeing:

(a) *Routeing*

A complex of measures concerning routes aimed at reducing the risk of casualties; it includes traffic separation schemes, two-way routes, tracks, areas to be avoided, inshore traffic zones and deep water routes.

(b) *Traffic separation scheme*

A scheme which separates traffic proceeding in opposite or nearly opposite directions by the use of a separation zone or line, traffic lanes or by other means.

(c) *Separation zone or line*

A zone or line separating traffic proceeding in one direction from traffic proceeding in another direction. A separation zone may also be used to separate a traffic lane from the adjacent inshore traffic zone.

(d) *Traffic lane*

An area within definite limts inside which one-way traffic is established.

(e) *Roundabout*

A circular area within definite limits in which traffic moves in a counterclockwise direction around a specified point or zone.

(f) *Inshore traffic zone*

A designated area between the landward boundary of a traffic separation scheme and the adjacent coast intended for coastal traffic.

(g) *Two-way route*

A route in an area within definite limits inside which two-way traffic is established.

(h) *Track*

The recommended route to be followed when proceeding between predetermined positions.

(i) *Deep water route*

A route in a designated area within definite limits which has been accurately surveyed for clearance of sea bottom and submerged obstacles to a minimum indicated depth of water.

2. The symbols in the following table are those recommended by the International Hydrographic Organization for representation of details of routeing measures on nautical charts. They are included in this publication for readers' information on what may be generally found in charts. Individual countries may, however, use on their charts symbols different from those given below.

Table 1

Detail	Presentation	Description
1. Outside limit of traffic lanes, two-way routes and inshore traffic zones	– – – – – – –	Dashed line – the symbol used for maritime limits in general
2. Outside limit of "roundabout"(1)		
3. Separation zone (2) (of any shape)		The zone shall be indicated by means of a tint light enough to reveal any hydrographic details
4. Separation line		A single tinted line

5. Centre of "roundabout" with no separation zone inside		A circle
6. Arrows indicating direction of traffic flow(3)	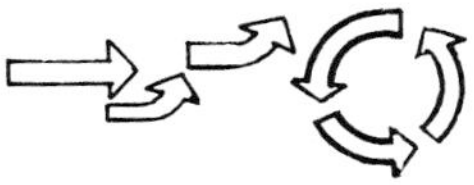	Open-outlined arrows so situated and shaped as to indicate general directions of traffic flow
7. Boundary of "areas to be avoided by ships of certain classes"(4)	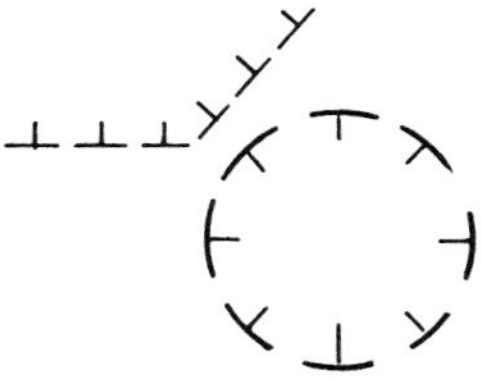	A line composed of a series of T-shaped signs, the cross-bar of the T being long and the down stroke short and pointing toward the area in question, within which a suitable legend may be inscribed
8. Limit of sea exploration and/or exploitation regions which may be dangerous for free navigation		
9. Recommended track when based on a system of fixed marks		A single or double continuous line
10. Recommended track when not based on a system of fixed marks	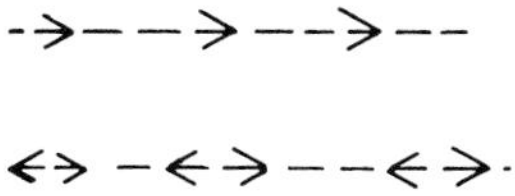	A single dashed line in which arrowheads are inserted at regular intervals, either singly to indicate a one-way track, or in opposing pairs to indicate a two-way track
11. Outside limit of deep water route, when depicted	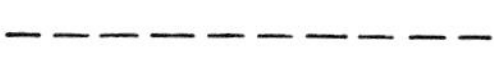	A dashed line
12. Deep water route when both outside limits are depicted	 DW DW	Dashed lines and the letters DW inserted at regular intervals between them. The minimum depth shall be inserted beside the abbreviation when considered critical
13. Deep water route, based on fixed marks	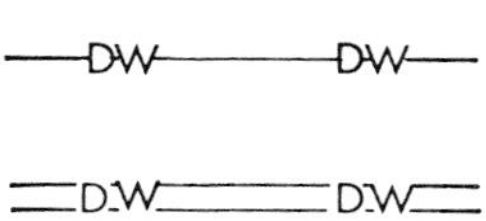 DW DW D.W D.W	A double or single continuous line with the abbreviation DW inserted at regular intervals. The minimum depth shall be indicated beside the abbreviation when considered critical. When using this symbol, the direction of traffic flow shall be indicated conventionally

14. Deep water route not based on fixed marks, direction of traffic flow	- - > -DW> —DW> — — <>DW< > —DW-	A single dashed line in which arrowheads are inserted at regular intervals, either singly to represent a one-way route, or in opposing pairs, to represent a two-way route, The abbreviation DW shall also be inserted at regular intervals along the symbol, and the minimum depth indicated beside the abbreviation when considered critical

Remarks
(1) The dashed line, representing outside limits of "roundabout" should be interrupted in places where ships are recommended to enter or to leave the scheme.
(2) In places where traffic is separated by natural features (islands, marked shoals, etc.) representation of the separation zone may be omitted.
(3). Dispersion of arrows, instead of placing them in a line is felt desirable.
(4) Notes on conditions of avoidance (classes and sizes of ships, nature of cargoes carried, etc.) may be given on charts and shall always be given in Sailing Directions.

General Observations
The routeing and traffic separation symbols to be used on charts should be printed in colour, preferably magenta.

Secondary details of routeing and traffic separation, such as figures indicating directions of traffic, schemes and their details, dimensions, distances from coast, etc., should not be shown on charts unless considered critical. These are given in this IMCO publication and may be given in Sailing Directions if so decided by hydrographic offices.

Methods of Routeing

1. When establishing routeing systems the following are among the methods which may be used;

(a) separation of traffic by separation zones or lines;

(b) separation of traffic by natural obstacles and geographically defined objects;

(c) separation of traffic by inshore traffic zones intended for keeping coastal traffic away from traffic separation schemes;

(d) separation of traffic by sectors at approaches to focal points;

(e) separation of traffic by roundabouts intended to facilitate navigation at focal points, where traffic separation schemes meet;

(f) routeing of traffic by deep water routes, two-way routes or tracks for ships proceeding in specific directions.

2. A description of methods (a) to (e) with drawings intended only to explain their function is given in the following:

(a) *By separation zones or lines* (Fig. 1)

In such cases, the separation of traffic is achieved by a separation zone or line between streams of traffic proceeding in opposite or nearly opposite directions. The outside limits in such a scheme are the outer boundaries of lanes intended for one-way traffic. Beyond such limits ships can navigate in any direction. A separation zone may also be used to separate a traffic lane from an inshore traffic zone.

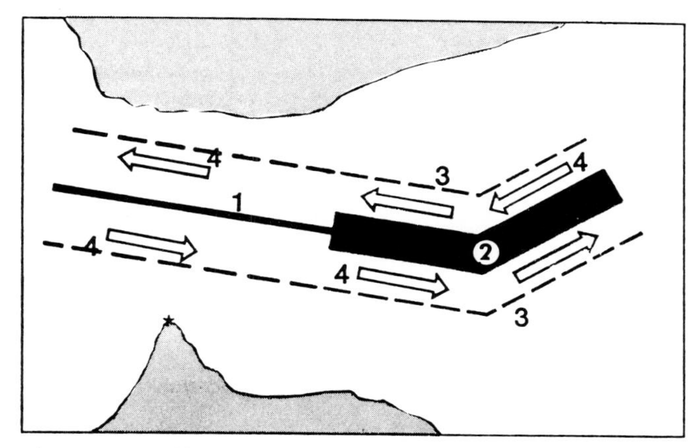

Fig. I 1. Traffic separation by separation line and zone.
1—Separation line
2—Separation zone
3—Outside limits of lanes
4—Arrows indicating main traffic direction

The width and length of separation zones and traffic lanes are determined after careful examination of local conditions, traffic density, prevailing hydrographic and meteorological conditions, space available for manoeuvring, etc., and generally their length is kept to the minimum necessary. In narrow passages and restricted waters a separation line may be adopted instead of a zone, for the separation of traffic, to allow for more navigable space.

(b) *By natural obstacles and geographically defined objects* (Fig. 2)

This method is used where there is a defined area with obstacles such as islands, shoals or rocks restricting free movement and providing a natural division for opposing traffic streams.

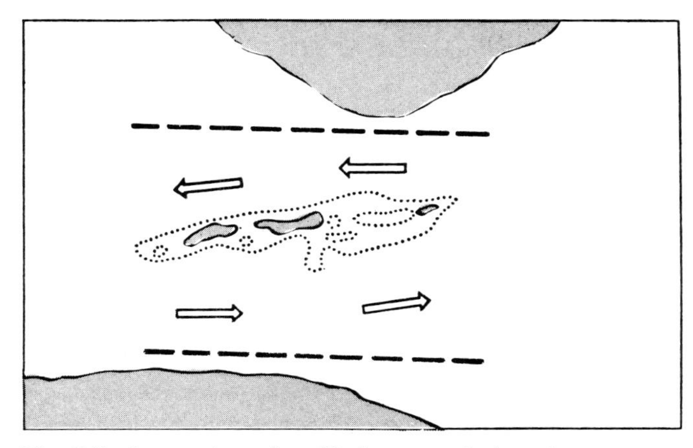

Fig. I 2. Separation of traffic by natural obstacles.

(c) *By inshore traffic zones* (Fig. 3)

By using inshore traffic zones coastal shipping can keep clear of through traffic in the adjacent traffic separation scheme. Ships navigating in any direction may be encountered in an inshore traffic zone.

(d) *By sectors at approaches to focal points* (Fig. 4)

Such a method is used where ships converge at a point or a small area from various directions. Port approaches, sea pilot stations, positions where landfall buoys or light vessels are fixed, entrances to channels, canals, estuaries, etc., may be considered as such focal points. The number of shipping lanes, their dimensions and directions depend mainly on the type of the local traffic.

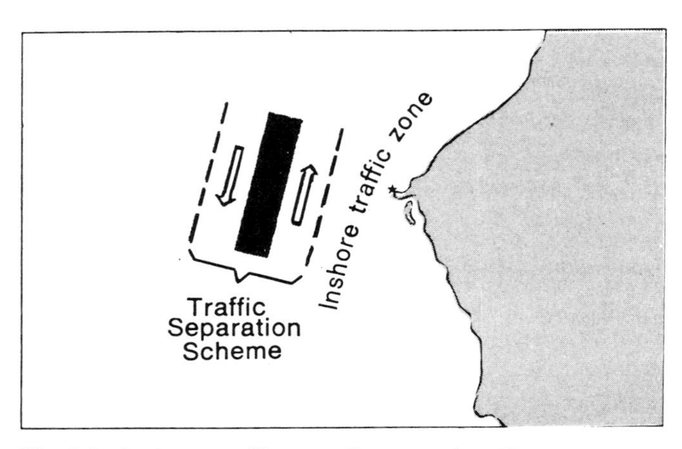

Fig. I 3. Inshore traffic zone for coastal traffic.

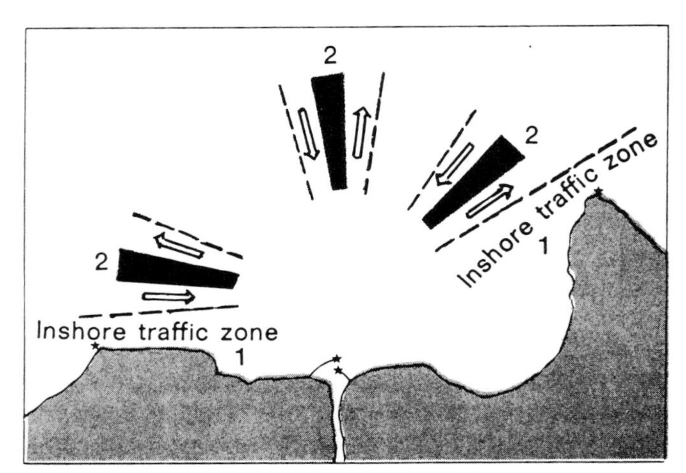

Fig. I 4. Sectorial traffic separation scheme at approaches to focal point.
1—Inshore traffic zone
2—Separation schemes for main traffic

(e) *By roundabouts* (Fig. 5)

To facilitate navigation at focal points where several traffic separation schemes meet, ships should move in a counter-clockwise direction around a specified point or zone until they are able to join the appropriate lane.

General Principles of Ships' Routeing

The Use of Routeing Systems

1. The International Regulations for Preventing Collisions at Sea apply to navigation in routeing systems.

2. Routeing systems are intended for use by day and by night in all weathers, in ice-free waters or under light ice conditions where no extraordinary manoeuvres or assistance by icebreaker(s) are required.

3. Routeing systems are recommended for use by all ships unless stated otherwise.

4. A deep water route is primarily intended for use by ships which because of their draught in relation to the available depth of water in the area concerned require the use of such a route. Through traffic to which the above consideration does not apply should, if practicable, avoid following deep water routes. When using a deep water route mariners should be aware of possible changes in the indicated depth of water due to meteorological or other effects.

5. A vessel using a traffic separation scheme shall:

(i) proceed in the appropriate traffic lane in the general direction of traffic flow for that lane;

(ii) so far as practicable keep clear of a traffic separation line or separation zone;

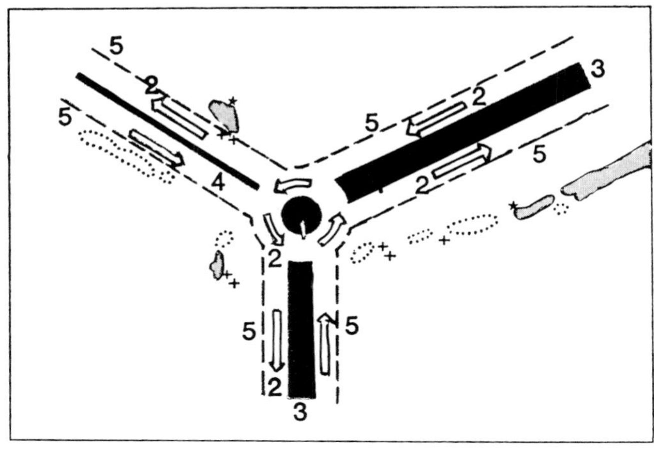

Fig. I 5. A roundabout where several traffic separation schemes meet.
1—Circular separation zone
2—Arrows indicating traffic direction
3—Separation zone
4—Separation line
5—Outside limits of lanes

(iii) normally join or leave a traffic lane at the termination of the lane, but when joining or leaving from the side shall do so at as small an angle to the general direction of traffic flow as practicable.

6. A vessel shall so far as practicable avoid crossing traffic lanes, but if obliged to do so shall cross as nearly as practicable at right angles to the general direction of traffic flow.

7. Inshore traffic zones shall not normally be used by through traffic which can safely use the appropriate traffic lane within the adjacent traffic separation scheme.

8. A vessel, other than a crossing vessel, shall not normally enter a separation zone or cross a separation line except:

(i) in cases of emergency to avoid immediate danger;

(ii) to engage in fishing within a separation zone.

9. A vessel navigating in areas near the terminations of traffic separation schemes shall do so with particular caution.

10. a vessel shall so far as practicable avoid anchoring in a traffic separation scheme or in areas near its terminations.

11. A vessel not using a traffic separation scheme shall avoid it by as wide a margin as is practicable.

12. The arrows printed on charts merely indicate the general direction of traffic; ships need not set their courses strictly along the arrows.

13. The signal "YG" meaning "You appear not to be complying with the traffic separation scheme" is provided in the International Code of Signals for appropriate use.

Part II Traffic Separation Schemes (Excerpts from)[2]

Navigation in the Vicinity of the Grand Banks of Newfoundland

Attention is drawn to Regulation 8 of Chapter V of the Convention for the Safety of Life at Sea, 1960. It directs that all ships proceeding on voyages in the vicinity of the Grand Banks of Newfoundland avoid as far as practicable the fishing banks of Newfoundland north of latitude 43°N. The reasons for avoiding the area are:

(a) high concentration of fishing vessels;

(b) prevailing adverse weather conditions;

(c) seasonal existence of icebergs.

In the Approaches to Chedabucto Bay

Description of the Traffic Separation Scheme

The traffic separation scheme of Chedabucto Bay consists of three parts.

Part I:

(a) A separation zone bounded by a line conecting the following geographical positions:

[2]The traffic separation schemes presented in this part are for general information only. Some geographic positions are expected to change as a result of port access studies being conducted by the Coast Guard. Mariners should refer to the latest edition of the applicable chart for current information.

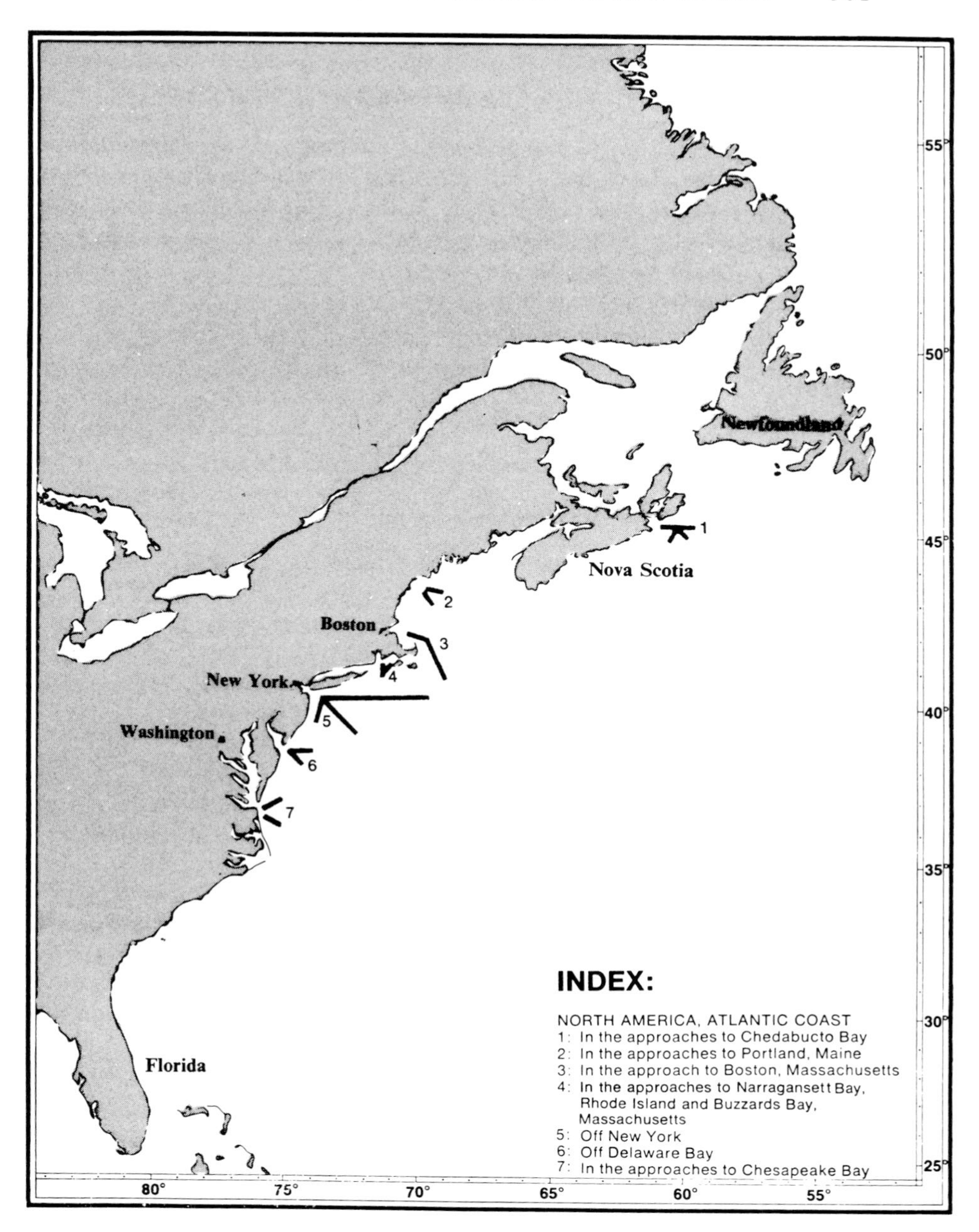

Fig. I 6. North America, Atlantic Coast.

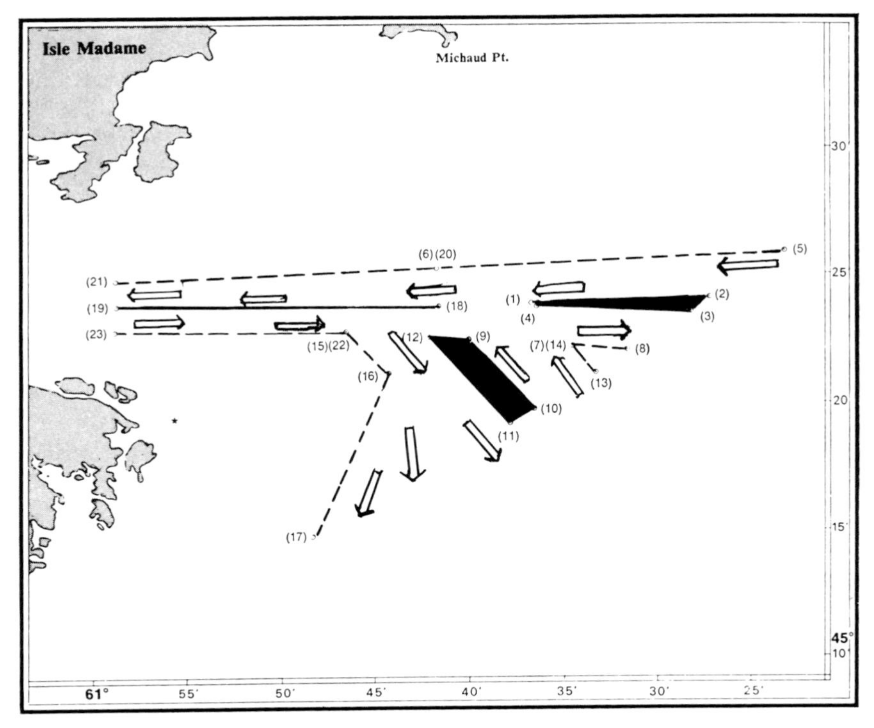

Fig. I 7. In approaches to Chedabucto Bay.

(1)	45°24′00″N.,	60°36′42″W.
(2)	45°24′12″N.,	60°27′10″W.
(3)	45°23′42″N.,	60°28′12″W.
(4)	45°23′49″N.,	60°36′29″W.

(b) A traffic lane for westbound traffic is established between the separation zone and a line connecting the following geographical positions:

(5)	45°26′00″N.,	60°23′12″W.
(6)	45°25′26″N.,	60°41′42″W.

(c) A traffic lane for eastbound traffic is established between the separation zone and a line connecting the following geographic positions:

(7)	45°22′18″N.,	60°34′30″W.
(8)	45°22′09″N.,	60°31′36″W.

The main traffic directions are;

092°–267°.

Part II:

(a) a separation zone bounded by a line connecting the following geographical positions:

(9)	45°22′34″N.,	60°40′00″W.
(10)	45°19′53″N.,	60°36′30″W.
(11)	45°19′18″N.,	60°37′48″W.
(12)	45°22′41″N.,	60°42′10″W.

(b) A traffic lane for north-westbound traffic is established between the separation zone and a line connecting the following geographical positions:

(13) 45°21′21″N., 60°33′18″W.
(14) 45°22′18″N., 60°34′30″W.

The main traffic direction is 318°.

(c) A traffic lane for southbound traffic is established between the separation zone and a line connecting the following geographical positions:

(15) 45°22′54″N., 60°46′30″W.
(16) 45°21′17″N., 60°44′24″W.
(17) 45°14′28″N., 60°48′23″W.

The main traffic directions are;

138° and 202°.

Part III:

(a) A separation line connects the following geographical positions:

(18) 45°23′54″N., 60°41′42″W.
(19) 45°23′54″N., 60°58′48″W.

(b) A traffic lane for westbound traffic established between the separation line and a line connecting the following geographic positions:

(20) 45°25′26″N., 60°41′42″W.
(21) 45°24′54″N., 60°58′48″W.

(c) A traffic lane for eastbound traffic is established between the separation line and a line connecting the following geographical positions:

(22) 45°22′54″N., 60°46′30″W.
(23) 45°22′54″N., 60°58′48″W.

The main traffic directions are:

090°–270°.

In the Approaches to Portland, Maine

Description of the Traffic Separation Scheme

The traffic separation scheme in the approaches to Portland, Maine, consists of two parts:

Part I—Eastern approach

A separation zone, one mile wide, is centred upon the following geographical positions;

(1) 43°30′.2N., 69°59′.4W.
(2) 43°24′.75N., 69°33′.0W.

A traffic lane, two miles wide, is established on each side of the separation zone.
The main traffic directions are:

107° and 287°.

Part II—Southern approach

A separation zone, one mile wide, is centred upon the following geographical positions:

(3) 43°26′.8N., 70°03′.5W.
(4) 43°07′.8N., 69°55′.3W.

A traffic lane, two miles wide, is established on each side of the separation zone.
The main traffic directions are:

162° and 342°.

Note:

Precautionary area

A precautionary area of radius five miles is centred upon geographical position 43°31′.5N., 70°06′.0W.

In the Approach to Boston, Massachusetts

Description of the Traffic Separation Scheme

A separation zone, one mile wide, is centred upon the following geographical positions:

(1)	42°21′.0N.,	70°40′.7W.
(2)	42°08′.5N.,	69°53′.6W.
(3)	40°49′.5N.,	69°00′.0W.

A traffic lane, two miles wide, is established on each side of the separation zone.

The main traffic directions are:

110°–290° and
153°–333°.

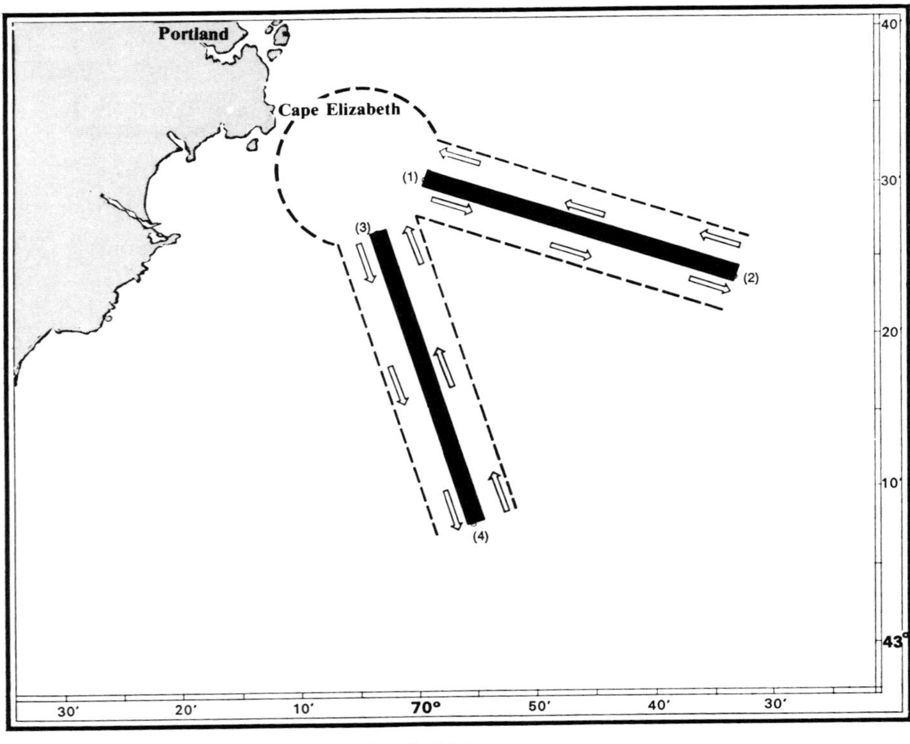

Fig. I 8. In approaches to Portland, Maine.

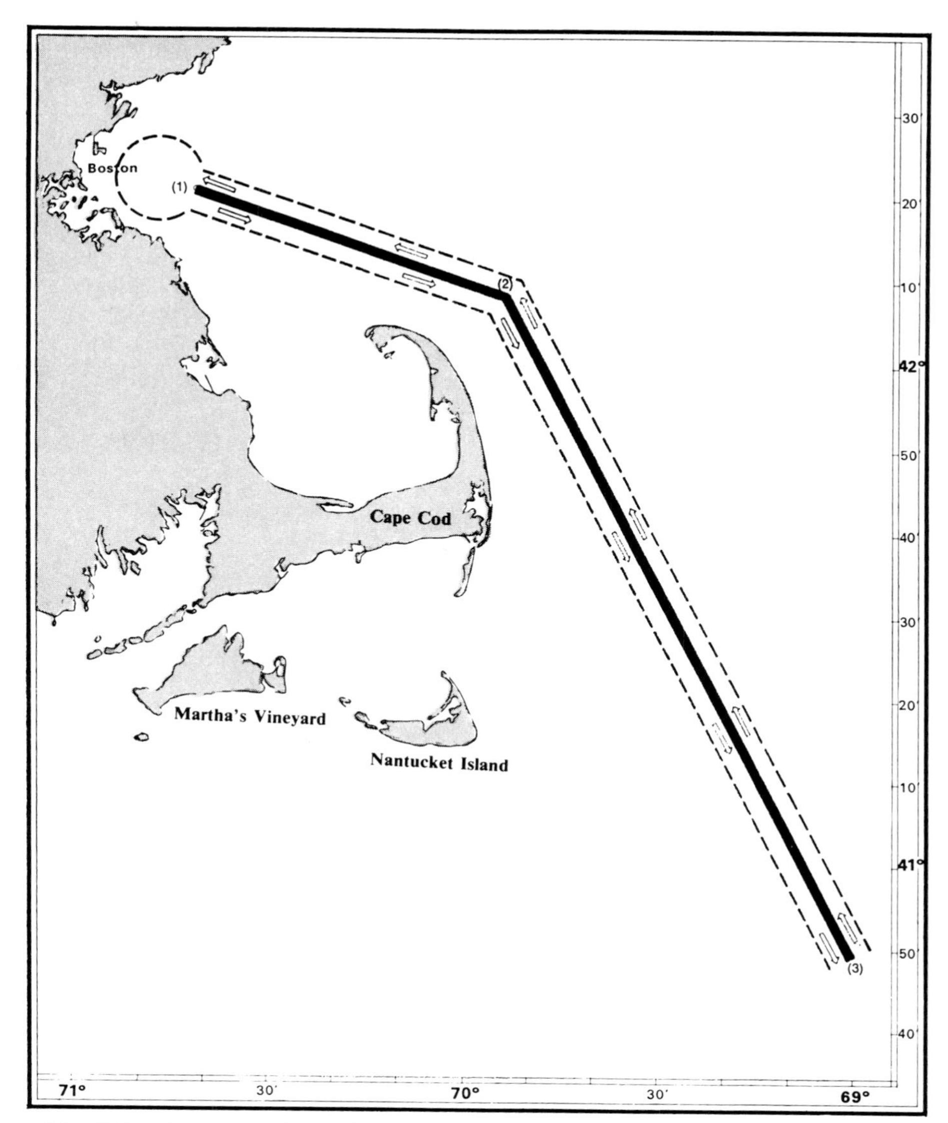

Fig. I 9. In approaches to Boston, Massachusetts.

Note:

Precautionary area

A precautionary area of radius five miles is centred upon geographical position 42°22′.7N., 70°48′.0W.

In the Approaches to Narragansett Bay, Rhode Island and Buzzards Bay, Massachusetts

Description of the Traffic Separation Scheme

The traffic separation scheme in the approaches to Narragansett Bay, Rhode Island, and Buzzards Bay, Massachusetts, consists of two parts:

Part I—Narragansett Bay approach

A separation zone, one mile wide, is centred upon the following geographical positions:

(1)	41°22′.7N.,	71°23′.4W.
(2)	41°11′.1N.,	71°23′.4W.

A traffic lane, one mile wide, is established on each side of the separation zone.

The main traffic directions are;

000° and 180°.

Part II—Buzzards Bay approach

A separation zone, one mile wide, is centred upon the following geographical positions:

(3)	41°10′.15N.,	71°19′.15W.
(4)	41°24′.9N.,	71°03′.9W.

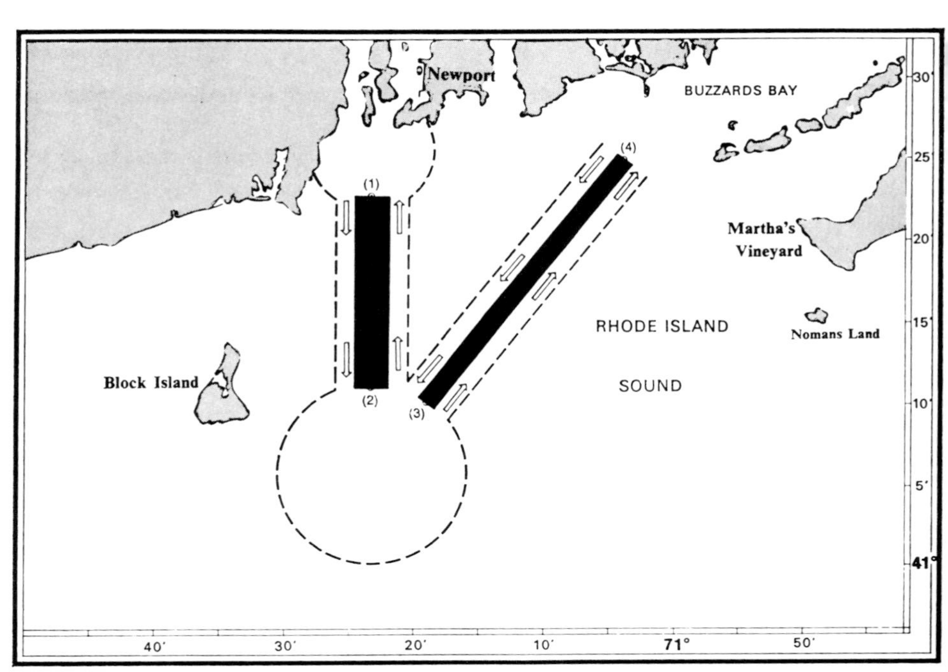

Fig. I 10. In approaches to Narragansett Bay, Rhode Island, and Buzzards Bay, Massachusetts.

A traffic lane, one mile wide, is established on each side of the separation zone. The main traffic directions are:

038° and 218°.

Note:

Precautionary areas

A precautionary area of radius 5.4 miles is centred upon geographical position 41°06′.0N., 71°23′.4W.

A precautionary area of radius 3.55 miles is centred upon geographical position 41°25′.6N., 71°23′.4W.

Restricted area

A restricted area, two miles wide, extending from the northern limit of the Narragansett Bay approach traffic separation zone to latitude 41°24′.7N. has been established.

The restricted area within the precautionary area will only be closed to vessel traffic by the Naval Underwater System Center during periods of daylight and optimum weather conditions for torpedo range usage. The closing of the restricted area will be indicated by the activation of a white strobe light mounted on Brenton Reef Light and controlled by a Naval vessel supporting the torpedo range activities. There would be no vessel restrictions expected during inclement weather or when the torpedo range is not in use.

Off New York

NOTE: Under review—possible insufficient navigational marking in the eastern and southeastern approaches.

Description of the Traffic Separation Scheme

The traffic separation scheme off New York consists of three parts.

Part I—Eastern approach

(a) A separation zone bounded by a line connecting the following geographical positions:

(1) 40°28′.5N., 69°27′.9W.
(2) 40°24′.2N., 73°11′.5W.
(3) 40°26′.0N., 73°40′.8W.
(4) 40°27′.0N., 73°40′.7W.
(5) 40°27′.0N., 73°11′.5W.
(6) 40°31′.5N., 69°28′.1W.

(b) A traffic lane for westbound traffic is established between the separation zone and a line connecting the following geographical positions:

(7) 40°36′.5N., 69°28′.2W.
(8) 40°32′.2N., 73°11′.5W.
(9) 40°27′.9N., 73°40′.6W.

(c) A traffic lane for eastbound traffic is established between the separation zone and a line connecting the following geographical positions:

(10) 40°25′.0N., 73°41′.2W.
(11) 40°19′.2N., 73°11′.5W.
(12) 40°23′.5N., 69°27′.8W.

Part II—South-eastern approach

(a) A separation zone bounded by a line connecting the following geographical positions:

(13) 39°20′.7N., 72°18′.0W.
(14) 40°06′.3N., 73°22′.7W.
(15) 40°22′.4N., 73°43′.5W.
(16) 40°23′.0N., 73°42′.7W.
(17) 40°08′.6N., 73°20′.1W.
(18) 39°23′.0N., 72°15′.2W.

(b) A traffic lane for north-westbound traffic is established between the separation zone and a line connecting the following geographical positions:

(19) 39°26′.7N., 72°10′.8W.
(20) 40°12′.2N., 73°15′.7W.
(21) 40°24′.0N., 73°41′.9W.

(c) A traffic lane for south-eastbound traffic is established between the separation zone and a line connecting the following geographical positions:

(22) 40°21′.7N., 73°44′.5W.
(23) 40°02′.7N., 73°27′.2W.
(24) 39°17′.0N., 72°22′.4W.

Part III—Southern approach

(a) A separation zone bounded by a line connecting the following geographical positions:

(25) 39°45′.7N., 73°48′.0W.
(26) 40°20′.5N., 73°48′.3W.
(27) 40°20′.7N., 73°47′.0W.
(28) 39°45′.7N., 73°44′.0W.

(b) A traffic lane for northbound traffic is established between the separation zone and a line connecting the following geographical positions:

(29) 39°45′.7N., 73°37′.7W.
(30) 40°21′.2N., 73°45′.8W.

(c) A traffic lane for southbound traffic is established between the separation zone and a line connecting the following geographical positions:

(31) 40°20′.4N., 73°49′.6W.
(32) 39°45′.7N., 73°54′.4W.

Note:

Precautionary area

A precautionary area of radius seven miles is centred upon the Ambrose Light in geographical position 40°27′.5N., 73°49′.9W.

Off Delaware Bay

Description of the Traffic Separation Scheme

The traffic separation scheme of Delaware Bay consists of two parts.

Part I—Eastern Approach

(a) A separation zone bounded by a line connecting the following geographical positions:

(1) 38°46′.8N., 74°34′.6W.
(2) 38°46′.8N., 74°55′.7W.
(3) 38°47′.8N., 74°55′.4W.
(4) 38°47′.8N., 74°34′.6W.

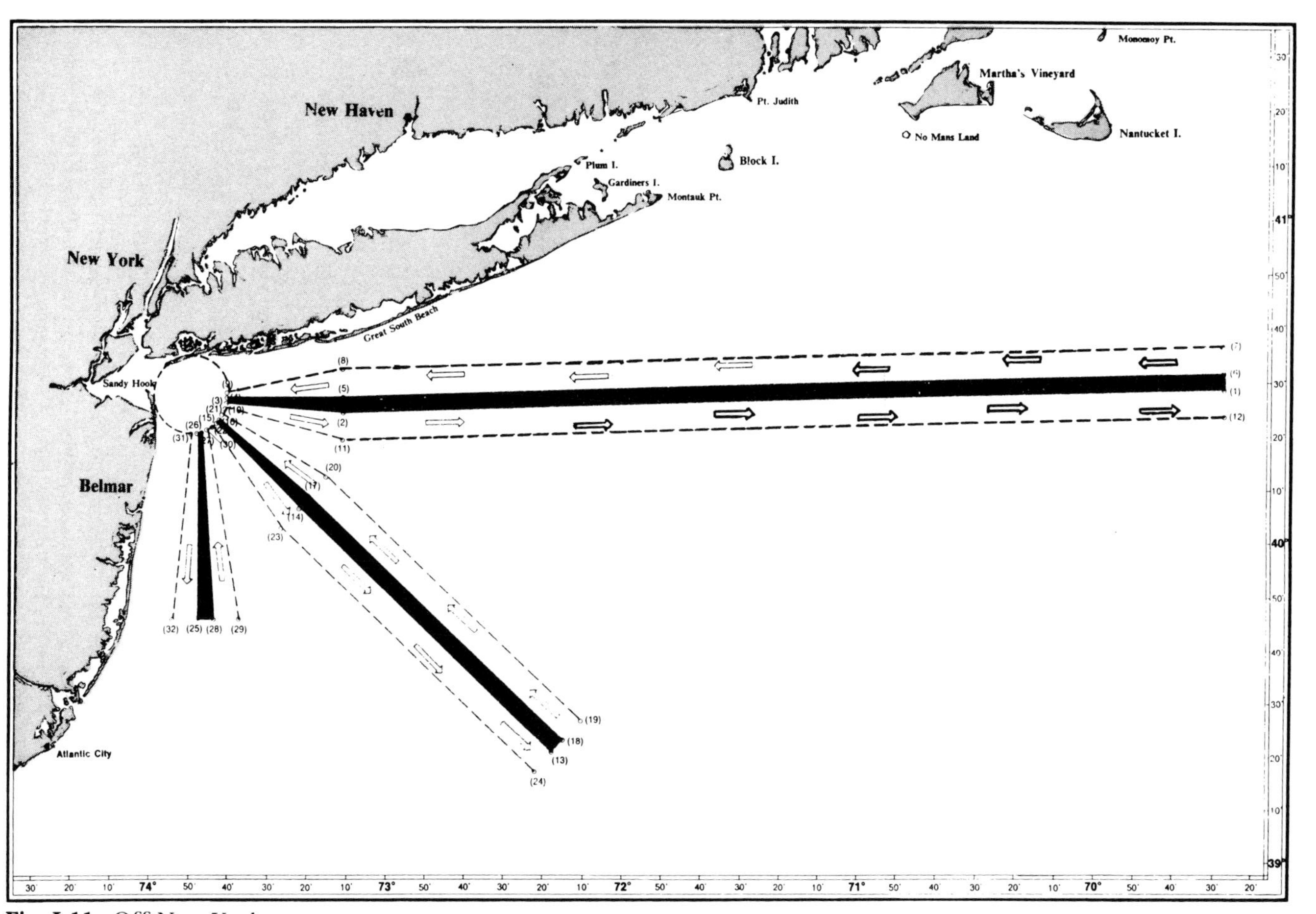

Fig. I 11. Off New York.

(b) A traffic lane for westbound traffic is established between the separation zone and a line connecting the following geographical positions:

(5) 38°49′.8N., 74°34′.6W.
(6) 38°48′.8N., 74°55′.3W.

(c) A traffic lane for eastbound traffic is established between the separation zone and a line connecting the following geographical positions:

(7) 38°45′.8N., 74°56′.1W.
(8) 38°44′.8N., 74°34′.6W.

Part II—South-eastern approach

(a) A separation zone bounded by a line connecting the following geographical positions:

(9) 38°27′.0N., 74°35′.6W.
(10) 38°43′.4N., 74°58′.0W.
(11) 38°44′.2N., 74°57′.2W.
(12) 38°27′.6N., 74°34′.6W.

(b) A traffic lane for north-westbound traffic is established between the separation zone and a line connecting the following geographical positions:

(13) 38°29′.1N., 74°32′.9W.
(14) 38°45′.1N., 74°56′.6W.

(c) A traffic lane for south-eastbound traffic is established between the separation zone and a line connecting the following geographical positions:

(15) 38°42′.8N., 74°58′.9W.
(16) 38°27′0N., 74°39′.2W.

Note:

Precautionary area

A precautionary area of radius eight miles is centered upon Harbour of Refuge Light in geographical position 38°48′.9N., 75°05′.6W.

In the Approaches to Chesapeake Bay

Description of the Traffic Separation Scheme

The traffic separation scheme in the approaches to Chesapeake Bay consists of two parts.

Part I—Eastern approach

A separation line connects the following geographical positions:

(1) 36°58′.7N., 75°48′.7W.
(2) 36°56′.5n., 75°56′.3W.

A traffic lane, half a mile wide, is established on each side of the separation line.

The main traffic directions are:

070° and
250°.

Part II—Southern approach

A separation line connects the following geographical positions:

(3) 36°51′.3N., 75°50′.9W.
(4) 36°55′.5N., 75°56′.6W.

A traffic lane, half a mile wide, is established on each side of the separation line.

The main traffic directions are:

132° and
312°.

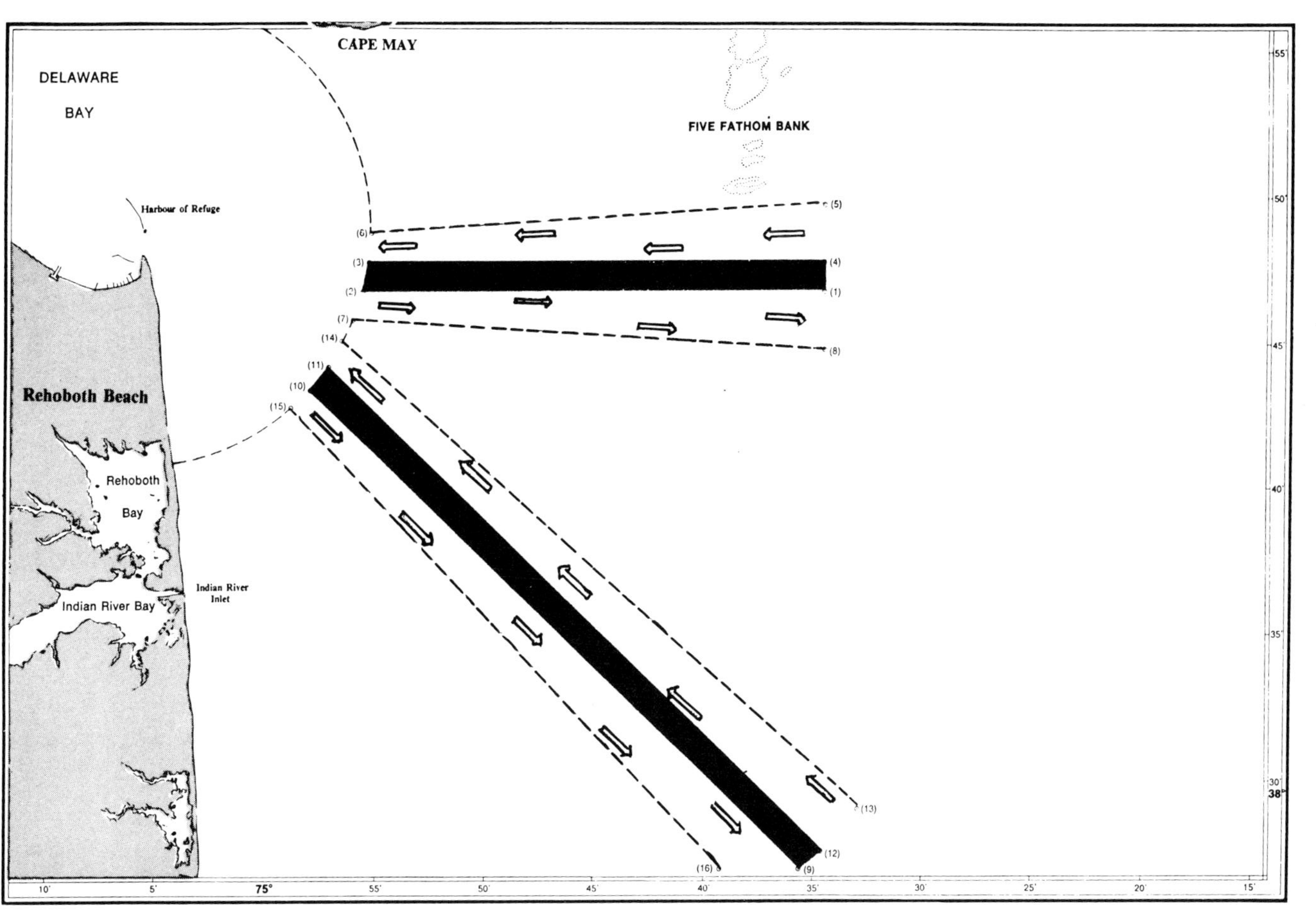

Fig. I 12. Off Delaware Bay.

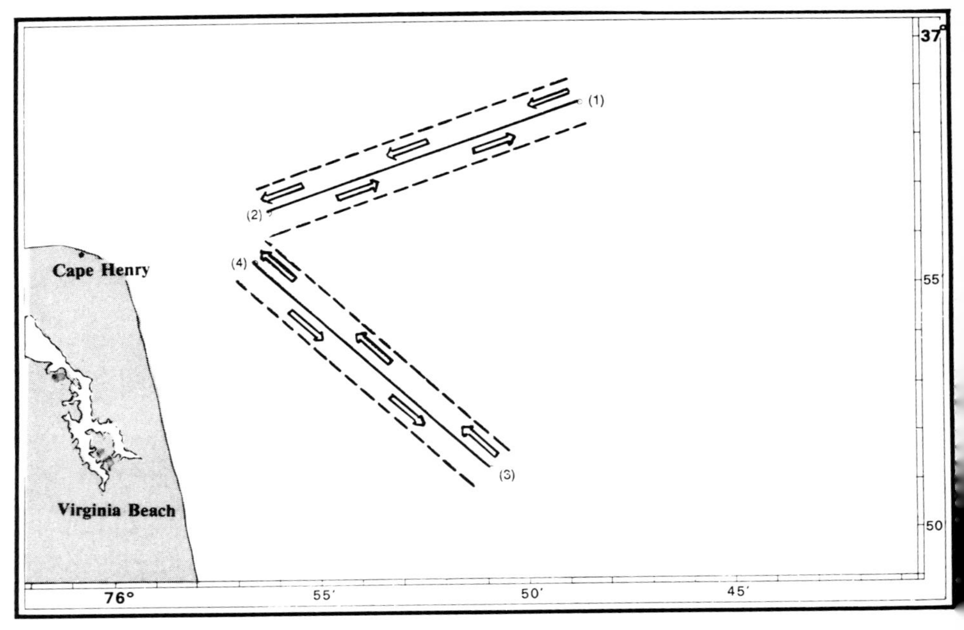

Fig. I 13. In approaches to Chesapeake Bay.

Off San Francisco

Description of the Traffic Separation Scheme

The traffic separation scheme off San Francisco consists of three parts.

Part I—Northern approach

(a) A separation zone bounded by a line connecting the following geographical positions:

(1)	37°48′.6N.,	122°47′.5W.
(2)	37°57′.1N.,	123°03′.5W.
(3)	37°55′.7N.,	123°04′.6W.
(4)	37°47′.8N.,	122°48′.2W.

(b) A traffic lane for north-westbound traffic is established between the separation zone and a line connecting the following geographical positions:

(5)	37°49′.4N.,	122°46′.6W.
(6)	37°58′.5N.,	123°02′.3W.

(c) A traffic lane for south-eastbound traffic is established between the separation zone and a line connecting the following geographical positions:

(7)	37°54′.3N.,	123°05′.7W.
(8)	37°46′.8N.,	122°48′.7W.

Part II—Southern approach

(a) A separation zone bounded by a line connecting the following geographical positions:

(9)	37°39′.1N.,	122°40′.3W.

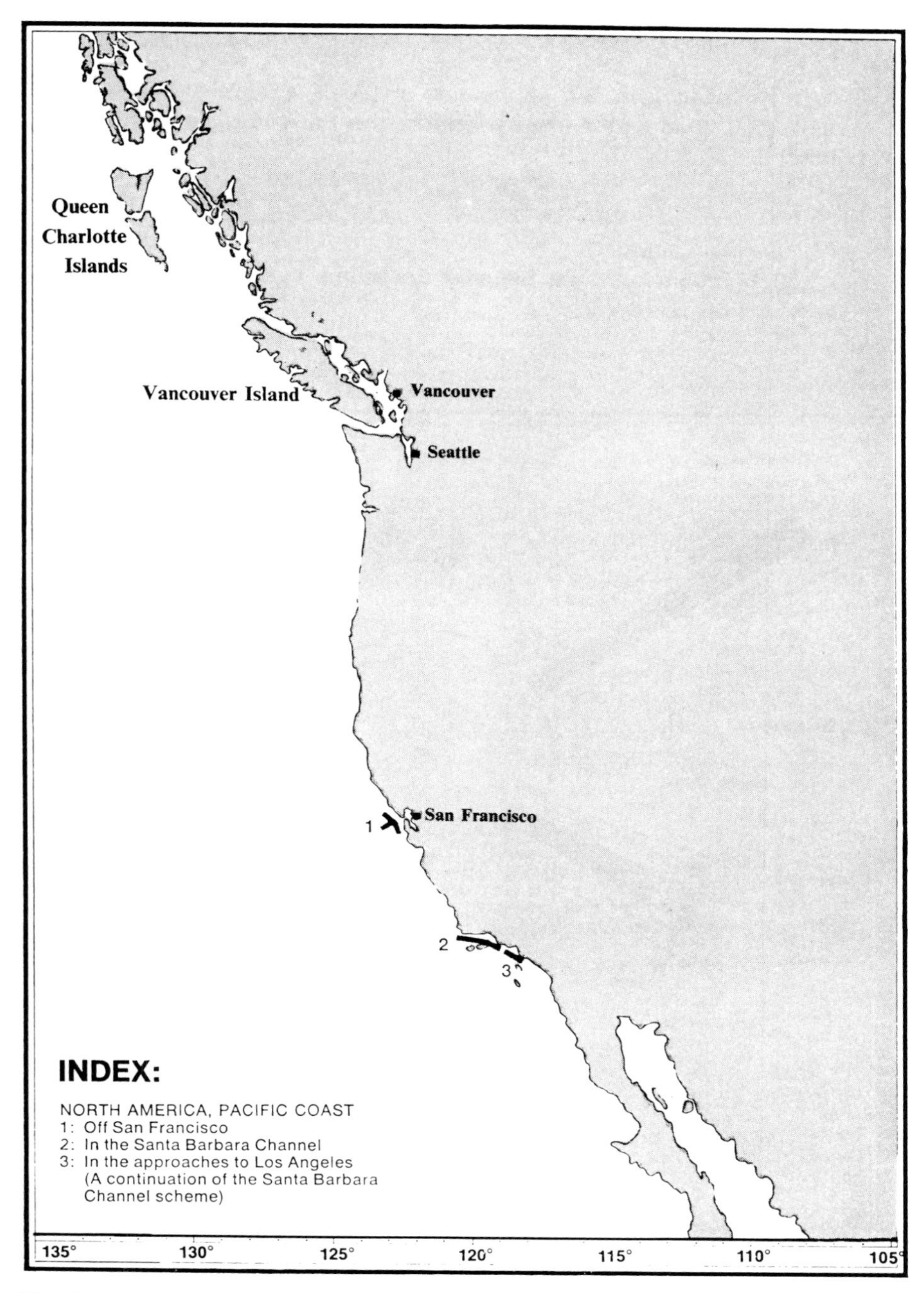

Fig. I 14. North America, Pacific Coast.

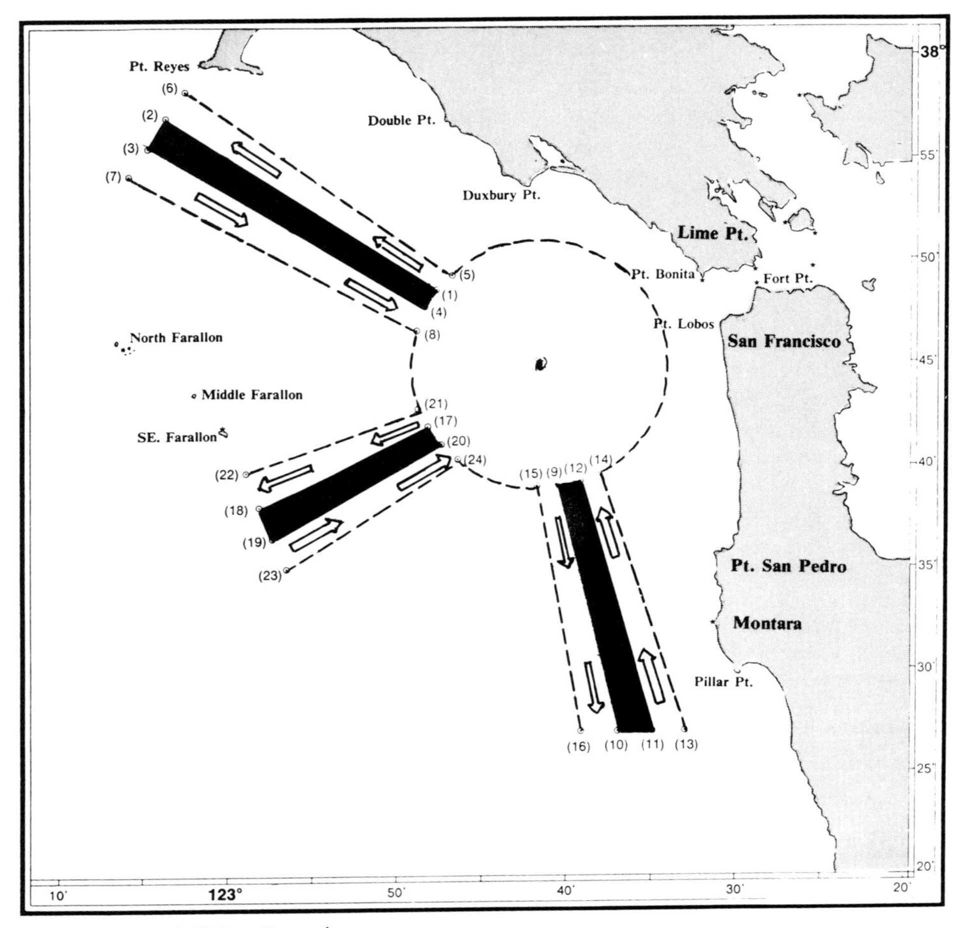

Fig. I 15. Off San Francisco.

(10) 37°27′.0N., 122°36′.9W.
(11) 37°27′.0N., 122°34′.8W.
(12) 37°39′.3N., 122°39′.1W.

(b) A traffic lane for northbound traffic is established between the separation zone and a line connecting the following geographical positions:

(13) 37°27′.0N., 122°32′.6W.
(14) 37°39′.7N., 122°37′.9W.

(c) A traffic lane for southbound traffic is established between the separation zone and a line connecting the following geographical positions:

(15) 37°39′.0N., 122°41′.6W.
(16) 37°27′.0N., 122°39′.0W.

Part III—Main approach

(a) A separation zone bounded by a line connecting the following geographical positions:

(17) 37°41′.9N., 122°48′.0W.
(18) 37°38′.1N., 122°58′.1W.
(19) 37°36′.5N., 122°57′.3W.
(20) 37°41′.1N., 122°47′.2W.

(b) A traffic lane for south-westbound traffic is established between the separation zone and a line connecting the following geographical positions:

(21) 37°42′.8N., 122°48′.5W.
(22) 37°39′.6N., 122°58′.8W.

(c) A traffic lane for north-eastbound traffic is established between the separation zone and a line connecting the following geographical positions:

(23) 37°35′.0N., 122°56′.5W.
(24) 37°40′.4N., 122°46′.3W.

Note:

Circular traffic separation zone

A circular traffic separation zone of radius half a mile is centred upon geographical position 37°45′.0N., 122°41′.5W.

Precautionary area

A precautionary area of radius six miles is centered upon geographical position 37°45′.0N., 122°41′.5W.

In the Santa Barbara Channel

Description of the Traffic Separation Scheme

A separation zone, two miles wide, is centred upon the following geographical positions:

(1) 34°20′.1N., 120°30′.4W.
(2) 34°04′.6N., 119°19′.6W.
(3) 33°44′.1N., 118°36′.3W.

A traffic lane, one mile wide, is established on each side of the separation zone.

The main traffic directions are:

105° – 285° and
120° – 300°.

Note:

Port Hueneme Fairway

The fairway at Port Hueneme is extended to meet the eastern edge of the northbound lane.

In the Approaches to Los Angeles—Long Beach

(A continuation of the Santa Barbara Channel scheme) (new scheme in southern approach)

Description of the Traffic Separation Scheme

Part I:

Western approach

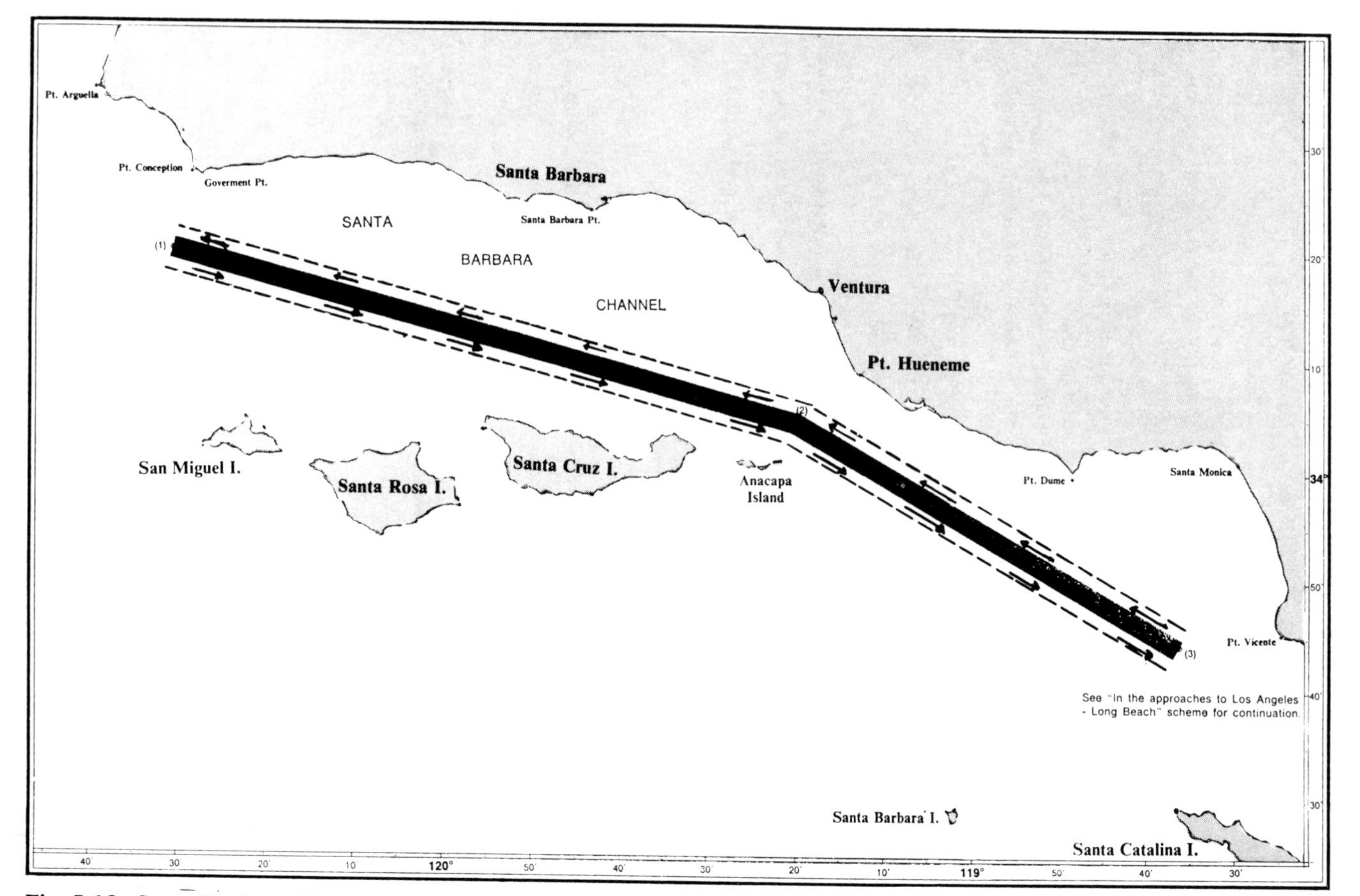

Fig. I 16. Santa Barbara Channel.

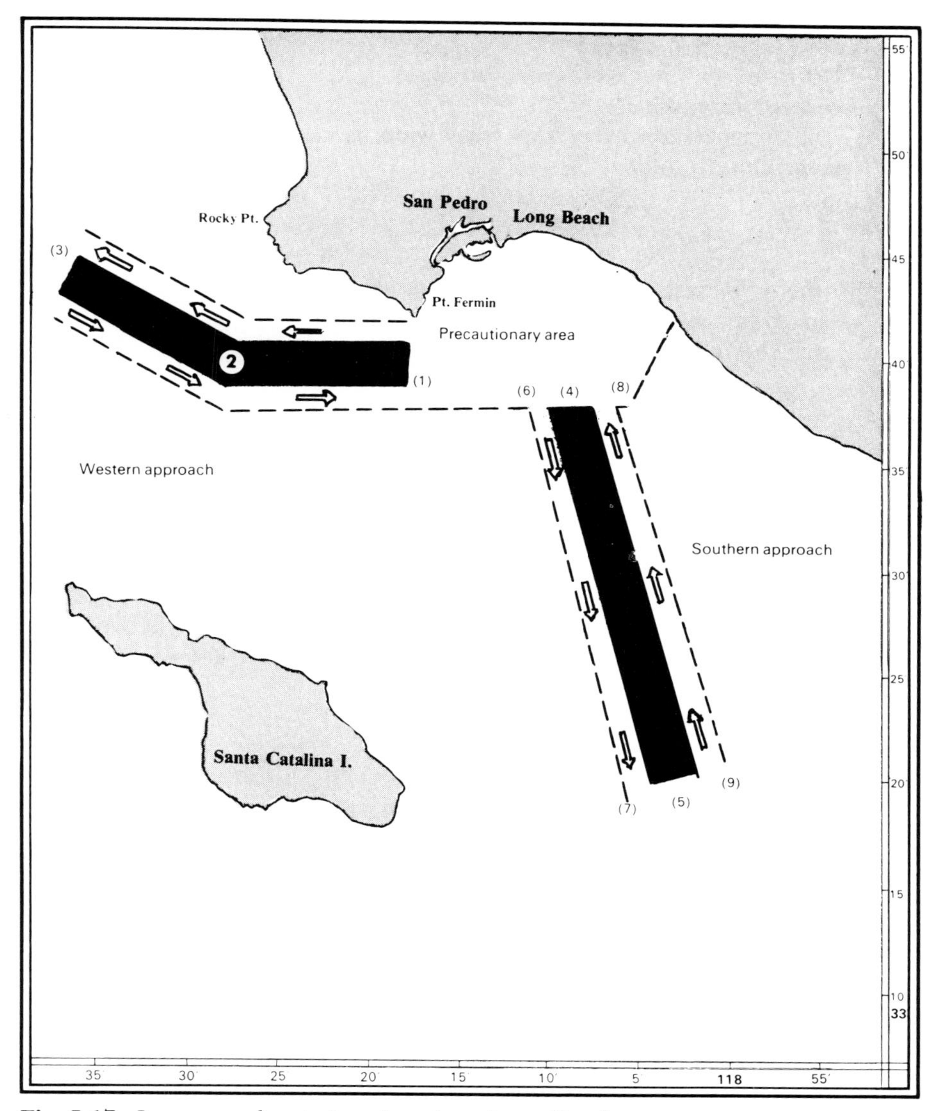

Fig. I 17. In approaches to Los Angeles—Long Beach.

(a) A separation zone, two miles wide, is centred upon the following geographical positions:

(1)	33°39′.7N.,	118°17′.6W.
(2)	33°39′.7N.,	118°27′.3W.
(3)	33°44′.1N.,	118°36′.3W.

(b) A traffic lane, one mile wide, is established on each side of the separation zone.

(c) The main traffic directions are:

090° – 270°
120° – 300°

Part II:

Southern approach

(a) A separation zone, two miles wide, is centred upon the following geographical positions:

(4) 33°37′.7N., 118°08′.9W.
(5) 33°19′.7N., 118°03′.4W.

(b) A traffic lane for southbound traffic is established between the separation zone and a line connecting the following geographical positions:

(6) 33°37′.7N., 118°11′.3W.
(7) 33°19′.1N., 118°06′.3W.

(c) A traffic lane for northbound traffic is established between the separation zone and a line connecting the following geographical positions:

(8) 33°37′.7N., 118°06′.5W.
(9) 33°20′.3N., 118°00′.5W.

(d) The main traffic directions are:

167° and 345°.

Precautionary area

The Los Angeles—Long Beach precautionary area consists of the water area enclosed by a line connecting Point Fermin Light at 33°42′.3N., 118°17′.6W. to 33°37′.7N., 118°17′.6W., thence to 33°37′.7N., 118°05′.4W., thence to the shoreline at 33°41′.7N., 118°02′.8W.

APPENDIX J

IMCO Recommendations on Navigational Watchkeeping[1]

SECTION I

Basic Principles to Be Observed in Keeping a Navigational Watch

Introduction

1. The master of every ship is bound to ensure that the watchkeeping arrangements are adequate for maintaining a safe navigational watch. Under his general direction, the officers of the watch are responsible for navigating the ship safely during their periods of duty when they will be particularly concerned to avoid collision and stranding.

2. This Section includes the basic principles which shall at least be taken into account by all ships.

Watch Arrangements

3. The composition of the watch, including the requirement for lookout(s), shall at all times be adequate and appropriate to the prevailing circumstances and conditions.

4. When deciding the composition of the watch on the bridge the following points are among those to be taken into account:

(a) at no time shall the bridge be left unattended;

(b) the weather conditions, visibility and whether there is daylight or darkness;

(c) the proximity of navigational hazards which may make it necessary for the officer in charge to carry out additional navigational duties;

(d) the use and operational condition of navigational aids such as radar or electronic position-indicating devices and any other equipment affecting the safe navigation of the ship;

(e) whether the ship is fitted with automatic steering;

(f) any additional demands on the navigational watch that may arise as a result of special operational circumstances.

Fitness for Duty

5. The watch system shall be such that the efficiency of the watchkeeping members of the crew is not impaired by fatigue. Accordingly, the duties shall be so

organized that the first watch at the commencement of a voyage and the subsequent relieving watches are sufficiently rested and otherwise fit when going on duty.

Navigation

6. The intended voyage shall be planned in advance taking into consideration all pertinent information and any course laid down shall be checked.

7. On taking over the watch the ship's estimated or true position, intended track, course and speed shall be confirmed; any navigational hazard expected to be encountered during the watch shall be noted.

8. During the watch the course steered, position and speed shall be checked at sufficiently frequent intervals using any available navigational aids necessary to ensure that the ship follows the planned course.

9. The safety and navigational equipment with which the ship is provided and the manner of its operation shall be clearly understood; in addition its operational condition shall be fully taken into account.

10. Whoever is in charge of a navigational watch shall not be assigned or undertake any duties which would interfere with the safe navigation of the ship.

Look-out

11. Every ship shall at all times maintain a proper look-out by sight and hearing as well as by all available means appropriate in the prevailing circumstances and conditions so as to make a full appraisal of the situation and of the risk of collision, stranding and other hazards to navigation. Additionally, the duties of look-out shall include the detection of ships or aircraft in distress, shipwrecked persons, wrecks and debris. In applying these principles the following shall be observed:

(a) whoever is keeping a look-out must be able to give full attention to the task and no duties shall be assigned or undertaken which would interfere with the keeping of a proper look-out;

(b) the duties of the person on look-out and helmsman are separate and the helmsman should not be considered the person on look-out while steering; except in small vessels where an unobstructed all round view is provided at the steering position and there is no impairment of night vision or other impediment to the keeping of a proper look-out;

(c) there may be circumstances in which the officer of the watch can safely be the sole look-out in daylight. However, this practice shall only be followed after the situation has been carefully assessed on each occasion and it has been established without doubt that it is safe to do so. Full account shall be taken of all relevant factors including but not limited to the state of weather, conditions of visibility, traffic density, proximity of navigational hazards and if navigating in or near a traffic separation scheme. 'Assistance must be summoned to the bridge when any change in the situation necessitates this and such assistance must be immediately available.'

Navigation with Pilot Embarked

12. Despite the duties and obligations of a pilot, his presence on board does not relieve the master or officer in charge of the watch from their duties and obligations for the safety of the ship. The master and the pilot shall exchange information regarding navigation procedures, local conditions and the ship's characteristics.

Protection of the Marine Environment

13. The master and officer in charge of the watch shall be aware of the serious effects of operational or accidental pollution of the marine environment and shall take all possible precautions to prevent such pollution particularly within the existing framework of existing international regulations.

SECTION II

Operational guidance for officers in charge of a navigational watch.

Introduction

1. This Section contains operational guidance of general application for officers in charge of a navigational watch, which masters are expected to supplement as appropriate. It is essential that officers of the watch appreciate that the efficient performance of their duties is necessary in the interest of safety of life and property at sea and the avoidance of pollution of the marine environment.

General

2. The officer of the watch is the master's representative and his primary responsibility at all times is the safe navigation of the vessel. He must at all times comply with the applicable regulations for preventing collisions at sea (see also paragraphs 23 and 24).

3. The officer of the watch should keep his watch on the bridge which he should in no circumstances leave until properly relieved. It is of especial importance that at all times the officer of the watch ensures that an efficient look-out is maintained. In a vessel with a separate chart room the officer of the watch may visit this, when essential, for a short period for the necessary performance of his navigational duties, but he should previously satisfy himself that it is safe to do so and ensure that an efficient look-out is maintained.

4. There may be circumstances in which the officer of the watch can safely be the sole look-out in daylight. However, this practice shall only be followed after the situation has been carefully assessed on each occasion and it has been established without doubt that it is safe to do so. Full account shall be taken of all relevant factors including but not limited to the state of weather, conditions of visibility, traffic density, proximity of navigational hazards and if navigating in or near a traffic separation scheme.

When the officer of the watch is acting as the sole look-out he must not hesitate to summon assistance to the bridge, and when for any reason he is unable to give his undivided attention to the look-out such assistance must be immediately available.

5. The officer of the watch should bear in mind that the engines are at his disposal and he should not hesitate to use them in case of need. However, timely notice of intended variations of engine speed should be given when possible. He should also keep prominently in mind the manoeuvring capabilities of his ship including its stopping distance.

6. The officer of the watch should also bear in mind that the sound signalling apparatus is at his disposal and he should not hesitate to use it in accordance with the applicable regulations for preventing collisions at sea.

7. The officer of the watch continues to be responsible for the safe navigation of the vessel despite the presence of the master on the bridge until the master informs him specifically that he has assumed responsibility and this is mutually understood.

Taking Over the Watch

8. The officer of the watch should not hand over the watch to the relieving officer if he has any reason to believe that the latter is apparently under any disability which would preclude him from carrying out his duties effectively. If in doubt, the officer of the watch should inform the master accordingly. The relieving officer of the watch should ensure that members of his watch are apparently fully capable of performing their duties and in particular the adjustment to night vision.

9. The relieving officer should not take over the watch until his vision is fully adjusted to the light conditions and he has personally satisfied himself regarding:

(a) standing orders and other special instructions of the master relating to the navigation of the vessel;

(b) the position, course, speed and draught of the vessel;

(c) prevailing and predicted tides, currents, weather, visibility and the effect of these factors upon course and speed;

(d) the navigational situation including but not limited to the following:

(i) the operational condition of all navigational and safety equipment being used or likely to be used during the watch;

(ii) errors of gyro and magnetic compasses;

(iii) the presence and movement of vessels in sight or known to be in the vicinity;

(iv) conditions and hazards likely to be encountered during his watch;

(v) the possible effects of heel, trim, water density and squat on underkeel clearance.

10. If at the time the officer of the watch is to be relieved a manoeuvre or other action to avoid any hazard is taking place, the relief of the officer should be deferred until such action is completed.

Periodic Check of Navigational Equipment

11. The officer of the watch should make regular checks to ensure that:

(a) the helmsman or the automatic pilot is steering the correct course;

(b) the standard compass error is established at least once a watch and when possible, after any major alteration of course. The standard and the gyro compasses should be frequently compared; repeaters should be synchronised with their master compass;

(c) the automatic pilot is tested in the manual position at least once a watch;

(d) the navigation and signal lights and other navigational equipment are functioning properly.

Automatic Pilot

12. Officers of the watch should bear in mind the need to station the helmsman and to put the steering into manual control in good time to allow any potentially hazardous situation to be dealt with in a safe manner. With a vessel under automatic steering it is highly dangerous to allow a situation to develop to the point where the officer of the watch is without assistance and has to break the continuity of the look-out in order to take emergency action. The change-over from automatic to manual steering and vice versa should be made by, or under the supervision of, a responsible officer.

Electronic Navigational Aids

13. The officer of the watch should be thoroughly familiar with the use of electronic navigational aids carried, including their capabilities and limitations.

Echo-sounder

14. The echo-sounder is a valuable navigational aid and should be used whenever appropriate.

Navigational Records

15. A proper record of the movements and activities of the vessel should be kept during the watch.

Radar

16. The officer of the watch should use the radar when appropriate and whenever restricted visibility is encountered or expected and at all times in congested waters having due regard to its limitations.

17. Whenever radar is in use, the officer of the watch should select an appropriate range scale, observe the display carefully and plot effectively.

18. The officer of the watch should ensure that range scales employed are changed at sufficiently frequent intervals so that echoes are detected as early as possible and that small or poor echoes do not escape detection.

19. The officer of the watch should ensure that plotting or systematic analysis is commenced in ample time, remembering that sufficient time can be made available by reducing speed if necessary.

20. In clear weather, whenever possible, the officer of the watch should carry out radar practice.

Navigation in Coastal Waters

21. The largest scale chart on board, suitable for the area and corrected with the latest available information, should be used. Fixes should be taken at frequent intervals; whenever circumstances allow, fixing should be carried out by more than one method.

22. The officer of the watch should positively identify all relevant navigation marks.

Clear Weather

23. The officer of the watch should take frequent and accurate compass bearings of approaching vessels as a means of early detection of risk of collision; such risk may sometimes exist even when an appreciable bearing change is evident, particularly when approaching a very large vessel or a tow or when approaching a vessel at close range. He should also take early and positive action in compliance with the applicable regulations for preventing collisions at sea and subsequently check that such action is having the desired effect.

Restricted Visibility

24. When restricted visibility is encountered or suspected, the first responsibility of the officer of the watch is to comply with the relevant rules of the applicable regulations for preventing collisions at sea, with particular regard to the sounding of fog signals, proceeding at a moderate speed and he shall have the engines ready for immediate manoeuvres. In addition, he should:

(a) inform the master (see paragraph 25);

(b) post look-out(s) and helmsman and, in congested waters, revert to hand steering immediately;

(c) exhibit navigation lights;

(d) operate and use the radar.

It is important that the officer of the watch should have the manoeuvring capabilities including the 'stopping distance' of his own vessel prominently in mind.

Calling the Master

25. The officer of the watch should notify the master immediately under the following circumstances:

(a) if restricted visibility is encountered or suspected;

(b) if the traffic conditions or the movements of other vessels are causing concern;

(c) if difficulty is experienced in maintaining course;

(d) on failure to sight land, a navigation mark or to obtain soundings by the expected time;

(e) if land or a navigation mark is sighted or a change in soundings occurs unexpectedly;

(f) on the breakdown of the engines, steering gear or any essential navigational equipment;

(g) in heavy weather if in any doubt about the possibility of weather damage;

(h) in any other emergency or situation in which he is in any doubt.

Despite the requirement to notify the master immediately in the foregoing circumstances, the officer of the watch should in addition not hesitate to take immediate action for the safety of the ship, where circumstances so require.

Navigation With Pilot Embarked

26. Despite the duties and obligations of a pilot, his presence on board does not relieve the officer of the watch from his duties and obligations for the safety of the ship. He should co-operate closely with the pilot and maintain an accurate check on the vessel's positions and movements. If he is in any doubt as to the pilot's actions or intentions, he should seek clarification from the pilot and if doubt still exists he should notify the master immediately and take whatever action is necessary before the master arrives.

The Watchkeeping Personnel

27. The officer of the watch should give the watchkeeping personnel all appropriate instructions and information which will ensure the keeping of a safe watch including an appropriate look-out.

Ship at Anchor

28. If the master considers it necessary a continuous navigational watch should be maintained. In all circumstances, however, the officer of the watch should:

(a) determine and plot the ship's position on the appropriate chart as soon as practicable and at sufficiently frequent intervals check when circumstances permit, by taking bearings of fixed navigational marks or readily identifiable shore objects, whether the ship is remaining securely at anchor;

(b) ensure that an efficient look-out is maintained;

(c) ensure that inspection rounds of the vessel are made periodically;

(d) observe meteorological and tidal conditions and the state of the sea;

(e) notify the master and undertake all necessary measures if the vessel drags the anchor;

(f) ensure that the state of readiness of the main engines and other machinery is in accordance with the master's instructions;

(g) if visibility deteriorates notify the master and comply with the applicable regulations for preventing collisions at sea;

(h) ensure that the vessel exhibits the appropriate lights and shapes and that appropriate sound signals are made at all times;

(i) take measures to protect the environment from pollution by the ship and comply with the applicable pollution regulations.

APPENDIX K

Conversion Table for Meters, Feet, and Fathoms

Meters	Feet	Fathoms	Meters	Feet	Fathoms	Feet	Meters	Feet	Meters	Fathoms	Meters	Fathoms	Meters
1	3.28	0.55	61	200.13	33.36	1	0.30	61	18.59	1	1.83	61	111.56
2	6.56	1.09	62	203.41	33.90	2	0.61	62	18.90	2	3.66	62	113.39
3	9.84	1.64	63	206.69	34.45	3	0.91	63	19.20	3	5.49	63	115.21
4	13.12	2.19	64	209.97	35.00	4	1.22	64	19.51	4	7.32	64	117.04
5	16.40	2.73	65	213.25	35.54	5	1.52	65	19.81	5	9.14	65	118.87
6	19.68	3.28	66	216.54	36.09	6	1.83	66	20.12	6	10.97	66	120.70
7	22.97	3.83	67	219.82	36.64	7	2.13	67	20.42	7	12.80	67	122.53
8	26.25	4.37	68	223.10	37.18	8	2.44	68	20.73	8	14.63	68	124.36
9	29.53	4.92	69	226.38	37.73	9	2.74	69	21.03	9	16.46	69	126.19
10	32.81	5.47	70	229.66	38.28	10	3.05	70	21.34	10	18.29	70	128.02
11	36.09	6.01	71	232.94	38.82	11	3.35	71	21.64	11	20.12	71	129.85
12	39.37	6.56	72	236.22	39.37	12	3.66	72	21.95	12	21.95	72	131.67
13	42.65	7.11	73	239.50	39.92	13	3.96	73	22.25	13	23.77	73	133.50
14	45.93	7.66	74	242.78	40.46	14	4.27	74	22.56	14	25.60	74	135.33
15	49.21	8.20	75	246.06	41.01	15	4.57	75	22.86	15	27.43	75	137.16
16	52.49	8.75	76	249.34	41.56	16	4.88	76	23.16	16	29.26	76	138.99
17	55.77	9.30	77	252.62	42.10	17	5.18	77	23.47	17	31.09	77	140.82
18	59.06	9.84	78	255.90	42.65	18	5.49	78	23.77	18	32.92	78	142.65
19	62.34	10.39	79	259.19	43.20	19	5.79	79	24.08	19	34.75	79	144.48
20	65.62	10.94	80	262.47	43.74	20	6.10	80	24.38	20	36.58	80	146.30
21	68.90	11.48	81	265.75	44.29	21	6.40	81	24.69	21	38.40	81	148.13
22	72.18	12.03	82	269.03	44.84	22	6.71	82	24.99	22	40.23	82	149.96
23	75.46	12.58	83	272.31	45.38	23	7.01	83	25.30	23	42.06	83	151.79
24	78.74	13.12	84	275.59	45.93	24	7.32	84	25.60	24	43.89	84	153.62
25	82.02	13.67	85	278.87	46.48	25	7.62	85	25.91	25	45.72	85	155.45
26	85.30	14.22	86	282.15	47.03	26	7.92	86	26.21	26	47.55	86	157.28
27	88.58	14.76	87	285.43	47.57	27	8.23	87	26.52	27	49.38	87	159.11
28	91.86	15.31	88	288.71	48.12	28	8.53	88	26.82	28	51.21	88	160.93
29	95.14	15.86	89	291.99	48.67	29	8.84	89	27.13	29	53.04	89	162.76
30	98.42	16.40	90	295.28	49.21	30	9.14	90	27.43	30	54.86	90	164.59
31	101.71	16.95	91	298.56	49.76	31	9.45	91	27.74	31	56.69	91	166.42
32	104.99	17.50	92	301.84	50.31	32	9.75	92	28.04	32	58.52	92	168.25
33	108.27	18.04	93	305.12	50.85	33	10.06	93	28.35	33	60.35	93	170.08
34	111.55	18.59	94	308.40	51.40	34	10.36	94	28.65	34	62.18	94	171.91
35	114.83	19.14	95	311.68	51.95	35	10.67	95	28.96	35	64.01	95	173.74
36	118.11	19.68	96	314.96	52.49	36	10.97	96	29.26	36	65.84	96	175.57
37	121.39	20.23	97	318.24	53.04	37	11.28	97	29.57	37	67.67	97	177.39
38	124.67	20.78	98	321.52	53.59	38	11.58	98	29.87	38	69.49	98	179.22
39	127.95	21.33	99	324.80	54.13	39	11.89	99	30.18	39	71.32	99	181.05
40	131.23	21.87	100	328.08	54.68	40	12.19	100	30.48	40	73.15	100	182.88
41	134.51	22.42	101	331.36	55.23	41	12.50	101	30.78	41	74.98	101	184.71
42	137.80	22.97	102	334.64	55.77	42	12.80	102	31.09	42	76.81	102	186.54
43	141.08	23.51	103	337.93	56.32	43	13.11	103	31.39	43	78.64	103	188.37
44	144.36	24.06	104	341.21	56.87	44	13.41	104	31.70	44	80.47	104	190.20
45	147.64	24.61	105	344.49	57.41	45	13.72	105	32.00	45	82.30	105	192.02
46	150.92	25.15	106	347.77	57.96	46	14.02	106	32.31	46	84.12	106	193.85
47	154.20	25.70	107	351.05	58.51	47	14.33	107	32.61	47	85.95	107	195.68
48	157.48	26.25	108	354.33	59.06	48	14.63	108	32.92	48	87.78	108	197.51
49	160.76	26.79	109	357.61	59.60	49	14.94	109	33.22	49	89.61	109	199.34
50	164.04	27.34	110	360.89	60.15	50	15.24	110	33.53	50	91.44	110	201.17
51	167.32	27.89	111	364.17	60.70	51	15.54	111	33.83	51	93.27	111	203.00
52	170.60	28.43	112	367.45	61.24	52	15.85	112	34.14	52	95.10	112	204.83
53	173.88	28.98	113	370.73	61.79	53	16.15	113	34.44	53	96.93	113	206.65
54	177.16	29.53	114	374.02	62.34	54	16.46	114	34.75	54	98.76	114	208.48
55	180.45	30.07	115	377.30	62.88	55	16.76	115	35.05	55	100.58	115	210.31
56	183.73	30.62	116	380.58	63.43	56	17.07	116	35.36	56	102.41	116	212.14
57	187.01	31.17	117	383.86	63.98	57	17.37	117	35.66	57	104.24	117	213.97
58	190.29	31.71	118	387.14	64.52	58	17.68	118	35.97	58	106.07	118	215.80
59	193.57	32.26	119	390.42	65.07	59	17.98	119	36.27	59	107.90	119	217.63
60	196.85	32.81	120	393.70	65.62	60	18.29	120	36.58	60	109.73	120	219.46

Index of Cases

Index